Plant Ecology

EDITED BY
Michael J. Crawley
Department of Pure and Applied Biology
Imperial College, London

Blackwell Scientific Publications

OXFORD LONDON EDINBURGH

BOSTON PALO ALTO MELBOURNE

This book is dedicated to
the memory of
E.J.SALISBURY FRS &
H.A.GLEASON
two of the founders of
modern plant ecology

© 1986 by
Blackwell Scientific Publications
Editorial offices:
Osney Mead, Oxford, OX2 0EL
8 John Street, London, WC1N 2ES
23 Ainslie Place, Edinburgh, EH3 6AJ
3 Cambridge Center, Suite 208
  Cambridge, MA 02142, USA
107 Barry Street, Carlton
  Victoria 3053, Australia

First published 1986
Reprinted 1989

Photoset by Enset (Photosetting),
Midsomer Norton, Bath, Avon
and printed in Great Britain by
Redwood Burn Limited,
Trowbridge, Wiltshire

DISTRIBUTORS

USA
  Publishers' Business Services
  288 Airport Industrial Drive
  Ypsilanti
  Michigan 48197
  (Orders: Tel: (313) 487 9720)

Canada
  Oxford University Press
  70 Wynford Drive
  Don Mills
  Ontario M3C 1J9
  (Orders: Tel: (416) 441–2941)

Australia
  Blackwell Scientific Publications
  (Australia) Pty Ltd
  107 Barry Street
  Carlton, Victoria 3053
  (Orders: Tel: (03) 347 0300)

British Library
Cataloguing in Publication Data

Plant ecology.
  1. Botany—Ecology
  I. Crawley, Michael J.
  581.5   QK901

ISBN 0–632–01363–X

Library of Congress
Cataloging in Publication Data

Plant ecology.
  1. Botany—Ecology
  I. Crawley, Michael J.
  581.5   QK901

ISBN 0–632–01363–X

Plant Ecology

Plant Ecology

EDITED BY
Michael J. Crawley
Department of Pure and Applied Biology
Imperial College, London

Blackwell Scientific Publications

OXFORD LONDON EDINBURGH

BOSTON PALO ALTO MELBOURNE

This book is dedicated to
the memory of
E.J.SALISBURY FRS &
H.A.GLEASON
two of the founders of
modern plant ecology

© 1986 by
Blackwell Scientific Publications
Editorial offices:
Osney Mead, Oxford, OX2 0EL
8 John Street, London, WC1N 2ES
23 Ainslie Place, Edinburgh, EH3 6AJ
3 Cambridge Center, Suite 208
    Cambridge, MA 02142, USA
107 Barry Street, Carlton
    Victoria 3053, Australia

First published 1986
Reprinted 1989

Photoset by Enset (Photosetting),
Midsomer Norton, Bath, Avon
and printed in Great Britain by
Redwood Burn Limited,
Trowbridge, Wiltshire

DISTRIBUTORS

USA
  Publishers' Business Services
  288 Airport Industrial Drive
  Ypsilanti
  Michigan 48197
  (*Orders*: Tel: (313) 487 9720)

Canada
  Oxford University Press
  70 Wynford Drive
  Don Mills
  Ontario M3C 1J9
  (*Orders*: Tel: (416) 441–2941)

Australia
  Blackwell Scientific Publications
  (Australia) Pty Ltd
  107 Barry Street
  Carlton, Victoria 3053
  (*Orders*: Tel: (03) 347 0300)

British Library
Cataloguing in Publication Data

Plant ecology.
  1. Botany—Ecology
  I. Crawley, Michael J.
  581.5   QK901

ISBN 0–632–01363–X

Library of Congress
Cataloging in Publication Data

Plant ecology.
  1. Botany—Ecology
  I. Crawley, Michael J.
  581.5   QK901

ISBN 0–632–01363–X

Contents

Preface

This book is about the factors affecting the distribution and abundance of plants. It aims to show how pattern and structure at different levels of plant organization (communities, populations and individuals) are influenced by abiotic factors like climate and soils, and by biotic interactions including competition with other plants, attack by herbivorous animals and plant pathogens, and relationships with mutualistic organisms of various kinds. One further aim has been to convey something of the excitement and dynamism of modern plant ecology. For too long the subject has been regarded as the poor relation of animal ecology. It has been transformed over the last two decades into a veritable growth industry, thanks chiefly to the inspiration of John Harper, and to the wide dispersal of the students he trained.

The present work differs from other textbooks on plant ecology in stressing dynamics rather than statics, and by espousing an experimental rather than a descriptive methodology. Throughout the book, the patterns of plant abundance which we see in the field are interpreted as the outcome of dynamic processes involving gains and losses. For example, we view the species richness of communities as the resolution of immigration and extinction, population density as the balance of recruitment and mortality within single species, and the size of individual plants as the result of births and deaths of modular component parts.

This book is based on a plant-centred view of ecology rather than the traditional, quadrat-centred view. Traditional quadrat-based measures like 'percentage cover' consign individual plants to oblivion, and discourage thinking about the evolutionary ecology of individuals. Quadrat based tests of association may detect positive statistical associations between species, even when the individual plants of the two species never grow next to one another, and never influence one another's population dynamics! The plant-centered view of ecology is intended to rectify some of these defects by focusing attention directly on the interactions between a plant and its immediate neighbours, and between plants and their mycorrhizal associates, pollinators and natural enemies. Throughout the book we are at pains to stress that our ultimate objective is to measure the impact of the processes we describe on the *fitness* of individual phenotypes (difficult though we recognize this task to be).

If there is a single philosophical strand running through the following chapters, it is this: 'seek simplicity but distrust it' (Lagrange). Our approach to ecology is to develop theory by the recognition of patterns in the field and the development of simple models to describe them. As Lewontin and Levins have said, only slightly tongue-in-cheek, 'Things are similar—this makes science possible. Things are different—this makes science necessary'! The role of theory in ecology is rather different than in some other sciences. It is not, for example, intended to make accurate predictions, as in astronomy or physics. Rather it attempts to separate the expected from the unexpected, the possible from the impossible and the surprising from the unsurprising (Lewontin). In the past, however, plant ecologists have tended to emphasize the complexity of ecological systems and to decry simple, theoretical models as being naïve and unrealistic. So prevalent, indeed, is this viewpoint amongst practising ecologists, that it has now been elevated to the position of a formal school of thought. During a recent botanizing trip to a species-rich area of serpentine soils, we had recourse to the local botanists' guide book. It promised enlightenment of the most profound kind. It began 'the factors responsible for the richness of the serpentine flora are . . .' we held our breath in anticipation . . . 'many, complex and interacting'. On that day the 'MC&I School' of ecology was born. Many and various have been its subsequent publications!

Of course this is not the only modern school of plant ecologists. Other recognizable disciplines include the 'Nothing's Happening School' (who believe in null models, random processes and non-equilibrium communities), the Biochemical Ecologists (the 'Find 'em and Grind 'em School'), the Mathematical Modellers (quantifying the bleeding obvious), the Agricultural Ecologists (spray and pray), and the Phytosociologists (ignore that plant, it shouldn't be here). There are as many schools of thought represented in this book as there are different contributors to it. Your editor, however, is a confirmed adherent to the Experimentalist School who 'suck it and see'!

The layout of the chapters is upside down in terms of the conventional, atom–to–universe textbook. I have chosen to begin with communities and work progressively downwards through populations and individuals for three reasons. First, it is with communities that we gain our initial impressions of plants and of plant ecology. Second, the theory of community ecology is relatively straightforward, so the theoretical material becomes more, rather than less challenging as the book goes on. Third, it makes a change!

It is worth pointing out explicitly what the book is *not* about. First and foremost, it is not about ecological methods. There is a vast literature on methodology which we have made no attempt to précis.

Students would do well to study carefully the problems of sampling plant populations, and to acquaint themselves with the statistical analysis and presentation of data. Only then, perhaps, should the more rarified quantitative methods like pattern analysis and the plethora of multivariate techniques be addressed. Three particularly good introductory papers for those about to embark on ecological experimentation are Lewontin (1974), Hurlbert (1984) and Bender, Case & Gilpin (1984). Other important aspects of plant ecology not covered by this book are micrometeorology, soil science, plant anatomy, plant physiology, plant evolution, plant geography and ecological biochemistry. Not only do these fields provide vital background information, but they also afford a variety of different perspectives from which plant ecology can be viewed.

Finally I should like to thank the many people who have helped in the preparation and production of this book.

Ascot, 1985   Michael J. Crawley

Chapter 1   The Structure of Plant Communities

MICHAEL J. CRAWLEY

1.1   Introduction

The stature, colour and texture of plants give landscape its unique character. As Darwin wrote, a 'traveller should be a botanist, for in all views plants form the chief embellishment'! The cast of the vegetation's features (its physiognomy) is determined by the size of the dominant plants (whether they are trees or bushes, herbs or mosses), by their spacing (whether they form continuous cover or are widely spaced-out), and by their seasonal prospect (whether the plants are deciduous or evergreen, and whether they undergo striking seasonal colour changes).

The first generation of plant ecologists (Warming, 1909; Raunkiaer, 1934) dedicated themselves to understanding why certain structures of vegetation are restricted to certain combinations of climate and soil. While these problems are far from resolved, modern plant community ecologists are occupied by questions involving species richness (why do so many (or so few) species of plants grow here?), species abundance (why is a single species dominant in one place, but many species co-dominant in another?), and patterns of spatial and temporal change (what determines the observed gradients in species composition we see as we climb up a mountain side, and what factors influence the succession of species we observe after disturbance?). The purpose of this chapter is to introduce the various structural attributes of plant communities, and Chapters 2 and 3 will consider the dynamics of how these patterns come about.

1.2   The definition of plant community

The plant 'community' is an abstraction of exactly the same kind as the 'population'; a community simply consists of all the plants occupying an area which an ecologist has circumscribed for the purposes of study. This definition draws attention immediately to the two key issues involved in studying plant communities; 1) how large should the area be? and 2) where, precisely, should the sample-area be put? These apparently trivial questions have fuelled the most heated controversy ever since the study of quantitative plant ecology began, towards the end of the nineteenth century (Goodall, 1952). Failure to standardize the size of study areas, and failure to agree on

1

whether the areas should be subjectively positioned in the most 'clearly typical' parts of the vegetation cover, or more objectively stationed by some form of randomization, have led to a great divergence of experimental approaches, and meant that clear comparisons between the findings of different studies are difficult or even impossible. Superimposed on these practical difficulties was a fundamental difference of opinion about the nature of the plant community itself. The two most polarized positions in this debate are represented by the views of the American ecologists Frederic Clements and H.A. Gleason. Clements believed that the plant community was a closely integrated system with numerous emergent properties, analogous to a 'super-organism'. In contrast, Gleason saw plant communities as random assemblages of adapted species exhibiting none of the properties of integrated organisms like homeostasis, repair, and predictable development alleged by Clements.

1.2.1   Clements' view of community structure

Clements (1916, 1928) studied plant communities throughout North America and was struck by the vast extent of the rather uniform vegetation types he found. He called these 'climax communities', and proposed that the nature of the climax was determined principally by climate. Clements felt that the developmental study of vegetation rested on the assumption that the climax formation was an organic entity which arose, grew, matured and died. Each climax was seen as being able to reproduce itself, 'repeating with essential fidelity the stages of its development', so that the life history of a community was a complex, but definite and predictable process, comparable with the life history of an individual plant.

Clements recognized three major classes of climax vegetation in North America; grasslands, scrub and forests and he subdivided each climax into a number of 'formations'. The grassland climax was divided into true grasslands (dominated by *Stipa* and *Bouteloua*) and sedgelands (*Carex* and *Poa*); the scrub climaxes into sagebrush (*Atriplex* and *Artemisia*), desert scrub (*Larrea* and *Franseria*) and chaparral (*Quercus* and *Ceanothus*); the forest climaxes into woodland (*Pinus* and *Juniperus*), montane forest (*Pinus* and *Pseudotsuga*), coast forest (*Thuja* and *Tsuga*), subalpine forest (*Abies* and *Picea*), boreal forest (*Picea* and *Larix*), lake forest (*Pinus* and *Tsuga*), deciduous forest (*Quercus* and *Fagus*), isthmian forest and insular forest. Each formation was then further subdivided into 'associations' (today, these are simply called communities).

Clements has been accused of espousing a static plant community in which the climax was a fixed, and spatially invariable organism, where individual plants were replaced faithfully by recruits of their own species. In fact, Clements was adamant that 'the most stable association is never in complete equilibrium, nor is it free from disturbed areas in which secondary succession is evident. Even when the final community seems most homogeneous and its factors uniform, quantitative study . . . reveals a swing of population and a variation in the controlling factors'.

Clements' view of succession was of a relatively orderly, predictable approach to a dynamic equilibrium. He identified six stages to the process: 1) nudation, 2) migration, 3) establishment, 4) competition, 5) reaction and 6) stabilization. Each of these might be 'successive or interacting', but the early processes tend to be successive and the later ones are more likely to be interacting (i.e. the frequency, strength and complexity of ecological interactions tend to increase through succession). Thus, in the course of development from bare substrate, through lichens and mosses to the final trees, there would be a series of recognizable (though ephemeral) communities before the climax was achieved (he called these 'seres'). He also distinguished clearly between primary successions, which were essentially soil-forming processes (and where there was no seed bank, and no reserve of vegetative propagules in the substrate at the outset), and secondary successions where the soil initially contained the propagules of many species characteristic of different stages of succession. Primary successions are dominated by the immigration of species from other areas, and occur on lava flows, dunes, rocks and in lakes, and tend to be associated with the accumulation of nutrients and organic matter in the soil. Secondary successions occur in fallow fields, drained areas, clear-cut forests, in the aftermath of fires, and so on. They involve only slight changes in soil (often nutrient *depletions*), and are less dependent on immigration of species.

1.2.2 Gleason's view of community structure

Gleason's name is associated with the 'individualistic concept' of plant community structure (Gleason, 1917, 1926, 1927). He freely admitted the *existence* of plant associations; 'we can walk over them, we can measure their extent, we can describe their structure in terms of their component species, we can correlate them with environment, we can frequently discover their past history and make inferences about their future'. In his view, however, recognizable plant communities owed their visible expression simply to the juxtaposition of individual plants, of the same or of different species, which may or may not interact directly with one another. The structure of the plant community was the result of continuously acting causes, chiefly 'migration and environmental selection' which operate independently in each place, and have 'no relation to the process on any other area; nor are they related to the vegetation of any other area, except as the latter may serve as a source of migrants or control the environment of the former'.

Where Gleason differed most from Clements was in his belief that the community 'is not an organism, but merely a coincidence'. He saw every species of plant as a law unto itself, and denied the emergent properties attributed to plant communities by Clements. He stressed the heterogeneity of community structure caused by accidents of seed dispersal, minor variations in environment, differences in the abundance of parent plants which could act as sources of seed, and the brevity of periods between disturbances.

Another fundamental difference from Clements' view was Gleason's denial of the determinism and directedness of succession. Gleason (following Cooper, 1926) emphasized the random elements of seed immigration and seedling establishment, highlighted the different durations of similar successional stages in different places, and pointed out that different initial conditions and different histories of disturbance could lead to different endpoints. (Clements stressed convergence to a single climatic climax.) Gleason also believed that disturbance was so frequent, and so patchily distributed in space, that most ecological communities should not be regarded as equilibrium assemblages.

Gleason's views are best summarized by one of his own examples. Assume a series of artificial ponds has been created in farmland. "Annually the surrounding fields have been ineffectively planted with seeds of *Typha* and other wind-distributed hydrophytes, and in some of the new pools *Typha* seeds germinate at once. Water-loving birds bring various species to the other pools. Various sorts of accidents conspire to the planting of all of them. The result is that certain pools soon have a vegetation of *Typha latifolia*, others of *Typha angustifolia*, others of *Scirpus validus*; plants of *Iris versicolor* appear in one, and *Saggitaria* in another, of *Alisma* in a third, of *Juncus effusus* in a fourth. Only the chances of seed dispersal have determined the allocation of species to different pools, but in the course of three or four years each pool has a different appearance, although the environment, aside from the reaction of the various species, is precisely the same for each. Are we dealing here with several different associations, or with a single association, or with merely the embryonic stages of some future association? Under our view, these become merely academic questions, and any answer which may be suggested is equally academic" (Gleason, 1927).

1.2.3   The modern synthesis

The modern synthesis is very close to Gleason's view of community structure and dynamics (see Chapters 2 and 3, and the studies of recruitment in plant populations in Chapters 4 and 5). The issue is not whether there are identifiable (if vague) kinds of communities (no one seriously disputes this); the question is, to what extent do *biological* interactions (between one plant and another, between plants and their herbivores, or between herbivores and their natural enemies) influence community structure, compared with limitations imposed by the physical environment (abiotic conditions like soil, weather and exposure).

Despite the severe criticisms levelled against most of the underlying assumptions of the climax concept (Cain, 1947; Egler, 1947; Mason, 1947), the term 'climax community' is still quite widely used. There may, indeed, be a place for a word to describe those communities that have been left alone long enough to pass through several generations of the dominant plants. Many ecologists, however, feel that the word 'climax' is so steeped in religious and ethical prejudices (continuous improvement, directed progress towards an ultimate goal, etc.), not to mention Freudian imagery, that its use is probably best avoided!

The legacy of Clements' 'super-organism' is to be found in Tansley's (1935) definition of the ecosystem ('the whole system . . . including not only the organism-complex, but also the whole complex of physical factors. These ecosystems . . . form one category of the multitudinous physical systems of the universe, which range from the universe as a whole down to the atom'). It also provides the philosophy behind 'whole-systems ecology', as epitomized by the International Biological Programme of 1964–1974. This 'systems-thinking' has underpinned a good deal of the conservation ethic (Hardin, 1968) and has inspired some terrible poetry (e.g. 'thou canst not stir a flower without troubling of a star'; Thompson, 1918)!

A denial of this holistic approach does not imply that ecological interactions are simple; far from it. As the following chapters will show, most processes really are 'Many, Complex and Interacting' (see page xii)! There is nothing in the reductionist approach which precludes the possibility of plant communities displaying emergent properties. Indeed, there is a marriage to be achieved between the realistic aspects of the climax notion (context-specific interactions between species, multi-species effects, etc.), and the experimental approach of the individualists (Levins & Lewontin, 1980; see Section 1.8).

1.3   The niche concept

The niche is a multidimensional description of a species' resource needs, habitat requirements and environmental tolerances (Hutchinson, 1957). For every vital attribute of a species' ecology it should be possible (at least in principle) to draw an axis to describe the range of possible values for the attribute, and then to plot the performance of the species at different positions along the axis. For example, a plant may not survive at very low or very high levels of soil moisture, but prosper only at intermediate levels (Fig. 1.1).

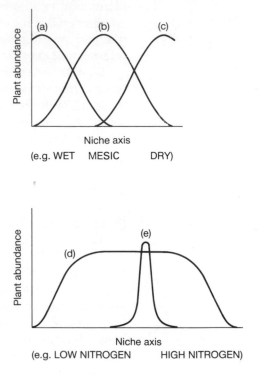

Fig. 1.1. The niche. Idealized relationships between plant abundance (cover, population density or biomass) and distance along a niche axis, of the kind that might be obtained from field transects. (*a*) A species with its peak abundance close to the minimum measured value. (*b*) The 'classic' bell-shaped niche response curve showing sub-optimal, optimal and super-optimal regions of niche space. (*c*) A species with its peak abundance close to the maximum measured value. (*d*) A species with a 'broad niche' growing abundantly over a wide range of niche conditions. (*e*) A species with a 'narrow niche' (a 'specialist') found only under a restricted range of conditions. It is important to distinguish between those broad-niched species which consist of rather uniform, but highly plastic individuals, and those comprising genetically polymorphic individuals, each having a rather narrow niche. Similarly, it is important to distinguish between plants where a narrow niche is caused by narrow tolerances of environmental factors, and those with broader fundamental niches that are restricted to a narrow realized niche as a result of competition with other species.

The niche concept serves as a shorthand summary of a species' complex suite of ecological attributes, including its abiotic tolerances, its maximum relative growth rate, its phenology, its susceptibilities to various enemies, and its relative competitive abilities with other plant species. Plant ecologists, however, have been rather slow to take this originally zoological definition of niche to their hearts. For whilst it is easy to see how resources might be divided up between animal species specializing on diets of different types (e.g. feeding on seeds of different sizes), it is much less obvious how plant resources like light, water or nitrogen could be apportioned between species. Habitat variables like soil moisture, pH or heavy metal concentration make obvious (and easily measurable) niche axes, and species attributes like temperature tolerances, germination requirements, root depth, flowering phenology or disease resistance are amenable to straightforward interpretations. But niche specialization in relation to plant resource partitioning is much less readily incorporated in Hutchinson's original model. In Chapter 2 Tilman presents an elegant solution to this problem by focusing on the performance of plants at different *ratios* of essential resources.

As every gardener knows, many plants can grow perfectly well under most soil conditions if they are freed from competition with the native vegetation by careful weeding. But of course the native vegetation, and the competitive milieu which it creates, is itself a prime component of the environment. The important point is not whether a species can grow in a particular soil, but whether it can hold its own on that soil in competition with the other members of the local flora (both as a seedling and as an established plant). Competitive ability, therefore, is not a species' attribute, but depends upon both the environmental conditions *and* the other plant species involved.

The distinction between what a plant could do on its own, and what it actually does in the presence of other plants and natural enemies, is formalized in the distinction between a species' fundamental and realized niches (Hutchinson, 1957). One of the earliest experiments on realized plant niche was carried out by Ellenberg (1953). He grew five grass species in monoculture and monitored their performance in soils with different depths of water table. He found that all the species grown on their own did best at the same water table. He then grew the species in competition with each other at different water tables. Not one of the species did best at the same depth of water table in the presence of competitors as it had when growing on its own! This important result illustrates that the realized niches of these grasses were distinctly different from their fundamental niches. In general, the 'physiological optimum' of a species is not necessarily the same as its 'ecological optimum' (measured as the position on the niche axis at which the species achieve maximum abundance under field conditions; see Chapter 8).

Niche breadth is the range of values along an axis at which the species can persist. Broad-niched (generalist) species can grow almost anywhere along the axis, while narrow-niched (specialist) species are restricted to a narrow band of values (Fig. 1.1). Species may be broad-niched in relation to some axes (e.g. soil pH) but narrow-niched in relation to others (e.g. germination temperature requirements). Whether a wide niche breadth derives from a high degree of phenotypic plasticity among rather uniform genotypes, or from a high degree of genetic polymorphism in populations is an important question. The problem is addressed further in Chapters 4, 9 and 10.

There are two main ways in which the niche concept is employed. The first is in classical 'autecological' studies like those reported in the *Biological Flora of the British Isles* (published in the *Journal of Ecology*). Individual accounts aim to explain the full range of biotic and abiotic conditions which influence the distribution, survival and reproduction of a particular species. A second use is in plant community ('synecological') studies, where certain niche parameters are fundamental to understanding community attributes like species

richness and to explainations of such dynamic processes as successional change (Whittaker, 1956). For instance, on a given length of
niche axis, a community may contain a small number of broad-niched
generalists, or a large number of narrow-niched specialists. Species
richness could be increased either by: 1) increasing the length of the
niche axis (so that more niches could be 'tacked on at the ends'); 2)
reducing the width of the existing niches; or 3) increasing the degree
of overlap tolerable between adjacent niches (so that more species
could be packed into a fixed length of niche axis). Perhaps the most
valuable aspect of the niche concept is that it requires us to distinguish clearly between the range of *conditions* under which a species
can survive, and the range of *resources* which it is capable of exploiting. These topics are expanded in Chapter 2.

1.4   Species richness

For every combination of soil, climate, altitude, slope and aspect
there will be one species which grows better than any other, so that it
produces more seeds or occupies more space by vegetative spread. In
a spatially uniform and temporally constant world this single species
would come to dominate the community to the exclusion of all
others. This is known as the Competitive Exclusion Principle, or
Gause's Principle (see Box 1.1), and it leads us to the expectation that
species richness should be uniformly low.

The reason that most communities contain so many species of
plants is that the competitive exclusion principle simply doesn't
work! Its logic is impeccable, but its assumptions are generally
wrong. Environments are neither spatially uniform nor temporally
constant; they are seasonal, spatially patchy, periodically disturbed,
and their constituent plant populations are subject to fluctuating
competition from other species and variable levels of impact from
herbivores, pathogens, pollinators and dispersal agents. Our task, in
attempting to understand why there are so many species of plants, is
to find what conditions invalidate the competitive exclusion principle. In doing so, we shall see why certain habitats are rich in species
and others are relatively poor.

Explanations of species richness hinge upon whether or not the
community is in equilibrium. There are two extreme schools of
thought on this, and the truth probably lies somewhere in between.
The stochastic school believe that most communities exist in a state of
non-equilibrium, where competitive exclusion is prevented by
periodic population reductions and environmental fluctuations. The
equilibrium school believe that co-existence is possible, even in
uniform environments, if certain criteria are met (for example, the
conditions derived from the Lotka-Volterra competition equations;
see Begon, Harper & Townsend, 1986).

List of Contributors

M.J. CRAWLEY *Imperial College at Silwood Park, Ascot, Berkshire, SL5 7PY*

A.H. FITTER *Department of Biology, University of York, York, YO1 5DD*

R.B. FOSTER *Department of Zoology, University of Iowa, Iowa City, Iowa, 52242, U.S.A.*

D.E. GILL *Department of Zoology, University of Maryland, College Park, Maryland, 20742, U.S.A.*

H.F. HOWE *Department of Zoology, University of Iowa, Iowa City, Iowa, 52242, U.S.A.*

S.P. HUBBELL *Department of Zoology, University of Iowa, Iowa City, Iowa, 52242, U.S.A.*

M.J. HUTCHINGS *School of Biological Sciences, University of Sussex, Falmer, Sussex, BN1 9QG*

D.A. LEVIN *Department of Botany, University of Texas, Austin, Texas, 78713–7640, U.S.A.*

H.A. MOONEY *Department of Biological Sciences, Stanford University, Stanford, California, 94305–2493, U.S.A.*

D. TILMAN *Department of Ecology and Behavioural Biology, University of Minnesota, Minneapolis, 55455, U.S.A.*

D.M. WALLER *Department of Botany, University of Wisconsin, Madison, Wisconsin, 53706, U.S.A.*

A.R. WATKINSON *School of Biological Sciences, University of East Anglia, Norwich, NR4 7TJ*

L.C. WESTLEY *Department of Zoology, University of Iowa, Iowa City, Iowa, 52242, U.S.A.*

In order to understand the origin and maintenance of species richness we can investigate several processes which contravene one or more of the assumptions of the Competitive Exclusion Principle. I shall deal with these in turn, but bear in mind that they are *not* mutually exclusive.

a The Lottery Model

This model, first proposed by Sale (1977), is a non-equilibrium explanation which involves no competition between the component species. Because its assumptions are so simple, it forms a useful starting point in constructing 'null models' of community structure. For example, there is no merit in invoking interspecific competition in the explanation of species richness if observed patterns can be explained adequately without it (Harvey *et al.*, 1983).

The Lottery Model assumes that establishment microsites appear unpredictably in an otherwise uniform environment, and that propagules establish in vacant microsites at random. Once in a site the plant holds tenure of it for a long time. Individuals produce seeds over an extended period, and the seeds are dispersed widely. The chance of a species establishing in a vacant microsite is simply a function of the proportion of all propagules produced by that species in that year. Where competition does occur, it is highly asymmetric and usually decided in favour of the prior resident. The rate of appearance of unoccupied microsites is a function of the longevity of the constituent species.

Computer simulations of this system show that when longevity and fecundity are high, species richness can be maintained for protracted periods, and exclusion is unlikely to occur even when there are substantial differences in fecundity or survivorship between the species. Analytical solution of the Lottery Model, however, shows that it cannot promote species richness in equilibrium systems (Hanski, 1986).

In the Lottery Model, each species has requirements for space that are general enough for there to be some chance of finding a suitable microsite (i.e. the species all have similar regeneration niches; see Chapter 8). This is in complete contrast to the Niche Differentiation Model (below) which assumes that species coexist by being different.

b Spatial heterogeneity

This hypothesis assumes that the competitive exclusion principle *does* operate, but that apparently uniform areas actually consist of many, subtly different microhabitats, in each of which one species excludes all others (see Chapter 2). This hypothesis is attractive,

Box 1.1 The competitive exclusion principle

The principle states that one species (the best competitor) will, given long enough, oust all others. This leads to the prediction that each separate community should contain just one species of each growth form. The logic is as follows. Two species must, by definition, be different. No matter how slight the difference, individuals of one species will produce more seeds (or occupy more microsites by vegetative spread) than individuals of the other. Since the total number of microsites is assumed to be limited, these differences, no matter how tiny, will eventually lead to the dominance of one species and the extinction of the other. For example, suppose 1000 seeds of two annual species (A and B) are sown each year; we start with an equal mixture of 500 A and 500 B. The seeds are harvested at the end of each growing season, mixed thoroughly, and 1000 randomly selected seeds are replanted. This is repeated year after year. Even though species A has only a small advantage of species B, the second species is virtually extinct after 100 generations (Fig. 1.2).

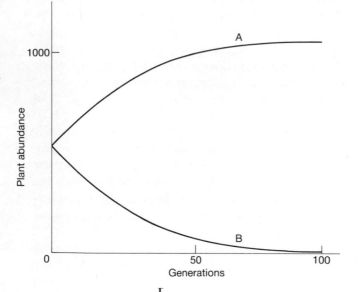

Fig. 1.2. Competitive exclusion. In a uniform environment one species will eventually exclude all others, no matter how small the differences in their reproductive output. The net reproductive rate of species A is only 5% greater than that of species B ($\lambda_A = 2.1$; $\lambda_B = 2.0$). By generation 100 species B is virtually extinct. The population is microsite-limited to a total of 1000 individuals in each generation.

More formally, let λ_A be the net reproductive rate of species A, and λ_B of species B; we define $\lambda_A > \lambda_B$ by some arbitrarily small amount. At planting there are N_A seeds of species A and N_B of species B. Each species produces S_A and S_B seeds respectively, where $S_A = N_A \cdot \lambda_A$ and $S_B = N_B \cdot \lambda_B$ to give a total of $T = S_A + S_B$ seeds. The number of seeds of species A planted in the next generation is therefore:

$$N_{A(t+1)} = 1000\,(S_A/T)$$

and

$$N_{B(t+1)} = 1000 - N_{A(t+1)}$$

In the limit, N_A tends to 1000 and N_B tends to zero; this is the competitive exclusion principle.

Experimental tests have confirmed the model's major prediction. For example, Harlan & Martini (1938) sowed 11 varieties of barley in equal proportions, then gathered random samples of seed for resowing. Within 4–11 years, the less suitable varieties were completely eliminated at any given site.

because spatial heterogeneity is so conspicuous in all real habitats (Rickleffs, 1977; Grubb *et al.*, 1982). However, as a model of species richness it is flawed, because by explaining everything, it explains nothing. It simply says that there must be as many different microhabitats (regeneration niches) as there are species in the community. Thus unless we know exactly what constitutes a suitable microhabitat for each species, and can measure the frequency with which different kinds of microhabitats occur, we can not predict species richness or species' abundance distributions.

c   The Musical Chairs Model

This model is based on migration between patches in a heterogeneous environment. Following Hutchinson's (1951) initial lead, a series of papers in the 1970s showed that coexistence of quite similar species is possible when, for example, the dispersal ability of the inferior competitor is sufficiently great that its rate of successful migration between patches is higher than the rate of extinction of populations within patches. The inferior competitor must be able to stay one jump ahead of the superior species (Levins & Culver, 1971; Horn & MacArthur, 1972; Levin, 1974; Slatkin, 1974). The limit to species richness is reached when migration rates become so great that the environment is no longer treated as patchy. It is the patchiness of the environment, rather than the number of different kinds of patch which allows coexistence to occur (Levin, 1974).

d   Niche separation and resource partitioning

This is an equilibrium explanation of species richness, and assumes that the niches of coexisting plant species are sufficiently different that competitive exclusion does not occur. Species-rich communities may be composed of: 1) species with narrower niches; 2) species with more broadly overlapping niches; 3) habitats providing 'longer' niche axes; or 4) a combination of these.

Just how different two species must be in order to allow them to coexist is a question of the greatest theoretical and practical interest. There is a substantial literature on the topic of the 'limiting similarity of species' (May & MacArthur, 1972; Abrams, 1976; Roughgarden,

1976) but the main findings of the theoretical work are flawed by the limitations of the simple models on which they are based. For example, generalizations which are sound for one-dimensional niche separation break down for multi-dimensional niches. We do not yet have any simple generalizations about the limiting similarity of many dimensional niches. However, if plant communities *are* equilibrium assemblages structured by competition, then there will be rules for limiting similarity, and we shall need to work them out (Sugihara, 1984).

Where the niche structure of complex plant communities has been investigated in some detail (e.g. limestone grasslands in northern England; Al-Mufti *et al.*, 1977; Grime & Curtis, 1976; Grime *et al.*, 1981; Sydes & Grime, 1984), niche differences have been found in germination behaviour, root depth, temperature thresholds, grazing tolerance, phenology (early growing versus late growing, early flowering versus late flowering) and many other factors. Whether or not these differences are necessary and sufficient to explain the maintenance of species richness remains to be discovered. It appears, however, that niche specificity is greater under more extreme environmental conditions. For example, plants of the extremely species-rich communities which occur on the ancient, nutrient-poor soils of Australia and South Africa, show a much higher degree of soil-specificity than species from more nutrient-rich communities (Shimda & Whittaker, 1984). However, there is very little information to suggest that resource partitioning (in its original, zoological sense) is a significant process contributing to the maintenance of plant species diversity, at least in grasslands (Braakhekke, 1980; but see Chapter 2).

e   Selective herbivory

Feeding by herbivores is one of the most obvious ways in which frequency dependent advantage might accrue to rare species, and it is well documented that selective feeding by polyphagous vertebrate herbivores on plant species which would otherwise be dominant can lead to increased plant species richness (the herbivore acts like Paine's (1966) 'keystone predator'); examples are given by Harper (1977) and Crawley (1983).

Vertebrate herbivores may, in fact, have a range of effects on plant species richness. They may eat preferred species to extinction, thereby reducing species richness. Alternatively, by suppressing certain plants, they may increase diversity by allowing the ingress of grazing tolerant or avoided species. Whether species richness increases or decreases under grazing therefore depends upon the balance between the number of species ousted by selective feeding, and the number of species able to establish under the altered competitive regime of the grazed community. Subtle, sub-lethal levels of feeding by inconspicuous invertebrate herbivores may alter the relative competitive abilities of the component plant species in such a way that frequency dependent reductions in abundance are brought about (Connell *et al.*, 1984). Similarly, selective attack on the seeds and seedlings of dominant tree species by insects or pathogens may have an impact in maintaining the species richness of tropical forests (Janzen, 1970; Connell, 1971; see Chapter 4).

f Disturbance

Disturbance of the soil surface and/or destruction of established plants may provide recruitment microsites, which allow the community to be invaded by new species, leading to increased species richness (Chapter 8). At very high rates of disturbance, however, the pool of adapted species is small, so species richness would be low (e.g. on mobile sand, or on frequently burned heathland). At very low rates of disturbance, competitive exclusion would occur, and species richness would again be low. Thus, species richness should be greatest at moderate levels of disturbance, because dominance is prevented, and the pool of potential colonists is relatively large (e.g. Armesto & Pickett, 1984). This has been termed the 'intermediate disturbance hypothesis' (Grime, 1973; Connell, 1979), and is another non-equilibrium explanation of species richness (and see Huston, 1979; Underwood & Denley, 1984).

g Refuges

There is almost certainly some truth in *all* the foregoing models. We can be confident that species richness is enhanced whenever the inferior competitors have some kind of refuge in which their numbers can be maintained. The refuge acts as a source of propagules which are dispersed to those parts of the habitat where, in equilibrium, the species would be excluded.

Refuges may be provided by one or more of the following: 1) tolerance of extreme conditions where no other species can survive (Proctor & Woodell, 1975); 2) spatial aggregation by the superior competitor which leaves a refuge of 'competitor free space' (Atkinson & Shorrocks, 1981; Pacala & Roughgarden, 1982; Hanski, 1983); 3) the existence of a long-lived seed bank (Grime, 1979); 4) herbivore resistance of one form or another (e.g. genetic polymorphism in chemical defences, see Crawley, 1983); 5) seasonal effects whereby the inferior competitor grows much earlier (or much later) in the season than the superior competitors (Al-Mufti *et al.*, 1977); 6) long life and protracted reproduction of individual plants (Kelly, 1985); 7) good years for one species may be bad years for another, and vice versa (Chapin & Shaver, 1985); 8) other aspects of niche differentiation (Grubb *et al.*, 1982).

In any event, if communities really are equilibrium assemblages, then the maintenance of high species richness requires that the rare species obtain an advantage over common ones, either by: 1) the common species suffering density dependent reductions in survival or fecundity (see Box 1.2); or 2) the rare species obtaining some frequency-dependent advantage in recruitment (Chapters 4 and 5).

h   Alpha and beta diversity

Whittaker (1975) distinguished two different kinds of species diversity which he called alpha and beta diversity. Alpha diversity refers to the number of species *within* sample areas, whilst beta diversity relates to the differences in species composition *between* samples (see Fig. 1.3). It is easy to see why different plants should be found on different substrates, and understanding beta diversity presents rather few problems. It is extremely difficult, on the other hand, to present a simple explanation of alpha diversity (see Models (a) to (g) above).

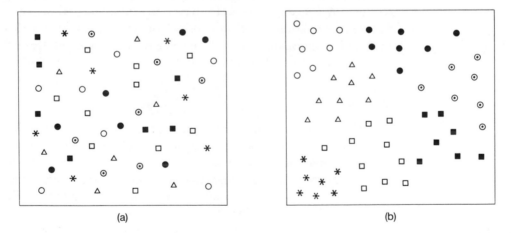

(a)   (b)

Fig. 1.3. (*a*) Fine grained and (*b*) coarse grained mixtures of species. In the fine grained mixture, each individual is equally likely to have neighbours belonging to any of the other species, in proportion to their relative abundances. In the coarse grained mixture, each individual is likely to have *neighbours of its own species*, despite the fact that species richness and relative abundance are the same in both cases.

Areas with the very highest levels of species richness are those where there are many species per quadrat, but no two quadrats are alike in their species composition (i.e. those with high alpha *and* beta diversities; Naveh & Whittaker, 1980). In regions of the lowest species richness there are few species in any sample, and the vegetation is spatially monotonous (low alpha and beta diversities).

Two communities with exactly the same *number* of species, S, may differ from one another in many other ways, and it is important that we recognize how this may influence studies of species richness. The most important differences which affect the interpretation of species richness are: 1) the relative abundance of the different species; 2) the degree of aggregation in the spatial distribution of each species; and 3) the degree to which the spatial distributions of the different species are correlated with one another.

1.4.1   Species area effects

Since the earliest studies of plant community structure it has been clear that the size of the quadrats employed was a vital consideration. The notion of 'minimal area' originated in floristic studies (Moravec, 1973) and referred to the quadrat size which was likely to include some asymptotic percentage of the 'total flora' of the community (90% was a figure commonly employed; see Goodall, 1952). Two thoroughly erroneous ideas are incorporated in this concept of minimal area: 1) that there is a fixed total flora for a community; and 2) that there is a unique relationship between species richness and area. First, there is the tacit assumption that plant communities are made up from species which are essentially 'community faithful' (Section 1.6.3). This is clearly not the case, and increasing quadrat size will almost always lead to increased species richness. Since the total cannot be known then nor can some arbitrary fraction (say 90%) of the total be known. Second, the concept of minimal area ignores the important effects which the spatial patterns of species can have on the measurement of species richness based on frequency counts (see Box 1.3). We cannot use frequency data to estimate species richness unless we know both the degree of aggregation of each species, and the extent of correlation between species' aggregation patterns (whether they tend to clump in the same places, or clumps of one species tend to preclude clumps of the other).

Plots of log-species-richness against log-area-sampled ('species-area curves') are often presented as summaries of vegetation survey work, and both the slope and the intercept of such graphs have been imbued with great ecological significance (Goodall, 1952; Grieg-Smith, 1983; Connor & McCoy, 1979). If species abundances can be described by the canonical log normal distribution (Section 1.5) of N individuals distributed between S species, and if N is directly proportional to the size of the sample area, A, then, in the standard species area relationship $S = cA^z$ the power z is roughly equal to 0.25 (May, 1975). There is nothing mysterious, therefore, in the fact that most log–log plots of species richness versus area have slopes of about 1/4 (Sugihara, 1981). In fact, interpretation of these curves is fraught with difficulties. For example, counts using political boundaries often give strange species-area curves, for the simple reason that large administrative areas tend to be large *because* they are barren (Crawley, 1986)!

Box 1.2 Density dependence

What is density dependence?

The adjectival phrase 'density dependent' is applied to rates of birth, death, migration, growth and development. So, for example, when we say that a particular death rate is density dependent, we mean that the percentage of plants dying from this cause increases as population density rises. Thus, if 80% of the seedlings die when there are 10 seedlings/m² but 95% die when there are 100 seedlings/m², then seedling death rate is density dependent.

Why does density dependence matter?

Populations without any density dependent processes are doomed to extinction. Try the experiment of tossing a coin to generate random changes in population size. Start with, say, 64 plants and *double* population size when you throw a head, and *half* it when you throw a tail. If the population drops below 1 it is deemed to have gone extinct. Repeat the experiment several times. When population sizes are plotted against time, it will become clear that in very few cases do numbers remain close to the starting value of 64; they either go extinct or increase without bounds. This experiment has a very important hidden assumption. Since heads and tails are equally likely, then halfing and doubling are equally likely, so, on average, the population should stay at its starting value. We have to ask, however, what ecological process could ensure that births were exactly matched by deaths on average over many generations? The answer is that no density independent process could achieve such a result, and the heads/tails model of population regulation is therefore fatally flawed. But even *with* the built-in assumption of balanced births and deaths, the populations still drift to extinction, or increase to levels where competition (the archetypal density dependent process) would occur. Density dependence matters, therefore, because without it populations could not persist.

What kinds of density dependence are there?

Most examples of density dependence are negative in their action. When we say, for example, that seed production is density dependent, we mean *negatively* density dependent (i.e. each plant produces *fewer* seeds at high population density than at low). In some cases, we may find positive density dependence (so-called 'Allee effects') where the number of seeds per plant *increases* with population density (though only at low densities). This might occur if pollination rates were extremely low when plants were widely spaced. Thus, 'density dependent' means negatively density dependent, and we use the explicit phrase 'positively density dependent' to describe Allee effects. The logistic model of population growth incorporates a very specific, simple kind of density dependence that is linear, continuous, and acts immediately (without a time lag). More realistic models may need to include density dependence which is strongly non-linear (it may not act at all until very high densities are reached), intermittent (only acting in certain seasons), or time-lagged (as when growth in one season influences seed production in the following year). These different kinds of density dependence have very different implications for population dynamics (see Chapter 5).

Key factors and regulating factors

A fundamental (and widely misunderstood) distinction is between 'key factors' and 'regulating factors'. Key factors are those which are principally responsible for *fluctuations* in plant abundance. They tend to be abiotic factors (like rainfall or fire), and to cause substantial year to year changes in population density. However, key factors are usually density independent, and, as we have seen, density independent factors *can not regulate* population density.

Regulating factors are, by definition, density dependent. They may not cause large amounts of mortality (or severe reductions in fecundity); all that is required is that they increase in intensity as plant population density rises. Because regulating factors need not cause massive mortality or dramatic reductions in growth, they are usually difficult to detect and can easily be overlooked. For example, in a population of desert annuals, the key factor influencing year to year fluctuations in abundance might be early summer rainfall, whereas the regulating factor might be density dependent predation of seeds by small mammals.

Detecting density dependence

The most widespread misconception about density dependence is seen in the tendency to equate density dependence with intraspecific competition. This is encapsulated in the view that if the mature plants are spaced so far apart that they could never compete with one another, then density dependence cannot be operating. This overlooks two vital, though elementary forms of density dependence. First, when plant recruitment is microsite-limited (e.g. there is only a small number of suitably sized gaps in the plant canopy), then the rate of seed establishment will be density dependent. The larger the number of seeds produced, the lower the percentage of seeds which will become established as seedlings. Second, if the plants are subject to attack by mobile, polyphagous herbivores, then density dependent seed predation can result when the animals aggregate in areas of relatively high plant density. In short, the fact that the mature plants are small or widely spaced out *does not* mean that the plant population is not regulated, nor that density dependent processes are not operating.

Since density dependence is most likely to occur during the seed and seedling stages, it is almost always going to be difficult to detect, since these stages are the most difficult and laborious to study! Furthermore, density dependence is likely to vary spatially, within each generation. Thus, detailed life-table studies need to be performed in several places each year. Different rates of birth, death, growth and development will be detected in different places, and some of these differences may be density dependent. When demographic rates are *averaged* over several patches, in order to produce a single figure for each generation, it may be impossible to detect the regulating factors.

Box 1.3 Quadrat size and plant frequency

Data on plant frequency are extremely easy to gather. We simply put down a quadrat and score every plant species as either present or absent. No attempt is made to estimate plant density. The percentage of quadrats in which a species is found is called the frequency of that species. Common species will have frequencies approaching 100%, and rare ones may have frequencies of 5% or less. Species with large individuals will register higher frequency scores than species at the same population density but with smaller individuals. It is important to understand how the measure of frequency depends upon the size of the quadrats used in sampling. At one extreme, very large quadrats will contain the vast majority of species present in the community, no matter where the quadrat is placed. Therefore, with big quadrats, most species have frequency values approaching 100%. At the other extreme, a point quadrat (a long, metal pin with a sharp point) will only touch *one* species in a single canopy layer. The percentage of touches may give a good estimate of plant cover for the common species, but unless sampling effort is very high, only a small proportion of the plant species present in the community will be detected.

It is common practice to estimate species richness by counting the plant species in quadrats of a fixed size. The hope is that the quadrat is sufficiently large that most of the plant species will be found within it. The size of the quadrat which will contain 'most of the species' has been called the 'minimal area'. The lower the average plant population density, the larger this quadrat will need to be. This is not the only consideration, however, because the spatial pattern of the plants also influences the likelihood of their turning up in quadrats of a given size.

In the following graph we plot the probability of finding a plant in a quadrat for three different species. Each species has the *same* average population density. They differ only in the degree to which the individuals are aggregated in their spatial distributions.

For a quadrat of area A we are virtually certain to find a species whose individuals are distributed at random (i.e. A is a reasonable 'minimal area' in this case). If the individuals are densely clumped, however, we only have a 30% chance of finding a species in quadrats of this size, despite the fact that its mean density is the same. Indeed, at this level of aggregation, we would need a quadrat of size 100 (ten times the scale of this graph) to have better than a 50:50 chance of obtaining the aggregated plant species in a randomly placed quadrat. To have better than a 75% chance of finding it, the quadrat would have to have an area of 100,000 units!

The calculations for obtaining the graph are as follows. Assume we can describe the spatial pattern of the plant by a negative binomial distribution (Elliot, 1977). The parameter k measures the degree of clumping; small values (like 0.1) mean tight clumping, while large values (like 10) mean the distribution of plants is random. In order to mark the species as 'present' we require one or more plants in the quadrat. The probability of getting one or more

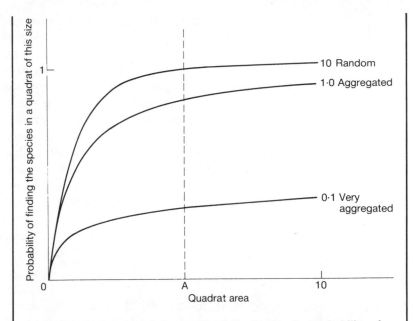

individuals is calculated most simply by finding the probability of getting *no* individuals (the so-called 'zero-term'), and taking this probability away from one. The zero-term of the negative binomial distribution is

$$P_0 = (k/(\mu+k))^k$$

where μ is the mean number of plants in a quadrat of size A. Thus, the probability of 'presence' is $1-P_0$. The graph was drawn for 3 values of k (10, 1 and 0.1) by working out $1-P_0$ as the area of the quadrat was increased. Increasing the area increases the mean number of plants per quadrat (μ), and we write $\mu = N.A$ where N is the population density of plants per unit area (of course, this is *not* affected by quadrat size or sampling method). For simplicity, N has been set to 1.0.

Because frequency depends so critically upon quadrat size, and upon the size of individual plants and their spatial arrangement, the interpretation of plant frequency data requires considerable care (see Goddall, 1952; Grieg-Smith, 1983).

Let us consider what shapes we would expect species-area curves to take under a variety of simple assumptions. First, let us assume that all the species are equally abundant, and that they are independently randomly distributed. We call this a 'fine grained' mixture of species (Fig. 1.4a). In such a case, the number of species in a sample (S) increases with quadrat area (A, the *fraction* of the total area sampled) according to the equation:

$$S = S_{max}(1-(1-A)^N)$$

where S_{max} is the maximum species richness (the total species pool) and N is the number of individuals of each species in the entire area (Fig. 1.4b; Arrhenius, 1921). If the species are not equally abundant we must specify a statistical distribution to describe the relative abundances of the different species; it is traditional to use a truncated log normal distribution for this purpose (see Section 1.5). Now S depends upon A in a manner influenced by the variance and skew of the distribution. In communities where species have more or less equal abundances (i.e. when 'equitability' is high and the Ns are all roughly equal), the result is the same as in Fig. 1.4a. When the communities demonstrate strong dominance, however, the tail of the distribution is extremely long, with the consequence that species richness increases rather uniformly with quadrat size over an extended range of areas (Fig. 1.4b). The appropriate equation in this case is:

$$S = \sum_{i=1}^{S_{max}} (1-(1-A)^{N_i})$$

where the N_is are distributed according to (say) a log normal distribution. If several equally abundant species are distributed in a 'coarse grained' manner (i.e. in dense, non-overlapping aggregates), then the species-area curve will increase in a more or less linear, step-wise manner; species are simply added as more patches are included within a single quadrat (Fig. 1.4c). Where the patches are of very different sizes, the model is a little more complex, and the slope of the curve depends upon whether the survey begins in a small or a large patch (Fig. 1.4d). In short, unless we know a great deal about the spatial distributions and relative abundances of the species, we can infer little from the species-curve area.

Similarly, on a much larger scale, comparisons of island or continental floras are confounded by processes affecting within- and between-area differences in species richness. For example, two regional floras with identical species richness will have very different species-area curves if one flora has high beta diversity and the other low. There are four basic configurations for such regional species-area curves; 1) alpha and beta diversity are both low (e.g. uniform, species-poor areas like central Australian spinifex); 2) alpha and beta diversity are both high (e.g. patchy, species-rich communities like tropical rain forests); 3) alpha diversity is high but beta diversity low (e.g. uniform but species-rich areas of goat-grazed Mediterranean maquis); and 4) alpha diversity is low but beta diversity high (e.g. some of the Cape heathlands of South Africa). The slope of these species-area plots is proportional to the beta diversity, while the intercept reflects alpha diversity. An intriguing, but as yet unanswered, question concerns the ecological processes responsible for the apparent convergence in the number of species in fixed area quadrats (e.g. 50×20 m) in comparable communities on different continents (Rice & Westoby, 1983); i.e. what processes regulate alpha diversity?

(a) Fine-grained; high equitability

(b) Fine-grained; low equitability

(c) Coarse-grained; equal patch sizes

(d) Coarse-grained; unequal patch sizes

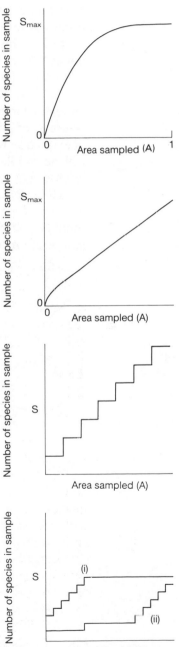

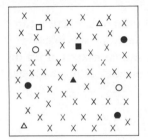
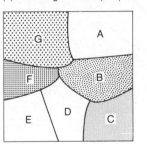
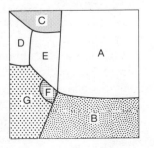

Fig. 1.4. Species-area curves for communities of different structures. (*a*) Fine grained mixture with high equitability; the species-area curve follows the classic Arrhenius asymptotic form. (*b*) With the same species richness, but pronounced dominance, the curve rises more or less linearly as the rare species are added. (*c*) Coarse grained mixtures produce stepped species-area curves. (*d*) When the patch sizes are very unequal, the shape of the curve depends upon where the first quadrat is placed: (*i*) starting in a small patch; (*ii*) starting in a large patch and encountering another large patch at the first increase in quadrat size.

1.4.2   Biogeography

The study of biogeography was revolutionized by the publication of MacArthur and Wilson's Equilibrium Theory in 1967. Botanists had known for years that distant islands supported fewer plant species than islands close to continents, and also that, for a given degree of isolation, large islands supported more plant species than small ones. The genius of MacArthur and Wilson was in proposing a single, simple hypothesis to explain both these phenomena.

MacArthur and Wilson saw the number of species on an island not as a fixed, immutable quantity, but as a dynamic balance between gains and losses—as an equilibrium between immigrations and extinctions. They argued that the rate of immigration will be higher on near islands than on far, because of their proximity to continental 'sources' of potential immigrants. Further, they argued that the rate of extinction will be greater on small islands because population sizes will be smaller and the intensity of competition likely to be more intense.

The rates of immigration or extinction can be plotted against the number of species, with separate sets of curves for large islands or for small (Fig. 1.5a), and for near islands or for distant (Fig. 1.5b). The immigration curve falls because as more species become established,

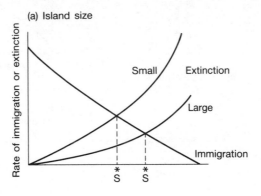

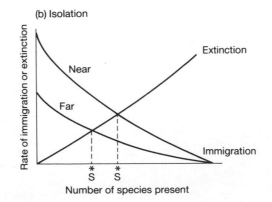

Fig. 1.5. MacArthur & Wilson's theory of island biogeography. (*a*) The effects of island size; larger islands have higher equilibrium species richness ($\overset{*}{S}$) because the extinction rate is lower. (*b*) The effects of isolation; near islands have higher equilibrium species richness because the immigration rate is higher. From MacArthur and Wilson (1967).

fewer immigrants will belong to new species. The curve is likely to be concave, because the more rapidly dispersing species are established first, and this causes a rapid drop in the overall immigration rate. The extinction curve rises with the number of species because the more species present, the more there are to become extinct, and the more likely any one species is to become extinct because of reduced population size. The immigration and extinction curves intersect at the equilibrium species richness, where species gains are exactly balanced by losses. The equilibrium is high on large, near islands, and low on small, distant ones.

A great many factors influence species richness on islands in addition to those considered by the Equilibrium Theory. For example, large islands usually have a greater variety of habitats than small ones, and most large islands have a wider range of altitudes than small ones (i.e. their beta diversity is higher). Species may also tend to become extinct faster on small islands because disturbances are more frequent or more devastating there (McGuiness, 1984). Furthermore, the floras of many islands contain ancient, relic species. For instance, the island of Crete in the Eastern Mediterranean contains about 10% endemic species, many of which were once distributed over much larger areas of southern Europe. These now persist in only a few rocky gorges where they have survived the changing climates of several ice ages, and escaped the attentions of the ubiquitous goats (Turrill, 1929). In this case, the species survive on the island because of the *lower* rate of extinction there. The rapid spread of alien plants through modern plant communities (Crawley, 1986), forcibly reminds us that species which grow together did not necessarily evolve together. The invasion of communities by pre-adapted species which evolved elsewhere may account for a considerable proportion of modern floras, compared with the number of species which actually co-evolved with their current coinhabitants.

The value of the Equilibrium Theory is that it provides a base line for explaining interesting patterns; the species-area curves which result can be used when comparing plant communities to highlight biologically interesting differences.

1.5   Species abundance distributions

For all the wealth of descriptive data on vegetation structure, there is still much to be learned about the factors controlling floristic composition, 'one of the most fascinating as well as urgent problems confronting the plant ecologist' (Watt, 1971). The question of the determination of relative abundance is extremely complex, of course, for until we know what regulates the abundance of *single* species, it is unlikely that we shall be able to understand the determination of the relative abundance of all species!

Two communities composed of the same plant species can differ greatly in structure depending upon their relative abundance distributions. Communities in which the species are all more or less equal in abundance are said to exhibit high 'evenness' (or high 'equitability'). Communities with one or two very abundant species and many rare ones are said to exhibit pronounced dominance (Fig. 1.6). The standard techniques for displaying relative abundance data are either to: 1) plot log abundance (measured as frequency, cover, biomass or primary productivity) ranked from the most to the least abundant species (Fig. 1.7a); or 2) group the species into abundance 'octaves' (1, 2, 4, 8, 16, 32, 64 etc., individuals per species), and to

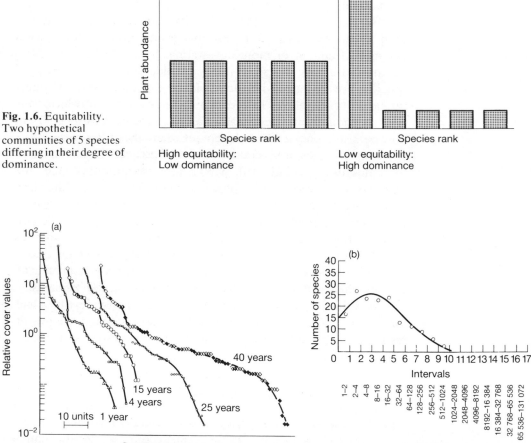

Fig. 1.6. Equitability. Two hypothetical communities of 5 species differing in their degree of dominance.

Fig. 1.7. Presentation of relative abundance data. (*a*) Rank abundance (or importance value) curves showing the relative abundance of plant species in old fields in 5 stages of abandonment in southern Illinois, USA, expressed as the percentage of total cover made up by each species, ranked from most abundant on the left, to least abundant on the right. Open symbols = herbs; half-open = shrubs; closed = trees. From Bazzaz (1975). The slope of the rank abundance curve always decreases as species richness increases. (*b*) Log normal distribution showing the frequency of species whose abundance lies in different 'octaves'. Relative abundance of diatom species in a sample taken from a stream in Pennsylvania, USA. From Patrick (1973).

plot the frequency of species in each class (Fig. 1.7b). The resulting histogram often approximates a truncated log normal distribution (Preston, 1962; see May, 1981, for details). There are many ways of characterizing the evenness of species' abundance distributions (essentially the 'flatness' of the curves in Fig. 1.7a), but perhaps the best measure of dominance is simply the proportion of total abundance made up by the single most abundant species ($p_1 = n_1/T$, where n_1 is the abundance of the commonest species and T is the total abundance of all species; May, 1981).

Since it only takes one species to fluctuate in order to alter relative abundance within the entire community, relative abundance is only as constant as the community's least constant species. Where relative abundance *is* observed to be constant, this may be due to low rates of turnover (e.g. individual plants are long-lived relative to the study period). Alternatively, relative abundance may be constant despite high rates of turnover of relatively short-lived individuals, in which case the ecological mechanisms responsible for stability are of considerable interest. What appear, superficially, to be rather stable abundance distributions, may prove to be highly variable after long-term study. A classic example is provided by the Breckland plots studied by Watt (1981), in which a series of three different dominant species succeeded one another over the course of 44 years (see Chapter 5). Short term study (three or four years) would no doubt have concluded that any one of these three communities exhibited a rather stable relative abundance of species! The quadrats in this study were very small (only 10 cm²), however, and the results might be criticized as showing little more than the waxing and waning of individual genets of different plant species. Relative abundance studies should ideally be carried out in quadrats large enough to contain *hundreds* of genetic individuals.

Most relative abundance distributions are drawn up with data on mature plants. However, the structure of the distribution reflects the outcome of a *series* of population dynamics processes: seed immigration, germination, establishment, growth and mortality. Differential seed production between species, coupled with differential recruitment, growth and mortality determine the final distribution of mature plants which we observe. The balance of relative performance may therefore fluctuate between species during the course of development; for example one species may have relatively high rates of seed immigration, another might have high germination, a third could exhibit high seedling survival, another show low rates of herbivore-induced mortality, a fifth might suffer less from shading as a mature plant, and so on. Torssell *et al.* (1975) describe an example of this kind of fluctuating relative performance for *Stylosanthes humilis* and *Digitaria ciliaris* in annual grasslands in the Northern Territory of Australia (Fig. 1.8).

Fig. 1.8. Changes in relative abundance of two species from an annual pasture in a dry monsoonal climate in northern Australia. The legume *Stylosanthes humilis* (A) and the grass *Digitaria ciliaris* (B) show differing rates of loss in the seed, seedling, vegetative and reproductive stages of their life cycles. The legume shows higher rates of establishment, but the grass shows higher rates of seed production and seed survival. After Torssell *et al.* (1975).

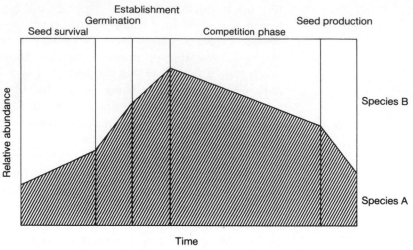

The long-term average distribution of relative abundance may reflect a dynamic equilibrium based on niche specialization of the component species, or it may be a non-equilibrium pattern reflecting the disturbance rate, species' longevities, site pre-emption, and so on (Section 1.4). In the first case, different species may have very specific microsite requirements ('regeneration niches'; Grubb, 1977), and the relative abundance of mature plants may simply reflect the relative abundance of suitable microsites at the time of recruitment. Alternatively, microsites may be more or less equally abundant for all species, and niche attributes such as maximal growth rates, shade tolerance, competitive ability for limiting soil nutrients, phenology, longevity, pest-resistance and so on, may play the vital role. If the community is not in equilibrium, then the frequency and intensity of disturbance, and the timing of plant reproduction relative to the availability of suitable microsites will determine the relative abundance of species. The species composition of non-equilibrium communities will not tend to recover following disturbance caused by the experimental addition or removal of plants, and nearby sites will often show different patterns of change (Sale, 1977).

As we have seen (Section 1.4), plants which are consistently rare must have periods (however brief) when they are advantaged relative to the commoner species if they are to persist in an equilibrium community. The rarer species of a Missouri prairie, for example, have been found to possess more dispersible seeds (Rabinowitz & Rapp, 1981), and to show a greater capacity for interference in greenhouse experiments (Rabinowitz, 1981). These matters are discussed in more detail in Chapters 2 and 3, but the question as to which plant communities, if any, can be regarded as equilibrium assemblages remains open to debate.

1.6   The physical structure of plant communities

Plant communities exhibit a wide variety of three-dimensional structures, and show spatial heterogeneity on many scales. The causes and consequences of this physical structure are felt throughout the animal and plant communities (see Lawton, 1983).

1.6.1   Life-forms in plant communities

The great biomes of the world show a remarkable degree of convergence in their physiognomy, despite wide differences in the taxonomic affinities of their floras. Thus, experts aside, we would be hard-pressed to distinguish between the chaparral of California and Chile (Mooney, 1977), or the nutrient-poor heaths of South Africa and Australia (Milewski, 1983). Arctic fell-fields look much like Antarctic (Smith, 1984), and rain forest in Brazil resembles rain forest in south east Asia (Leigh, 1975). This convergence of community physiognomy is vivid testimony of the importance of climatic factors as agents of natural selection. While there are numerous subtle differences in species richness and in the relative abundances of different forms, it is clear that there are broad trends in the dominant life-forms comprising plant communities under different climatic conditions.

Raunkiaer (1934) classified plants according to their adaptations for surviving the unfavourable season (winter cold or summer drought). These 'life-forms' (see Chapter 8) can be summarized into spectra describing the proportion of the total flora made up of different kinds of plant. Figure 1.9 shows the life form spectra of five different climates, compared with the spectrum of the world's flora as a whole. Several points emerge from these comparisons: 1) Under ideal growth conditions (constant warmth and moisture) trees are dominant, simply because the competitive spoils go to the tallest individuals; 2) Where there is no unfavourable season, or the less favourable season is not too severe, then tree-like plants (phanerophytes) predominate in the flora as a whole; 3) In less equable climates, trees may still be the dominant plants in most communities, but the flora as a whole is made up predominantly of other life forms (e.g. hemicryptophytes in northern temperate latitudes); 4) Where the summer is arid (as in deserts and Mediterranean climates), there is a preponderance of annual plants (which *avoid* drought by passing the dry period as dormant seeds) and geophytes (which avoid drought by die-back of their above ground parts, and survive by means of underground storage organs). Other desert plants which *tolerate* drought may appear to be dominant if the vegetation is surveyed during the dry season (e.g. xerophytic shrubs and stem succulents); 5) Where extreme cold and exposure characterize the

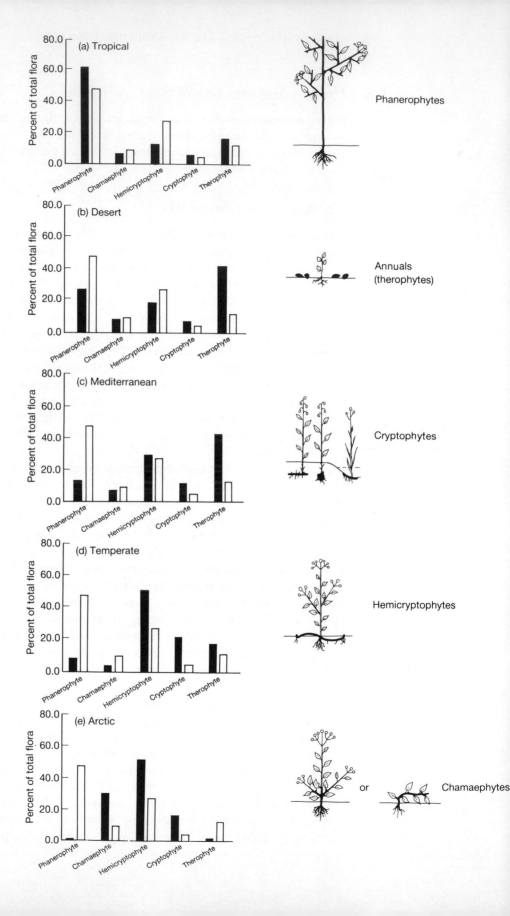

unfavourable season, there is a shift towards cushion-forming plants and other species whose buds are held close to the soil surface (chamaephytes). Many arctic plants have their perennating buds protected by the dead leaf-bases of last year's shoots, because both the exposed aerial environment and the frozen soil are extremely inhospitable conditions for bud survival.

We should be careful not to read too much into the percentage make-up of entire floras. In terms of community dynamics it is much more important to know that trees dominate the vegetation than that hemicryptophytes are predominant in the flora! Again, the length of time a community has remained undisturbed, and its history of human influence can have profound effects on the relative abundance of different life forms (compare certain northern hemisphere Mediterranean communities (e.g. Californian chaparral) with their southern counterparts (e.g. South African fynbos; see Shimda & Whittaker, 1984). Nevertheless, Raunkiaer's system does serve a valuable purpose in focusing our attention on the important (and as yet unanswered) question of why certain life forms coexist in fairly predictable proportions in communities from quite different floristic regions.

1.6.2   Vertical structure of plant communities

The maximum height to which vegetation can grow is determined largely by climate. Wherever water is continuously available, and there is shelter from strong, drying winds, then tall trees are found. In the continuously warm, wet conditions of tropical rain forest, plant communities reach their pinnacle of vertical structuring, with a closed canopy at 25–45 m, a mosaic of light-gaps at differing heights (Chapter 3), understory species, regenerating canopy trees, vast buttressed trees emerging 20 m above the canopy top, festoons of lianes and epiphytes, and so on. Richards (1983) cautions against too broad an extrapolation of a multi-strata, 'text book' model of rain forest structure. He points out that while single-dominant tropical rain forests do show clearly defined strata, mixed forests usually do not, and that, more important than the stratification of the trees is the boundary between the euphotic zone in which the crowns are more or less fully exposed to sunlight, and the shaded zone (the undergrowth) beneath. Similarly, much of the vertical structuring is a result of gap-formation and the dynamics of gap-recolonization (Chapter 3),

Fig. 1.9. Life forms in different biome types. Solid bars show the percentages of 5 life forms in a range of biomes. For comparison, the open bars show the composition of the world's flora as a whole (note that these figures are biased somewhat by the fact that most of the world's vascular plant species are tropical in origin). Each life form reaches its peak representation in a different biome: (*a*) phanerophytes in the tropics; (*b*) annuals in deserts and Mediterranean ecosystems; (*c*) cryptophytes in Mediterranean and temperate systems; (*d*) hemicryptophytes in temperate ecosystems; and (*e*) chamaephytes in arctic and alpine habitats.

rather than a consequence of the coexistence of specialized sub-canopy, or 'shrub layer' species beneath the crowns of the dominant trees.

The world's tallest plants are found in the temperate rain forests of the west coast of North America, from northern California to the Olympic penninsula of Washington. Species like *Sequoia semper-virens* and *Pseudotsuga menziesii* reach extraordinary heights (over 80 m), and achieve massive girths (up to 20 m). Despite their great height, the vertical stratification of these communities does not approach the complexity of tropical rain forest. In areas of sporadic rainfall, the maximum height of the trees is reduced and the spacing between one tree and the next is increased. Vertical development of the woodland becomes progressively impoverished as aridity increases, until the community passes into savanna with sparsely scattered, short trees surrounded by open grassland. Eventually, if rainfall is too low, fires too frequent, or exposure too great, tree growth is precluded altogether. In such cases, vertical community development is restricted to dwarf woody shrubs (as in Northern ericaceous heathlands), herbaceous or woody cushion plants (alpine and certain arctic environments with prolonged snow cover), sclerophyllous scrub (in desert regions), or lichen heath (high latitude coastal communities).

The other major factor limiting the vertical development of communities is feeding by vertebrate herbivores. Prolonged exposure to browsing herbivores can lead to the destruction of existing woodland (e.g. elephants in *Terminalia* woodlands in east Africa; Laws *et al.*, 1975), or to the prevention of woodland regeneration (e.g. by sheep and rabbits in British oak woodlands; Pigott, 1983). Browsing by animals like giraffes can sculpture isolated plants into the most peculiar topiaries, whilst deer can destroy the entire shrub and ground layers of the forests in which they overwinter (Crawley, 1983).

One of the most profound effects of vertical community structure is on total plant species richness. In tropical forests, for example, increased vertical complexity allows substantial increases in species richness by creating new habitats for plants like lianes (Putz, 1984) and vascular epiphytes (Benzig, 1983), growth forms which are rare or absent in other kinds of communities (see Chapter 8). The influence of vertical structure is also felt indirectly through its effects on microclimate (temperature and gas profiles, air flow and turbulence, wind pollination; Wellington & Trimble, 1984), and through effects on the structure of the animal community. There are close correlations, for example, between foliage height diversity and the number of bird species found in woodlands (MacArthur & MacArthur, 1961), and between insect species richness and the vertical complexity of grasslands (Morris, 1981).

1.6.3 Spatial structure of plant communities

Wherever we travel, we find that the environment alters from place to place, indeed often from step to step; correspondingly, the vegetation alters in both species composition and in the frequency of the component species. Figure 1.10 shows plant communities with two extreme spatial structures. In the first case there are distinct zones of different plant species, as we might see in the emergent vegetation surrounding shallow lakes. In the second, there are indistinct patterns, one grading imperceptibly into the other, as we might find in a transect up a forested mountain-side. We need to determine the extent to which these spatial patterns in plant community structure reflect changes in abiotic conditions (soil nutrients, water depth, etc.), and to what extent the patterns are the result of biotic interactions (interspecific plant competition, seed dispersal, herbivorous animals, etc.). Also, to what extent do changing abiotic conditions alter the outcome of biotic interactions? For example, changing ratios of limiting soil nutrients may alter the relative competitive abilities of different plant species (Chapter 2).

One of the most fundamental questions about the nature of plant communities concerns the way in which species respond to gradients in environmental conditions. Do whole recognizable *sets* of species appear and disappear at points along a gradient (as in Fig. 1.10a), or do species come and go more or less independently of one another along the gradient (Fig. 1.10b)? In the first case we would be in no doubt that there were distinct, recognizable communities. In the second case, one mix of species grades imperceptibly into another as we pass along the gradient; what we define as a community is therefore simply a matter of convenience (an identical community would not occur anywhere else). This distinction forms the basis for the long-running (and still unresolved) dispute between those who wish to classify and name discrete plant communities (the phytosociologists—e.g. Braun-Blanquet, 1932; Becking, 1957; Soo, 1964–73) and those who see plant communities as continuously variable (the continuum school—Curtis & McIntosh, 1951; Whittaker, 1967). At this stage it is sufficient to note that, as in so many ecological disputes, it is the kind of system one studies that colours one's view of the ecological world as a whole.

In Europe, where man has influenced all the plant communities for so long, and the scale of semi-natural vegetation is so limited, the attention of the ecologist has been focused on very abrupt changes in environmental conditions in very small patches of vegetation. For example, the sharp transition from a bracken clad slope to a *Juncus* dominated valley bottom, or the immediate change from heather moor to base-rich fen where a calcareous stream cuts through an acid peat-bog, define quite unequivocal plant communities where there is

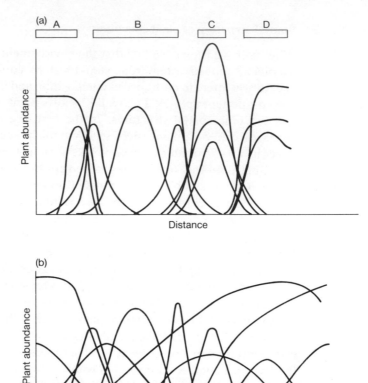

Fig. 1.10. The spatial structure of plant communities. Transects through two hypothetical plant communities: (*a*) where there are clear, discrete communities (A, B, C, D) separated by narrow intermediate zones known as ecotones (shown by the gaps between the open bars); (*b*) where there are no clear boundaries between communities, and plant species come and go along the transect more or less independently of one another. After Whittaker (1972).

very little overlap in species composition between one and the other (Tansley, 1939). In short, abrupt environmental transitions create distinctive plant communities. Even so, making these communities sufficiently uniform to permit the construction of a taxonomy requires the subjective elimination of a great many vegetation samples, including: 1) communities which are 'obviously unstable'; 2) small fragments; 3) ecotones; and 4) mosaics (Poore, 1955)!

American ecologists, working on vast expanses of almost virgin forest, took a different view. They saw a world structured by rather smooth, gradual changes in environmental conditions. For example, the alluvial floodplain forest of the Mississippi stretches all the way from Minnesota to Louisiana; no ecologist would refer the forests of the upper and lower Mississippi to the same plant community, yet there is no place along the entire length where one can logically mark a boundary between them (Gleason, 1926). In such a world it was impossible to erect a taxonomy of plant communities because no two sample areas contained the same species in the same relative abundances. These areas had to be described in terms of a continuum (Curtis, 1959), and effort spent in trying to define fixed communities was effort essentially wasted (Davis, 1981). Of course, the fact that vegetation changes as a continuum does not preclude our using sensible names to describe different communities. After all, electromagnetic radiation changes continuously in wave-length through the visible spectrum, but this does not stop us describing colours as red or blue. In the same way, the names we give to plant communities are labels of convenience, rather than descriptions of rigidly structured, highly integrated, deterministic associations of plants.

Even where recognizable boundaries between communities do occur, their positions are not necessarily fixed. For example, Gleason (1927) describes the mosaic of prairie and alluvial floodplain forest found in Illinois. Prairie can not invade forest because the majority of prairie species are intolerant of shade. On the other hand, the density of the prairie turf, the action of vertebrate herbivores, and the high competitive ability of grasses for water, preclude the invasion of prairie by tree seedlings. The comings and goings of forest and prairie therefore reflect the slow advance of the forest due to the shade cast by its leading, overhanging edge, abetted by the slow advance caused by erosion which gradually widens the river banks and makes more habitat available to forest. However, forest plants suffer much more severely from fire than do the hemicryptophytes of the prairie (see Chapter 8), so forest tends to be replaced by prairie in areas where fires are frequent.

Modern methods of multivariate analysis involving gradient analysis, ordination and classification are described by Gauch (1982). This proliferation of multivariate techniques for the analysis of spatial variation in plant community structure has not, however, led to great advances in our understanding of the processes underlying these patterns. As May (1985) observes, "the wilderness of meticulous classification and ordination of plant communities, in which plant ecology has wandered for so long, began in the pursuit of answers to questions but then became an activity simply for its own sake"!

1.6.4   Allelopathy and spatial patterns

In addition to competing with one another through the exploitation of shared resources, plants may also compete by interference (Box 1.4). This is most likely to take the form of suppression of the

Box 1.4 Competition

Competition is an interaction between two individual plants that reduces the fitness of one or both of them. Some ecologists insist that the word competition be reserved for mutually deleterious interactions ($-/-$), and that asymmetric interactions ($0/-$), where one party suffers little or no reduction in fitness be called 'amensalisms'. In this book we use the one word, competition, to cover both kinds of interactions, because: 1) a great many plant–plant interactions are asymmetric; and 2) we often do not know the fitness implications of particular interactions in advance, so that calling them amensal rather begs the question.

Inter or intra?

Interactions between individuals of the *same* species are called intraspecific competition, while interactions between individuals of *different* species are called interspecific competition.

Interference or exploitation?

Exploitation occurs when plants compete by reducing the availability of shared, limiting resources such as light, water, soil nutrients or germination microsites. Exploitation may lead to more or less asymmetric competition (see below).

Interference occurs when fitness is reduced by essentially 'behavioural' mechanisms, which do not directly involve limiting, shared resources. For example, plants interfere when they compete via the production of allelopathic chemicals. Interference is usually strongly asymmetric.

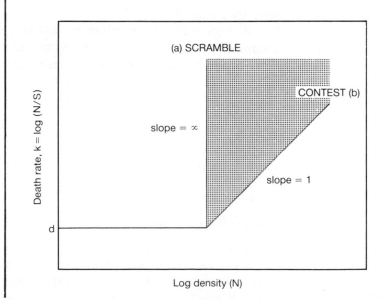

Contest or scramble?

There is a continuum of exploitation competition from highly asymmetric (contest) to rather symmetric (scramble). In contest competition there are winners and losers; one individual gets all the resources it requires while another gets only the 'left-overs'. For example, plants contest for light, since the taller individual can pre-empt the incoming radiation. In scramble, the limited resources are divided equally between the competitors. Plants with similar root morphologies might scramble for soil nutrients. These two kinds of competition have different effects on population dynamics. If we plot the death rate (or some other measure of fitness loss) against the density of competing plants (on log axes), we obtain graphs of the form shown above.

At low densities there is no competition at all, but above a certain threshold the rate of resource supply is insufficient for every plant to meet its full growth potential, and competition begins. In scramble, all plants suffer equally, and in the limit no plants obtain sufficient resources to survive, so the death rate (k) is infinite (curve a). In pure contest, a *fixed number* of plants can be supported at a given rate of resource supply, so the death rate rises linearly with log density with a slope of 1.0 (curve b). Most real competitive interactions will fall between these extremes. We call the interactions scramble when their curves are close to a in shape, and contest when they are more like b. In terms of dynamics, scramble competition is destabilizing compared with contest (see Begon, Harper & Townsend, 1986, for details).

germination or growth of neighbouring plants by chemicals which are leached, exuded or volatalized from the plant (so called 'allelo-chemicals'; see Rice, 1984).

Some of the most pronounced patterns of zonation in Californian annual grasslands are thought to be due to allelopathic interactions between long-lived shrubs like *Salvia leucophylla* and the annual herbs and grasses (Muller & del Moral, 1966). There is a bare zone of up to 2 m around each *Salvia* plant, and a zone from 3–8 m containing stunted plants of *Bromus mollis*, *Erodium cicutarium* and *Festuca megalura*. Muller & del Moral suggest that the zonation is initiated and perpetuated by volatile terpenes, evolved into the air from the leaves of the *Salvia* during periods of high temperature, adsorbed by the dry soil, and held there until the early part of the following growing season. Seeds and seedlings in contact with terpene-containing soils extract some of the terpenes by solution in cutin, and the terpenes are transported into the cells via phospholipids in the plasmodesmata.

As with all hypotheses of this kind, it is exceptionally difficult to devise experiments which show unequivocally that the observed pattern is due to allelopathic interference between plants, rather than to competition for water, light or nutrients, to subtle soil interactions, or to the influence of herbivorous animals (Harper, 1975). Bartholomew (1970), for example, suggested on the basis of fencing experiments, that the bare zones were caused by grazing animals rather than by allelochemicals. In one of the most thorough attempts to eliminate other possible explanations, McPherson and Muller (1969) examined the causes of the bare zones around the Californian shrub *Adenostoma fasciculatum*. Fertilizer and shading experiments ruled out the likelihood of competition for nutrients or light. Competition for water was dismissed because *Adenostoma* has a very deep root system, and the phenology of the excluded annual plants ensured that they were not subject to drought (their germination was triggered by rainfall). *Adenostoma* produced no volatile terpenes, but toxins did leach from its leaves after each rain. These accumulated in the soil and inhibited the germination of annual plants. On fenced plots where the shrub was cut back to ground level, more than 1000 seedlings m² appeared, compared with only 40/m² in the uncut controls. Small mammals did reduce recruitment beneath the shrubs (fencing increased seedling density from 40 to 70/m²), but the effect was small in comparison with the removal of the *Adenostoma* foliage and the consequent reduction in the input of toxic leachates.

The production of allelopathic compounds in living plants can be influenced by light quality (e.g. many plants produce less under glass), nutrient deficiency, drought, extremes of temperature, or by pathogen and herbivore attack (see Chapter 12). Allelochemicals are typically taken up by roots and translocated through xylem in the transpiration stream. In addition to volatilization, root exudation (Foy *et al.*, 1971) and leaching from leaves (del Moral & Muller, 1969), allelochemicals may move from plant to plant via natural root grafts (Graham & Bormann, 1966), fungal bridges (Harley & Smith, 1983) or via the haustoria of parasitic vascular plants (Rice, 1984).

The modes of action of allelochemicals include: 1) inhibition of cell division in root meristems; 2) inhibition of oxidation of plant hormones like indoleacetic acid (IAA); 3) altered membrane permeability; 4) interference with mineral uptake (ash roots, for example, inhibit the uptake of labelled ^{32}P by oak roots); 5) reduced stomatal aperture leading to impaired photosynthetic rate; 6) interference with mitochondrial function; 7) reduced protein synthesis; 8) inhibition or stimulation of specific enzymes; 9) clogging of the xylem leading to impaired water balance; 10) various indirect effects, including inhibition of soil micro-organisms, leading to reduced rates of decomposition, or altered chemical composition of the breakdown products (Rice, 1984). Subtle allelopathic effects are likely to occur in many (if not most) plant communities (e.g. Newman & Rovira,

1975), and the experimental challenge lies in determining their importance in community dynamics in comparison with other ecological processes like exploitative competition, natural enemy attack, mutualistic relationships, and so on.

1.6.5   Quantitative methods for describing spatial patterns

Perhaps more energy has been devoted to this aspect of plant ecology than to any other (Goodall, 1952; Pielou, 1977; Grieg-Smith, 1983). The spatial patterns described in these studies vary in scale from a few millimetres squared to many kilometres squared and are the result of very different ecological processes. On the smallest scales, the patterns reflect the interactions between individual plants (Chapter 4) and are therefore the outcome (at least in part) of processes involved in plant population dynamics. Beyond the zone of immediate, first order neighbours, direct plant–plant interactions are relatively unimportant (Chapter 5), and pattern analyses on larger scales deal either with patterns of underlying soil conditions (water table, nutrient availability, heavy metal contamination), or with the growth patterns of spreading, clonal plants (e.g. Kershaw, 1963). Spatially variable biotic processes do occur on larger scales (e.g. the foraging of large vertebrate herbivores, or the activities of pollinating and seed dispersing animals; Chapter 6), but these are rarely addressed in traditional pattern analyses.

The desirable properties of a measure of plant spatial pattern include: 1) that the *degree* of non-randomness can be estimated (it is not enough simply to say that the pattern is not random); 2) that the measure is insensitive to the size of sampling unit chosen; 3) that the measure is easy to estimate from data and readily applied in theoretical models; and 4) that the measure is not a function of plant density (so that density-independent mortality does not alter the estimate of pattern). In fact, it is impossible to combine all these features in one measure. For instance, the parameter k of the negative binomial is a useful descriptor of the degree of aggregation, but the value of k is affected by density (see Elliot, 1977, for a fully worked example). Similarly, quadrat methods are easy to apply in the field, but *all* fixed-area quadrats influence the kind of pattern they detect (see Section 1.6.6 below). Lloyd's (1967) index of 'mean crowding' is useful because it is not sensitive to density independent mortality, but it is cumbersome to use in theoretical models (Pielou, 1977). Plotless (nearest neighbour) methods do not suffer the disadvantages of fixed area quadrats, but they only work well for plants with discrete individuals (like annuals and trees), and are difficult to incorporate in simple analytical models (Cottam & Curtis, 1949; Goodall, 1965; Byth & Ripley, 1980). For a full discussion of the pros and cons of the different methods, see Grieg-Smith (1983) and Bartlett (1975). A clear explanation of more advanced models of spatial patterning is given by Diggle (1983).

Many of the early pattern analyses were attempts to correlate obvious surface features of the vegetation with underlying soil conditions (Knapp, 1974). These studies suffered from the age-old problem that correlation does not imply causation, and it was always difficult to know whether or not the environmental factor they measured was the direct cause of the pattern. Many of the studies also failed to define a sensible null hypothesis; they did not ask 'what pattern would be expected if the hypothesized environmental factor were *not* operating'? In other words, they had no 'null pattern' with which to compare their field measurements (Harvey *et al.*, 1983). All too often a random (Poisson) distribution was taken as the null hypothesis without any firm grounds for believing it to be appropriate. For example, soil conditions are unlikely to be uniform, with the consequence that plants are likely to be clumped in the 'better' patches. Similarly, plants might be aggregated due to limited seed dispersal around parents, clonal spread, and so on. Thus, finding that field data are not described by the Poisson distribution tells us nothing that we did not already know! On the other hand, tests of complete spatial randomness (CSR) can form a useful starting point, if only for the reason that when a pattern does *not* depart significantly from CSR, it scarcely merits any further formal statistical analysis (Diggle, 1983).

In discussing the use and usefulness of pattern analysis, it is well worth tracing the development of the controversy between those who believed that regular spacing in desert shrubs was indicative of competition for water (Woodell *et al.*, 1969; Yeaton & Cody, 1976) and those who challenged this view (Anderson, 1971; Barbour, 1973; Ebert & McMaster, 1981). The acid test of this kind of hypothesis is a manipulative field experiment. For example, the water potentials of plants could be measured before and after their neighbours were removed. Such experiments demonstrate that some species (e.g. *Larrea tridentata*) do compete for water and therefore benefit when their neighbours are cut down. Other species (e.g. *Ambrosia dumosa*) have only a brief growing season following rain, and water does not appear to be a limiting resource, despite the general aridity (Fonteyn & Mahall, 1978, 1981). The important point is that it was the *experiments* and not the pattern analyses which provided the vital evidence for testing the hypothesis of competition between the individual shrubs.

1.6.6   Spatial pattern and quadrat size

Since we recognize spatial patterns in plant distribution on several natural scales (shoots within individuals, separate individuals, groups of individuals, etc.), it is not surprising that the *size* of quadrat we choose, influences the *kind* of pattern we detect (Aberdeen, 1958). For instance, if the quadrats are big enough for many individuals to fit within one quadrat, we should be able to detect clumping of individuals (caused, for example, by restricted seed dispersal about parent plants, or by underlying spatial heterogeneity in soils). If, on the other hand, the individual plants are about the same size as the quadrats (e.g. clonal plants of grasslands) then pattern analysis measures nothing more than the proportion of the habitat occupied (the analysis is essentially a description of frequency; see Box 1.3). If the quadrats are smaller than the individual plants (e.g. they may contain several to many shoots of the same clonal individual), then pattern analysis only tells us that shoots of the same individual tend to be clumped! This dependence of the pattern on quadrat size is illustrated in Fig. 1.11. The limitations of quadrat-based pattern analysis are so severe, and the alternative methods of nearest-neighbour analysis are so time-consuming and technically so difficult, that we should only undertake pattern analyses when they are absolutely necessary in order to answer very clearly formulated questions.

Quadrat size can also affect both the detection and the interpretation of interspecific association (Grieg-Smith, 1983). If the quadrats are too small, then species will appear to be negatively associated simply because two individuals cannot occupy the same space. In

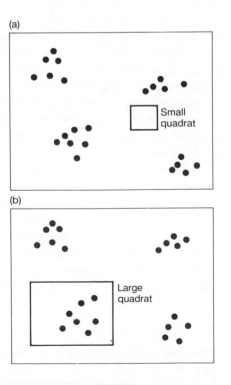

Fig. 1.11. The effect of quadrat size on pattern determination. Varying the size of the sampling quadrat leads to interpretation of the same plant distribution in quite different ways: (*a*) small quadrats detect the pattern of plants as being aggregated; (*b*) large quadrats detect the same pattern as being regular.

medium-sized quadrats, large individuals can cause spurious positive associations between smaller plant species, since all smaller plants are constrained to grow in the gaps between the larger ones. Large quadrats may fail to detect genuine negative associations between species if the scale of spatial heterogeneity is small (i.e. one quadrat may contain several different patches, some with species A and some with B, even though A and B cannot coexist in any one kind of patch).

Quadrat size also influences the measurement of plant species richness. There may be *apparent* changes in species richness when an experimental treatment leads to dramatic changes in the size of individual plants. For example, cessation of mowing leads to greatly increased size of a few individuals, and thus to a reduction in the number of plants, and hence of species, found in small quadrats (as on Darwin's celebrated lawn (1859)).

1.6.7   Spatial patterns reflecting temporal changes

A great many of the spatial patterns which we observe in plant communities today reflect processes of recovery from disturbances which occurred at different times in the past. Sometimes, these patterns are caused by localized, external forces like violent storms, land-slips or lava-flows. When the disturbance does not kill the plants (as in many communities regularly affected by fire) the patterns may represent nothing more than stages in the *regrowth* of the dominant species. In other cases, the changes are driven by the life histories of the individual plants themselves, as they pass through their pioneer, building, mature and degenerate phases of growth (e.g. the 'stand cycles' described by Watt (1947, 1955) for *Calluna vulgaris* and *Pteridium aquilinum*). Aubreville (1971) suggested that spatial variation in species richness in tropical rain forest trees is maintained by temporal variation in species composition at a given location. He saw the forest as a mosaic of patches at different stages in a cyclic succession. Hubbell & Foster (1986) tested Aubreville's mosaic hypothesis of tropical forest regeneration for their Barro Colorado Island plot (described in Chapter 3), and conclude that there is little evidence to support it. In very few cases was the probability of self-replacement significantly lower than of replacement by any other species. The *apparent* mosaic is simply a result of the high tree species richness and the consequently low probability of replacement of a tree by an individual of its own kind.

The classic studies of succession (Cowles, 1899; Clements, 1928; Crocker & Major, 1955) used observations on the structure of plant communities growing at different points in space (e.g. at different distances from the snout of a glacier, or from the front of the youngest sand dune) in order to *infer* the changes which occurred in plant communities through time. These, and other direct, long-term studies of successional changes, form the subject matter of the next section.

1.7 Succession

Succession is the process whereby one plant community changes into another. It involves the immigration and extinction of species, coupled with changes in the relative abundance of different plants. Succession represents community dynamics occurring on a time scale of the order of the life-spans of the dominant plants (in contrast to much slower, evolutionary changes, occurring over hundreds or thousands of generations, or the much more rapid seasonal or annual fluctuations in species' abundances). Succession occurs because, for each species, the probability of establishment changes through time, as both the abiotic environment (e.g. soil conditions, light intensity) and the biotic environment (e.g. the abundance of natural enemies, the nature and competitive ability of neighbouring plants) are altered. The degree to which these changes follow a predictable sequence determines the extent to which succession can be viewed as an orderly process. Some successions converge to rather uniform, predictable end-points, independent of the initial conditions. Others are non-converging or cyclic, or have alternative stable end-points, with their dynamics completely dominated by the history of disturbance and immigration. Thorough reviews of succession are to be found in Drury & Nisbet (1973); Connell & Slayter (1977); Noble & Slayter (1980) and Crawley et al. (1986).

The great debate in the 1920s centred on whether the changes during succession were orderly and 'goal-directed' (headed towards a invariable climax community), or whether they were essentially random, with many possible end-points determined by initial conditions, accidents of immigration and by periodic, patchy episodes of disturbance (see Section 1.2). The modern debate is no less intense, but concentrates on rather different processes. Nowadays, the role of chance in succession is universally admitted, and the central question concerns whether or not it is sensible to regard succession as a single process at all.

1.7.1 Interglacial cycles

As the polar ice-caps advanced and retreated during the Pleistocene, so individual plant species migrated southwards and northwards across the continents. As they migrated, so the composition of different plant communities altered (e.g. different North American tree species have advanced at average rates of between 100 and 400 m per year since the end of the last ice age; Davis, 1981). During a 'typical' interglacial period we can recognize four broad phases of climate, soil and community development (Godwin, 1975). Immediately following the retreat of the ice, there is an open community characterized by low temperatures and high exposure, supporting scattered arctic–alpine plants on soils of high pH. Next, as temperatures rise and leaching begins to reduce pH, a neutral grassland or scrub develops. By the height of the interglacial, weathering and leaching have further reduced the pH of the soil, and shade-casting, broadleaved woodlands have developed on slightly acid, brown forest soils. As the ice re-advances, declining temperatures coupled with continued leaching of already acid soils, lead to the replacement of broadleaved woodland by coniferous woodland and heathland on very acid, podsolized soils. Similar cycles can be reconstructed from fossil plant remains preserved in lake sediments throughout the north-temperate regions (Pennington, 1967; Godwin, 1975).

In regions that have not been glaciated at all during the Pleistocene (e.g. Australia), the process of leaching continues for millennia. Without the periodic input of nutrients that comes from glacial deposition, soils tend to be exceptionally low in nutrients such as phosphorus. These areas support unusual floras, and tend to undergo successions of a rather different kind (Noble & Slatyer, 1980; Rice & Westoby, 1983).

The impact of glacial cycles on the floristic composition of tropical plant communities is the subject of considerable debate (Sutton *et al.*, 1983). It may be that the advancing ice sheets are a prime factor promoting the high species diversity of tropical rain forest trees, because they forced large numbers of trees to coexist in small tropical refuges during the dry periods associated with the maximum advance of the glaciers. The trees may have evolved in isolation, only to be forced into cohabitation during later glacial episodes. In addition, it is likely that during longer periods of genetic isolation in fragmented forest refuges, speciation occurred which further increased species diversity. Far from representing long-term stable communities, therefore, it is possible that current tropical rain forests owe their high species richness to the fact that they have *not* remained stable long enough to cause the competitive exclusion of many of the tree species thrown together during the ice ages (Section 1.4).

1.7.2   Primary succession

On August 27 1883, telegrams reached Singapore from Jakarta, capital of the Dutch East Indies; 'During the night terrific detonations from Krakatoa (volcanic island, Straits of Sunda) audible as far as Soerakarta—ashes falling as far as Cheribon—flashes visible from here'. By the end of the eruption, only one-third of the entire island of Krakatau remained, and that was covered several metres deep in ash and pumice. New islands appeared where the sea had previously been 36 m deep. The explosion was heard 4652 km away and ash fell on ships 6076 km away. Giant waves, 40 m high, devastated everything in their path and caused the deaths of 36,417 people, washing away 165 coastal villages (Simkin & Fiske, 1983).

No plant life at all survived the eruption, but the island was invaded by a series of plant species over the next 45 years (Fig. 1.12; Docters van Leeuwen, 1929). This kind of curve of species accumulation is probably typical of the early stages of primary successions, and reflects a change in both species richness and life form through time. The first colonists were cryptogams, and the last were shade-loving forest species. No doubt the curve would reach an asymptote if followed for a longer period.

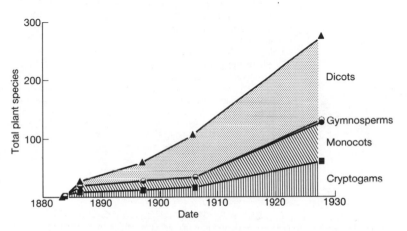

Fig. 1.12. Primary succession following the volcanic eruption which destroyed the island of Krakatoa. Plant species richness increased rapidly over the first 40 years, associated with changes in the relative abundance of different life forms (see text for details).

The life-history of the initial colonists of primary successions is determined largely by the nature of the substrate. On smooth bare rock, they will be lichens; on ribbed rock, bryophytes will appear first; on block scree the first plants might be trees; on gravel we find perennial herbs; on sandy substrates the colonists will be grasses, and so on (Grubb, 1986).

Most primary successions (following glacial retreat, volcanic eruption, earthquake, sediment deposition or reduction of sea level) are associated with: 1) increasing soil nitrogen; and 2) increasing height of the mature plants (leading to shading of low-growing species). Perhaps the most important process is the accumulation of nitrogen in the soil. Mature ecosystems support soil nitrogen pools of between 5000 and 10,000 kg N/ha[1] in the topsoil (Table 1.1). During primary succession, this pool must be built up virtually from scratch. Experiments on a variety of virgin substrates have shown that woody plants cannot invade successional communities until 400–1200 kg N/ha[1] have built up in the soil; this may take anything from 20–100 years or more (Table 1.2). Many of the early vascular plant colonists

Table 1.1. Organic carbon and nitrogen storage in ecosystems classified according to Holdridge's system of life zones (from Post *et al.*, 1985).

Life Zone	Carbon (kg m^{-3})	Nitrogen (g m^{-3})	C/N ratio
Moist tundra	10.9	638.5	18.3
Wet tundra	20.7	1251.3	18.4
Rain tundra	36.6	2226.0	15.6
Boreal dry bush	10.2	631.0	16.0
Boreal moist forest	15.5	1034.1	16.0
Boreal wet forest	15.0	980.1	16.9
Boreal rain forest	32.2	1512.2	25.8
Cool temperate desert bush	9.9	779.8	12.5
Cool temperate steppe	13.3	1032.2	15.1
Cool temperate moist forest	12.0	626.1	22.5
Cool temperate wet forest	17.5	930.6	20.7
Cool temperate rain forest	20.3	1210.3	15.7
Warm temperate desert	1.4	106.5	15.3
Warm temperate thorn steppe	7.6	538.0	15.9
Warm temperate dry forest	8.3	644.8	14.0
Warm temperate moist forest	9.3	648.3	25.1
Warm temperate wet forest	26.8	2806.6	15.6
Subtropical desert bush	29.1	2282.6	12.7
Subtropical thorn woodland	5.4	379.3	14.3
Subtropical dry forest	11.5	1070.4	10.2
Subtropical moist forest	9.2	987.9	10.3
Subtropical wet forest	9.4	2853.4	3.3
Tropical thorn woodland	2.6	264.6	9.2
Tropical very dry forest	6.9	597.2	13.7
Tropical dry forest	10.2	885.9	13.3
Tropical moist forest	11.4	802.9	14.9
Tropical wet forest	15.0	655.2	30.2

Table 1.2. Comparison of nitrogen content and compartmentation between major ecosystem pools in naturally colonized and reclaimed china clay wastes. (n.d. not detectable.) (From Marrs *et al.*, 1983.)

Ecosystem	Age	Total N (kg ha^{-1})	% Shoots	% Roots	% Litter	% Soil
(a) Naturally colonized china clay wastes						
Pioneer						
(*Lupinus arboreus*)	16–18	291	37	1	6	56
(*Calluna vulgaris*/*Ulex europaeus*)	30–55	823	13	5	3	79
Intermediate						
(*Salix atrocinerea*)	40–76	981	8	18	6	68
Woodland						
(*Betula pendula*/ *Rhododendron ponticum*/ *Quercus robur*)	40–116	1770	30	3	0	67
(b) Reclaimed china clay wastes						
Sand tips	3–84	211	11	59	n.d.	30
Mica dam walls	3–84	441	8	61	n.d.	31

are legumes and other nitrogen-fixing species which add substantially to the soil pool. For example, on china clay waste, tree lupins *Lupinus arboreus* added 72 kg N/ha[1]/year[1], and gorse *Ulex europeus* added 27 kg. These plants add nitrogen at a much faster rate than occurs via weathering (often negligible) or atmospheric input from rain (about 10 kg N/ha[1]/year[1]; Marrs *et al.*, 1983).

In a mature temperate ecosystem with a productivity of 10 t/ha[1]/year[1] (of which 1.5% is nitrogen), there must be an annual uptake of about 150 kg N/ha[1]. This nitrogen supply is achieved by mineralization of the capital accumulated in the soil organic matter, and the rate of uptake is limited both by the size of the pool and by the rate of mineralization. It appears that young ecosystems function on a greatly reduced soil capital by virtue of high rates of mineralization coupled with a preponderance of nitrogen-fixing species (Marrs *et al.*, 1983; Vitousek & Mattson, 1984).

1.7.3   Secondary succession

Secondary successions begin with a more or less mature soil containing a sizeable bank of seeds and vegetative propagules. One school of thought holds that secondary succession is nothing more than the expression of the life histories of the species already present at the outset. Thus short-lived plants with high relative growth rates flourish and achieve dominance first, followed by herbaceous perennials, then by rapidly growing, short-lived trees, followed eventually by slow-growing, long-lived trees. The observed 'succession' of species is nothing more than the replacement of small, short-lived plants by large, long-lived ones. This is known as the Initial Floristic Composition model. At the other extreme is the model of Relay Floristics which emphasizes a more or less strict sequence of plant species and stresses the importance of 'facilitation' (Egler, 1954).

Facilitation is the process whereby species A paves the way for species B; the community cannot be invaded by species B until environmental conditions have been altered by the activities of species A. The most obvious examples of facilitation come from primary successions; for example, on freshly exposed glacial debris, mosses facilitate the invasion of grasses by providing them with roothold, shelter and moisture (Smith, 1984). On a rather different note, ericaceous plants like *Vaccinium* cannot enter a post-glacial succession until substrate pH has been reduced from 8 to about 4.5 (Crocker & Major, 1955). Rather more subtle examples occur when immigration of seed depends upon the services of dispersing animals; for instance, primary successions on bare chalk are not invaded by bird-dispersed shrubs like *Crataegus* until woody plants like *Betula* have grown to sufficient height to provide perches for the birds (Finegan, 1984). Whether these last two processes represent facilitation or simply 'enablement' is a moot point. In any case, it seems clear that genuine *interspecific* facilitation is the exception rather than the rule. Most examples of facilitation are a consequence of community-level processes affecting soils (increased nitrogen and organic matter, reduced pH) or light availability (increased shade), and are *not* the consequence of specific interactions between one plant species and another.

Between the extremes of Initial Floristic Composition and Relay Floristics are the 'Tolerance' and 'Inhibition' models (Connell & Slatyer, 1977). The Tolerance Model assumes that succession progresses by the replacement of early, fast-growing species, by plants capable of regenerating in the conditions of depleted light and nutrient resources which these early species create (Chapter 2). In contrast, the Inhibition Model assumes that early species simply inhibit the establishment of later species by site pre-emption. The longer-lived the initial colonists, the more slowly the succession proceeds. For instance, in south-east Asia certain tropical forest successions following storm damage get 'stuck' after the development of a dense cover of ferns, and there is little change in the vegetation for a protracted period until the ferns are eventually shaded out (Whitmore, 1985).

Other models of succession stress the importance of the proximity of seed parents, and the fact that immigrant species differ greatly from place to place, depending upon what particular mature plants happen to grow nearby (Horn, 1981). Noble and Slatyer (1980) consider that certain 'vital attributes' of species determine their place in succession. These include their mode of arrival at a site, their persistence, and their ability to establish at different stages of succession. These, in turn, can be related to the duration of their dormancy in the seed bank, their age at first fruiting, and their longevity as adults (see also van der Valk, 1981; Hils & Vankat, 1982; Pickett, 1982). In all secondary successions, the precise timing of the disturbance which initiates the succession is vital in determining the sequence of events that follows. For example, Keever's (1950) classic study showed the importance of the phenology of germination and early growth in relation to the time of year at which old fields are abandoned, in determining which plant species would dominate first, and how long that dominance would last.

Several quantitative descriptions of secondary succession have been attempted. One set of models has used multi-species versions of the Lotka-Volterra equations to address the question of what makes communities invasible (see Crawley, 1986). Other models involve drawing up matrices of 'replacement probability' for individuals of each species in the community (Horn, 1976). The simplest way to do this is to observe the distribution of saplings of different species beneath the canopies of mature individuals of each tree in a forest, and then to assume that replacement probability is equal to the proportion of saplings belonging to particular species. For example, a pioneer species is likely to be replaced by a mid successional species in a given time period, whereas a late successional plant is more likely to be replaced by another individual of its own kind. These replacement probability matrices define 'Markov chains' which predict the convergence of the community to an equilibrium state (Usher, 1986). Usher's analysis suggests that: 1) successional processes are not 'first-order' (they depend upon the history of the system, as well as on its present configuration), so that complicated Markov chain models will be needed for realism; and 2) successional processes are not stationary in time or in space. Particular local conditions influence the replacement probabilities, and it is unlikely that predictions based on simple, small-scale counts of saplings will generalize to large-scale plant communities. Perhaps the most damning criticism of replacement probability models, however, is that profound change in community structure is often determined by very rare events (which tend, of course, not to occur during detailed studies!). For example, very rare runs of years with rainfall favourable for germination and seedling growth, or a rapid sequence of unusually hot fires, may shape the vegetation of arid lands (Noble & Slatyer, 1980), while rare disease epidemics may eradicate vertebrate herbivores, allowing the establishment of seedling trees (e.g. the recruitment of oaks in England following the virtual extermination of rabbits by myxomatosis; Crawley, 1983).

The problem, as usual, is that *observational* studies are unlikely to distinguish between the different models of succession. All of the models are roughly consistent with the observed trends and exceptions can always be explained away. It will take long-term, manipulative field experiments to expose the relative importance of inhibition, tolerance and facilitation in particular ecosystems. Nor should we be surprised if the results of these experiments show that individual species have their own, unique sets of rules governing invasion and extinction, rather than whole successional series of communities being of one kind or another (see Section 1.5).

1.8   The dynamics of plant communities

If we reject the concept of the climax community in favour of the individualistic approach (Section 1.2.3), how are we to understand the community's emergent properties of structure and dynamics? We need a framework which brings together the individual populations comprising the community in a way which enables us to understand the processes of extinction, species immigration, and fluctuation in relative abundance, without invoking mysterious, super-organismic imperatives.

Fig. 1.13. Community structure and the dynamics of community change. The N = nutrients, A = plants or other autotrophic species, and H = herbivore. For simplicity, other herbivore species and herbivore enemies have been omitted. Interactions ending in circles are negative in their effects; pointed arrows are positive. The semicircular negative loops indicate 'self-regulation' terms due to intraspecific, density dependent processes. The directions of response of community variables to parameter changes entering the system at different nodes are shown below. The responses are those of a slowly moving equilibrium after transient effects are damped.

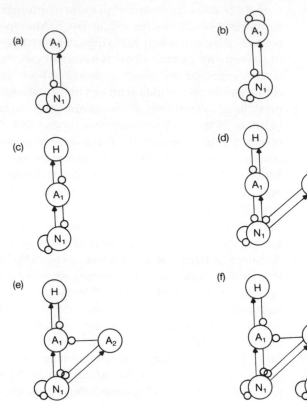

Model	Change entering through	Effect on:				
		N1	N2	A1	A2	H
a	N1	0		+		
	A1	−		+		
b	N1	+		+		
	A1	−		+		
c	N1	+		0		+
	A1	0		0		+
	H	+		−		+
d	N1	0		0	+	0
	A1	0		0	0	+
	A2	−		0	+	−
	H	0		−	+	0
e	N1	0		0	+	−
	A1	0		0	0	+
	A2	−		0	+	−
	H	0		−	+	−
f	N1	+	−	0	+	?
	N2	−	+	0	+	−
	A1	0	0	0	0	+
	A2	−	−	0	+	−
	H	+	−	−	+	?

From Levins & Lewontin (1980).

Levins and Lewontin (1980) suggest that we reject both mechanistic reductionism and idealist holism. They recognize the following properties in communities: 1) The community is a contingent whole in reciprocal interaction with lower and higher level wholes, and not completely determined by them. 2) There are properties at the community level (like diversity, equitability, biomass, primary production, invasibility and the patterning of food webs) exhibiting patterns which are both striking and which show some kind of geographical regularity. 3) There are many configurations of populations which preserve the same qualitative properties of communities (e.g. species can be substituted, and communities persist despite their component populations constantly fluctuating). 4) The species of a community interact (plants with other plants, with their herbivores and pathogens, with their pollinators, dispersal agents and symbionts, with the enemies of their herbivores, and so on), so that the environment is structured from *interacting* biotic and abiotic factors. 5) The essential asymmetry of many interactions means that it is impossible for all species to be most abundant where their environment is most 'favourable'. For example, a plant may be driven to low density by a specialist herbivore under the abiotic conditions where it would thrive best in the absence of the herbivore. Conversely, the plant may be able to defend itself against attack under good soil conditions, so that herbivore numbers are at their lowest where the host plant is at its most vigorous. 6) The way a change in some physical character or genetic characteristic of a population affects the other populations in a community depends upon the way the community is structured. This last point is perhaps the most important. If species *do* interact, then community structure determines the consequences of their interaction. Figure 1.13 shows some hypothetical equilibrium communities made up of few species, and describes the direction of change in each species in response to an increase in the level of the other. Some important generalizations emerge from this simple analysis: 1) the response of a species to the direct impact of the external environment depends upon the way that species fit into the community (for example, species A1 increases with N1 in (a) and (b) but is unaffected in all other systems); 2) some species respond to changes arising almost anywhere in the system (A2 or H) while others (A1) are insensitive to most inputs; 3) some species affect most other species in the community (A_2 or H) whereas others have very restricted effects (changes in Al are felt only in H); 4) a change may not be detectable at the point of action but only elsewhere in the community (changing the A_1-A_2 link between (d) and (e) only affects the way H responds to N_1); 5) a positive input into a species can have a positive, negative or zero effect on its population size; 6) the same pair of species may show positive or negative correlations in their abundances depending upon where the variation enters the system (compare N_1 and N_2 in (f)).

This approach to the community allows for a greater richness than a rigidly reductionist view. It allows us to work with the relative autonomy and reciprocal interaction of systems on different levels, it stresses the inseperability of the physical environment and biotic factors, and highlights the origins of correlations between species' abundances. The community is seen as consisting of many loosely-coupled sub-systems dominated by asymmetric interactions, rather than as a tightly and symmetrically interacting system of mutually dependent elements (see also Cohen, 1978; Pimm, 1982).

As plant ecology matures, we can begin to develop a rather more sophisticated approach to ecological relationships than the simple, two-species competitive and predator-prey models which have dominated our thinking for so long. A consideration of multiple-species interactions permits a massive variety of dynamic behaviour. For example, theoretical models of two coexisting but non-competing species can manifest all the behaviour expected of competing species if they share a common (but perhaps invisible) species of pathogen (Holt & Pickering, 1985). Thus increasing the abundance of species A leads to a reduction in the abundance of species B, and vice versa. Unless we knew that they shared a common pathogen and that this pathogen influenced their population dynamics, we would conclude, quite wrongly, that the species were competitors (see also Holt, 1977, for the potential effects of shared predators).

Field studies also provide a rich variety of multi-species interactions. For example, the work of Davidson and her colleagues on seed-feeding desert animals has highlighted a number of cases of indirect mutualisms which arise between apparent competitors. In a guild of three ants, the larger species facilitates the smallest by excluding the medium-sized species. In the absence of the largest ant, the medium-sized species excludes the smallest one (Davidson, 1985). Again, voles may facilitate ants via their effects on plant recruitment. Voles eat larger seeds than ants, and when voles are experimentally excluded, the large-seeded plants out-compete the smaller, small-seeded plant species upon which the ants rely. Thus when voles are removed the vegetation changes to such a degree that ants can no longer be supported in their former abundance (Davidson *et al.*, 1985).

Recognizing which communities are equilibrium assemblages dominated by biotic interactions such as these, and which are non-equilibrium assemblages dominated by chance, or by chronic abiotic disturbances, represents a major challenge for plant community ecologists. We need to understand how both these kinds of dynamic systems work, a task taken up from quite different, but complementary angles in the next two chapters.

Chapter 2   Resources, Competition and the Dynamics of Plant Communities

DAVID TILMAN

2.1   Introduction

As discussed in Chapter 1, there are many factors that influence the structure of plant communities. A central goal of plant ecology is to understand the processes and mechanisms that cause the patterns we see. This is no easy task, however. There are more than 300,000 species of terrestrial vascular plants world-wide, and more than 1,000,000 species of animal. Any given habitat may contain from a few plant species to hundreds of different species, all of which are potentially interacting with each other, with their abiotic environment, with soil micro-organisms, and with herbivores, pathogens, predators, pollinators and dispersal agents. Thus, plant community structure is likely to be influenced by interspecific competition, by herbivory, by predation, and by mutualism. Additionally, the habitat in which plants live is not constant. Rainfall, temperature, and the population densities of various herbivores, predators and mutualists change through time, on both short and long time scales. The natural, ecological world may be one of the most complex systems that scientists have tried to understand. One way to approach such a complex system is to start with a few important processes, and see what features of the system can and cannot be explained using those processes.

Numerous observational and experimental studies, reviewed later in this chapter, have shown strong effects of the availabilities of nutrient resources and light on the species diversity, species composition, and species dominance of both aquatic and terrestrial plant communities. In this chapter, I shall explore the possible effects of interspecific competition for resources on plant community structure. I shall first present a simple model of plant competition for resources, and then compare its theoretical predictions with the patterns observed in a variety of plant communities worldwide. This is not meant to imply that other processes, such as herbivory, are unimportant. Rather, it is done in the hope of finding a simple theory that can predict the broad, general patterns we observe in plant communities. Such a simple theory could then be expanded, as needed, to describe nature more realistically.

51

2.2   Theory

Ecologists define interspecific competition (i.e. competition between different species) as an interaction in which increases in the population density of one species lead to decreases in the per capita growth rate and population density of another species (see Box 1.4). This definition, however, does not specify the mechanisms whereby each species inhibits the other. At least for most plants, it seems unlikely that inhibition is a direct result of changes in density as such (but see Section 1.6.3). Rather, changes in the population density of one plant species are likely to affect the availabilities of various resources, such as nitrogen, water, phosphorus and light, and thus influence the growth of other species indirectly.

2.2.1   Competition for a single resource

The basic information needed to predict the outcome of competition for resources is the dependence of the rates of population growth of

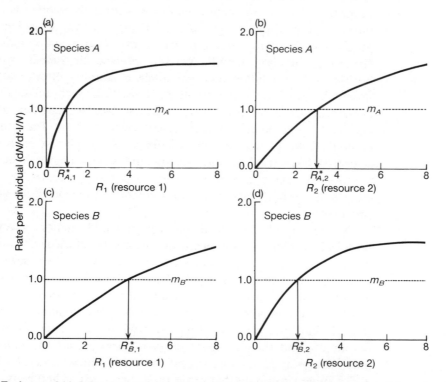

Fig. 2.1. Each part of this figure shows the dependence of the per capita rate of reproduction (solid curve) and the per capita rate of mortality (broken line) of a species on resource availability. For each species and each resource, the resource availability at which reproductive rate equals mortality rate is indicated by the arrow pointing at R. Note that Species A has a lower R* for resource 1 but a higher R* for resource 2 than Species B. The R* of a species is the amount of a resource the species requires to survive in a habitat. The four R* values from this figure are used to construct Fig. 2.2.

each species on resource availability. In the following discussion, N refers to the population density of a species, i.e. to the number of individuals per unit area. Consider, for instance, the dependence of the per capita rate of reproduction, expressed as $dN/dt \cdot 1/N$, of species A on the availability of a limiting resource, R_1 (Fig. 2.1a). For a species to exist in a habitat, its reproductive rate must be greater than or equal to its mortality rate. If species A lived in a habitat in which it experienced the mortality rate indicated by the broken line, its mortality rate (m_A) would equal its reproductive rate only in habitats with $R_{A,1}^* = 1.0$ of the resource (Fig. 2.1a). If a habitat had less R_1 than this, the population density of species A would decline. As the population density declined, the rate of consumption of the resource would decline, and resource availability would increase. If a habitat had more R_1 than $R_{A,1}^*$, the population density of species A would increase, leading to an increased rate of consumption of the resource, and thus to a lower availability of the resource. In the absence of interspecific competition, the population density of species A would equilibrate at the point at which its rate of resource consumption equalled the rate of resource supply, and its reproductive rate equalled its death rate. At equilibrium, species A would thus maintain R_1 at a level of $R_{A,1}^*$ (called the 'requirement' of species A for R_1). Similarly, based on the curves in Fig. 2.1c, species B has a requirement of $R_{B,1}^* = 4.0$. If species A and B competed for this resource, and if this interaction went to equilibrium, theory predicts that the single species with the lower requirement for the resource would displace all other species (Tilman, 1976; Hsu, Hubbell & Waltman, 1977). Thus, species A should displace species B when both are limited by R_1 because species A can reduce R_1 below the level required for the existence of species B.

There have been several experimental studies of plant competition for a single limiting resource. The first attempt to test this theory was an experimental study of nutrient competition between two species of freshwater algae (Tilman, 1976, 1977). This work showed that the species with the lower R^* for the limiting resource (either phosphorus or silicate) competitively displaced the other species. Many additional studies have been performed since then, with most of them showing that R^* differences corresponded with differences in competitive ability as predicted by theory (Tilman, 1981; Holm & Armstrong, 1981; Hansen & Hubbell, 1980; Tilman *et al.*, 1982).

2.2.2   Competition for two resources

In order to predict the outcome of plant competition for two or more resources, it is necessary to know the dependence of the growth of each species on the availability of *all* limiting resources. A resource is limiting if increases in its availability bring about an increase in plant growth rate. Information on the dependence of plant growth on resources which are not limiting under natural conditions is of no ecological importance. Plants require many different resources, including H_2O, light, and various forms of N, P, K, C, Ca, Mg, S, Co, Cu, Mo, Mn and Zn. Each of these are essential to plant survival. Thus the growth rate of a nitrogen-limited plant is not increased by addition of P or K, or any resource other than nitrogen. For essential resources, the growth rate of a plant is determined by the availability of the *one* resource in lowest supply relative to need (Chapter 12). However different forms of an essential resource, such as the ammonia and nitrate forms of nitrogen, can be substituted for each other. This chapter will only consider competition for limiting, essential resources. A more complete development of theory can be found in Tilman (1982).

2.2.3   Resource isoclines

Although resource competition theory is best expressed using differential equation models (Tilman, 1977, 1982, 1985), its major features can be understood graphically (Tilman, 1980). Consider species A of Fig. 2.1. It requires $R_{A,1}{}^* = 1.0$ of resource 1 and $R_{A,2}{}^* = 3.0$ of resource 2 in order to maintain a stable population. These requirements define the resource-dependent growth isocline of Fig. 2.2a. For all environments that have availabilities of R_1 and R_2 that fall *on* this isocline, the reproductive rate of species A will equal its mortality rate; there will be no change in its population density (it is in equilibrium and $dN/dt = 0$). If resource availabilities fall outside this isocline (further from the origin), population density will increase ($dN/dt > 0$). Population density will decrease ($dN/dt < 0$) if resource availabilities fall inside the isocline. The right angle corner in the isocline means that species A is limited by *either* R_1 or R_2, whichever is in lower supply relative to its need. It is equally limited by R_1 and R_2 at the corner of the isocline. Above this point, changes in R_2 have no effect on its growth rate, and it is limited by R_1. On the horizontal portion of the isocline, it is limited by R_2, and changes in R_1 do not affect its growth. To see this, choose several different points on the growth isocline, and see, for each point, how addition of R_1 or R_2 would affect the growth of species A.

2.2.4   Resource consumption vectors

Equilibrium will occur when resource consumption equals resource supply, and reproduction equals mortality. Reproduction equals mortality for any point on the growth isocline. The actual point on the

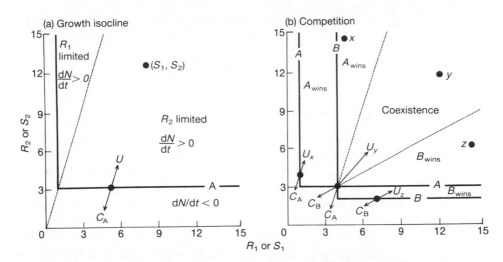

Fig. 2.2.(*a*) The solid curve with the right angle corner is the resource-dependent growth isocline of Species A. At all points on the isocline, the reproductive rate of Species A equals its mortality rate. Above the isocline it increases in abundance, and below the isocline it declines. Note that the position of the isocline is determined by the values of $R_{A,1}*$ and $R_{A,2}*$ from Fig. 2.1(*a*) and 2.1 (*b*). The broken line from the origin through the corner of the isocline shows the amounts of R_1 and R_2 at which the species is equally limited by both resources. The dot on the isocline is the one-species equilibrium point, where the reproductive rate equals the mortality rate, and the resource consumption rate (C_A) equals the resource supply rate (U). (*b*) The isoclines for Species A and B, determined by the $R*$ of Fig. 2.1., and their consumption vectors (C_A and C_B), define the types of habitats in which Species A is dominant (e.g. supply point *x*), both species coexist in stable equilibrium (supply point *y*), and Species B is dominant (supply point *z*). Modified from Tilman (1982).

isocline that will be the equilibrium point in a given habitat depends on the rates of resource consumption and supply. Optimal foraging theory (Rapport, 1971; Tilman, 1982) predicts that a plant should consume essential resources in the proportion in which it is equally limited by them. This is illustrated with the consumption vector, C_A, which is parallel to the broken line from the origin through the corner of the isocline (Fig. 2.2a). The broken line shows the optimal ratio of $R_1 : R_2$ for species A (i.e. the $R_1 : R_2$ ratio it requires for growth).

2.2.5 Resource supply vectors

Let S_1 and S_2 be the maximal amounts of all forms of resources 1 and 2 in a given habitat. The point (S_1, S_2) is called the resource supply point of that habitat. Each habitat is considered to have a particular resource supply point. The rate of supply of a resource should be proportional to the amount of the resource that is not already in the available form. This would give

$$dR_j/dt = a(S_j - R_j),$$

where a is a rate constant, and j refers to resource j. This equation defines a resource supply vector, U, that always points toward the supply point. Thus, the resource supply point, (S_1, S_2) of Fig. 2.2a, leads to the equilibrium point shown with a dot. At the equilibrium point, the population density of species A is such that its total rate of resource consumption equals the total rate or resource supply, and its reproductive rate equals its mortality rate.

2.2.6   Coexistence and displacement

To predict the outcome of competition between two or more species for two limiting resources, it is only necessary to superimpose their isoclines and consumption vectors. The R^* of species A and B of Fig. 2.1 give the isoclines shown in Fig. 2.2b. The point at which these isoclines cross is a two-species equilibrium point; that is, a point of potential coexistence of these two species. At this point, the reproductive rates of both species A and B equal their mortality rates. However, these species will only be able to coexist in habitats with certain resource supply points (Fig. 2.2b). If habitats have low supply rates of R_1 and high supply rates of R_2, such as at supply point x, both species will be limited by R_1. Species A, which is the superior competitor for R_1, will reduce the level of R_1 down to a point on its isocline at which there is insufficient R_1 for the survival of species B. Thus, species A will competitively displace species B from habitats with low $S_1 : S_2$ ratios (Fig. 2.2b). Comparably, species B is a superior competitor for R_2, and it will displace species A from habitats in which both species are limited by R_2. Such habitats have high $S_1 : S_2$ ratios.

The two species can coexist in intermediate habitats in which *each species is limited by a different resource*, species A by R_2 and species B by R_1. For habitats with supply points within the region defined by the consumption rates of these species at the two-species equilibrium point (Fig. 2.2b), the consumption of the two species will eventually reduce resource levels down to the equilibrium point. Thus, the resource requirements of these two species define habitats in which one species is dominant, both coexist, or the other is dominant. Along a resource ratio gradient, such as from supply point x to supply point y to supply point z, there is a smooth transition from dominance by species A, to coexistence of A and B, to dominance by species B. This resource ratio gradient is a gradient from low $S_1 : S_2$ ratios to high $S_1 : S_2$ ratios. Such separation along the $S_1 : S_2$ gradient only occurs if the species are differentiated in their requirements for R_1 and R_2, with the superior competitor for one resource being the inferior competitor for the other resource.

Several experimental studies of algal competition for limiting phosphate and silicate have shown that this simple theory can predict the outcome of interspecific competition, including stable coexistence (Tilman, 1976, 1977, 1982). Additionally, the distributional patterns of algal species along natural resource ratio gradients in Lake Michigan are consistent with their requirements for the limiting resources and the outcome of laboratory competition experiments among the species (Tilman, 1982).

2.2.7 Multispecies competition

Finally, let us consider a case in which five different species compete for two essential resources. Again, let us assume that these species

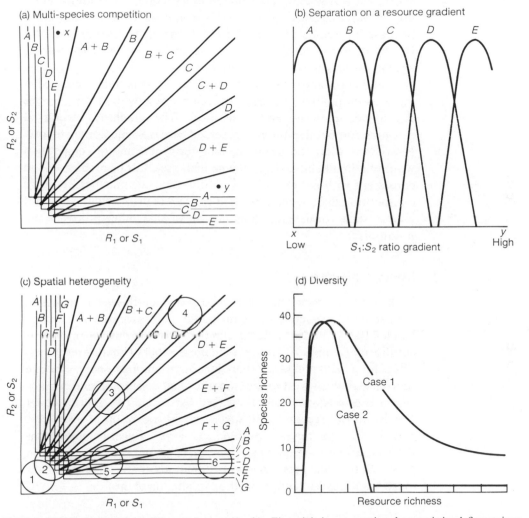

Fig. 2.3. (a) The isoclines for 5 different plant species (A–E), and their consumption characteristics define regions in which various species or pairs of species may exist. For instance, Species A and B could coexist stably in any habitats that had supply points (S_1, S_2) that fell within the region labelled 'A+B'. (b) Competition between the 5 species would lead them to be separated along a resource supply ratio gradient as illustrated. This could be considered to be a gradient from low to high $S_1 : S_2$ ratios, such as from point x to point y in Fig. 2.3(a). (c) Seven species (A to G) are shown competing for two resources. Spatial heterogeneity in the supply rates of the resources is illustrated by the circular resource supply regions. This could allow many species to coexist in resource-poor habitats (circle 2), but would not allow more than two species to coexist in resource-rich habitats (circle 4). (d) Simulations of cases in which 40 species competed for two resources in a spatially heterogeneous habitat led to these diversity curves. Note that diversity falls more rapidly with resource enrichment when all resources but one are added (case 2), than when the supply rates of all resources are increased (case 1). Modified from Tilman (1980; 1982).

are differentiated such that each species is a superior competitor for a particular ratio of the limiting resources, and that each species consumes the resources in the ratio in which it is equally limited by them. This gives regions in which various pairs of species can coexist (Fig. 2.3a), and predicts that the species should, at equilibrium, be separated along a $S_1 : S_2$ resource ratio gradient (Fig. 2.3b). The point at which each species reaches its greatest abundance along the resource ratio gradient is determined by its requirements for the limiting resources. Thus, species A is dominant at low $S_1 : S_2$ ratios because it is the best competitor for R_1 but the worst competitor for R_2. At all points along the gradient, R_1 and R_2 are important limiting resources for some of the species.

Let us now apply this theory to a variety of patterns observed in plant communities, including species diversity patterns, species dominance patterns and succession. The graphical version of resource competition theory presented here is an equilibrium theory. It predicts what the ultimate, long-term outcome of competition would be in a habitat once change had ceased. If its predictions are consistent with patterns observed in natural communities, resource competition may be an important process in these communities. Of course such consistency does *not* mean that other processes are unimportant.

2.3   Species diversity

In the form in which it has been discussed so far, this theory makes an almost trivial prediction about the number of species that can coexist in a habitat. It predicts that the number of species that can coexist, at equilibrium, can be no greater than the number of limiting resources. Thus, for a habitat in which there is a single limiting resource, the one species with the lowest requirement for that resource should displace all other species at equilibrium (see Box 1.1). Two species could coexist on two resources, 10 species could coexist on 10 resources, and so on. This raises a problem, however. Experimentation has shown that there are, at most, four or five resources limiting plants in most habitats, and yet there may be hundreds of species coexisting in these habitats. What might be the explanation of this?

2.3.1   Spatial heterogeneity

One explanation could be that natural habitats are not uniform. So far the model has assumed that all plants in a habitat experience the same rate of resource supply. However, there is much spatial heterogeneity in the levels of nutrients in both soils and lakes (e.g. Burgess & Webster, 1980; Lehman, 1982; Yost *et al.*, 1982). This means that a habitat has both an average supply point and variance within it.

Different plants, depending on where they are located, experience different rates of resource supply. Such variance could be illustrated graphically by showing all of the different supply points experienced by the individual plants in a heterogeneous habitat. Let the circles in Fig. 2.3c include 99% of the point-to-point spatial heterogeneity in the (S_1, S_2) as experienced by individual plants in a habitat. Such spatial heterogeneity can allow many more species to coexist than there are limiting resources. All that is required for a species to exist in a habitat is that there be some site with a suitable (S_1, S_2), i.e. that the circle overlap its region of existence. For instance, all seven species could stably coexist in a habitat with the heterogeneity represented by the circle number 2 (Fig. 2.3c), because there are microsites in which each species can survive.

2.3.2   Resource availability

How should species richness depend on resource availability? Consider two different ways in which resource levels might change. First, it could be that richer habitats are richer in all resources. If that were so, this theory would predict that there would be maximal species richness in habitats that were moderately resource poor, providing that spatial heterogeneity of the resources—the diameter of the circles—did not increase more than linearly with the resource level (Tilman, 1982). As shown in Fig. 2.3c no species could exist in habitat 1, seven species could coexist in the slightly more resource rich habitat 2, four species could coexist in the even richer habitat 3, and two species could coexist in the even richer habitat 4. Simulations of similar cases in which 40 species were assumed to be competing for two essential resources gave the diversity curve in Fig. 2.3d (case 1). This theory thus predicts that maximal species richness should occur in moderately resource poor habitats, habitats that have just enough resources for species survival in the absence of interspecific competition.

Consider, also, a second case in which S_1 increases but S_2 remains constant, such as for habitats 2, 5 and 6 of Fig. 2.3c. Again, there is maximal plant diversity in resource poor habitats. In this case, all resources but R_1 are being added, and diversity falls more rapidly with resource richness because R_1 becomes the only limiting resource (Fig. 2.3d, case 2).

2.3.3   Observational studies

This prediction agrees well with many observational and experimental studies of both planktonic algal communities and terrestrial plant communities. For instance, Beadle (1966) observed that the maximal diversity of Australian xeromorphic and rainforest genera

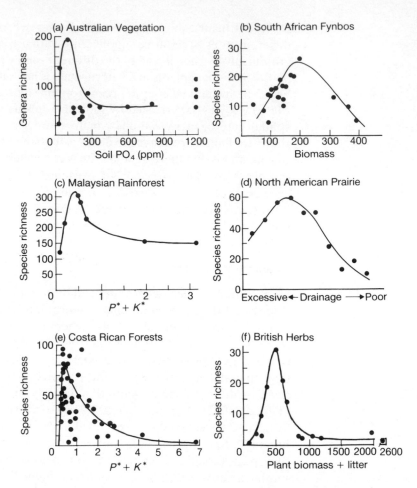

Fig. 2.4. The dependence of plant community diversity on resource richness (or some measure of resource richness) for a variety of habitats. Figures redrawn from: (a) Beadle (1966); (b) Bond (1983); (c) Tilman (1982) based on data in Ashton (1977); (d) Dix & Smeins (1967); (e) Tilman (1982) based on data in Holdridge et al. (1971); and (f) Al-Mufti et al. (1977). Note that Bond estimated above-ground plant biomass (in arbitrary units) by using the area under a plant foliage-height profile. Tilman normalized the data of Ashton & Holdridge et al., on soil phosphorus (P) and potassium (K) levels, and then used the summation of these normalized variables as a measure of soil resource richness. Dix & Smeins divided their study areas into drainage types which they ranked from those with the best drainage to those with the poorest. Al-Mufti et al., estimated productivity by summing the total above ground plant biomass and the soil litter of each area. For all the figures it is assumed that the x axis is a measure of the availability of various limiting soil resources, with habitats with the greatest availability being at the right hand side of each graph.

occurred in very phosphorus-poor soils, and that areas with richer soils were more species poor (Fig. 2.4a). Shmida et al. (1984) found that the regions with the highest plant diversity within Israel had fairly low rainfall; a graph of species richness versus regional rainfall gave a humped curve.

Two of the worlds most diverse plant communities, the fynbos vegetation of South Africa and the heath scrublands of Australia, occur on very nutrient poor soils. In both of these areas, nearby areas with more resource-rich soils have much lower diversity (Goldblatt, 1978; Kruger & Taylor, 1979; Specht & Rayson, 1957; Specht, 1963). This is illustrated by a comparative study performed in South African vegetation by Bond (1983; Fig. 2.4b).

The number of species of Malaysian tropical rainforest woody species had a similar dependence on the availability of two important soil resources, phosphorus and potassium (Ashton, 1977; Fig. 2.4c). The work of Holdridge *et al.* (1971) in the New World tropical forests of Costa Rica (Fig. 2.4e) showed that the most diverse forests occurred on moderately resource poor soils, and that rich soils supported many fewer species. For North American prairie, Dix & Smeins (1967), reported that the greatest plant diversity occurred in areas with low water availability (Fig. 2.4d). A humped diversity curve was also reported by Al-Mufti *et al.* (1977) for British vegetation; maximal diversity occurred in habitats with low productivity, as measured by the total of plant standing crop and litter (Fig. 2.4f). Other studies in other habitats have also shown that maximal species richness occurs in moderately resource poor terrestrial habitats, as predicted by theory (e.g. Young, 1934; Huston, 1980; Mellinger & McNaughton, 1975; Whittaker & Niering, 1975). Additionally, similar patterns have been reported for the planktonic algae of a variety of lakes and the oceans (e.g. Steeman-Nielsen, 1954; Fischer, 1960; Dugdale, 1972; Smayda, 1975; Blasco, 1971; Nelson & Goering, 1978; Williams, 1964; Patrick, 1963, 1967; Patten, 1962; Schelske & Stoermer, 1971). Consistent with theoretical predictions, for both terrestrial and aquatic plant communities, maximal species diversity occurs in moderately resource poor habitats. Very resource rich habitats and extremely resource poor habitats can support the long-term persistence of many fewer species.

2.3.4 Experimental studies

Because few habitats are sufficiently resource poor to be below the diversity peak (Fig. 2.4), and because most additions of nutrient resources represent major increases in resource supply, most experimental additions of limiting resources should lead to decreased diversity. Such fertilization experiments have been performed in a variety of habitats world-wide, and have shown just this pattern. For instance, fertilization of British pastures led to dramatic decreases in species richness during a 17 year experiment (Milton, 1940, 1947). In experiments performed on turf transplanted from British coastal sand dunes, Willis & Yemm (1961) and Willis (1963) found that the number of coexisting species fell from 22 to 5 following the addition of a complete mineral fertilizer, whereas the species richness of controls did not decrease. They observed similar, but less dramatic decreases in species richness in undisturbed natural vegetation. Bakelaar & Odum (1978) found that the species diversity of an eight year old field (abandoned from agriculture eight years previously) fell significantly following application of a fertilizer containing N, P and K. Kirchner (1977) found that the diversity of North American short-grass prairie decreased significantly after three years of watering and fertilization.

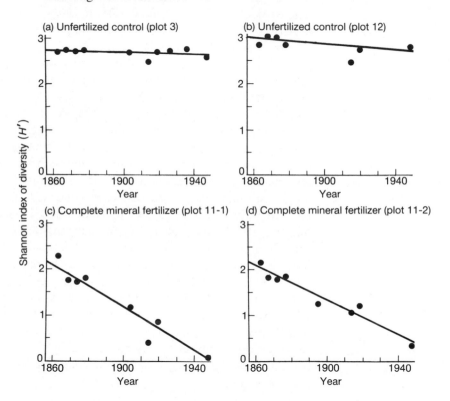

Fig. 2.5. Effects of fertilization on the species diversity of the Park Grass Experiments at Rothamsted. England. The Shannon Index of species diversity, H', measures both the total number of species in the community and the evenness of their relative abundance distributions. The more equally dominant species are, and the more species there are. the greater the value of H'. (a) and (b) show that there were no significant changes in diversity during a 100 year period in the unfertilized control plots. (c) and (d) show that the diversity of plots fertilized with a complete mineral fertilizer containing N, P, K, Ca and Mg fell dramatically through time. Many of the grasses and forbs living in this regularly mowed pasture are long-lived perennial plants which have a large proportion of their biomass in below-ground storage organs. This may account for the slow rate of species displacement. Modified from Tilman (1982).

The most long-term fertilization study that has been performed (and, indeed, the most long-term research in ecology) is the Park Grass Experiment. In 1856 in Rothamsted, England, a mowed pasture was divided into 20 plots, each receiving a different combination of mineral fertilizers (Lawes & Gilbert, 1880). Although this research was started to determine which mineral elements could improve the yield of hay, Lawes & Gilbert became fascinated with the effects of the treatments on the diversity and species composition of the plots. From 1862 onwards, there have been several censuses of the abundances of the species in these plots (Thurston, 1969). Throughout the experiments, there were no significant changes in the species diversity (Fig. 2.5a and b) or the species composition (Fig. 2.6a) of the *unfertilized* control plots. In contrast, the species diversity of the plots receiving all mineral fertilizers, with a high rate of nitrogen application (as ammonium) fell dramatically (Fig. 2.5c,d). These plots became dominated by a single species that comprised 90–99% of the total plant biomass (Fig. 2.6c,d). The grass *Holcus lanatus* dominated the plot which was unlimed and had a pH less than about 4.0. Another grass, *Alopecurus pratensis*, dominated the next higher pH plot, and two grasses, *Alopecurus* and *Arrhenatherum elatius*, coexisted in the plot with the highest pH (Fig. 2.6b,c,d).

The effect of fertilization on diversity was not limited to the plots receiving unusually high rates of nitrogen addition. For every year in which the Park Grass Experiments have been sampled, there has been a negative correlation between the species richness of the plots and their total above-ground biomass (Silvertown, 1980; Tilman, 1982). To the extent to which the addition of particular nutrients led to increased plant biomass, those nutrients also led to a lower number of coexisting species. Soil pH, which changed in response to fertilization, also had a great effect on diversity (Silvertown, 1980; Tilman, 1982).

In summary, a simple theory of plant competition for limiting resources predicts that plant communities should have maximal diversity in moderately resource poor habitats. Both observational and experimental studies performed in a wide variety of terrestrial and aquatic habitats are consistent with this prediction.

2.4   Patterns in species dominance

2.4.1   Observational studies

Anyone who has walked through a pasture, a prairie, or an abandoned field will have noticed patterns in the vegetation. Often a field will seem to be a mosaic made up of clumps of various species. Alternatively, there may be fairly distinct zones in which each species is dominant, as is frequently the case along even minor elevational gradients. Some species seem to reach their greatest abundance on hill tops, some on the slopes, and others at the bottom of the hills

(Chapter 1). There have been numerous attempts to explain such patterns in vegetation. Most of these have shown a strong correlation between some soil characteristic, such as moisture content (water holding capacity), nitrogen concentration or pH, and the observed

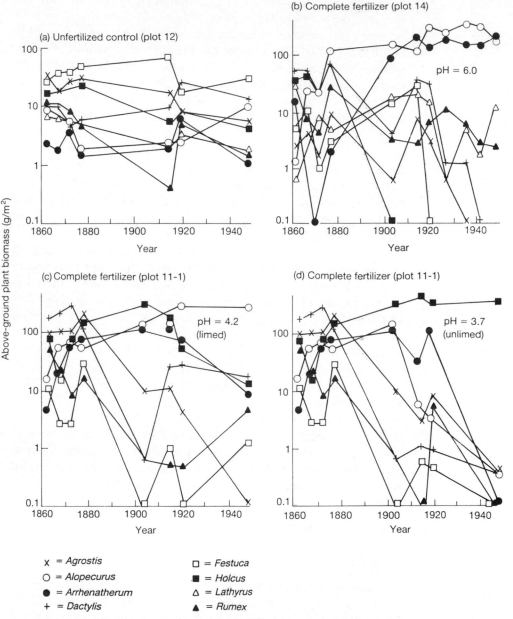

Fig. 2.6. The dynamics of change in four of the Park Grass Experiment plots at Rothamsted, England. (*a*) Plot 12 was unfertilized. Plots 14, 11–1 limed and 11–1 unlimed received complete mineral fertilizer. (*b*) Because Plot 14 received its nitrogen as nitrate, its soil became more alkaline, having a pH of 6.0 by the 1960s. Plot 11–1 received its nitrogen as ammonia, and its soils became very acidic (see Chapter 12). (*c*) The portion of Plot 11–1 that was limed to raise its pH was dominated by a different grass species (*Alopecurus pratensis*) than (*d*) the unlimed portion (*Holcus lanatus*). Modified from Tilman (1982).

pattern (e.g. Box, 1961; Beals & Cope, 1964; Zedler & Zedler, 1969). For instance, for understory woodland herbs, Pigott & Taylor (1964) found that *Deschampsia caespitosa* was most abundant on the bottom of dales, that the nettle *Urtica dioica* was most abundant just up from the base of the slopes, and that the sides of the slopes were dominated by *Murcurialis perennis*. This pattern seemed to be caused by differences in the nitrogen and phosphorus contents of these soils and in the requirements of these species for N and P. Snaydon (1962) looked at the distribution of white clover, *Trifolium repens*, in relation to several soil characteristics in both a 10×10 m area and a 130 km² area. In both areas, he observed that white clover was most common in soils with high calcium and high phosphorus levels. Within the 10×10 m area, he found that calcium and phosphorus levels varied by a factor of 3 between spots as close together as one metre, and this local variation in soil chemistry corresponded with the distributional pattern of white clover. On broader scales, white clover was only present in local areas that were rich in calcium and phosphorus and had soils of higher pH.

In the vegetation of the upper Galilee of Israel, there is considerable variation in the composition of mature, stable plant communities, even though they experience similar climatic conditions. Rabinovitch-Vin (1983) found that this variation corresponded closely with the original parent material on which these soils had formed. Similar correspondence between original parent material and the ultimate composition of the vegetation on a site have been reported in a variety of areas (e.g. Olson, 1958; Lindsey, 1961; Hole, 1976). Such patterns strongly suggest that the availabilities of various soil resources are an important determinant of the composition of terrestrial plant communities. Similar patterns have been reported for a variety of algal communities by Smith (1983), who found that the nitrogen to phosphorus (N:P) ratio of lake water was a major determinant of the abundance of various nitrogen-fixing blue-green algae. These blue-green algae reached their greatest abundance at low N:P ratios, as might be expected because of their ability to use atmospheric N_2 as a nitrogen source and their high requirements for P (Fig. 2.7).

2.4.2 Experimental studies

There are other possible explanations for at least some of the patterns summarized above. Just as soils influence plants, so plants influence soils. Although the N:P ratios of lakes may correlate well with the relative abundances of blue-green algae, they may also correlate well with some other factor which is the true cause of the pattern. Cause and effect are difficult to separate in any purely observational study. This can be at least partially overcome through experimentation.

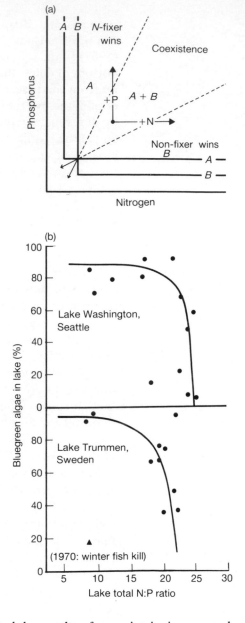

Fig. 2.7. (*a*) A hypothetical case of competition between a nitrogen-fixing plant (Species A) and a plant incapable of nitrogen fixation (Species B). Note that the N-fixer should be the superior competitor in very N-poor habitats (where the N : P supply ratio is low). In N-rich habitats (with high N : P ratios), both species should be P-limited, and Species B, the non-fixer, should displace Species A. If P were added to a habitat where both species were coexisting, it would favour the N-fixer, as illustrated by the vertical arrow. The addition of N to such a habitat would favour the non-fixer, as shown by the horizontal arrow. (*b*) In both Lake Washington (northwest USA) and Lake Trummen (Sweden) the greatest abundance of blue-green algal species capable of nitrogen fixation occurred during years in which the ratio of total nitrogen to total phosphorus was low, as predicted above. Modified from Tilman (1982) and Smith (1983).

Schindler (1977) reported the results of some intriguing research in the Experimental Lakes Area of Canada in which whole lakes were fertilized with different ratios of nitrogen to phosphorus. For instance, when Lake 227 was fertilized with an N : P ratio of 31:1, the lake was dominated by the green alga *Scenedesmus*, which is not capable of nitrogen fixation. When the N : P fertilizer was changed to give a ratio of 11:1, a nitrogen-fixing blue-green alga, *Aphanizomenon*, became dominant. After fertilization of another lake, Lake 226NE, with an N : P ratio of 12:1, there was a dramatic increase in the abundance of another blue-green alga, *Anabaena* (see also Kilham & Kilham, 1982). Similar results have been obtained when natural algal communities were allowed to compete for nutrients in laboratory experiments (Tilman & Kiesling, 1983). When the N:P ratio was low, various species of blue-green algae capable of nitrogen-fixation became dominant. When N:P ratios were high, and there was ample silicon, various species of diatoms (shell-bearing algae), became dominant. Thus, there is direct experimental evidence that the correlations reported by Smith (1983) for a wide variety of lakes world-wide are indicative of differences in the competitive abilities of blue-green algal species and other algal species for nitrogen and phosphorus.

Just as nitrogen-fixing blue-green algae reach their greatest abundance in nitrogen-limited lakes, so legumes and other terrestrial vascular plants capable of symbiotic nitrogen fixation, reach their greatest abundance in habitats with nitrogen-poor soils (e.g. Campbell, 1927; Young, 1934; Foote & Jackobs, 1966). If legumes are superior competitors for nitrogen, but inferior competitors for other resources such as phosphorus, calcium or light, legumes should decrease in abundance when nitrogen is added (see Fig. 2.7a), but become more abundant when all resources except nitrogen are added.

The Park Grass Experiment at Rothamsted provides several opportunities to test this hypothesis (Tilman, 1982). Averaging over all samples collected in a plot during the 130 years of these experiments, those plots receiving no fertilizer contained 7.9% legumes by biomass. The plot receiving all nutrients except N contained 21.4% legumes. Thus, adding all nutrients except nitrogen led to an approximately three fold increase in the abundance of legumes. In contrast those plots receiving nitrogen averaged less than 0.5% legumes. In addition, several plots were fertilized with nitrogen for a period, and then fertilized with all nutrients but nitrogen thereafter. During the period that it was fertilized with N, plot 15 averaged 0.07% legumes but averaged 20% of the total biomass during the subsequent period when all nutrients except nitrogen were added. Thus, both the whole lake fertilization experiments and the Park Grass Experiments have shown dramatic changes in the relative abundances of nitrogen-fixing plants. These changes are consistent with predictions based on the assumption that nitrogen-fixing plants are superior competitors for nitrogen but are inferior competitors for other resources, such as phosphorus and light (Fig. 2.7a).

In addition to their effects on legumes, each pattern of fertilization in the Park Grass Experiment has tended to favour a different plant species. For instance, several of the plots received different amounts of two limiting soil nutrients, nitrogen and phosphorus. Looking only at those plots with similar pH (from 4.2–6.0), different plant species were dominant at different relative avail-

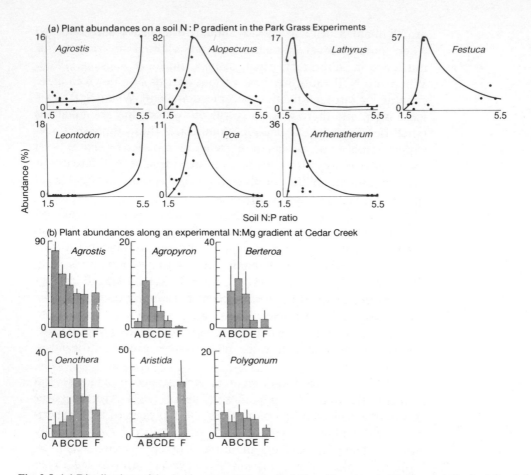

Fig. 2.8. (*a*) Distributions of the dominant species in the Park Grass Experiment at Rothamsted, England, in relation to soil N : P ratios. Note that pH ranges from 4.2–6.0 between different plots. Modified from Tilman (1982). (*b*) Distributions of the major species in a N/Mg fertilization experiment performed at the Cedar Creek Natural History Area, Bethel, Minnesota, USA. Letters A–F refer to points ranked along an experimental N : Mg fertilization gradient. Treatment A received the highest rate of application of N and the lowest rate of Mg, and there was a smooth gradation to Treatment E, which received the lowest rate of supply of N and the highest of Mg. Treatment F was a control receiving neither N nor Mg. Modified from Tilman (1984).

abilities of N and P (Fig. 2.8a). Thus, changes in the relative rates of supply of two limiting resources led to separation of the major species along a gradient, just as theory predicts should happen when plants compete for essential resources (Fig. 2.3).

Tilman (1984) established an experimental N:Mg gradient by fertilizing a newly disturbed field with different ratios of nitrogen and magnesium. By the second year of the experiment, five of the six most abundant species were distinctly separated along the gradient (Fig. 2.8b). Similar experiments, performed in a variety of plant communities around the world (e.g. Milton, 1934; Willis & Yemm, 1961; Specht, 1963; Ginzo, *et al.*, 1982; Tilman, 1982), have shown major changes in plant dominance following fertilization. Such results are consistent with the hypothesis that competition for resources is an important process influencing the local abundance of species. They demonstrate experimentally that different supply rates of soil resources lead to dominance by different species. More research is needed, however, before we will be able to predict which species will become dominant in a particular habitat in response to a particular set of environmental conditions or manipulations.

2.5   Primary succession

When bare ground is newly exposed following such events as glacial recession, sand dune formation, landslide or volcanic eruption, the ground is colonized by a series of species in a process that is called primary succession (Chapter 1). The species replacements observed during primary succession often seem to correspond with changes in the availabilities of soil nutrients and light at the soil surface. For instance, recession of glaciers in Glacier Bay, Alaska, has exposed 105 km of bare gravel and sand substrate during the 225 year warming period since the Little Ice Age (Lawrence, 1979). When initially exposed, the substrate had very low levels of nitrogen, and the initial colonists to the habitat showed signs of extreme nitrogen deficiency (Lawrence et al., 1967). Nitrogen is the main limiting soil resource in all primary successions, because the mineral substrates in which soils form contain little nitrogen, but do contain sufficient amounts of P, K, Ca, Mg, and other minerals required by plants (Jenny, 1980). At Glacier Bay, nitrogen levels increased 5 to 10 fold within 100 years (Crocker & Major, 1955; Fig. 2.9a). As soil nitrogen levels increased, total plant biomass increased, and the proportion of light penetrating to the soil surface decreased. As these changes occurred, the dominant species changed from short, nitrogen-fixing species such as black crust-forming soil algae, lichens *Stereocaulon* and *Lempholemma* and a rose *Dryas drummondi*, to such taller species such as nitrogen-fixing alders, to the even taller cottonwood trees, and finally the even taller spruce and fir trees (Cooper, 1923, 1939; Lawrence, 1958; Lawrence et al., 1967; Reiners et al., 1971; Worley, 1973). A qualitatively similar pattern has been reported for primary succession following sand dune formation along the southern coast of Lake Michigan (Cowles, 1899; Olson, 1958; Robertson & Vitousek, 1981).

What could cause such patterns? Although numerous factors are involved, plant competition for limiting resources may play a central role in these and other primary successions and in secondary successions on poor soils (Tilman, 1982, 1985). Just as species that differ in their competitive abilities for two essential resources should become separated along a spatial gradient in the relative availabilities of the resources (Fig. 2.3b), so should species be separated through time along a temporal gradient of the relative availabilities of resources.

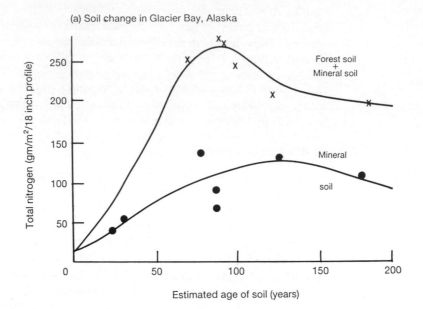

(a) Soil change in Glacier Bay, Alaska

(b) Glacier Bay succession

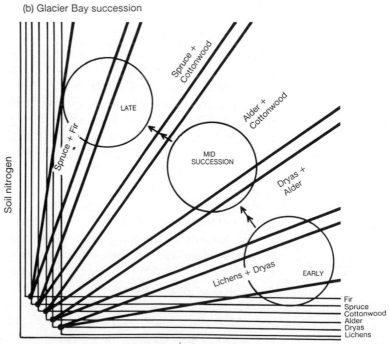

No matter what the limiting resources might be, theory predicts that there can be a directional and predictable change in the species composition of a plant community *only to the extent that there are regular and predictable changes in the relative availabilities of the limiting resources.* Because the relative availability of two resources can be approximated by their ratio, this has been called the resource ratio hypothesis of succession (Tilman, 1982, 1985). A major temporal gradient in primary successions is from habitats that have high light at the soil surface but low soil nitrogen, to habitats that have low light at the soil surface but high availability of nitrogen. If some of the major species involved in succession at Glacier Bay, Alaska, differ in their requirements for light and nitrogen as illustrated in Figure 2.9b, then the observed changes in soil nitrogen and light could explain at least the broad pattern of succession observed there.

2.5.1   The soil-resource: light gradient

During primary succession in Glacier Bay, there is a major and obvious change in the structure of the vegetation. The initial dominants are short, and as succession proceeds, these species are replaced by others which are increasingly tall.

Why are taller species dominant later in primary succession? As plant biomass increases, less light reaches the soil surface. Light is a directional resource, supplied from above, so that taller plants can obtain a greater proportion of the light. The simplest indicator of the ability of plants to compete for light is their height at maturity (Givnish, 1982). During at least the first few hundred years, primary succession tends to be a directional process in which the richness of the soil and plant height both increase. The increased height of the mature dominant plants is consistent with the hypothesis that competition for light is an important process on richer soils.

Similar patterns have been reported in mature, non-successional vegetation around the world. If soils within an area are ranked from

Fig. 2.9. By sampling sites in a primary succession at Glacier Bay, Alaska, USA, which had been deglaciated at different times during the past 200 years, a time sequence of total nitrogen levels can be obtained. Both the nitrogen in the mineral portion of the soil and the total nitrogen in the mineral portion plus the forest floor layer increased during succession. After Crocker & Major (1955). (*b*) The pattern of species replacement in succession may depend on the ability to compete for nitrogen and for light. This figure shows light at the soil surface decreasing through succession while total soil nitrogen levels increase. The hypothesis is that species which are superior competitors for nitrogen (such as lichens and *Dryas*) are poor competitors for light, and vice versa. The major species (or species groups) are assumed to be separated in their requirements for nitrogen, with lichens the best competitors, followed by *Dryas*, alder, cottonwood, spruce and then fir, the worst nitrogen competitor. The species are assumed to be ranked in reverse order in their competitive abilities for light. Modified from Tilman (1982, 1985).

those that have the lowest availability of a limiting soil resource to those that have the highest availability, the dominant plant species are found to change from plants that are short at maturity to plants that are tall at maturity (Tilman, 1986). Such gradients can be considered to be gradients in the relative availability, or ratio, of soil resources versus light. I call them soil-resource:light gradients (Tilman, 1986). For instance, on sandy soils of Michigan, Wisconsin and Minnesota, the most nitrogen-poor soils are dominated by lichens, forbs and grasses, with only scattered individuals of cherry bushes and oak trees. Richer soils have an oak savannah vegetation. Still richer soils have closed canopy oak forests, and the richest soils are dominated by closed-canopy maple forests (Hole, 1976). The species dominant on poor soils are short at maturity, while the species dominant on richer soils are increasingly taller at maturity along this soil resource:light gradient.

Similarly, in the areas of California, Chile, Italy, South Africa, and Australia that have a Mediterranean-type climate, the vegetation ranges from open, low scrublands with plants about 0.3 m tall in areas with poor soils to dense scrub dominated by different species reaching 10 m at maturity on the most moist soils (Cody & Mooney, 1978; Beard, 1983). In the case of the British flora, Grime (1979) reports that taller species tend to be found in areas with richer soils, whereas shorter plants tend to be found in habitats with poor soils. For the new world tropics, Beard (1944, 1955) observed that the height of the dominant species depended on the number of months per year that an area received rainfall. The most productive regions, which received 12 months of rainfall a year, were dominated by species that ranged from 40–50 m in height. Increasingly less productive regions of the new world tropics were dominated by species that were progressively less tall. The driest area that he recognized was a scrubland-cactus assemblage which averaged 2–4 m in height.

Thus, along a gradient from resource-poor (dry) to resource-rich (wet) sites, there is a replacement of species, with the species dominant in the most resource-rich sites being much taller than those dominant in the resource-poor areas. Marks (1983) reported a similar pattern for the areas in which native North American plants were dominant before European disturbance. The shorter species, common in secondary succession today, were dominant in such resource poor habitats as sand plains, gravel bars, limestone outcrops, talus slopes, rock crevices, eroded ridgetops, and eroded stream banks, whereas the taller species, such as maple, were dominant on rich soils.

This strong tendency for species dominant on rich soils to be taller at maturity than those dominant on poor soils is consistent with the hypothesis that taller species are superior competitors for light.

However, why are species that are shorter at maturity favoured on resource-poor soils? To answer this question, it is necessary to consider the forces that control the evolution of plant life histories. Put simply, the theory of evolution by natural selection states that a trait that leads to a greater net reproductive rate will be favoured over other traits. In a habitat in which individuals of a species are not light-limited, there is no selective advantage for increases in height (except, perhaps, for seed dispersal), or for delayed reproduction. Thus, shorter individuals, capable of earlier reproduction would be favoured. In a habitat in which individuals are light-limited, those which delayed reproduction, and allocated more of their potential growth to increases in height, could be favoured, since they would acquire a greater proportion of the limiting resource, light.

There is another potential cost, in addition to reduced or delayed reproduction, that plants face in order to be superior competitors for light. Taller plants require a stronger stem for the support of their photosynthetic tissues (Chapter 9). However, these plants must still transport essential soil resources from their roots to their leaves. Thus, they must allocate a greater proportion of their potential growth to the production of structural and transport tissues, and they must bear the energetic cost of the respiration of these tissues. Such costs cause woody perennial plants to have much lower maximal rates of weight gain, as seedlings, than herbaceous (non-woody) perennial plants, and cause herbaceous perennial plants to have lower seedling growth rates than annual plants (Grime & Hunt, 1975).

Thus, several major life history traits of plants may be strongly influenced by the position along a soil resource:light gradient at which a species is a superior competitor. In general, plants that are good competitors for nutrients, but poor competitors for light, should be shorter at maturity, reproduce sooner, and have a greater maximal rate of weight gain than species that are superior competitors for light (Tilman, 1985, 1986). There should be a smooth gradation in these traits along a soil-resource:light gradient, and although there are many other factors which influence the evolution of plant life histories, a major axis for differentiation of plant life histories and speciation has been the gradient from nutrient-poor to nutrient-rich habitats (i.e. the soil resource:light gradient; Tilman, 1985, 1986).

2.6   The dynamics of plant communities

So far in this chapter, I have discussed broad relationships between the species composition and diversity of plant communities and the availabilities of various resources. I have focused on patterns in mature vegetation and on the rather slow dynamics of primary succession. I did this because these are some of the more obvious and clear patterns, and because a simple, equilibrium theory of plant competition for resources, like that illustrated in Figs 2.1–2.3, has the potential to predict such patterns. However, many of the dynamic changes in plant communities with which we are familiar occur on much shorter time scales. Many of us have seen dramatic changes within a one or two year period following abandonment of a garden plot or a farmer's field. To predict such dynamics, it is necessary to use explicit, differential equation models of plant competition for resources (Tilman, 1985). Although I will not present such models here, I do want to point out that the short-term dynamics observed following the perturbation or manipulation of a plant community can be *quite different* from the long-term, equilibrium effect of that same manipulation.

This makes it extremely difficult to interpret the results of short-term, manipulative experiments. Consider, for instance, the primary succession at Glacier Bay, Alaska. What might happen to the composition of an early successional plant community if it were fertilized with nitrogen? Figure 2.9b predicts that the long-term effects would be a more rapid succession, with spruce and fir becoming dominant sooner. However, the immediate effects could be quite different. If the early successional, shorter species do have greater maximal growth rates (as discussed above), the immediate response to the addition of nitrogen could be an increase in their abundance. Just after nutrients are added, the habitat will have unusually high availabilities of nutrients for the ambient light level. The species capable of growing most rapidly under those situations would dominate initially. Those species should be the ones with the greater maximal growth rates. Eventually, as these species grew, they would reduce light to a lower level, and conditions would then begin to favour later successional species. Computer simulations of such cases have shown that the initial effect of fertilizers is to increase the abundance of early successional species, whereas the long-term effect is to favour late successional species (Tilman, 1985). Thus, because there are differences in the maximal growth rates of species, it is very difficult to interpret the immediate effects of any manipulation unambiguously. Long-term experimentation is required.

Probably because richer land is more commonly used for agriculture, the vast majority of documented secondary successions have occurred on rich soils, where much of the initial dynamics may reflect nothing more than differences in the maximal growth rates of the species (Egler, 1954, 1976). However, it has often been observed that the species sequence of secondary succession is similar to, but much faster than, the species sequence in nearby primary successions. If the life histories of species have evolved to correspond with the point along a soil-resource:light gradient where they are the superior competitors, a dynamic model of plant competition for these resources would predict such similarity. The initial dominance by species that are shorter, and better nutrient competitors, however, would not result from competition for nutrients on these nutrient rich soils, but come from the greater maximal growth rates that such species have evolved (Tilman, 1985). As I believe is illustrated by much of the work on secondary successions on rich soils, too detailed a view of the short-term dynamics of plant communities may obscure the simple, general underlying causes.

2.7   Conclusions

Many different processes control the structure of plant communities. This chapter has taken a broad look at one of these processes, namely plant competition for limiting resources. A variety of observational and experimental studies, performed in an assortment of aquatic and terrestrial habitats world-wide, have shown that the species richness, species composition and species dominance of plant communities are greatly affected by the availabilities of limiting resources. At least in broad outline, many of these patterns are consistent with the predictions of a simple theory of plant competition for essential resources. As predicted by theory, the plant communities with the greatest diversity within a region occur in resource-poor habitats. Increases in the supply rates of limiting resources lead to decreased diversity. Much of the local spatial heterogeneity of plant abundances may be caused by local spatial heterogeneity in the availability of soil resources.

The broad patterns of primary succession and of secondary succession on poor soils correspond with changes in the relative availabilities of limiting resources. Theory predicts that succession should be a predictable and directional process only to the extent that the processes controlling the supply rates of limiting resources are *predictable* and *directional*. In many of these successions, soil resources increase and the availability of light at the soil surface decreases, leading to a gradient through time in the soil resources:light ratio. The separation of successional species along these temporal gradients is very similar to their separation along comparable spatial gradients in mature communities. Both of these patterns are potentially explained by a simple model of resource competition.

Thus, many of the broad, long-term patterns observed in plant communities may be the result of interspecific competition for various limiting resources. Such correspondence does not mean that other processes are unimportant in structuring these plant communities, but the simple theory that predicts them may provide a framework for more detailed studies.

Chapter 3 Canopy Gaps and the Dynamics of a Neotropical Forest

STEPHEN P. HUBBELL & ROBIN B. FOSTER

3.1 Introduction

When a tree falls in the forest, the canopy hole or 'light gap' it creates sets in motion a chain of events known as 'gap-phase regeneration' which culminates in the replacement of the previous canopy tree by a new one. Gap disturbances provide the principal or only means by which most tree species can maintain their representation in closed-canopy forests. Therefore understanding the gap disturbance regime and knowing the rules for tree-by-tree replacement processes are fundamental to understanding the structure and dynamics of closed-canopy tree communities. This chapter summarizes what we have recently learned about the abundance, distribution, and geometry of canopy gaps in a neotropical forest on Barro Colorado Island (BCI), Panama, and how the gap-disturbance regime of the forest influences the structure and species composition of the canopy tree community.

Such seemingly simple events as tree replacement and gap-phase regeneration are actually complex processes that are still poorly understood. Biology, chance, and history all play a role in the replacement process (Hubbell & Foster, 1986a). Biology plays a role because tree species vary, for example, in their capacity to disperse seeds, their powers of seed dormancy, their longevity as supressed saplings in shade, and in their tolerance for gap microclimates. Chance and history also play a role because the precise location of the next treefall gap is unpredictable, so that successful growth and maturation of a sapling may depend as much on being in the right place at the right time as on the species to which it belongs.

3.2 Gap-phase regeneration: the current paradigm

Before presenting our Panamanian forest results, it is helpful to outline some basic elements of the current paradigm for gap-phase dynamics, and consider some questions and hypotheses. A recent review by Brokaw (1985a) gives a more detailed treatment of the following points.

3.2.1   Gap structure

The idealized treefall gap made by an uprooted tree is a dumb-bell shaped area with three major zones of biological importance (Oldeman, 1978). Zone one is the area around the base of the tree, the area with maximal soil disturbance due to the uprooting of the tree, and maximal light due to the new opening in the canopy directly overhead. Zone two lies along the midsectional axis of the gap, next to the fallen trunk. In this zone there is relatively little disturbance of the soil or of the understory vegetation of saplings except in the immediate area hit by the fallen trunk. Zone three is the area struck by the falling tree crown, the area suffering the greatest destruction of the understory vegetation. Gap size varies considerably, depending upon the size of the tree that falls, the orientation of the falling tree in relation to neighbouring trees, and the number of other trees that fall with it. Lianas (woody climbers) may link trees together, so that when one tree falls, it sometimes pulls other trees down with it (Putz, 1984).

3.2.2   Gap creation

Theory aside, actual gaps are diverse in structure because of the many causes and circumstances of gap creation. Many trees die not by uprooting but by trunk-snapping during a windstorm. Other trees may die standing up, and drop their branches over a long period of time. Species differ in their likelihood of dying in these various ways. Gap creation is seasonal in many tropical forests, with a peak frequency during the rainy months (Brokaw, 1982); wet trees are heavier and less streamlined in wind, and wet or loosened soil provides less stable rooting for trees. Treefall seasonality may influence the success of establishment of tree species which depend on dispersal and germination to colonize gaps (Brokaw, 1982; Garwood, 1983). Rates of treefall also vary with soils and topography (exposure to wind).

3.2.3   Gap environment

The great structural diversity of gaps produces similarly great diversity in gap microclimates. Gap size, shape, compass orientation, and the height of surrounding vegetation, as well as the vegetation and treefall debris in the gap itself, all influence gap light, temperature, moisture, and wind regimes (Brokaw, 1985a; Denslow, 1980). Light in gaps is higher in intensity, lasts longer during the day, and is of different spectral quality than in the understory of the closed-canopy forest. Soil and the air layer near the ground are often much hotter in gaps during the day, and cooler at night. Humidity is often lower in gaps than in surrounding forest, but soil water content may actually be higher because of reduced root uptake and lower transpirational water loss from gaps (Lee, 1978). Soil nutrients in gaps are enriched in local patches, particularly in the areas adjacent to the decaying fallen tree and in the disturbed soil of the uprooted area.

3.2.4   Regeneration guilds

In discussing gap-phase regeneration, it is useful to distinguish at least three guilds of canopy tree species on the basis of their life history patterns and requirements for regeneration. These are primary or mature forest tree species, early pioneers, and late secondary species (Hartshorn, 1978; Brokaw, 1985a), but names are rather misleading because all three kinds of trees are represented in mature forest.

Primary tree species germinate in the shade, or in both sun and shade, and can persist as suppressed juveniles for some time until a gap opens overhead. As a group these species have moderate to high shade tolerance (low compensation point), low photosynthetic and growth rates, moderate to high wood density, and relatively complex branching architecture. They have rather brief (or no) periods of seed dormancy (less true for species in seasonal tropical forests, as Garwood, 1983, has shown), variable powers of seed dispersal and diverse seed dispersal mechanisms. Their seeds are generally large and few compared to the seeds of pioneers (see Chapter 6).

Early pioneer species have a different strategy for waiting for gaps to open; they frequently persist for long periods as dormant seeds in the soil. These seeds require specific gap disturbance cues to germinate. High temperatures or light intensities may trigger germination, or a specific light quality may be required. For example, a high ratio of red to far red light, characteristic of moderate to large light gaps, stimulates seed germination in *Cecropia obtusifolia* (Moraceae) (Vasquez-Yanes & Smith, 1982), whereas low ratios characteristic of small gaps and the forest interior inhibit germination. Early pioneer species have high maximal photosynthetic rates and growth rates, and growth responds much more dramatically to increased light than in primary species. They are much less tolerant of shade than primary species, and their juveniles are found mainly, if not exclusively, in gaps. Pioneers typically exhibit rapid elongation of a monopodial shoot, simple branching architecture, and low wood density. They also mature rapidly, often flowering while they are still small, and as a group have relatively short life spans. They generally produce large numbers of small seeds, widely disseminated by birds and bats.

Late secondary species are typically shade-intolerant heliophiles (sun-loving plants) that have high maximal rates of photosynthesis and growth like early pioneer species. However, they often grow to much larger stature and persist for long periods in the canopy. Many of the emergent species (trees which grow taller than the general forest canopy) are of this type. Their wood density is variable, but generally lower than primary species. Many late secondary species, particularly the emergents, do not have seed dormancy, and rely on wind or ground mammals to disperse their seeds to new gaps.

3.2.5   Gap regeneration

Canopy gaps fill in with regrowth from three sources: seeds, plants established prior to gap formation, and lateral ingrowth of branches from trees on the gap periphery. In large gaps, pioneers and late secondary species, germinating from seed arriving before or after gap formation, grow rapidly and overtop the slower-growing established juveniles of primary species which survived the treefall.

Brokaw (1980, 1985b) studied the first six years of gap-phase regeneration in 30 gaps on BCI. In large gaps, he found that species numbers and stem densities increased rapidly, then levelled off or declined in 2–3 years, due to reduced recruitment and increased mortality, especially among the pioneer species. Pioneers colonized gaps of all sizes but as a group were denser in large gaps, where they grew faster than primary species. In small gaps, pioneers suffered heavy mortality, and in most cases only the primary species persisted. The growth rate of pioneer species increased with gap size, but it did not do so in primary species, which nevertheless grew faster in gaps than in the shaded understory. The density of stems of primary species was unrelated to gap size.

After considerable thinning of stems among pioneers as well as late secondary and primary species, the gap-phase regeneration enters a building phase in which survivors grow into larger sub-canopy sub-adults. As the short-lived early pioneers die out, they are replaced by primary species which were continuously present in the understory of the pioneers. In the larger gaps, persistent late secondary species in the developing canopy of the maturing gap may continue to suppress the primary species for much longer. Growth during this phase may be episodic as juvenile canopy trees are alternately suppressed and released from shade competition.

3.3   Some hypotheses and predictions

The relative importance of biology, chance, and history in gap-phase regeneration and subsequent canopy tree replacement is of considerable interest in plant community ecology because it touches the heart of the debate over whether forest communities are in species-equilibrium (Hubbell, 1984; and see Chapter 2). At issue is the strength if not the existence of equilibrating forces in forest communities, tending to stabilize particular taxonomic assemblages of tree species, and to resist invasion by other species (Hubbell & Foster, 1985). One potentially important mechanism for stabilization could be specialization for conditions of regeneration (Grubb, 1977).

Arguing that there is a greater range in microclimate from the forest interior to the centre of a large gap in tropical forests than in temperate forests, Ricklefs (1977) suggested that niche specialization along this gap microclimatic gradient could help to explain the higher tree diversity of tropical forests. Brokaw (1982, 1985b), Denslow (1980), Hartshorn (1978, 1980) and Whitmore (1974) have independently argued that the size, and possibly the seasonal timing, of gaps might be critically important variables in the establishment of particular tree species. Brokaw (1982, 1985a) has provided strong observational data in support of the gap-size factor for pioneers; but for primary species, the evidence is still equivocal.

Alternatively, specialization into regeneration niches may not involve explicit microsite requirements, but take the form of epidsodic, asynchronously-cued reproduction among tree species (the variable recruitment hypothesis). In theoretical models, a multi-species stochastic equilibrium exists when each tree species has a time period during which its per capita recruitment is greater than the average per capita recruitment of all the other species in the community (Chesson & Warner, 1981). In space-limited communities in which the total support capacity for all species is finite, this mechanism results in a frequency-dependent reproductive advantage for rare species. Unfortunately, existing data on gap disturbance regimes and tree replacement processes are not yet adequate to evaluate most of these hypotheses rigorously.

Hartshorn (1978) made predictions about tropical forest dynamics in relation to the gap disturbance regime, at least one of which can be evaluated with present data, however. Among other things, he predicted that there should be a positive relation between turnover rate of tropical forests and the proportion of shade-intolerant species which depend on gaps for successful regeneration. Turnover rate is estimated by dividing the total forest area by the gap area created per unit time. Denslow (1980) made a similar prediction by arguing that a high proportion of the tree species in a tropical forest should be adapted for regeneration under the most common disturbance regime in the forest. If large gaps are the rule, for example, then heliophilic (sun-loving) species doing best in large gaps should predominate in terms of species richness and relative abundance.

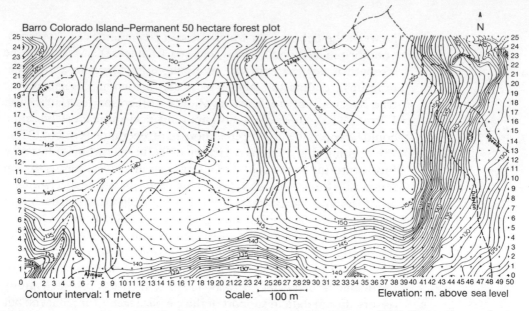

Barro Colorado Island–Permanent 50 hectare forest plot

Contour interval: 1 metre   Scale: ⊢ 100 m ⊣   Elevation: m. above sea level

Fig. 3.1. Topographic map of the 50 hectare plot on Barro Colorado Island, Panama (BCI). Contour lines are drawn for each metre of elevational change. Dotted lines mark permanent trails through the plot. Grid marks indicate the corners of 1250, 20×20 metre quadrats. A stream in the lower left corner drains a seasonal swamp (left of centre). Grades on the steeper slopes range between 10 and 20%.

3.4   A neotropical forest case study

In 1980 we began a long-term study of a half-square kilometre plot of tropical forest on Barro Colorado Island (BCI), Panama (Hubbell & Foster, 1983), a 15 km² nature reserve in Gatun Lake. In this plot (Fig. 3.1), we have completely identified, measured, and mapped to the nearest metre all free-standing woody plants with a 1 cm or greater stem diameter at breast height (dbh). There are nearly a quarter of a million trees, saplings, and shrubs in the 50-hectare plot, representing approximately 300 species. The plot is located in 'old-growth' forest (Foster & Brokaw, 1982). Recent palaeoecological surveys of the site reveal that the forest has never been clearcut for agriculture: fossil corn phytoliths and weedy pasture grasses are absent from plot soil cores, in contrast to permanent pre-Columbian settlement areas along the Rio Chagres several kilometres away (D. Piperno, pers. comm.). However, radiocarbon dating of cores indicate that from 500 A.D. until 545 years ago, small temporary clearings were occasionally made in the forest, probably for seasonal hunting camps. Thus the forest has been essentially undisturbed by man for the past 500 years or more.

3.4.1   Methods

In 1983 and 1984 we took a complete census of the canopy and light gaps in the plot. The canopy census is straightforward: the plot is gridded into 20,000 permanent 5×5 m subplots, and a measurement of canopy height and layering is taken at the corner of each of these subplots. The mean intercensus interval is about a year and a half.

The sampling method was as follows. Using a pole with a surveyor's rod level, we took a vertical sighting over each point and estimated canopy height with a range finder. Canopy heights were classified as follows: 0–2 m (henceforth called '0 m'), 2–5 m ('2 m'), 5–10 m ('5 m'), 10–20 m ('10 m'), 20–30 m ('20 m'), and 30 m and above ('30 m'). These classes were chosen because they provided useful height benchmarks for early to middle stages of gap-phase regeneration. The higher categories are broader because canopy heights at these distances could not be determined with high precision. We also recorded which height classes below the canopy contained vegetation layers.

This method for measuring the canopy has advantages and disadvantages. The advantages are that the method is simple, reproducible, and can be done extensively in a reasonable period of time. Probably the most important advantage of the method is that it eliminates virtually all subjectivity from the problem of defining treefall gaps. This has been a major problem when estimating gap area and using this estimate to calculate forest turnover rates (Brokaw, 1982). Finally, unlike previous methods used in gap studies, our method also reveals the presence of old gaps in the canopy, which no longer penetrate down to ground level. Thus it is possible to unveil the recent history of gap disturbance in the forest, beyond a simple description of present ground-level gaps. These older gaps at higher levels in the canopy may be very important for the successful maturation of sub-adult trees during the building phase of forest regrowth. The principal disadvantages are: 1) gaps smaller than 5×5 m in size cannot be accurately censused; 2) the method does not identify which trees are responsible for creating either canopy or gap; and 3) the true light environment for each plant is not precisely measured.

3.4.2 Questions

The discussion which follows is divided into two main parts. The first part focuses on a description of gaps and gap dynamics over a 17-month period in the 50-hectare plot. Some organizing questions are these: 1) how abundant are gaps in the plot? 2) how does gap abundance change with gap size? 3) what are the horizontal and vertical cross-sectional shapes of the gaps? 4) how are the gaps spatially distributed in the forest? 5) do new gaps open up at random locations in non-gap areas, or are they more likely to occur adjacent to pre-existing gaps, or on certain soils and terrain? 6) once a gap is created, how rapidly does the canopy fill in? 7) is the canopy height distribution in the forest in equilibrium with the current gap disturbance regime? 8) are different areas in the forest experiencing different gap disturbance regimes?

The second part of the chapter summarizes the responses of 81 canopy tree species to gaps, focusing on the distribution of saplings in gaps, gap edges, and forest interior, and addressing the following questions: 1) how far does a tree have to disperse its seeds to reach a gap? 2) how do canopy tree species differ in the proportion of their saplings which are located in gaps or in the forest interior? 3) are guilds of ecologically similar species in terms of gap requirements evident? 4) what is the relationship between the abundance of a tree species and its apparent degree of gap dependence? 5) are there more heliophilic species in the canopy in areas of the forest which have experienced greater gap disturbance in recent history? The evidence outlined in this section is circumstantial at present because we do not yet have the recensus data for the plot which will allow us to estimate growth, reproduction, and mortality rates.

3.5   Distribution, abundance and geometry of canopy gaps

3.5.1   Gap distribution and abundance

From the canopy data collected at points every 5 m, we can reconstruct horizontal cross-sections of the canopy over the entire plot. The 1983 horizontal canopy cross-sections at the ground (< 2 m) and at 5, 10, 20 and 30 m above ground, permit several immediate conclusions about gaps in the BCI forest:

(i) The total gap area increases steadily with increasing height above ground. At ground level, total gap area is 0.68 ha (1.36%) increasing to 31.99 ha (63.98%) at 30 m.

(ii) Small gaps are more abundant than large gaps. A nearly linear negative log–log relationship exists between gap abundance and gap size for canopy cross-sections up to a height of 10 m (Fig. 3.2).

(iii) The number of canopy gaps is maximal at intermediate cross-sectional height (10 m), but mean gap size increases monotonically with height in the canopy. Large gaps become ever commoner relative to small gaps as one moves upwards through the vegetation profile. At 10, 5 and 2 m, a doubling in gap size results in a 2.2-, 3.2- and a 5.4-fold reduction in gap number, respectively (Fig. 3.2). At cross-sections above 10 m, gaps begin to merge and coalesce, resulting in an increase in mean gap size and a reduction in total gap number.

(iv) At ground level up to the 10 m cross-section, gap dispersion in the forest is indistinguishable from random (based on nearest neighbour analysis of gap centres).

(v) However, above 10 m, gaps coalesce to form non-random

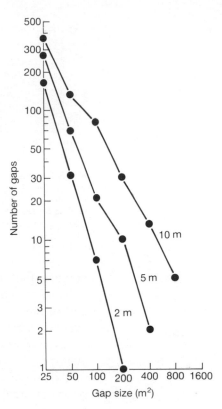

Fig. 3.2. The linear log–log relationship between gap abundance and gap size at 3 cross-sectional heights in the canopy (2, 5, and 10 m). The line becomes progressively steeper at lower levels in the canopy, indicating that large gaps are relatively rarer near the forest floor than they are higher in the forest canopy.

patches, with noticeably less canopy cover in the western half of the plot. At 20 m, mean per hectare canopy cover is 4.7% lower in the west (p < 0.004, rank sum test); at 30 m, mean per hectare canopy is 16.9% lower in the west (p < 0.001).

(vi) The linear track of a violent windstorm that caused extensive treefalls in the early 1960s is still visible in canopy cross-sections at higher levels (20 and 30 m). The track runs from the east central area of the plot southwest to the plot edge, extending over a distance of at least 350 m (Fig. 3.4). The storm track is precisely aligned with the direction of prevailing trade winds in the northeast.

3.5.2 Gap geometry

In east-west profile, the average gap reaching the forest floor is shaped like an inverted triangle (a cone in three dimensions) (Fig. 3.3). At the ground, a mean of only about 5 m of soil is exposed, widening to 25 m or more in the 20–30 m height level. The average gap is asymmetrical with lower canopies on its eastern margin in the direction of prevailing winds.

Some gaps in the forest have columnar vertical cross-sections, but these are a distinct minority. The cone shape is due in part to the

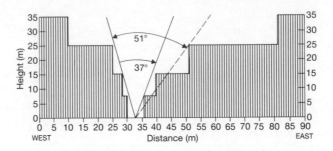

Fig. 3.3. Vertical cross-section in an east–west direction through the average gap in the BCI forest which reaches the forest floor. The gap is triangular in cross-section, standing on its vertex. Steps in the vegetation profile were placed at the midpoint of each canopy height class as the best estimate of mean vegetation height in the class. The aperture of the gap to the sky from gap centre at ground level is between 37 and 51°. Note the asymmetry of the sides of the average gap: canopy height rises more slowly to the east, the direction of the prevailing winds, than to the west.

damage pattern caused by the falling tree crown, in part to the survival of saplings of primary species not struck by the falling tree, and in part to averaging the vertical cross-section of many gaps in various stages of inward regrowth from the gap edge.

The aperture of the gap to the sky, viewed from ground centre, average 37–51 degrees, depending upon the particular tangent drawn to crowns of trees surrounding the gap. Thus, gap centres are exposed to significant periods of direct sunlight during the day. The cone-shape of gaps generates a range of gap-edge microsites having increasing canopy height as one moves away from the centre of the gap. These gap edges extend at least 10 m into the forest from central gap areas. In the analysis of sapling spatial distributions which follows, we distinguish two levels of gap-edge environment: less than 5 m, and 5–10 m, from gaps. Forest interior is defined as any site more than 10 m from any gap.

Gap perimeters become increasingly irregular in shape at higher levels in the canopy, and gap shapes are elongate. Gap perimeters must fall between a maximum perimeter (linear, with all gap 5×5 plots in a line) and a minimum perimeter (a circle); actual gap shapes are more linear than circular.

3.6   Canopy and gap dynamics

At the moment we have only two canopy censuses, roughly a year and a half apart, on which to base an analysis of canopy and gap dynamics in the BCI forest. However, because a large area has been censused in detail, a number of important conclusions can be drawn which seem unlikely to change.

(i) Most 5×5 m plots showed no change in canopy height class over the intercensus interval, except at 0–2 m, where a greater number graduated to the next higher height class than stayed the same (Table 3.1). Nevertheless, transitions were possible between all height classes. A few gap plots made transitions in 17 months to high canopies, but these transitions were invariably due to ingrowth of high branches from neighbouring trees.

Table 3.1. Transitions between canopy height categories from 1983–84 in the 50-hectare forest plot on Barro Colorado Island, Panama. The rows are canopy heights in 1983, and columns are the corresponding heights in 1984. Entries in the table give the numbers of 5×5 m plots whose canopy underwent the indicated height transition, and the observed probabilities of that transition (in parentheses).

		Canopy Height Class in 1984 (metres)					
		0–2	2–5	5–10	10–20	20–30	30 up
Canopy	0–2	62 (0.228)	81 (0.298)	49 (0.180)	48 (0.176)	28 (0.103)	4 (0.015)
Height	2–5	25 (0.062)	186 (0.463)	96 (0.239)	54 (0.134)	35 (0.087)	6 (0.015)
Class	5–10	17 (0.013)	54 (0.043)	710 (0.563)	354 (0.281)	95 (0.075)	30 (0.024)
in 1983	10–20	60 (0.013)	60 (0.013)	142 (0.030)	3460 (0.737)	822 (0.175)	153 (0.033)
(metres)	20–30	67 (0.011)	49 (0.008)	71 (0.012)	575 (0.093)	4977 (0.807)	426 (0.069)
	30 up	27 (0.004)	10 (0.001)	43 (0.006)	176 (0.025)	530 (0.074)	6418 (0.891)

(ii) The canopy gap disturbance regime is near equilibrium over the plot as a whole, but slightly more high-canopy forest is currently falling than is being replaced. This conclusion is drawn by analyzing the matrix of transitions occurring between height classes between 1983 and 1984, which is a quantitative description of the disturbance regime for a year and half (Table 3.1). Using linear algebra, it is straightforward to determine the equilibrium percentage of each canopy height class expected in the forest if the matrix of canopy transition probabilities (i.e. the disturbance regime) remains unchanged. The observed percentages of 5×5 m plots in each height class in 1983 were as follows: 0–2 m (1.32%), 2–5 m (2.01%), 5–10 m (6.30%), 10–20 m (23.49%), 20–30 m (33.25%), and over 30 m (36.02%). The equilibrium percentages based on the current gap disturbance regime of the plot are: 0–2 m (1.32%), 2–5 m (2.29%), 5–10 m (4.80%), 10–20 m (23.14%), 20–30 m (36.74%), and over 30 m (31.71%). Thus, the percentage of area in ground-level and 2–5 m gaps is very near equilibrium. The sub-canopy layer, from 20–30 m, however, can be expected to increase at the expense of the highest canopy class (> 30 m) over the next decade or so if the present disturbance regime continues.

(iii) Local differences in gap disturbance regime exist within the plot. Although the canopy structure is near equilibrium for the plot as a whole, this does not imply that the gap disturbance regime must be identical from one area to the next. Gap area per hectare varies considerably, especially at higher levels in the canopy. For example, gap area per hectare open above 5 m varies between 2.50% and 22.75%. For each hectare we can repeat the type of equilibrium analysis done for the plot as a whole, computing the expected equilibrium canopy structure by hectare. This analysis reveals considerable local heterogeneity in disturbance regime. The mean equilibrium percentages of area predicted for each height class per hectare, and their standard deviations, are: 0–2 m (1.47±1.42%), 2–5 m (2.99±3.59%), 5–10 m (5.10±4.58%), 10–20 m (23.98± 13.32%), 20–30 m (37.13±18.71%), and over 30 m (29.33± 29.79%).

(iv) The two height classes representing forest areas in a building-phase have the lowest coefficients of variation (standard deviation/ mean) of equilibrium values from hectare to hectare: 10–20 m (0.56) and 20–30 m (0.50). In contrast, gap-phase areas in a late stage of thinning, and forest areas with the highest canopy, both exhibit wider variation: 2–5 m (1.20) and over 30 m (1.02). These observations suggest instability and relatively fast-paced dynamics in the canopy of gap-phase regeneration sites and in high-canopy forest, and stability with slow change in the canopy of regenerating sites in a building-phase.

(v) Trees adjacent to gaps, or which stand above the surrounding canopy, suffer a greater risk of falling than trees which are sur-rounded by plants as tall or taller than themselves. The spatial pattern of 5×5 m plots with 30 m canopies in 1983 and less than 20 m canopies in 1984 shows that drops in canopy height are much more likely to occur on the edges of gaps (Fig. 3.4). We can quantify this risk by computing the probability that a plot with a 30 m canopy in 1983 will fall to a lower canopy class in 1984, as a function of the number of adjacent plots that had lower canopies in 1983. In the grid, each

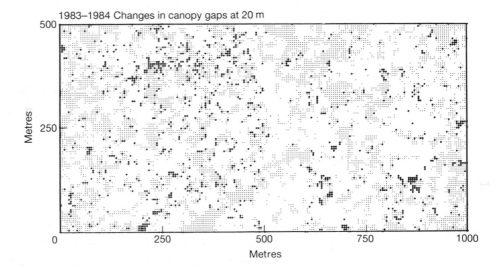

Fig. 3.4. Changes in high-canopy gaps (20–30 m height class) between the 1983 and 1984 canopy censuses. Small dots indicate 5×5 m plots with canopies below 20 m in 1983. Large dots indicate 5×5 m plots with canopies 30 m or above in 1983, but with canopies lower than 20 m in 1984. Note that almost invariably these new gaps occur along the edges of existing gaps. The track of a violent storm in the early 1960s is still visible in a line of gaps running north-east from the southern edge of the plot.

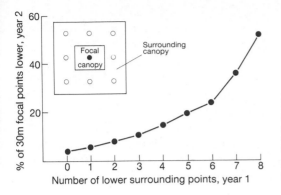

Fig. 3.5. Relationship between the probability of a canopy fall in a 5×5 m plot as a function of the number of surrounding 5×5 m plots with *lower* canopies in the preceding year. Data are for 5×5 m plots at 30 m in 1983. The curve indicates a higher probability of falling in the interval 1983–84 if more of the 8 surrounding plots have canopies lower than 30 m in 1983.

5×5 m plot has eight 5×5 m plot neighbours. If the focal plot in the centre was at 30 m in 1983, and all neighbouring plots were at 30 m as well, the probability of a fall in canopy height in the focal plot was only 0.039 (Fig. 3.5). However, focal plots taller than all 8 neighbouring plots had more than a 50:50 chance of falling (0.515).

3.7   Canopy gaps and patterns of canopy tree recruitment

In the following analysis, attention is restricted to gaps occurring the 5–10 m canopy height class. This height class is appropriate because the smallest saplings in the census (1 cm dbh) are already approximately 4–5 m tall, and many new treefall gaps retain a low canopy of saplings of primary species, present prior to gap formation, which survive the treefall. The data refer to the distribution of saplings having diameters of 1–4 cm dbh, few of which exceed the top of this canopy class (10 m) in height.

3.7.1   Dispersal and gap colonization

The dispersal problems of primary forest tree species, early pioneers, and late secondary species are different in each case. While most species in each regeneration guild require gaps to mature, they differ in their degree of shade-tolerance as juveniles. Therefore, shade-tolerant primary species would be expected to survive and grow better in smaller gaps than shade-intolerant species. Because there are more small gaps than large, the distance a parent tree of a primary species must disperse its seeds to encounter a suitable gap is less than for an early pioneer or late secondary species. We can illustrate the problem by throwing 200 points at random in the plot and computing the mean distance to the nth nearest gap as a function of gap size (Fig. 3.6). In this hypothetical but realistic example, a shade-tolerant species with 200 adults and a minimum critical gap size of 25 m² would encounter 20 gaps within about 100 m. On the other hand, seeds of a less shade-tolerant species which minimally requires 100 m² gaps must travel over 300 m to encounter 20 gaps.

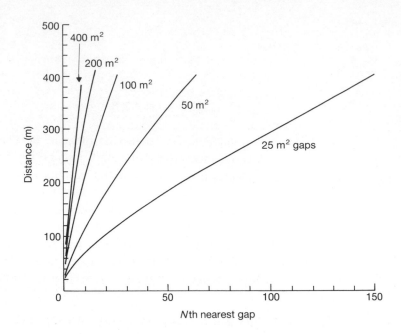

Fig. 3.6. Mean distance to the *n*th nearest neighbour gap as a function of gap size for a hypothetical tree species with 200 adults distributed at random throughout the 50 ha plot. Distances were computed using actual coordinates of gap centres for each size class. Because there are many more smaller gaps than large ones, the travel distance required to reach the *n*th nearest gap is much greater for large gaps than for small.

The spatial pattern of adult trees of canopy species in the forest will also affect mean distances to gaps, but the effect is not very strong because the gaps themselves are nearly randomly dispersed. However, adult dispersion will strongly affect the intensity of intraspecific competition for gaps. For example, many of the BCI canopy tree species are clumped as adults (Hubbell & Foster, 1983). In these species, competition among the progeny of aggregated adults will be especially intense because the number of nearby gaps per adult is reduced. Unfortunately, a thorough discussion of the dispersion of adult canopy trees in relation to competition for gap space among their offspring is a major subject in its own right and beyond the scope of the present chapter.

3.7.2   Sapling distributions in gap, edge and understory

If primary, early pioneer, and late secondary species constitute valid regeneration guilds with differing requirements for gaps, these requirements should be reflected in differences in sapling distribution over gap and non-gap areas of the forest. At present we can mount only a rather crude analysis because we lack dynamic information on comparative species performance in gaps compared with the shaded forest interior. Thus we cannot presently separate the effects of differential growth, survival, and recruitment on juvenile spatial patterns. However, we can address a simpler question, namely whether species differ in the proportion of 1–4 cm dbh juveniles found in gaps. In fact, we can test for significant gap association or avoidance because we have mapped not only the saplings that occur in gaps, but also those that occur elsewhere. For this analysis, we have divided the 50-hectare plot into 4 classes of microsite: gap, interior gap edge (less than 5 m from gap), exterior gap edge (5–10 m from gap), and non-gap or canopy understory (more than 10 m from gap). Analysis is restricted to the 81 canopy tree species.

A well-developed guild of heliophilous, mainly late-secondary species exists and is readily distinguished from other guilds by the strong skewing of juveniles into gaps and gap edges and away from the forest interior. The observed microsite distribution of juveniles for a representative species of heliophile is shown in Fig. 3.7, and compared with the expected distribution based on the fraction of total area in the 50-hectare plot in each microsite class. Of the 81 canopy species, 14 (17.3%) demonstrated significant gap association of their 1–4 cm dbh saplings.

In spite of significant association of juveniles with gaps in these heliophilous species, most of them still have a large number of saplings in non-gap areas. Thus the suggestion that gap requirements in these species are absolute is incorrect. However, the disparity between observed and expected numbers of saplings in forest interior microsites for these species does suggest that they suffer substantially increased rates of mortality in the shade.

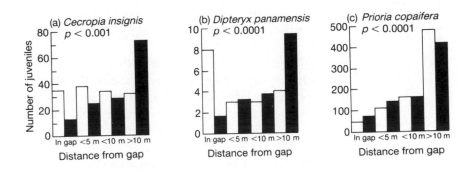

Fig. 3.7. Distribution of 1–4 cm dbh juveniles of canopy tree species in 4 microsite classes: in gaps (5×5 m plots classified as gaps in the 5–10 m canopy cross section): interior edges of gaps (< 5 m from a gap plot); exterior edge of gaps (5–10 m from a gap plot); and forest interior (> 10 m from a gap plot). The black bars give the expected juvenile distribution based on the proportion of total area in the 50 ha plot belonging to each microsite class. The open bars give the distributions of juveniles observed in each microsite class. The difference between observed and expected is tested by chi-square with 3 df. (*a*) *Cecropia insignis*, a species which is over represented in gaps and on interior edges of gaps. (*b*) *Dipteryx panamensis*, a species over represented only in gaps. (*c*) *Prioria copaifera*, a species over represented in the forest interior.

A smaller guild of canopy species demonstrated gap avoidance with significant skewing of 1–4 cm dbh saplings away from gaps and towards the forest interior. Six canopy tree species (7.4%) showed significant negative sapling association with gaps. Four other canopy species had no juveniles in gaps, but rarity of these species, rather than gap avoidance, could be the explanation of this.

Saplings of a still smaller guild of species were significantly associated with the gap edge microsites (4 species, 4.9%) rather than with gaps or forest interior.

The largest group of species (57 out of 81) showed *no* apparent specialization in their microsite requirements in so far as the present test could demonstrate. 70% of canopy species had sapling distributions among microsites which were indistinguishable from those expected by chance.

3.7.3   Gap disturbance regime and canopy tree composition

If Hartshorn (1978) and Denslow (1980b) are correct in suggesting that tropical canopy tree species are adapted to different gap disturbance regimes, then the following observations should hold: i) species with extreme light or shade requirements should be less abundant than species with intermediate requirements, relative to the mean gap disturbance regime; ii) there should be a greater number of species with such intermediate requirements, reflecting the mean gap disturbance regime, and no bimodality of gap specialists versus shade specialists; iii) the percentage of high light-requiring species should increase with an increase in gap disturbance rate, and shade-loving species should increase with a decrease in gap disturbance rate. Each of these expectations is upheld for canopy tree species in the BCI forest.

We can use the fraction of saplings 1–4 cm dbh in gaps as an index of heliophily for species, and ask how the abundance of a canopy species is related to its index score. High and low scores should be found in occasional or rare species and intermediate scores in common species; this is the pattern observed. If species are classified into three abundance categories, rare (10–99 individuals), occasional (100–999 individuals), and common (1000 or more individuals), we find that species with high or very low indices of heliophily are always occasional or rare.

Conversely, common species are tightly clustered about the index value of 9.6%, the expected proportion of saplings in gaps based on the proportion of the total plot in gaps. To analyse the clustering of common species, consider the following: there are 51 species that are within 5% of the expected percentage of juveniles in gaps (9.6%); there are 16 'common' species (more than 1000 individuals), and all of these are included in the 51 species within 5% of the expected values. The probability of this occurring by chance in a total of 81 species, is 51 take 16, divided by 81 take 16, or 0.0012.

We can conclude that a necessary, but not sufficient, condition for becoming an abundant canopy species in this forest is an ability to survive under the most prevalent microsite conditions and gap disturbance regime, which on BCI, produces mostly small gaps and rarely larger ones. This ability is not sufficient, because many occasional and rare species also exhibit the expected proportion of juveniles in gaps. The abundance of these species must be limited by something other than extreme gap or non-gap requirements.

In addition to species abundance, the number of species also shows the expected pattern. The number of species as a function of the index of heliophily is a unimodal distribution, with the mode centred over the expected fraction of juveniles in gaps (Fig. 3.8). Species exhibiting extremely high or low heliophily scores are a small fraction of the total, so that the distribution is quite peaked about the mode.

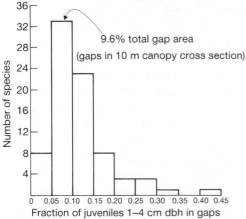

Fig. 3.8. Relationship between species richness and the index of heliophily (defined as the fraction of 1–4 cm dbh juveniles of a species that occur in gaps) for 81 canopy trees species in the BCI forest. The distribution is centred on the expected index of heliophily, suggesting that most species in the forest are adapted to the most common disturbance regime in the BCI forest which produces mainly small gaps.

Finally, the proportion of heliophilous trees over 10 cm dbh per hectare increases with canopy openness, as expected. We took the 16 species with the highest scores in terms of fraction of 1–4 cm saplings in gaps then determined the percentage of trees over 10 cm dbh per hectare that were among this set of 16. The results indicate a highly significant correlation between the percentage of heliophilous trees, and the percentage of canopy open at 30 m per hectare, although there was considerable scatter. Caution is needed in interpreting this result, however, because hectares with greater openness at 30 m also have more gaps in the 5–10 m height class in which the juveniles were counted. To eliminate this potential problem, we used the 25 hectares in the west half to estimate the index of heliophily, and the 25 hectares in the east half to evaluate independently the relationship between proportion of heliophilous trees over 10 cm dbh and canopy openness; the correlation remained just as strong ($p < 0.01$).

3.8   Paradigm reconsidered

For the moment at least, this study of gap dynamics and forest structure is the largest ever mounted for any forest, temperate or tropical. Data on the distribution, abundance, and geometry of over 600 gaps (5–10 m canopy height class) were recorded, along with the distribution of the saplings of 81 canopy tree species inside, outside and on the edges of these gaps. In the light of these results, it is useful to take a brief retrospective look at the current paradigm for gap dynamics. We are still at an early stage of analysis, because we lack information about tree survival, growth and reproduction in relation to gap dynamics and tree replacement processes in the BCI plot. Nevertheless, there are several modifications as well as supporting statements we can make concerning present views of gap-phase regeneration which seem likely to stand.

This study confirms earlier work showing that large gaps are spatially and temporally less frequent than small gaps. However, the number and proportion of gaps of a given size change as one moves up to higher vegetation layers in the forest. Mean gap size increases monotonically with height, but gap number is greatest at an intermediate height (10–20 m). Gaps measured near the ground (2–5 or 5–10 m) appear randomly dispersed throughout the forest. If these were the only gaps recorded, one would arrive at the erroneous conclusion that the forest has a homogeneous gap disturbance regime. At higher levels in the canopy, however, major differences in local gap disturbance regimes over the recent past are evident. Contrary to expectation, the forested areas on steeper slopes had greater percentages of high canopy than flat areas.

In classical theory, gaps are vertical-sided holes in the canopy leading straight to the forest floor, but very few gaps in the present study conform to this ideal. Most are shaped like an inverted cone standing on its point. This accounts for the expansion of gap area as one moves up to higher levels in the canopy. In horizontal cross section, most intermediate-sized and large gaps are more linear than round, but few exhibit a recognizable dumb-bell shape.

If the gap disturbance regime over the 17-month period of this study persists, then present canopy structure will change, especially in the high canopy. For the present disturbance regime, the equilibrium coverage by canopy over 30 m tall is 12% less than the coverage in the forest today. The current paradigm for gap-phase dynamics implicitly assumes a forest in equilibrium with its disturbance regime. However, the BCI evidence for canopy disequilibrium cannot be used to challenge the paradigm because the present disturbance regime is probably more severe than it was before BCI became an island in the artificial Gatun Lake 70 years ago. For example, there are indications that storm wind speed may have increased now that storms travel unimpeded by vegetation across the lake to the island (Foster & Brokaw, 1982).

Strong evidence for regeneration guilds was found, particularly for late secondary heliophiles and for primary species, the bulk of the 81 species in the sample. Low-stature, early pioneer species do not show up because the sample was restricted to high-canopy tree species. Evidence was not found, however, for extreme special-ization for regenerative niches of the sort proposed by Ricklefs (1977). A few species showed significant gap edge association or gap avoidance, but the majority of species showed statistical indifference to microsite differences in the plot.

Since we have examined sapling distributions in over 600 gaps, the widespread indifference to gap/non-gap distinctions in the BCI forest may be real, and not due simply to our inability to detect niche differences among these species. Elsewhere we have argued that one should not invariably expect fine-tuned partitioning of regenerative niches (Hubbell & Foster, 1986a). Instead, one should often find large guilds composed of ecologically very similar species in terms of their regeneration requirements. Adaptive convergence is expected from Denslow's (1980) argument that most species in the forest will be adapted to the prevalent gap disturbance regime. For example, BCI species should be selected to do well in small gaps because most regeneration sites are of that type. We observe that primary species are indeed dominant over extreme heliophiles, both in species rich-ness and in numerical abundance.

3.9   Concluding remarks

Other than to stress their importance, we have said nothing specific about tree replacement processes except to predict a shifting balance in representation of the major regeneration guilds in the canopy tree community as the gap disturbance regime changes. Herein lies the key, however, to a true understanding of tropical forest dynamics. Based on the data from the first census, we can only make educated guesses about tree replacement probabilities from the species com-position of the saplings of different size classes beneath the canopy of overstory and midstory trees. Common BCI tree species all appear capable of self-replacement from this analysis. However, because of the high species richness of the BCI forest, the probabilities of self-replacement, or of replacement by any other given species, are low (Hubbell & Foster, 1986b). The surprise result was that the calculated probabilities of self-replacement were statistically no lower as a group than the probabilities of replacement by other species, although saplings of the two commonest species did show avoidance of conspecific adult trees. Further descriptive and experi-mental studies of tropical tree reproductive performance in a spatial context are now needed to check these results. Competition from neighbouring plants, physical site factors, predators, pathogens, pollinators, and seed dispersal agents all have potentially important contributary roles in determining tree-by-tree replacement probabilities.

Chapter 4 The Structure of Plant Populations

MICHAEL J. HUTCHINGS

4.1 Introduction

The various structures which we can identify in populations of plants result from the action of biotic and abiotic forces to which their members, and in some cases their ancestors, have been exposed in the past. The most obvious impact of the forces experienced by the ancestors of a population is upon genetical structure, but the spatial structure of plants in a population is also a legacy of the spatial arrangement of parent plants, and of the interactions which have taken place between plants in the past. Plant interactions together with abiotic factors also mould the 'performance structure' of populations. This can be viewed as an expression of the opportunities for growth realized by each member of the population in the course of its development. Finally, the age structure of a population reflects both past opportunities for recruitment, and the mortality risks to which each recruit has subsequently been exposed. These four aspects of structure—performance, spacing, age and genetic structure—are the subject of this chapter.

All the facets of population structure are interrelated, so that an alteration in one facet generates alterations in the others. The growth of the individuals in a population, for example, sets in motion a chain-reaction of modifications to population structure. Therefore, while it is convenient to discuss each aspect of structure in isolation, it is important to remember the close links that exist between them. I shall begin with a description of plant performance, which is the most conspicuous aspect of population structure, and then discuss in turn the less conspicuous facets of spacing, age and genetical structure. Finally, some consideration is given to the influence of abiotic environmental variation upon population structure.

4.2 Performance structure in plant populations

4.2.1 Plant weights

In plant populations, the sizes of individuals are far from uniform. This is true even in monocultures such as cereal crops or forest plantations, which consist of uniformly-aged plants. Since many aspects of performance are correlated with size, they too will vary

Box 4.1 Measuring size inequality of plants

The simplest measures used to characterize plant size distributions are mean, variance, skew and kurtosis. The mean is simply the average plant size. Variance describes the variability in size (calculated as the average of the *squared* residuals; a residual is the difference between an individual plant's size and the mean plant size). Skew indicates whether the distribution has long 'tails' to the left (skew < 0) or to the right (skew > 0), or whether the distribution is symmetrically bell-shaped (skew $= 0$). Skew is calculated on the basis of the *cubes* of the residuals. Kurtosis measures the degree to which the distribution is more pointy (leptokurtic) or more flat-topped (platykurtic) than a normal distribution. Kurtosis is based on the *fourth powers* of the residuals, with negative values indicating platykurtosis and positive values showing leptokurtosis. Normal distributions have kurtosis $= 0$.

If y is the weight of an individual plant, and there are n different plants in a sample, the parameters are calculated as follows:

Mean $\quad \bar{y} = \Sigma y/n$

Variance $s^2 = (\Sigma y^2 - (\Sigma y)^2/n)/(n-1)$

Skew $\quad g_1 = (n\Sigma y^3 - 3\Sigma y\Sigma y^2 + 2(\Sigma y)^3/n)/(s^3(n-1)(n-2))$

Kurtosis $g_2 = ((n+1)(n\Sigma y^4 - 4\Sigma y\Sigma y^3 + (6(\Sigma y)^2\Sigma y^2/n) - 3(\Sigma y)^4/n^2)/(s^4(n-1)(n-2)(n-3))) - (3(n-1)^2/((n-2)(n-3)))$

where $\Sigma y \quad$ is the sum of all the plant weights
$\quad\quad \Sigma y^2 \quad$ is the sum of all the *squared* plant weights
$\quad\quad \Sigma y^3 \quad$ is the sum of all the *cubed* plant weights
$\quad\quad \Sigma y^4 \quad$ is the sum of all the *fourthed* plant weights
$\quad\quad (\Sigma y)^2$ is the *grand total* of the plant weights squared
$\quad\quad (\Sigma y)^3$ is the grand total of the plant weights cubed
$\quad\quad (\Sigma y)^4$ is the grand total of the plant weights fourthed
$\quad\quad s \quad$ is the standard deviation; the square root of s^2

For worked examples, see Sokal & Rohlf (1981, p 114–117).

The Gini coefficient

When the emphasis is to be placed on size inequality rather than on asymmetry (skewness) in describing the size hierarchy, then the Gini coefficient is an appropriate measure (Weiner & Solbrig, 1984). For n plants of mean size $\bar{y}$, this is given by:

$$G = \frac{\sum\limits_{i=1}^{n} \sum\limits_{j=1}^{n} [y_i - y_j]}{2n^2\bar{y}}$$

When all the plants are the same size, $G = 0$, while in a population of infinite size where all but one of the plants are infinitely small, $G = 1$.

Since data on plant dry weights often follow a log-normal distribution (Box 4.2), it is usually better to carry out statistical analyses on log weights rather than on weights. This has the added advantage of making the calculation of relative growth rates more straightforward (Box 4.3).

from plant to plant. While seedling populations may have symmetrical weight distributions (Obeid *et al.*, 1967; Harper *et al.*, 1970; Rabinowitz, 1979), a variety of factors combine to bring about an asymmetric, positively skewed distribution of adult plant weights (see Box 4.1). The factors responsible are considered below. The adult population therefore consists of a small number of large plants and a large number of small ones (Fig. 4.1). We call this the 'size hierarchy' as a convenient epithet, bearing in mind three important caveats: i) all plants in the population do *not* directly influence the performance of all others; ii) the strength of interaction between plants is not necessarily reflected by their respective ranks in the size hierarchy; and iii) the immediate, 'first order' neighbours of a plant have by far the greatest impact on its performance, almost independent of where they fall in the size hierarchy of the entire population. Weiner & Solbrig (1984) stress that one of the essential features of the size hierarchy is that the few, large plants often make up most of the population's biomass.

Many factors are influential in promoting variation in the sizes of plants in populations. The size of seeds is not constant for a given species, and seedling size is often correlated with the size of the seed from which it emerged (Black, 1957; Crawley & Nachapong, 1985). The relative growth rate (RGR) of the emerging seedling may also be controlled by its genotype. In addition, Ross and Harper (1972) have demonstrated that the time of germination relative to neighbours is a major determinant of the future growth of plants. In monocultures of the grass *Dactylis glomerata*, the growth of the earliest plants to germinate was hardly affected by inter-plant competition, whereas those germinating ten days later showed negligible increase in weight after 35 days of growth (Fig. 4.2a). Ross and Harper (1972) also showed that the size of the unoccupied space in which a seed germinated could significantly affect its subsequent performance (Fig. 4.2c), although its position in that space had little effect on its growth (Fig. 4.2b), while several authors have demonstrated a positive correlation between plant performance and the distance to close neighbours (Pielou, 1961; Yeaton & Cody, 1976; Liddle *et al.*, 1982). Indeed, Mack and Harper (1977) have shown that distance to neighbours, their size and their spatial arrangement can account for up to 69% of the variation in individual plant weight. In the words of

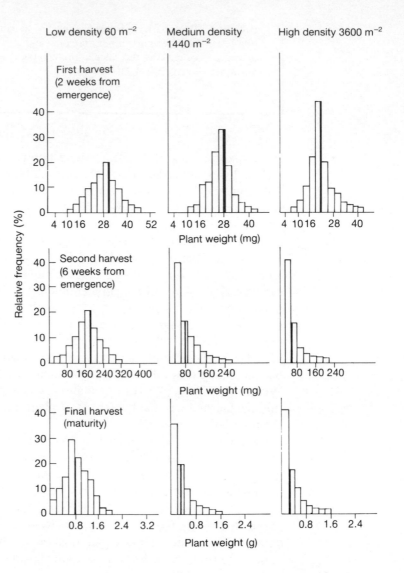

Fig. 4.1. Frequency distributions of plant weights in populations of flax, *Linum usitatissimum*, sown at 3 densities. Mean plant weight is shown by the black bars. From data in Obeid *et al.*, (1967).

Ross and Harper (1972) 'an individual's potential to capture resources is dictated by the number and proximity of neighbours already capturing resources' from the same resource pool. Herbivorous animals can also have a dramatic impact on the size hierarchy of plants; if they preferentially attack smaller, already weakened plants, they may accentuate the inequality in the sizes of plants, whereas a preference for larger plants would tend to reduce plant size variation (Crawley, 1983).

Whilst much variation in individual plant weight in populations

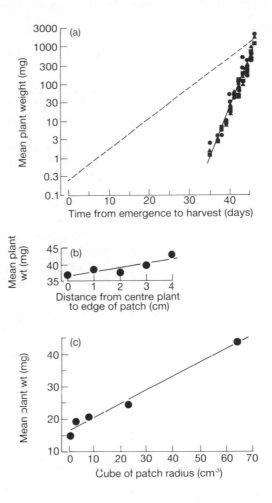

Fig. 4.2. Gap size and seedling establishment. (*a*) Mean weight of groups of plants emerging at different times in 3 populations. The dotted line indicates the weights of plants growing for different lengths of time in the *absence* of competition. (*b*) Response of individual plant weight to varying the distance of the centre plant from the perimeter of patches of a given size, showing the rather slight effect of distance. (*c*) Response of individual plant weight to varying sizes of patches, showing a much more pronounced effect. From Ross & Harper (1972).

may be caused by inter-plant competition (Koyama & Kira, 1956), it has long been known that competition is not an *essential* causative factor of a size hierarchy. Most plant ecologists believe that competition between plants is highly asymmetric (it is one-sided, in the sense that the larger plant has a much greater adverse effect on the smaller plant than vice versa), and that when competition is occurring, dominance and suppression in the population are major causes of the size hierarchy. While agreeing that the size hierarchy could develop in the absence of competition, however, Turner & Rabinowitz (1983) have argued that competition may actually *delay* the development of a hierarchy by slowing growth. In any event, the existence of a size hierarchy is not, in itself, evidence for intra- or inter-specific competition having taken place (see Box 4.2).

Most studies have shown that weight skewness in populations increases through time, and that it is higher by a given date in denser populations (Obeid *et al.,* 1967; Ford, 1975; Fig. 4.1). However,

Box 4.2 Plant size distributions

Assume that we begin with a normal distribution of seed sizes, and that these germinate to produce a normal distribution of shoot dry weights amongst the young seedlings. At any subsequent time, the shape of the frequency distribution of plant sizes reflects the interplay between size-specific growth and death rates, and such factors as plant competition and size-specific attack by natural enemies. The point to be made here is that strongly skewed distributions can arise without competition, so the existence of a size hierarchy is not *evidence* of competition. There could be a small number of large plants and a large number of small ones, even in the absence of competition (Hara, 1984).

The degree to which variance, skew and kurtosis of a plant size distribution change through time, depends upon the relationship between growth rate and size (specifically on the relationship between the increment of a size measure and the same size measure):

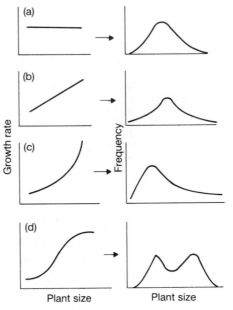

a) Regardless of plant size, when growth rate is constant, mean size increases without any change in variance. b) When growth rate increases linearly with size, then variance in plant size increases with time, but skew is unaltered. c) When growth rate increases faster than linearly with size, then variance and skew increase with time, resulting in a log-normal distribution of plant weights. Most data from even-aged monocultures exhibit this pattern. d) When growth rate is a sigmoid function of size, then bimodality can develop, and variance, skew and kurtosis all change through time (see Westoby (1982) and Hara (1984) for details).

there are exceptions to these generalizations (e.g. the work by Turner & Rabinowitz (1983) cited above). Mohler *et al.,* (1978) provide an excellent example of how size structure (measured as trunk diameter) changes, and skewness can *decrease,* with age, as populations of fir trees, *Abies balsamea,* decline in density from 11.5 plants/m² at three years to 0.2 plants/m² at 59 years (peak death rates occurred when the trees were between 19 and 35 years of age. Fig. 4.3). In cases like this, when competition causes mortality, the interpretation of changes in skewness becomes more difficult; for

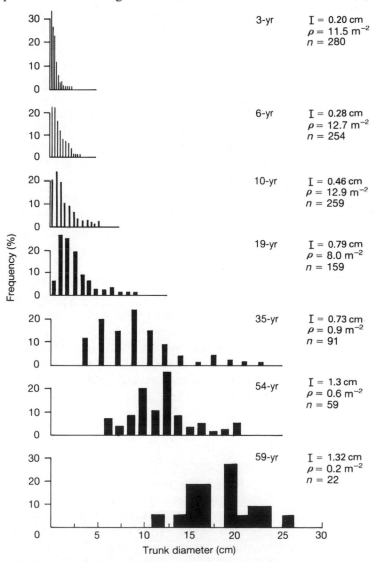

Fig. 4.3. Frequency distributions of trunk diameter for *Abies balsamea* stands arranged in age sequence. The interval of trunk diameter between successive bars (I) in the histograms was determined by dividing the range of observed diameters into 12 equal intervals. ρ indicates stand density and n shows the number of trees in the sample. From Mohler *et al.* (1978).

example, Begon (1984) has shown how different size-dependent survivorship probabilities could either exaggerate or reduce skewness over the course of a growing period. Typically, however, mortality will be concentrated in the smallest size classes, so that skewness often stabilizes once plants have begun to die (White & Harper, 1970). This is because deaths of the smallest individuals will truncate the left-hand end of the distribution of plant weights, whilst the rapid growth of the largest plants extends the right-hand end of the distribution.

The smallest plants in thinning populations are known to have low or even negative values of RGR (see Box 4.3). For example, Ford (1975) analysed the relationship between RGR, tree girth and mortality in sitka spruce *Picea sitchensis*. RGR was very low and independent of girth for small trees, but strongly correlated with girth for plants above median girth. All the plants which died in his population were small. In a similar analysis of RGR in monocultures of French marigolds *Tagetes patula* he found that RGR was positively correlated with plant weight early in growth; later on all large plants had similar high RGR's, while there was a positive correlation between RGR and weight for small plants. The slope of this relationship fell as time passed. Generally, a substantial proportion of the plants in competing populations appear to have RGR's close to zero.

In experiments on monocultures of sitka spruce and lodgepole pine *Pinus contorta* planted in hexagonal lattices, Cannell *et al.* (1984) analysed the influence of competing neighbours upon growth rates. Each tree has six first order neighbours in a hexagonal lattice, and every individual in the population was classified as having 0, 1, 2, 3, 4, 5 or 6 first order neighbours taller than itself. Those trees taller than all their neighbours had high RHGR's (relative height growth rates), while those with many taller neighbours had significantly lower RHGR's (Fig. 4.4a,b). Many trees died during the experiment; there was an inverse relationship between the probability of a tree surviving and the number of its first order neighbours which survived (Fig. 4.4c). In other words, trees were more likely to survive if their neighbours were dying.

Long-term field studies of *Astrocaryum mexicanum*, a dominant understorey palm of tropical rain forests in south-eastern Mexico, are providing detailed insights into the fate of individuals from different parts of the size hierarchy. For example, the probability of a tree surviving from the second to the fifth year increases steeply with the size of the plant when it is two years old (Fig. 4.5a), and both the frequency of flowering (Fig. 4.5b) and the mean annual fecundity (Fig. 4.5c) are much higher for plants from larger size classes (Sarukhán *et al.*, 1984).

The early literature on size hierarchies (Koyama & Kira, 1956; Ogden, 1970), and the results of elementary calculations from simple models of plant growth suggested that plant weight frequency distributions ultimately become log-normal. More recently, the generality of log-normal weight frequency distributions has been questioned, and a variety of skewed and kurtotic distributions have been described from field data (Box 4.2). It has been suggested that the development of a *bimodal* (i.e. two peaked) size distribution is more usual in circumstances where competition has taken place (Ford & Newbould, 1970; Ford, 1975). Ford (1975) argues that bimodality in weight frequency distributions is generated by the

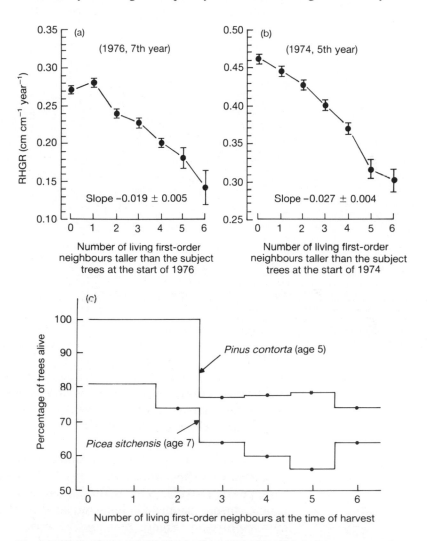

Fig. 4.4. Mean relative growth rates in height (RHGR) of: (*a*) *Picea sitchensis*; and (*b*) *Pinus contorta* trees in the year before harvest as a function of the number of the first-order neighbours *taller* than themselves. Vertical bars indicate 1 standard error. (*c*) Percentage of trees alive for *Picea* and *Pinus* with different numbers of living first-order neighbours; the fewer living neighbours, the higher the chance of surviving. From Cannell *et al.* (1984).

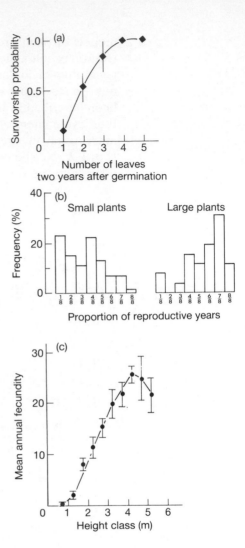

Fig. 4.5. Population structure and plant performance in the palm *Astrocaryum mexicanum*. (*a*) Probability of survival increases with seedling size. (*b*) Most large plants reproduce nearly every year, but most small plants reproduce rather infrequently. (*c*) Taller plants have significantly greater mean annual fecundity. From Sarukhán *et al.* (1984).

characteristic distribution of RGR values among the plants in competing populations, in which, as we have seen, many small plants have RGR's close to zero, but there is an abrupt change to higher RGR values in the larger plants.

A simulation model of population growth that incorporated differences in the intrinsic growth rate of plants and one sided competition (Aikman & Watkinson, 1980) produced size hierarchies which closely paralleled the bimodal distributions recorded for monocultures of French marigold by Ford (1975; see Fig. 4.6). Westoby (1982) has subsequently shown that only models involving one sided competition will produce these bimodal frequency distributions of plant sizes (Box 4.2). However, the sensitivity of the appearance of bimodality in weight histograms to the number of classes in the histogram, and the lack of direct statistical tests for

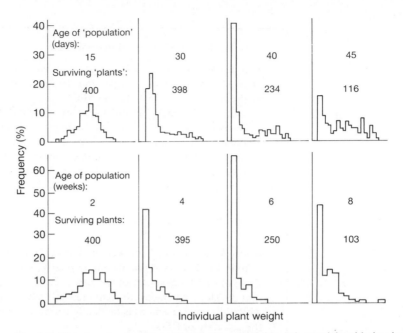

Fig. 4.6. The changing frequency distribution of individual plant weight with time in a self-thinning population. Above: results of a simulation model by Aikman & Watkinson (1980). Below: data from a laboratory population of *Tagetes patula* collected by Ford (1975).

bimodality, make its significance difficult to assess. Many populations which show signs of intraspecific competition do not display any hint of bimodality.

As the examples quoted above illustrate, most studies of plant size frequency distributions have been made on monocultures, and the few investigations of mixtures have generally involved only two species. Ogden's (1970) study of a mixed annual herb community is an exception. Each of the constituent species populations had positively skewed frequency distributions of shoot weight, many of them approximating log-normality. Butcher (1983) investigated size hierarchies in monocultures and two-species mixtures of wild oats *Avena fatua*, leafless pea *Pisum sativum* var. Filby, and conventional pea *Pisum sativum* var. Birte. In monoculture, wild oat developed the greatest size inequality, as measured by the Gini coefficient (Box 4.1), followed by the conventional pea and the leafless pea. In two-species mixtures, the degree of inequality appears to be related to the ability of one of the species to outcompete the other. In a mixture of wild oat and conventional pea, the pea considerably outgrew the oat, resulting in a virtually complete separation of their weight frequency distributions; the mixture displayed far greater inequality than either of its components in monoculture. With wild oat and leafless pea, however, the two weight frequency distributions did not separate, and the size inequality of the whole population was similar to those

Box 4.3 Expressing plant growth rate

The simplest way to understand plant growth rates is to plot a graph of log shoot dry weight against time. The slope of the graph changes as the plant ages. The steeper the slope, the faster the plant is growing. At maturity (or under severe competition) plants may stop growing (slope = 0) or even lose weight (in which case the slope is negative). The usual means of expressing the rate of growth is simply to determine the slope of the graph at any point in time (i.e. change in *log* weight per unit time). This is called the relative growth-rate (RGR) of the plant at that time (or at that size). In symbols, this is:

$$RGR = \frac{\log_e W_2 - \log_e W_1}{t_2 - t_1}$$

where W_2 is the shoot dry weight at time 2 (t_2), and W_1 is the weight at time 1 (t_1). To find the maximum relative growth rate we find the slope of the steepest part of the curve; this is called RGR_{max}. Note that the slope one measures is influenced by the length of time over which the measurement is made. The longer the period, the *lower* the estimate of RGR_{max}:

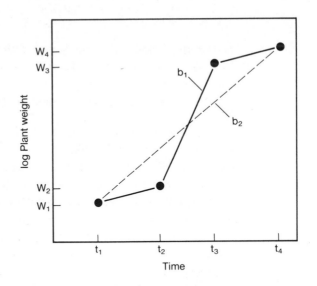

where the slope b_2 calculated over the longer period is shallower than the slope b_1 calculated over a single time period.

It is important to note the distinction between growth rate and relative growth rate. Growth rate (sensu Box 4.1) is the rate of change in plant weight, dW/dt (with dimensions g time^{-1}). Relative growth rate is the rate of change in log plant weight, $d(\log W)/dt$ (with dimensions time^{-1}).

Other important measures used in the quantitative analysis of plant growth (with their dimensions in parentheses) are:

U.L.R. Unit Leaf Rate.	The mean rate of interest of total plant dry weight per unit leaf area (g m^{-2} time^{-1}).
L.A.R. Leaf Area Ratio.	Ratio of leaf area to total dry weight (m^2 g^{-1}).
R.G.R. Relative Growth Rate.	= U.L.R.×L.A.R. (time^{-1}).
S.L.A. Specific Leaf Area.	The leaf area per unit leaf dry weight (m^2 g^{-1}). U.L.R. and S.L.A. tend to be negatively correlated (see Chapter 12).
L.A.I. Leaf Area Index.	The ratio of the area of leaf (measured on one surface only) to the area of ground beneath (dimensionless).
L.A.D. Leaf Area Duration.	The area beneath a graph of L.A.I. against time computed over the whole growing season (time).
R.W.R. Root Weight Ratio.	The ratio of total root dry weight to total plant dry weight (roots plus shoots) (dimensionless).

More detailed definitions can be found in Evans (1972).

from the individual monocultures (Fig. 4.7). Weiner (1985) also investigated size hierarchies in monocultures and mixtures of strawberry clover *Trifolium incarnatum* and ryegrass *Lolium multiflorum* (Fig. 4.8). In monocultures at low fertility and high density, the clover population had a lower mean weight per plant than ryegrass, but a far greater size inequality, because the largest clover plants were heavier than the largest ryegrass individuals. In a 50:50 mixture at the same density, the mean and maximum size of ryegrass plants far exceeded that of clover, and the size frequency distributions of the two species were markedly separated. While size inequality for all the plants in the mixture was intermediate between its values in the two monocultures, the component populations showed *opposite* changes in their inequality. The suppressed clover population had higher inequality than in monoculture, while the ryegrass showed reduced inequality. Under these experimental conditions, ryegrass is clearly a better competitor than clover. In comparison with the monocultures, the majority of clover plants in mixtures suffered as a result of imposing strong interspecific competition from ryegrass in place of weak intraspecific competition from clover plants. In contrast, ryegrass benefited in mixtures, since strong intraspecific competition was alleviated by replacing ryegrass individuals with less competitive clover plants.

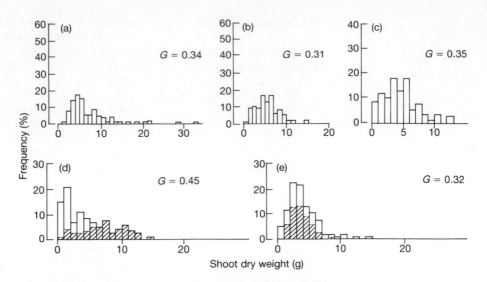

Fig. 4.7. The distribution of shoot dry weight in (*a*) monocultures of conventional pea, var. Birte; (*b*) monocultures of leafless pea, var. Filby; (*c*) monocultures of wild oat, *Avena fatua*; (*d*) a 50:50 mixture of Birte and *Avena*; (*e*) a 50:50 mixture of Filby and *Avena*. All populations were sown at the same density. The contribution made by the pea plants in mixture is shown by the hatched area. G = value of the Gini coefficient. After Butcher (1983).

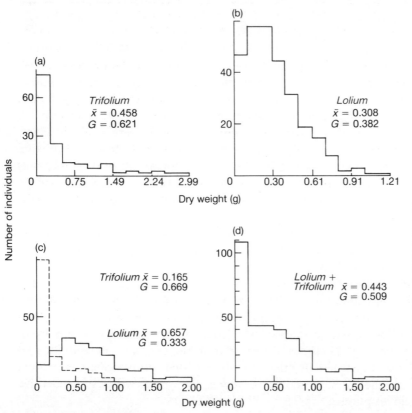

Fig. 4.8. Dry weight distributions for (*a*) monospecific populations of *Trifolium incarnatum*; (*b*) monospecific populations of *Lolium multiflorum*; (*c*) mixtures of *Trifolium* (hatched lines) and *Lolium* (solid lines) with their weight distributions shown separately; (*d*) the mixture as a whole, both species taken together. x̄ = mean weight, G = Gini coefficient. From Weiner (1985).

In mixed populations, the development of widely differing size distributions for the constituent species has dramatic consequences for subsequent performance and survival. Although the behaviour of the individual species in mixtures is not generally well understood, the mixtures themselves often conform broadly to patterns expected from a knowledge of the behaviour of monocultures. Bazzaz and Harper (1976) found that 50:50 mixtures of the two crucifers *Sinapis alba* and *Lepidium sativum* accumulated biomass and self-thinned according to the $-3/2$ power law (see Chapter 5), and had size frequency distributions which began close to normality and moved towards log-normality through time. However, *Sinapis* showed little thinning and accumulated much of the mixture's biomass, while *Lepidium* thinned considerably and made little growth.

4.2.2   Other aspects of performance

a) **Height Distribution** Most published data show that whereas population weight distributions become positively skewed as time passes, the distribution of plant heights either remains symmetrical or becomes negatively skewed (Koyama & Kira, 1956; Kuroiwa, 1960; Ogden, 1970). This is because many of the short plants die because they receive little direct light and are therefore at a severe competitive disadvantage. Furthermore, many of the suppressed plants undergo etiolation (rapid height extension towards the light) and so grow closer to the height of the surrounding vegetation. However, these plants increase very little in weight, and many ultimately die, partly because their increased height is not matched by an increase in the mechanical strength of their stems and trunks (Ogden, 1970; Hutchings & Barkham, 1976).

b) **Productivity** Recent studies on plant productivity have paid close attention to the structure of the population and to the dynamics of turnover of the different plant parts. This combination of the sampling methods from traditional energetics with the detailed census techniques of plant demography, has allowed considerably improved precision in the estimation of primary productivity (Flower-Ellis & Persson, 1980; Callaghan & Collins, 1981), and promises to revolutionize the study of the impact of herbivores on the demography of plant modules (Chapter 9).

c) **Seed sizes** Individual seed weights are now known to vary greatly both within and between individual plants (Chapter 8). In many cases the frequency distribution of individual seed weight approximates, if not to a normal distribution, at least to symmetry (Rabinowitz, 1979; Schaal, 1980; Pitelka *et al.*, 1983). Where more than one seed morph is produced, the weight distribution may take a more complex form (Harper *et al.*, 1970). While seed weight can be a strikingly invariate characteristic under a wide range of conditions, relative to other aspects of performance (Obeid *et al.*, 1967; Harper *et al.*, 1970), Thompson (1984) has commented that this impression is usually based on mean values calculated from bulk weights of many seeds. In the umbellifer *Lomatium grayi*, for instance, individual seed masses showed a 16-fold variation (Thompson, 1984); while this may be exceptional (but see Black, 1957, 1959), Thompson quotes several examples of seed masses varying from 2- to 6-fold within species. Positively skewed distributions of seed mass may be found when unfilled, unviable seeds form a major fraction of seed production (Gross & Werner, 1983; Thompson, 1984). This can occur when the number of seeds initiated by a plant exceeds the number which can be filled, either because of resource limitation, or because of a deterioration in environmental conditions between seed initiation and maturation (Fuller *et al.*, 1983; see Chapter 6). Variation in individual seed weights has rarely been considered in previous studies of population structure, but the importance of the correlation between seed weight and the size and competitive ability of the resulting seedlings (Black, 1957; Crawley & Nachapong, 1985) suggests that we should pay close attention to seed weight variation in future.

d) Fitness The ultimate measure of a plant's performance is the number of its offspring which reproduce in future generations. However, in field studies involving many plants, it is practically impossible to tell which parent plant produced which emerging seedling. In such cases, the best practical estimate of offspring contributed to future generations may simply be a direct count of the seed production of individual plants. Generally, seed production is closely correlated with the size of the reproducing plant (Solbrig, 1981). In many species, the probability of flowering is also size-dependent, so that only plants larger than some critical threshold size will produce any flowers at all (Werner, 1975; Harper, 1977; Waite, 1980). As expected from the correlation between seed output and plant size, and the positively skewed distribution of weights in plant populations, the distribution of seed output per plant is strongly and positively skewed (Levin & Wilson, 1978; Schaal, 1980; Mack & Pyke, 1983). This means that the great majority of seeds is produced by a very small proportion of the plants. The implications of this for population genetics are discussed in Chapter 7.

4.3   Spatial structure of plant populations

4.3.1   Spatial structure of seed and seedling populations

We use the term 'regular' to describe spatial patterns which are more uniform, and the term 'aggregated' to describe patterns which are more clumped than randomly distributed. The methods whereby these patterns can be quantified are explained in Chapter 1. The spatial distribution of seedlings of such a species is determined by the interplay between:

i) The distribution of seed-producing parent plants. For example, despite 'extensive' sampling of the seed rain, Rabinowitz & Rapp (1980) failed to find any seeds of certain species that had aggregated flowers, and the seedlings of such species would be expected to show a patchy, aggregated pattern.

ii) The pattern of seed rain about each parent plant. This is influenced by wind and water movements, or by the abundance and activity of animal dispersal agents (Chapter 6). Seeds and seedlings of many species exhibit secondary dispersal after reaching the ground, as a result of factors such as rain splash, frost heaving and animal movements (Watkinson, 1978; Liddle *et al.*, 1982), while the established vegetation intercepts many dispersing seeds, and by preventing them ever reaching the ground, acts as a potent mortality factor.

iii) The abundance and foraging behaviour of seed- and seedling-feeding herbivores. In some cases, seed-feeding animals modify the spatial pattern of dispersed seeds by caching or by scatter-hoarding, and these activities modify the spatial pattern of seedlings which germinate. In the case of caching, scattered seeds are collected into clumps. If there is a large crop of seeds and many caches are made, some may escape rediscovery and predation, and produce seedlings (Bossema, 1968; Abbott & Quink, 1970; Smith, 1975). In the case of scatter-hoarding, aggregated seeds are re-distributed in a more regular pattern. This may increase the probability of individual seeds escaping predation (Smith & Follmer, 1972).

iv) The spatial distribution of suitable germination sites (Harper *et al.*, 1965); this subject is considered next.

Rabinowitz and Rapp (1980) report that the total seed rain arrives at the soil surface in a spatially aggregated pattern. The seeds of some species such as thistles, dandelions and fireweed are dispersed in clumps, and many other passively dispersed species produce high local densities of deposited seeds. If all dispersed seeds were equally likely to germinate, it is clear that locally dense clumps of seedlings would commonly result, and this, in turn, would lead to intense competition between the seedlings. While this certainly happens in many species, all seeds are not equally likely to germinate under field conditions. Germination requires that the seed is both in the correct physiological state and in a suitable microsite. The accumulation of a bank of ungerminated, viable seeds in most soils (Roberts, 1981) testifies to the fact that many seeds remain in a dormant condition after dispersal, perhaps for a very long time (see Section 4.4.1).

The number of microsites (termed 'safe-sites' by Harper, 1961) which fulfil all the conditions required for germination of any species (providing sufficient water and oxygen, the correct light and temperature conditions, etc.) may well be limited in any given area of habitat, so that, once all these microsites are occupied by seeds, addition of further seeds results in reduced overall rates of germination (Palmblad, 1968). When microsites do limit recruitment, then the spatial pattern of seedlings may simply reflect the distribution of microsites; such control of recruitment patterns, linked to soil-surface disturbances, is observed with the ragwort *Senecio jacobaea* (Crawley & Nachapong, 1985) and with the crucifer *Cardamine pratensis* (Duggan, 1985). Some evidence, however, indicates that germinating seeds may themselves modify their local microenvironments in such a way that germination of other seeds of the same species is either promoted or inhibited (Linhart, 1976; Waite & Hutchings, 1978). These density dependent effects are apparently extremely local in nature, and may require physical contact between the seeds. While some advantages of density dependent promotion of germination were suggested by Linhart (1976) (e.g. better moisture retention, stabilization of the ambient microenvironment and easier soil penetration by roots when many individuals germinate together), it is likely that the intense clumping of seedlings which results would be disadvantageous to individual plants, and would tend to lead to density dependent mortality, through intraspecific competition or increased risk of herbivore or fungal attack. Indeed, Waite & Hutchings (1979) showed that whereas clumps of *Plantago coronopus* seeds displayed positively density dependent germination in the laboratory, this was less marked in the field, and conferred no greater establishment rate upon individuals than when the seeds were sown in isolation. It may be that seeds experiencing density dependent enhanced germination rates are just the hapless recipients of an external stimulus that confers on them no consistent selective advantage.

The major advantage of germination which responds *negatively* to increasing density of seeds would be that the production of dense, highly competitive populations of seedlings is avoided. Such a response would also regulate the population by ensuring that relatively few seeds germinate at any one time. A reservoir of ungerminated seeds may accumulate which buffers the populations against occasional seed failures, and may confer colonizing potential at a later date (see Section 4.4.1). Such control of the germination of neighbouring seeds is often achieved by the release of allelopathic chemicals from germinating seeds, growing plants or decaying remains of plants. An example of intraspecific allelopathic control occurs in reedmace *Typha latifolia*, which has a toxic effect on the germination and growth of *T. latifolia* seeds and seedlings (McNaughton, 1968). The principal chemicals involved are phenolic compounds which are most concentrated in decaying *Typha* leaves, but are also sufficiently concentrated in the soil to inhibit invasion by seedlings.

Established plants of many species are able to suppress seedlings in their immediate vicinity by a variety of means, including casting deep shade, vigorous competition for water and nutrients in the upper layers of the soil, production of inhibitory chemicals, or by supporting a large herbivore fauna which can eliminate seedlings as they appear. In such cases successful establishment of many species from seed requires that gaps are created in the established vegetation (Grubb, 1976; Fenner, 1978a; Orians, 1980). These gaps might be caused by the death of large individual plants (e.g. tree fall in a tropical forest; see Chapter 3), or by small scale disturbances due to animals (footprints, dung piles, digs or scrapes). Large scale disturbances through fire, flood or landslide, and the various disruptive activities of man, can provide extensive areas suitable for the establishment of many species from seed.

In the absence of disturbance, a more or less continuous canopy of vegetation can develop, in which gaps are small and few in number. Nearly all the species which are inhibited from germinating beneath the dense leaf canopy are typical of open habitats (King, 1975). Fenner (1978a) has shown that, in comparison with species which are characteristic of closed turf, the growth of colonizing species is more suppressed in shade, showing greater reductions in RGR, and greater percentage reductions in final weight. Conversely, not all species require gaps for germination. For example, many woodland species are stimulated to germinate beneath a leaf canopy which biases the spectral composition of the light they receive, increasing its far-red:red ratio in comparison with direct sunlight (see Chapter 3).

Given the complex of factors which combine to determine the spatial patterns of seeds and seedlings, we might ask if any sense can be made of the spatial and temporal distributions shown by different species in a given site? One study providing considerable insight was carried out by Gross and Werner (1982), who analysed the reasons why four short-lived monocarpic perennial species occupied different points in secondary succession on abandoned agricultural land. Two small-seeded species, *Verbascum thapsus* and *Oenothera biennis*, characterized recently abandoned fields (1–4 years old). They produced huge numbers of seeds which could remain viable in the soil for 80–100 years. Thus their seeds were often present in the soil seed bank and capable of germinating as soon as bare ground was available. *Daucus carota* produced fewer, large, but shorter-lived seeds that did not form a seed bank. *Daucus* normally reached abandoned fields by immigration some 4–7 years after abandonment. The seeds of *Tragopogon dubius* are larger still, they are viable for only a short time, and produced in much lower numbers; this species is likely to colonize sites between 15 and 40 years after abandonment. Normally, the availability of bare ground falls as the time since abandonment increases, so recruitment of *Verbascum* and *Oenothera*, which require bare ground, falls off rapidly. *Daucus* and *Tragopogon* do not need bare ground for establishment (because their larger seeds enable them to establish in shade), but they need to immigrate from other habitats. The sequence of establishment of these species reflects the conflicting benefits of either having small, persistent seeds in the soil bank, or being able to establish in shade. Similarly, there is a trade-off between seed size and seed number; the larger seeds of *Tragopogon* are produced in lower numbers, so that its probability of colonizing a site in any year is less than for the more prolific *Daucus*. Thus differences in germination microsites, together with differences in dispersal and survival probabilities, interact to determine the spatial distribution and temporal patterns of abundance of seedling populations of these species.

4.3.2   Spatial structure of established populations

The spatial distribution of mature plants reflects the spatial pattern of recruitment (Section 4.3.1) and the influence of mortality factors which differ in intensity from place to place. Thus, where density dependent mortality is important, the spatial distribution of mature plants becomes less aggregated than the distribution of seedlings. In contrast, when abiotic mortality factors are most important, mortality may be negatively density dependent, because most plant deaths are expected to occur at the edge of the population distribution, where conditions are least favourable, and where recruitment is likely to have been lower and more patchy. Such inverse density dependent mortality means that the spatial pattern of mature plants is *more* aggregated than the seedling populations. In this Section I describe how biotic factors, especially plant competition, influence the distribution of mature plants in space.

Strictly random spatial patterns are rare in plant populations. Regular distributions are also uncommon, and many reported cases are now disputed (see Chapter 1); these distributions are considered later in this section. Most studies of pattern in plant populations have revealed some degree of aggregation. Among the many factors responsible for aggregation are the following:

a) The majority of dispersing seeds land close to the parent plant, and the probability of mortality caused by seed predators, seedling herbivores or fungal pathogens may be highest near the parent plant, falling off as distance increases (this is the 'distance hypothesis'; Janzen, 1970). An alternative model suggests that the mortality risk of seeds and seedlings due to herbivores and pathogens is a function of the density of seeds, rather than of distance from parent plants as such (this is the 'density hypothesis'; Connell, 1979). By plotting the relationships of seed density and mortality rate against distance from the parent plant, a predictive graph of recruitment against distance can be generated. Janzen (1970) argued that this should produce a peak of recruitment some distance away from the parent because mortality beneath the parent is so high (Fig. 4.9), and that this process may help to explain why individual trees of a given species in tropical rain forests might be widely spaced out. Hubbell (1980) countered this argument by suggesting that: 1) when the axes of seed density and mortality rate are scaled realistically, peak recruitment might occur beneath the parent plant because of the relatively high seed density deposited there; and 2) at least in tropical dry forest, many species of trees are not spaced out at all, but show *clumped* distributions of seedlings beneath parents (Hubbell, 1979). Simple mathematical models (Geritz *et al.*, 1984) have tended to confirm Hubbell's predictions of progeny clumping.

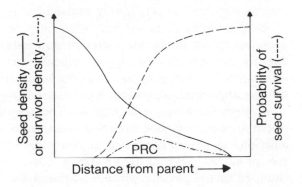

Fig. 4.9. Janzen's (1970) original 'distance hypothesis' model of tree recruitment. Seed density (solid line) is assumed to fall off with distance from the parent. Probability of seed survival (dashed line) is assumed to *increase* with distance from the parent as a result of reduced risk from natural enemies. Consequently, the population recruitment curve (PRC) peaks at some distance from the parent tree. Peak probability of seed survival is always much less than 1.0 (see Chapter 6).

Augspurger (1983) points out that predictions of offspring recruitment distances from parent plants generally ignore the fact that the probability of survival of recruits at different distances from parent plants may change with time. If this is the case, the position of the recruitment curve would alter as time goes by. In the tropical tree *Platypodium elegans*, the distances between surviving offspring and parent trees generally increased with time, showing density- or distance-dependent mortality. Median distances from parents to three month old plants were greater than parent to seedling distances; many seedlings near to the parent were killed by fungal pathogens. Subsequent survival of seedlings was dependent on the availability of light gaps on the forest floor (see Chapter 3). Finally, older saplings were found to be further from the parents than were one year old plants. Thus, for *P. elegans*, the spaced-out adult plants appear to conform to Janzen's distance model, whereas the distribution of the younger recruits is explained better by Hubbell's prediction of seedling clumping.

Experimental comparisons of the 'distance' and 'density hypotheses' (Wilson & Janzen, 1972; Connell, 1979; O'Dowd & Hay, 1980) do not give clear reasons for preferring one to the other; in some cases mortality of seeds and seedlings depends upon their density and is virtually independent of distance to parent plants, while in others, distance appears to be more important than density. There have been few field experiments in which both distance and seed density have been manipulated simultaneously, but in a study with acorns of *Quercus robur* at seven densities (from 1–64 seeds per cache) and five distances (1–16 m) from parent trees, there was no effect of either density or distance on seed survival; in each of the three years, *all* the seeds had been eaten or taken away by small mammals within 7 days (Crawley, unpublished results)!

b) Germination, as we have seen, may be positively density dependent, and suitable microsites for germination may be clumped due to the activities of digging animals, soil microtopographic factors, etc. On a larger scale, a heterogeneous habitat might afford suitable and unsuitable patches for the growth of particular species.

c) While there may be large distances between adjacent plants of a clonal species, the modular construction units (shoots, tillers) of individual clones are generally clumped within a habitat because they are produced vegetatively. Kershaw (1973) has discussed the clumped spatial patterns which arise from different kinds of clonal morphology. The distance at which a parent clone establishes a new module will have been subject to selection in response to several (potentially conflicting) considerations, including: 1) minimizing competition between connected modules of the same clone (the habitat volume from which they extract resources should not overlap extensively); 2) the investment and metabolic costs of differing lengths of stolons (typically annual structures) or rhizomes (typically perennial structures) on which new modules are produced; 3) the ability of daughter modules to gain resources directly from the parent (Chapter 12); 4) the need to counteract invasion by or avoid competition from other plants; and 5) the patchiness of the habitat. As the patch sizes alter, so the length of stolons or rhizomes which will maximize the probability of 'escaping' from an unfavourable patch, or remaining within a favourable patch will alter (Hutchings & Bradbury, 1986).

The clonal growth-forms which result can be described using the relative terms 'phalanx' (highly branched and slowly expanding clones with closely packed modules) and 'guerilla' (less branched, invasive clones with loosely packed modules; see Chapter 9). Whichever growth form is selected, the pattern of placement of modules must be sufficiently flexible to enable the plant to 'forage' efficiently for resources, and to alter its form in response to local environmental conditions. It should retain the ability to exploit favourable locations intensively and to move away from adverse conditions. Certain predictions follow from these simple considerations of foraging. For example, growth form might be of the guerilla-type to enable 'escape' from conditions of interspecific competition (e.g. *Trifolium repens* growing with grasses; Harper, 1983), whereas a phalanx growth form might efficiently exploit 'good' soil conditions. However, data to support such predictions are few, and predictions and results often conflict (Hutchings & Bradbury, 1986). The plasticity of growth form under different growing conditions is a subject which will repay much more extensive study.

The detailed architecture of individual plants is thus responsive to a variety of factors. These include abiotic conditions, the species and stature of neighbouring plants, and the degree and kind of herbivore attack (Lovett Doust, 1981; Harper, 1983; Crawley, 1983; Bülow-Olsen *et al.*, 1984). In some clonal species (but by no means all), the placement of modules may follow simple rules which might allow rather precise predictions of growth and form (see Chapter 9). In some situations, however, the basic growth rules may be so disturbed by edaphic variability (e.g. the presence of stones, patchily distributed nutrients), by neighbours or by herbivores, that the rules cannot be worked out from the plant's final form (Cook, 1983).

Whenever plants are irregularly spaced, differential spatial demands are made upon the resources which the habitat can supply. When heavy local demand for limited resources by dense aggregations of plants results in competition, at least some of the plants will suffer reduced performance, and some may die. This competition-induced mortality follows a density dependent pattern, so that local density differences are reduced, and clumped or random distributions tend to become more regular through time (Laessle, 1965; Hutchings, 1978; Schlesinger & Gill, 1978). By subdividing populations to analyse the spatial distribution of different sized plants, Ford (1975) was able to show that the plants which emerged as dominants were regularly distributed, even though the distribution of the whole population was far from uniform. In effect, competition for resources with uniform availability produces an arrangement of large plants which are as far apart as possible, with suppressed plants in the spaces between them. The large plants have RGR's higher than the population average, and Begon (1984) has proposed that they experience population density as being much lower than it really is. He suggests that the large plants are seriously affected only by competition from the other dominant plants, whereas the small plants experience density as being far higher than it really is, and suffer disproportionately intense competition. On the other hand, it could be argued that population density, either actual or experienced, is irrelevant as far as the individual plant is concerned, and that it is the *size* of the plant's immediate neighbours that is important.

Regular distributions of plants have often been reported in desert perennial communities, and it has been proposed that in these cases regular spacing is the result of competition for water (Woodell *et al.*, 1969), which may have converted an initially aggregated or random spatial pattern into a more regular pattern (see Chapter 1). Yeaton and Cody (1976), Yeaton *et al.* (1977) and Phillips and MacMahon (1981) have provided supporting evidence for the hypothesis that competition is occurring in such communities, by demonstrating positive correlations between the distances separating nearest neighbour plants and the sum of their sizes. In deserts, even though the above ground parts of perennial species might form less than 10% ground cover, the root systems of neighbouring plants often meet. Large, laterally spreading root systems are necessary to enable sufficient water to be absorbed to ensure survival between the rare falls of rain. Often, established plants are old, and recruitment is rare.

Not all studies of desert perennial communities reveal regular spacing and competition for water, however. For example, Gulmon *et al.* (1979) concluded that for the cactus *Copiapoa cinerea* there was no evidence of competition for water, and plants were randomly distributed or clumped. Their calculations suggested that water use by the established plants was far below that available annually. The water storage capacity of very small plants was low, and the time between successive rain storms could be very long, and these facts rather than competition prevented plant establishment. Clearly, even these relatively simple desert perennial communities do not lend themselves to straightforward generalizations about spatial structure and the factors which control it.

A final aspect of spatial structure concerns the non-random distribution of the sexes in species having separate male and female plants (dioecious species). Meagher (1980) has shown that for the lily *Chamaelirium luteum*, plants were more likely to have nearest neighbours of the *same* sex, and that males were in higher proportion where the local population density of the species was higher. Grant and Mitton (1979) observed changes in sex ratios of adult *Populus tremuloides* along an altitudinal gradient, with females predominating at lower elevations and males at higher. Freeman *et al.* (1976) reported marked spatial segregation of the sexes along a moisture gradient for a variety of dioecious species. In these and many other cases, it appears that males are found under 'poor' conditions and females under 'good', presumably because female reproductive function is more severely resource-limited (Willson & Burley, 1983); the precise definition of the conditions affecting sex expression, and the implications of labile sex expression for population dynamics and evolution, are discussed in Chapters 7 and 10.

4.4   Age structure in plant populations

4.4.1   The seed bank: dispersal in time

Detailed investigations of temporal changes in the seed banks of a variety of soils enabled Thompson and Grime (1979) to distinguish two classes of plants: i) species with transient seed banks, in which all seeds germinate or die within a year of dispersal, and ii) species with persistent seed banks, in which a fraction of the dispersed seed survives for more than a year in a dormant condition in the soil.

Three states of seed dormancy have been recognized (Harper, 1957, 1977). 1) **Innate dormancy**, which prevents seeds from germinating while still attached to the parent plant. Many species of temperate latitudes retain strong innate dormancy after dispersal, and this is broken only by a period of cold treatment, after which germination occurs in the spring. 2) **Induced dormancy**, in which inability to germinate has an endogenous physiological or biochemical cause. This form of dormancy is acquired as a consequence of some environmental experience to which the dispersed seed is exposed after ripening. If, for example, the seeds referred to above did not germinate in spring, induced dormancy might result, preventing germination until a further period of cold treatment has occurred. Thus, once again, germination could not occur until spring. 3) **Enforced dormancy**, in which an environmental restraint prevents the germination of otherwise germinable seeds. Thus, enforced dormancy has an exogenous rather than endogenous cause. Many seeds buried in soil are in a state of enforced dormancy and only require light in order to germinate. This explains the often spectacular abundance of seedlings appearing after soil is dug or ploughed. Prior to germinating, seeds can pass between different dormancy states as conditions change.

The rate at which seeds emerge from dormancy, and the length of time for which they remain viable, are subject to different kinds of selection in different environments. Rapid germination of all seeds and associated short-term viability may lead to a population explosion if germination coincides with favourable conditions, or to extinction if conditions deteriorate suddenly after germination. Conversely, periodic germination of a few seeds from a seed population with long-term viability limits the potential for population growth but increases the likelihood of at least some of the seeds germinating during favourable conditions at some time in the future. It also lessens the probability of mortality from intraspecific competition between seedlings (Section 4.3.1). Protracted germination of a cohort of seeds has been called 'dispersal in time' by Levins (1969). In many species which germinate in this way, dormant seeds (and dormant buds and meristems in clonal species) may greatly outnumber growing plants. Because of the practical problems of accurately recording the fate of seeds of species with long-term dormancy, investigations of age structure tend to ignore the fraction of the plant population in the seed bank (but see Roberts & Feast, 1973; Sarukhán, 1974; Sagar & Mortimer, 1976).

The effect of germination of different fractions of the seed bank on population growth and stability in randomly fluctuating environments has been analysed by Cohen (1966). Individuals of his model species either reproduced once or died without reproducing. Germinated seeds had a probability P of successful reproduction, and both P and the average number of seeds produced per germinated seedling, Y, depended upon environmental conditions. (Y was assumed to be independent of population density—an assumption only true in models!). Each year a fraction, G, of the seed bank germinated, and a fraction, D, of the remaining seeds died. If S_t is the size of the seed bank in year t, then the size in the next year will be:

$$S_{t+1} = S_t (1-G-D(1-G)+G . Y_t)$$

The growth rate of the seed bank will thus always be increased by decreasing the death rate D and increasing the seed output per germinating seed. In year t, an average of Y_t seeds is produced per seedling.

Cohen then set the problem of predicting how different germination rates will affect the growth rate of the seed bank, given different combinations of P, Y and D. Some of his results are shown in Fig. 4.10, where each of the variables P, Y and D has been given either a high or a low value. The sensitivity of population growth rate and optimal germination rate to different values of these three variables is clear. Given a high probability of reproduction and a high average seed output per plant, the proportion of the seed bank dying hardly affects the growth rate, and, not surprisingly, the highest growth rate is achieved when all seeds germinate. Under these conditions, therefore, *selection* for long term viability would not be beneficial, and no seed bank would accumulate.

When average seed output per plant is low, population growth rate is inevitably far lower, but when G is also small, the decay rate of the seed bank has a marked influence on growth rate. If the probability of successful reproduction is low, then extinction of the population is almost inevitable, unless the germination rate and death rate are both low. This combination of variables would lead to selection for long-term viability and the accumulation of a seed bank.

The influence of defoliating and seed-feeding herbivores on the dynamics of this model are discussed by Crawley (1983). The effects

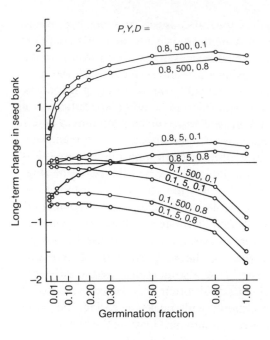

Fig. 4.10. Long-term expectation of growth rate expressed in terms of change in the size of the seed bank as a function of the fraction of seeds germinating. Triplets of parameter values for each curve indicate: P = probability of successful reproduction; Y = average number of seeds produced per germinated seedling; D = death rate of seeds remaining in the seed bank. After Cohen (1966).

of herbivory on the stability and equilibrium density of the plant population depend upon: 1) whether attack rate is dependent upon plant density; and 2) the degree to which herbivore numbers are determined by the plant species in question (e.g. whether the herbivore is a specialist or a generalist, and whether it is sedentary or mobile). Dormancy and seed bank accumulation are favoured under circumstances when all (or most) of the seed crop is destroyed by herbivores in certain years. Therefore, in addition to allowing recruitment in the face of uncertain germination and growth conditions, dormancy also enables escape from seed-feeding enemies (see also the phenomenon of mast-fruiting with its attendant 'predator satiation'; Chapter 5).

4.4.2 Age structure of the growing plants in populations

There are few short-cut methods for determining the age structure of populations of growing plants. Attempts to correlate size or other easily-measured aspects of performance with age give notoriously inaccurate results, since size variation increases rapidly as plants age, and this variation is increased by competition and habitat heterogeneity. Generally, three approaches can be used to determine population age-structures (Hutchings, 1985). First, if the population can be sacrificed, some type of annual marker can be counted (e.g. tree rings or bud scars). Second, individual plants can be recorded as they recruit to the population (as zero year olds), and uniquely tagged so that their future survival can be followed. This is laborious, time-consuming and requires several censuses, perhaps over many years; nonetheless it does provide the most valuable kind of data for plant demography. Third, with large trees, it may be possible to make non-destructive ring-counts using a core-borer (Newbould, 1967).

Age structures for various plants of widely differing taxonomy and ecology are shown in Figs 4.11–4.14. Law's (1981) study of a colonizing population of the grass *Poa annua* enabled him to determine age structures directly, because the arrival of each plant, and the time of each death, were recorded in monthly censuses. His results demonstrate changes in the age structure through time (Fig. 5.1b). The first four cohorts to colonize the site each consisted of few plants, but these, and the large cohorts which established soon after, showed high survivorship throughout the study. Thus, while the early stages of establishment were inevitably characterized by a high proportion of young individuals, the proportion of older plants increased during later stages. Law also demonstrated a wave-like pattern in the number of plants of different ages, reflecting seasonal patterns in rates of recruitment, with peaks in spring and early autumn, and troughs in winter and summer. Despite the fact that this was a population of a species which colonizes open habitats, recruitment continued for the entire duration of the study.

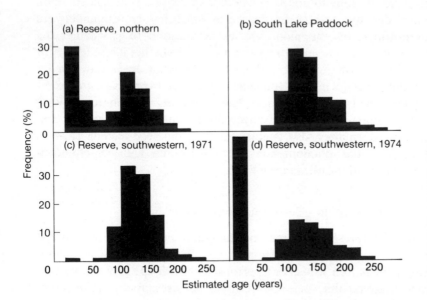

Fig. 4.11. Age structures of *Acacia burkittii* populations (*a*) sheep excluded since 1920; (*b*) sheep grazed continuously; (*c*) fenced against sheep but rabbit-grazed until 1969; (*d*) the same site, 3 years later, following rabbit eradication in 1969. After Crisp & Lange (1976).

While opportunities for recruitment of colonizing species like *Poa annua* tend to decline as a site becomes vegetated, site clearance is not always necessary to allow plant recruitment. Sometimes a change in management which alters the risks attending seedling establishment is sufficient, as found in a study of the arid zone shrub *Acacia burkittii* by Crisp & Lange (1976). Unfortunately, the ages of plants

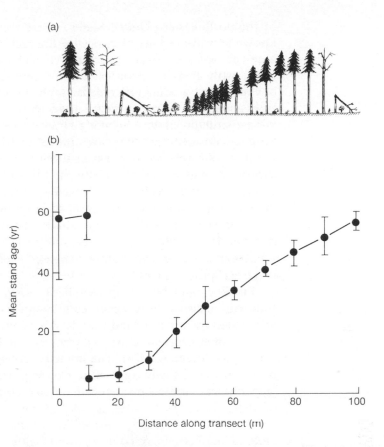

Fig. 4.12. (*a*) Wave regeneration in high altitude balsam fir, *Abies balsamea*. (*b*) Mean stand age plotted against distance along a transect across an average 'wave'. After Sprugel (1976).

could not be directly determined, but had to be estimated by regression using a combined measure involving height and canopy diameter. The age frequency histograms (Fig. 4.11) show that few shrubs live for over 200 years, and that while there has been little recruitment in some of the populations over the last century, others show substantial gains in the last 25 years prior to the study. In 1920, an area was fenced to prevent the grazing of sheep and the effect of this simple management expedient on recruitment is readily seen by comparing Figs 4.11a and b. Despite fencing against sheep, rabbit grazing prevented recruitment in the south western part of the reserve, but the effect of rabbit eradication in 1969 is shown in huge recruitment by 1974 (compare Figs 4.11c and d). These age structure data enabled Crisp & Lange to predict extinction of *A. burkittii* within a century in areas where current sheep grazing practices persist, and greatly reduced populations due to rabbit grazing, wherever sheep grazing alone is prevented.

Dynamic patterns in age structure can sometimes be inferred from analysis of community-wide spatial patterns of senescence and regeneration (Watt, 1947; and see Chapter 3). An interesting example occurs in the high-altitude balsam fir *Abies balsamea* forests of the north-eastern United States, where a striking phenomenon known as 'wave regeneration' occurs (Sprugel, 1976). Large areas of green canopy are broken by crescent-shaped bands of standing dead trees, with areas of vigorous seedling regeneration beneath and beyond them, grading into a dense sapling stand, then into mature trees about 100 m away on the far side of the wave. Figure 4.12 shows the age structure of the trees along a transect across a complete wave. Sprugel considers that the waves probably originate when a small group of old trees blows down or succumbs to fungal attack. This exposes a small crescent of plants, and the wave propagates down wind, away from this focus. Increased wind speeds, desiccation and increased deposition of rime ice at the exposed edge lead to high winter foliage loss, and production losses through foliage cooling in summer. It is possible that at 60 years of age the trees are beginning to senesce in any case, and that the increased exposure simply tips the balance (Sprugel, 1976).

Finally, populations of species characteristic of later successional stages often display little recent recruitment, and have age structures dominated by rather old individuals. An example can be seen in an analysis of oak age structure in a primary wood in Sussex, England (Fig. 4.13; Streeter, 1974). The limited seedling recruitment which occurred was virtually confined to clearings and to the 'decay' and 'gap' phases of the woodland regeneration cycle but even so, very few of these seedlings became established plants. In another study, conducted in the Colorado Rocky Mountains, Knowles and Grant (1983) identified climax, colonizing and fugitive species of coniferous trees on the basis of their age structures. The 'climax' species such as Engelmann spruce *Picea engelmannii* and ponderosa pine *Pinus ponderosa* live much longer (up to 500 and 300 years respectively) and have distinct points of inflection in their cumulative age distributions (at 220 and 120 years respectively; Fig. 4.14a). Knowles and Grant believe that this indicates a transition to a comparatively stable condition for a substantial fraction of the population. The

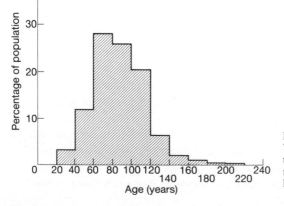

Fig. 4.13. Age frequency of an oak population at Nap Wood, Sussex, England, showing the percentage of trees in 20-year age classes. There has been very little recruitment in the past 60 years, and virtually none in the last 20. From data in Streeter (1974).

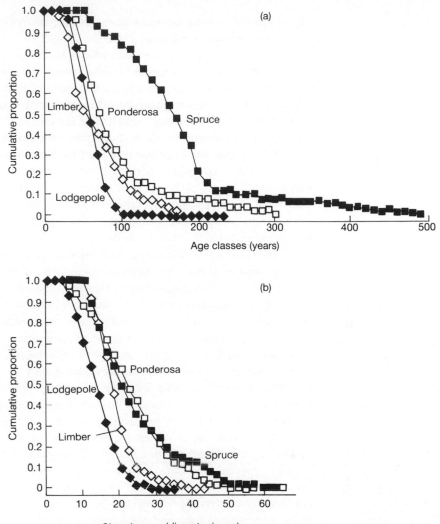

Fig. 4.14. Age and size structures of 4 coniferous tree species in the Rocky Mountains, Colorado, USA. (*a*) Cumulative age distributions for all species and all sites. (*b*) Cumulative size class distributions for all species at all sites. All pairwise comparisons of curves are significantly different by the Kolmogorov-Smirnov two-sample goodness-of-fit procedure, except for the comparison between spruce and ponderosa pine in (*b*). From Knowles & Grant (1983).

'colonizing' (early successional) lodgepole pine *Pinus contorta* has a much lower life expectancy (about 100 years), and no hint of an inflection in its age structure (despite a single datum point for a tree aged 230 years). The 'fugitive' species, limber pine *Pinus flexilis*, has an intermediate type of age structure. Note the similarity between the size distributions of ponderosa pine and Engelmann spruce (Fig. 4.14b), despite the wide differences in their age distributions. Knowles and Grant suggest that this similarity in size distribution reflects the fact that both are climax dominants, and argue that community level interactions, which are more or less independent of age, determine size structure.

The two hypotheses presented by Knowles and Grant, namely: 1) that an inflection in the cumulative age distribution is a characteristic of 'climax' species in mature stands of vegetation; and 2) that species characteristic of the climax have similar size distributions, have been strongly disputed by Parker and Peet (1984). Data are presented which fail to support the hypotheses, and Parker and Peet argue persuasively that the vegetation studied by Knowles and Grant has been subject to considerable disturbance in the past and is still in a successional stage, rather than at climax. In such cases, when the history of the site is imperfectly understood, the age structures of populations which have developed over long periods are clearly extremely difficult to interpret in detail.

Population age structures can be used to infer the survivorship through time of cohorts of plants recruited in the same time interval. Survivorship curves generated in this way will often contain considerable errors, however, since they are usually the product of observations made on the population at a single date, and the age structure observed must be assumed to have arisen as a result of constant annual recruitment, and age-specific mortality rates which do not change from year to year. As we have seen, such assumptions are often unjustified, as for example, when grazing pressure changes (Fig. 4.11). A good discussion of this use of age structures to generate 'static' ('time-specific' or 'vertical') life tables and age-specific survivorship curves, is provided by Begon and Mortimer (1981, pp. 12–14). Accurate determination of the survivorship curve of a cohort requires regular censusing from the recruitment of the cohort until the death of its last member. Depletion curves, which plot the survival through time of *multi*-aged populations also necessitate periodic census of the population (see Chapter 5). Both survivorship and depletion curves can be used to calculate half-lives—the length of time taken for half the cohort (or original population) to die (Harper, 1967). Characteristically, half-lives are longer for multi-aged populations than for cohorts, because multi-aged populations contain a proportion of established plants when observations begin, whereas all the plants in a cohort are observed from the time of their recruitment as seedlings, when they usually experience a higher than average mortality rate.

Brief mention should be made of the use of 'age states' to describe plant populations (Rabotnov, 1978; Gatsuk *et al.*, 1980; Sharitz & McCormick, 1973). Here, the stage of development which the plants have reached (juvenile, vegetative, reproductive, senescent, etc.) is recorded, rather than their actual ages. A given developmental stage may be reached by plants at widely different ages, and the attainment of a particular state may not preclude a plant from returning to a state it had held previously (e.g. reverting from sexual to vegetative). The value of determining the age state structure of a population is that the state of development of its members may be a much better indicator of the population's condition than its age structure. However, Gatsuk *et al.* (1980) recommend determination of both developmental stages *and* calendar ages wherever possible.

4.4.3 Age structure of populations of modules

Up to this point, age structures have been discussed only with respect to genets. However, valuable information can also be obtained by considering the age structure of plant parts such as shoots, tillers and leaves (see also Chapter 9). For example, Noble *et al.* (1979) analysed the age structure of shoot populations of the sand sedge *Carex arenaria* in two dune systems. At each site certain plots were fertilized with N, P and K and others left as controls; addition of nutrients produced an increased stable density of shoots and higher birth and death rates of shoots, so that the age structure of the population came to be dominated by younger shoots. This study emphasizes that if attention had not been paid to the detailed turnover of shoots, and if only their densities at the beginning and end of the experiment had been recorded, then the dramatic impact of fertilizers on the shoot dynamics of the sedge would have been overlooked completely.

For the sand sedge, abiotic conditions markedly affected shoot age structures. Turkington's (1983) research on cloned genets of *Trifolium repens* shows that the local *biotic* environment (specifically, the identity of neighbouring plant species) can also influence the age structure of leaf populations. In the presence of different neighbour species, a single cloned genet of *T. repens* exhibited widely different leaf birth and death rates, and accumulated different leaf population sizes. At any census date, leaf age structure and the phenology of leaf growth (as shown by the speed of resumption of leaf production after winter) differed when the genet was grown with different neighbour species (see Fig. 4.15 and Section 4.5).

4.5 The genetic structure of plant populations

Field studies from a wide variety of habitat types have uncovered genetic variation within populations over very small distances, caused by such factors as heavy metal contamination, local differences in nutrient level, grazing, disturbance, trampling, soil moisture and wind exposure (references in Turkington & Aarssen, 1984). This genetic variability manifests itself in many ways, including plant morphology (Watson, 1969; McNeilly, 1981), physiology (Shaver *et al.*, 1979), competitive ability (Lovett Doust, 1981), phenology (Lewis, 1976), demography (Cook, 1979) and patterns of life history (Solbrig & Simpson, 1974).

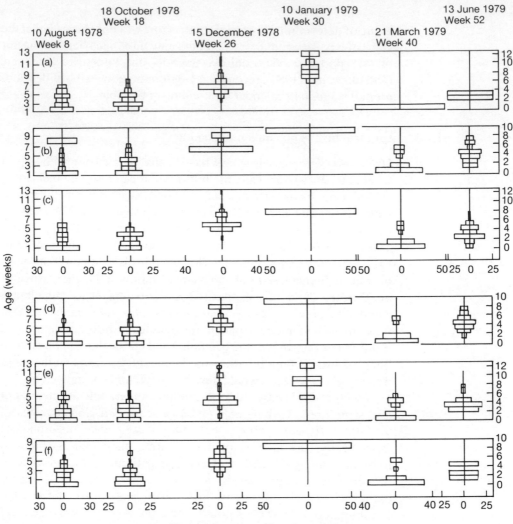

Age (weeks)

Percentage composition of population

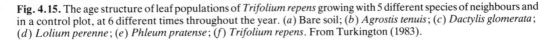

Fig. 4.15. The age structure of leaf populations of *Trifolium repens* growing with 5 different species of neighbours and in a control plot, at 6 different times throughout the year. (*a*) Bare soil; (*b*) *Agrostis tenuis*; (*c*) *Dactylis glomerata*; (*d*) *Lolium perenne*; (*e*) *Phleum pratense*; (*f*) *Trifolium repens*. From Turkington (1983).

Different genotypes confer different probabilities of survival and reproduction under different conditions, and as a result the population of plants which survives to reproduce may represent a very limited segment of the genotypic diversity which was present in the seedling population. Antonovics (1978) has suggested that the 'cost' of seedling thinning may be offset to some extent by greater speed and precision of genetic adaptation. This is not to say, of course, that selective mortality at the seedling stage necessarily produces the adult plants which would have exhibited maximum survival and fecundity.

The probability of an individual seed contributing offspring to a future generation is generally very low. For example, in a variety of habitats from experimental gardens to woodland shade, Barkham (1980) calculated that the probability of juvenile plants (daughter bulbs and seedlings) of wild daffodil *Narcissus pseudonarcissus* dying ranged from 0.09–0.43/year, the probability of subadults dying ranged from 0.08–0.11/year and the probability of adults dying ranged from 0.04–0.06/year. Without quantifying the mortality risks to which seeds themselves are exposed (Chapter 5), it is clear that very few seeds will evade all hazards and reproduce. The hazards encountered at various points during a plant's life cycle have been illustrated by Sagar (1974; Fig. 4.16).

Little is known about changes in genetic composition of plant populations during the seed phase of the life cycle, but there have been a number of recent studies of the genetic composition of populations of growing plants. Using electrophoretic techniques (Chapter 7), Gottlieb (1977) could show no significant genotypic differences between large and small plants in natural populations of the annual

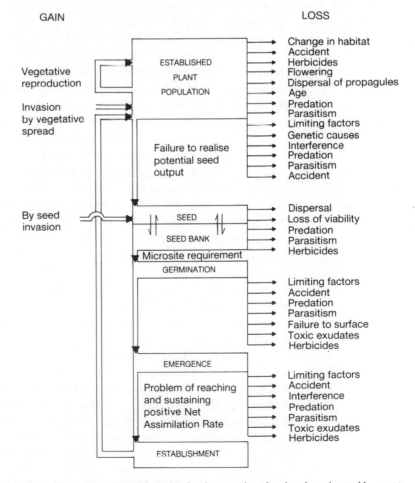

Fig. 4.16. A generalized life cycle of a plant species, showing the gains and losses at different stages. After Sagar (1974).

herb *Stephanomeria exigua*, thus apparently vindicating, at least for
this species, the ecologist's tendency to ignore genotypes when
studying fecundity differences between individual plants in populations. Unfortunately, there is no guarantee that the genetic loci
which Gottlieb investigated have any relevance to plant size, while
others, more linked to size might have demonstrated differences
between large and small plants. Gottlieb proposed that plant size and
fecundity were primarily controlled by local environmental factors,
and were unlikely to be the basis of directed evolutionary changes.
Trends in the genetic composition of plant populations over time
have been documented, however. Bazzaz *et al.* (1982) established
populations of *Phlox drummondii* from equal proportions of ten
cultivars which could be recognized by the colour and markings of
their petals. After 60 days of growth, the frequencies of cultivars
amongst the surviving plants differed significantly from their
frequencies after 38 days, and from their frequencies in seedling
populations. Cultivars showed different patterns of survivorship, and
relative survivorship changed with population age and with soil
nutrient status. Natural populations of *Echium plantagineum* also
show genotype-specific thinning. Burdon *et al.* (1983) used electrophoretic techniques to demonstrate that certain alleles altered in
their frequency in the surviving population with the passage of time;
plants heterozygous at particular genetic loci had characteristically
higher survival rates than plants homozygous at the same loci.

Detailed investigation of genetic structure in plant populations
has sometimes produced unexpected results, showing that genetic
variability was far greater than anticipated. For example, Jefferies &
Gottlieb (1983) predicted that *Puccinellia x phryganodes*, an
apparently sterile triploid with extensive clonal growth, would have
very little genetic variation, either within or between populations.
However, electrophoretic analysis revealed that most assayed
material was genetically unique! This led them to hypothesize that
fertile seed is occasionally produced even though it has never been
witnessed. (Flowering is uncommon, but small amounts of viable
pollen can be produced.) Alternatively, somatic mutations within the
clones may have increased genetic variability in the population (see
Chapter 10).

A second example concerns white clover *Trifolium repens* which
has been studied more intensively than any other herbaceous clonal
species (indeed Turkington & Aarssen (1984) have called it 'the
Drosophila of plant biology'!). While some clonal species develop
stable local patches of genetically monotonous vegetation (patches of
ground occupied by one genet to the exclusion of all others; e.g.
bracken fern *Pteridium aquilinum* (Oinonen, 1967), and the grasses
Festuca rubra and *Holcus mollis* (Harberd, 1961, 1967)), this is not
the case with *T. repens*. By analysing the distribution of clones with
different leaf markings in pastures up to about 900 years of age, Cahn

and Harper (1976a) consistently found between three and six closely intermingled clones of *T. repens* in each 10×10 cm square. We might have expected, however, that over such long periods of continuous, stable management, the fittest clone would have excluded all others and achieved local dominance, thus reducing genetic diversity. At least two reasons can be suggested as to why this has not happened. First, Cahn and Harper (1976b) demonstrated that sheep grazed preferentially on the commonest clover morph, and this frequency dependent feeding moderated any tendency for one morph to dominate the sward. Second, the guerilla growth form of *T. repens* makes it likely that any clone will make contact with many different neighbouring plants. Some of these will be competitively superior, others not. The rapid movement of guerilla clones of *T. repens* through the sward means that most of these interactions with neighbours will be fleeting (see below). Therefore it is unlikely that an established clone of *T. repens* will remain in contact with enough competitively superior neighbours long enough for it to be totally eliminated from the habitat.

There has been substantial progress in recent years in the analysis of genetic variability in populations of *T. repens*. Burdon (1980) sampled 50 plants from a regular grid established on a pasture in Wales, and screened the plants for a range of genetically determined vegetative and floral characters. On average, any two clones differed in more than three vegetative characters and more than five vegetative plus floral characters. Trathan (in Harper, 1983), using isoenzyme analysis, has recorded 48–50 genotypes of *T. repens* per square metre! Part of this genetic variability can be accounted for by the discovery that particular clones of *T. repens* perform best in the presence of specific kinds of neighbours. Turkington & Harper (1979) collected ramets of *T. repens* from sites in a permanent pasture which were dominated by one of the four grasses *Lolium perenne*, *Holcus lanatus*, *Cynosurus cristatus* and *Agrostis tenuis*. The ramets were multiplied clonally in the greenhouse, then samples of each clone were transplanted back into the field beneath each of the four grasses. The *T. repens* clones grew better in the presence of the grass from whose neighbourhood they had originally been collected, and performed less well with the other species. These results were verified in swards of the grasses grown in the greenhouse (Fig. 4.17). It appears, therefore, that *T. repens* has diversified on a micro-evolutionary scale, to the extent that specific clones grow best in the presence of particular species of neighbours. Therefore, in addition to the factors referred to above, the diversity of *neighbours* is a prime factor in maintaining the genetic diversity of these clover populations.

Another layer of complexity in the genetical structure of *T. repens* populations involves the persistence in the same populations of both cyanogenic and acyanogenic morphs. Cyanogenic morphs generate

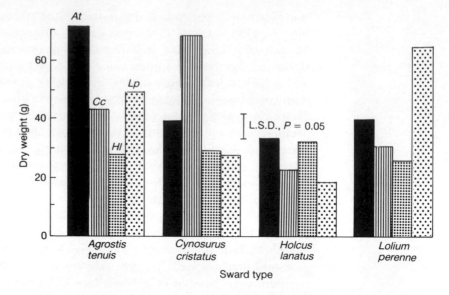

Fig. 4.17. The dry weight of plants of *Trifolium repens* from a permanent grassland sward, sampled from patches dominated by 4 different perennial grasses, and grown in all combinations of mixture with the four grass species. Clover 'types' *At = Agrostis tenuis*; *Cc = Cynosurus cristatus*; *Hl = Holcus lanatus*; *Lp = Lolium perenne*. Note that each clover 'type' grows best when grown with the grass species from which it was collected in the field. From Turkington & Harper (1979).

cyanide when they are damaged, and suffer little or no mollusc grazing; a field study by Dirzo and Harper (1982a) demonstrated that these morphs were over-represented in areas where there were high densities of active molluscs. Conversely, acyanogenic morphs were over-represented where molluscs were less abundant. Comparative studies of each morph under field conditions reveal the reason for the relative success of the acyanogenic morphs when grazing pressure is relaxed. Acyanogenic morphs showed consistent tendencies for greater survivorship, leaf and stolon production, and flowering, and lower frost damage to leaves than cyanogenic morphs (Dirzo & Harper, 1982b). These observations are consistent with the theory that the way in which an organism's resources are allocated to the competing activities of reproduction, growth and protection from herbivores, represents a compromise, and greater expenditure on one of these activities reduces the availability of resources to bestow on the others (Chapter 12). Dirzo and Harper (1982b) argue that the provision of herbivore protection in cyanogenic plants involves a metabolic cost which must be paid for in reduced competitive ability and/or reproduction.

4.6   Abiotic influences on population structure

As we have seen, each plant in a natural population experiences a microenvironment which is biotically unique, and which exerts a major influence on its performance and its likelihood of survival; the sizes, distances, spatial arrangement and specific identity of its neighbours differ from those of other individuals. Abiotic conditions also vary throughout the habitat, perhaps even over distances as small as the seeds themselves, and this variation exerts an effect on individual performance. Wellington and Trimble (1984), for example, graphically describe how, to a seedling, the two sides of a shallow hoof-print may differ as dramatically in their microclimates (in radiant cooling, frost incidence, and cold air drainage) as the two sides of a steep mountain valley, several kilometres apart!

The extent to which such small scale abiotic heterogeneity influences population structure and individual fitness has been investigated by Hartgerink and Bazzaz (1984) in a greenhouse experiment on the annual herb *Abutilon theophrasti*. Heterogeneity was established at three densities in experimental plots by either adding nutrients, sand or stones, or by compacting the soil in many small, seedling-sized patches. Single seedlings were grown in each of the patch types created. Seedling populations were also grown in the same spatial arrangement in similar plots in which every effort was made to eliminate microhabitat heterogeneity. Several aspects of performance were measured for the seedling and adult plants from each treatment. The effects of density and patch type on the biomass of seedlings are shown in Fig. 4.18. Both density, and the interaction of patch type with presence or absence of heterogeneity, had a significant effect on plant biomass; the pattern of response to different types of patch in the heterogeneous treatment was significantly different from that of plants in the same locations but in homogeneous environments. Seed production varied significantly with the type of patch in which the plants were grown. Much of the variance (47–62% depending on density) in seedling height after two days of growth could be explained by patch type differences, but as growth proceeded, the level of explanation due to patch type fell, reaching 20–25% at the final harvest. One third of the variance in individual biomass at final harvest was attributable to differences in patch type. Overall, at final harvest, 58–76% of the variance in plant biomass, depending on density, could be explained by a multiple regression in which most of the explained variance was due to patch type and seedling size early in growth. In contrast to the experiments discussed in Section 4.2.1, individual plant sizes were *not* correlated with the sizes of their immediate neighbours (see also Fowler, 1984).

The conclusions drawn from this study are highly relevant to a discussion of population structure. Hartgerink and Bazzaz (1984) propose that small-scale stochastic events which are unpredictable in time and space (including, for example, the availability of seedling-sized patches of differing suitabilities for plant growth), may strongly influence the position taken by a plant in the size hierarchy. Those components of the genotypes of plants which might determine fitness

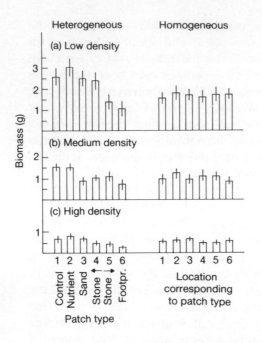

Fig. 4.18. Soil surface heterogeneity and plant performance. Mean above ground biomass of plants on each patch type on a heterogeneous substrate, or in the corresponding locations in the container on a homogeneous substrate, at three population densities; (*a*) 130 plants/m^2; (*b*) 410/plants m^2; (*c*) 770 plants/m^2. In Patch type 2 extra nutrients were added to the soil, whereas in Patch type 3 sand was mixed with the soil. Patch type 4 indicates a ceramic tile placed above the seed; type 5 indicates a tile below the seed. Patch type 6 is a 'footprint' created by compacting the soil. From Hartgerink & Bazzaz (1984).

under controlled conditions may have their influence completely nullified in the field as a result of these stochastic events. Consequently, selection against less favourable genotypes will probably not take place in any predictable fashion. This, ultimately, will contribute to the maintenance of genetic diversity in populations. Environmental heterogeneity at the scale of seeds and seedlings can exert an overwhelming effect on the fitness of the individual, and can also largely control the expression of variance in population parameters, as Hartgerink and Bazzaz (1984) demonstrate. In the past, plant population biology has tended to concentrate on mean values of demographic and performance parameters. Evolution, however, takes place at the level of the individual plant. Harper (1981) and Watkinson (in Chapter 5) emphasize that study of the *variance* in populations, rather than mean values, is most relevant to understanding evolution. The 'mean plant' is an almost meaningless and biologically misleading abstraction.

This chapter has described the variation between plants within populations which produces *structures* of different kinds—performance, spatial, age and genetic structures—and has drawn attention to the biotic and abiotic factors responsible for this variation. Now that these structures are becoming familiar, a major challenge facing plant population ecologists is to explain the dynamics of the processes responsible for the formation of these structures, and a major challenge for evolutionary ecologists is to understand natural selection in the context of individual plants that are members of structured populations.

Chapter 5 Plant Population Dynamics

ANDREW R. WATKINSON

5.1 Introduction

In studying the dynamics of populations we are concerned with the ways that various biological and physical factors interact to bring about changes in plant numbers through time and space. By quantifying births and deaths, immigrations and emigrations, we confront such questions as why some species are rare and others common, and what processes are responsible for fluctuation in their numbers? Related to these topics are questions about why the age and size distributions of individuals vary from one population to another (Chapter 4), and what factors determine the genetic structure of plant populations (Chapter 7). While these structural aspects of populations are objects of study in their own right, it should be remembered that the same processes that determine the dynamics of populations are also largely responsible for determining their size, age and genetic structures.

5.1.1 Population flux

Observations on natural populations show that when conditions are suitable and resources are freely available (as in the early stages of secondary succession), all populations have the potential to increase exponentially. With time, however, the rate of population growth slows and eventually a maximum population size is reached (Fig. 5.1). For example, in the Botanic Gardens in Liverpool, a derelict piece of land was laid bare in late December. Law (1981) found that the first colonists of the weedy grass *Poa annua* appeared in the following April. Most of these colonists survived and produced large numbers of seeds that germinated from August to September, leading to a massive increase in recruitment and a corresponding increase in population size. Following this initial very rapid increase in population size, the growth rate of the population declined. The growth rate is density-dependent (the rate of increase declines as the density of the population increases), because seedling recruits to the population at high densities germinate in dense aggregations around the parent plant where they have to compete for a limited supply of resources. As a consequence they grow much more slowly, produce fewer seeds and are more likely to die at an early age.

137

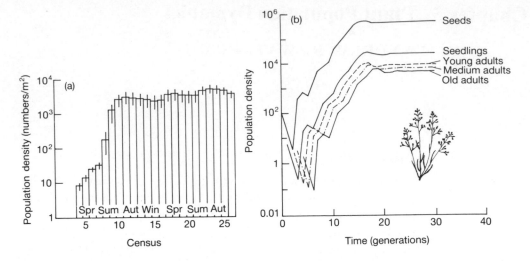

Fig. 5.1. The population density at each census during colonization of derelict land by *Poa annua*. (a) Field data from Law (1981). (b) A simulation of the density-regulated growth of 5 age-classes, based on the density-dependent Leslie Matrix model given in Box 5.3. From Law (1975).

In a different environment, the colonization of bare mud flats by the annual *Suaeda maritima* shows a similar pattern. In this case, however, the time taken for the population to reach its maximum density was longer, because *S. maritima* has discrete generations and produces only one seed crop per year (Joenje, 1978).

That population growth is *regulated* and limited by the availability of some resource can be seen clearly in populations of both *Poa annua* and *Suaeda maritima*. Observations on these populations, however, were carried out over a relatively short period. A much longer study on a number of rare plants in a grassland community in northern England (Bradshaw, 1981) also showed that despite year-to-year fluctuations, population sizes were frequently regulated within rather narrow bounds over a 12 year period (Fig. 5.2). However, not all populations show numerical constancy following an initial period of exponential growth; for example, populations of *Hieracium pilosella* (Davy & Jefferies, 1981; Watt, 1981) declined rapidly, immediately after reaching a peak of abundance in 1953 (Fig. 5.3), and were replaced by *Thymus praecox* as the dominant species. Watt (1981) interpreted the rise and fall of the dominant species over a period of 44 years in terms of climatic change, species' interactions, and the changing vigour of even-aged populations. The clear message from this classic long-term study is that in order to understand the abundance of plants it is necessary to understand the patterns of birth and death with age, the interactions with other organisms and the effect of environmental variables such as climate on birth and death rates.

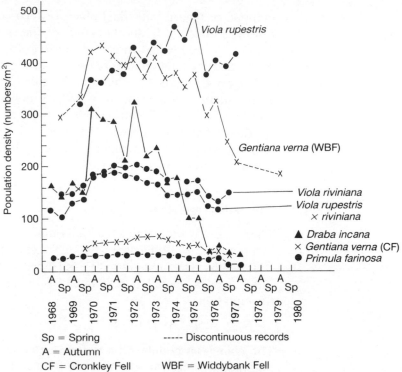

Simple monitoring of population size, however, is insufficient on its own. Even in those populations where the number of individuals remains relatively constant from one census to the next, there may be a rapid flux of births and deaths and a considerable turnover of individuals. This underlying flux can only be revealed if the fate of

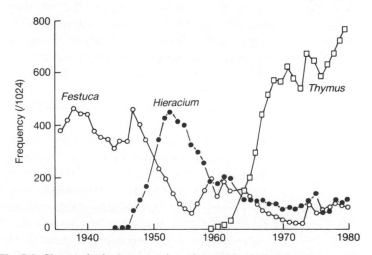

Fig. 5.3. Changes in the frequency (out of a possible 1024) of *Festuca ovina* (O), *Hieracium pilosella* (●) and *Thymus drucei* and *T. serpyllum* (□) in an enclosed grassland plot in the Breckland of eastern England between 1936 and 1980. From Davy & Jefferies (1981).

mapped or tagged individual plants or their component modules is followed in detail. In another study of *Hieracium pilosella* in which the fates of individual rosettes were followed, Bishop *et al.* (1978) found that the total number of rosettes in a grazed population varied little over 4.5 years, but that this concealed a rapid flux of rosettes, approximately equivalent to a 25% turnover per year. Very few seedlings established during this period, and most new rosettes resulted from clonal growth. The one major change in population size occurred in the summer of 1976, coincident with the worst drought in England for 200 years.

5.1.2   Population regulation

There are only four factors that determine the numbers of individuals in a population (Williamson, 1972). These are the numbers of births (B), deaths (D), immigrants (I) and emigrants (E). The number of individuals in a population at time $t+1$ is related to that at time t by the equation

$$N_{t+1} = N_t + B - D + I - E \tag{1}$$

The change in population size from one census to the next can be obtained by subtracting N_t from both sides of the equation to give

$$\Delta N = B - D + I - E \tag{2}$$

Clearly if the sum of the terms on the right-hand side of the equation is positive the population will increase in size, whereas if ΔN is negative the population will decline. If the numbers of births, deaths, immigrants and emigrants do not change with density, then a population will increase to infinity if ΔN is positive, or will decline to extinction if ΔN is negative. Although in theory the size of a population will remain stable if ΔN is exactly zero, the introduction of random variation again results in population density declining to zero (Moran, 1962). These three kinds of populations are said to be unregulated.

No factor which alters the growth rate of a population, independent of population density, can be invoked to explain why population densities persist within narrow bounds. Yet observations on a large number of populations (e.g. Harper, 1977; Silvertown, 1982) have shown that population size often remains more or less constant from year to year (see Fig. 5.2), while in others it has been shown that the growth rate of a population decreases as population size increases (see Fig. 5.1). Clearly the assumption that B, D, I and E are all independent of plant density is untenable. Unless one or more of the demographic variables in equation (1) are functions of N, population size cannot be regulated, and the population will not persist in the face of an uncertain and fluctuating environment. Considering the importance of density dependence in the regulation of population size, it is surprising how few plant demographic studies have investigated density phenomena in natural populations (Antonovics & Levin, 1980).

Whilst the density dependent control of one or all of the variables in equation (1) is necessary to explain population regulation, it is the interaction between density dependent and density independent processes that determines the actual size of a population. Consider a population in which the birth rate (i.e. seed production) is density dependent, while the death rate is density independent but varies with the physical conditions. The rates of immigration and emigration are assumed to be equal. In Fig. 5.4a there are three equilibrium populations (n_1, n_2, n_3) corresponding to different death rates, which reflect the physical conditions in the three environments. In real populations, the actual rates will inevitably vary (Fig. 5.4b) and the populations will fluctuate round the equilibrium. Whilst it is naive to expect all populations to be at equilibrium, Fig. 5.4 nevertheless illustrates clearly how population size will vary depending on physical and biotic conditions. The relative importance of different density dependent and density independent processes will vary considerably from population to population. For example, in Fig. 5.4c a major change in the birth rate results in only a minor change in the equilibrium population size, whereas in Fig. 5.4d, the same shift in birth rate produces a major change in population size.

5.1.3   The individual and the population

Population biologists are interested in why populations are as large as they are, and why they vary in size through both space and time. Yet, what is a population? For a geneticist a population is generally considered to be a group of individuals genetically connected through parenthood or mating (Chapter 7), but for an ecologist a population can be defined simply as the total number of individuals of a single species in an area circumscribed for the purposes of study (see Chapter 1).

It is also necessary to consider exactly what an *individual* is. This is discussed in detail in Chapter 9, but it is necessary to make two points at this stage: 1) we should recognize that individual plants may occur in various life-states, for example as seeds or vegetative plants (for plants such as desert annuals, the large majority of individuals may occur below ground in the seed bank, only emerging occasionally above ground as vegetative plants to produce more seeds); 2) plant growth is modular, unlike the unitary growth of most animals (Harper, 1981). This means that for most plants the zygote develops into a modular organism where a basic structural unit is iterated (Chapter 9). In some modular organisms the products of growth

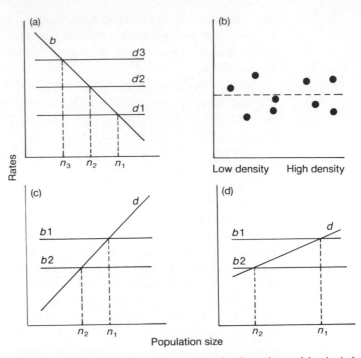

Fig. 5.4. A graphical model showing how density-dependent and density-independent processes interact to determine population size. In (a) the birth rate, b, is density-dependent. Three different levels of density-independent mortality (d_1, d_2 and d_3) lead to three different average population sizes (n_1, n_2 and n_3). The higher the death rate, the lower the equilibrium population size. (b) Equilibrium populations exist where mean birth rate equals mean death rate. In reality the average birth and death rates will vary and the population will fluctuate around the equilibrium. The existence of fluctuation does *not* mean that the populations are unregulated. (c) and (d) show how the same change in density-independent birth rate can lead to very different changes in population size when the *form* of the density-dependent death rate varies between populations.

fragment into physiologically independent parts, as with the fronds of *Lemna minor* or the rosettes of *Hieracium pilosella*, whereas in others, the modules may remain part of a more or less closely integrated physiological whole (e.g. *Poa pratensis;* Nyahoza *et al.*, 1973).

Populations of modular organisms are composed of N individual genets (individuals that develop from zygotes), each made up of a subpopulation of z modules. The number of genets in a population can be described by equation (1), while the modular growth of the *i*th genet can be described by the equation

$$z_{i,t+1} = z_{i,t} + B' - D' \tag{3}$$

where B′ is the number of module births and D′ is the total number of module deaths (Noble *et al.*, 1979; Harper, 1981). Equation (3) allows the growth of individual plants to be studied at a demographic level (Chapter 9), and the total number of modules (n) in a population to be calculated from the equation

$$n = \sum_{j=1}^{N} z_j \tag{4}$$

Few studies have attempted to measure the flux in the number of genets *and* the modular flux within individual genets (but see Soane & Watkinson, 1979). Most studies have been concerned with the flux of either genets (annuals, facultative biennials, trees) or the total number of modules (herbaceous, perennials) in a population

$$n_{t+1} = n_t + b - d + i - e \tag{5}$$

where b, d, i and e define the numbers of births, deaths, immigrants and emigrants of modules irrespective of the genet which produced them, or whether they arise from reproduction or clonal growth.

5.2   The fates of individuals

In order to understand what factors determine the abundance of plants, it is necessary to understand how the demographic parameters in equations (1), (3) and (5) are affected by the age and size of plants, and by population density, competitors, herbivores, pathogens, weather, soil conditions and various hazards like fire, or burial by sand. Monitoring the fate of individuals from birth can give us some insight into these processes, but the amount of useful information gained from such studies is limited, and it is disappointing that so much research has simply monitored the fate of individuals. Observational studies rarely reveal the causes of death, and generally only provide substantial information on how fecundity and mortality vary with age and size.

Ideally we would monitor the fates of individuals from the time of zygote formation, since it is at this point that the life of a genet begins, but this is an extremely difficult stage at which to begin counting (Harper, 1977). Most studies follow the fates of genets from the time of seed maturation or (much less satisfactorily) from seedling emergence. A rare exception is provided by the study of De Steven (1982) who monitored the fates of witch-hazel (*Hamamelis virginiana*) fruits from initial fruit set to seed maturation and dispersal. Only 14–16% of the fruits set in each of two years survived to maturation and dispersal (Fig. 5.5a). Physiological abortion resulted in the death of approximately 60% of the fruits. Other major causes of mortality included damage by caterpillars, infestation by a host-specific seed weevil and consumption by squirrels (Fig. 5.5b). De Steven speculated that much of the physiological abortion was the result of foliage-feeding caterpillars reducing leaf area, and thus

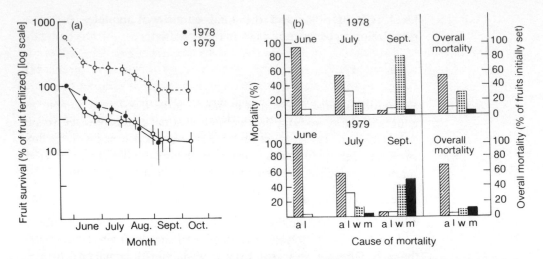

Fig. 5.5. (a) The survival of fruits of witch hazel (*Hamamelis virginiana*) from the time of fruit set to the time of fruit maturation (1978) or the time of seed dispersal (1979), in Michigan, USA. The two lower curves show percentage survival. The upper curve is scaled to allow for the higher fruit set in 1979 (5.6 times the 1978 crop). The greater fruit production in 1979 resulted in many more seeds being dispersed than in 1978. Values are means of four plots with 1 standard deviation; (b) The relative contributions of physiological abortion, a; feeding by Lepidoptera, l; weevil infestation, w; and mammal damage, m, to the loss of fruits. From de Steven (1982).

limiting the internal resources for fruit maturation (see also Crawley, 1985).

Factors other than resource limitation may influence the percentage of fertilized ovules developing into seeds. The seed-ovule ratio is about 85% in many annuals and approximately 50% in perennials. Wiens (1984) has speculated that such consistent post-zygotic abortion might result from selection against unfit genetic recombinants.

5.2.1   Patterns of mortality

Following seed maturation, most plants stand still and wait to have their seeds dispersed by animals, wind or water (Chapter 6). Undoubtedly some seeds are transported long distances from the parent plant, but the large majority of seeds are deposited close to the parent (Chapter 4). The density of seeds then declines sharply with distance. For a variety of species, the curves relating seed density to distance can be described by inverse square or inverse cube relationships (Chapter 6). Where the dispersal curve approximates an inverse cube relationship it is likely that the population will colonize as an advancing front, whereas species with more specialized mechanisms of dispersal, and dispersal curves approximating the inverse square, will colonize an area of land more as isolated individuals over a greater distance (Harper, 1977). Examples of plant invasions into new areas are provided by Mack (1981), Nip-van der Voort *et al.* (1979) and Usher (1973).

The fates of seeds are difficult to follow once they reach the ground and are incorporated in the surface litter or soil. Occasionally it is possible to follow the fates of individually labelled seeds (Lawrence & Rediske, 1962; Watkinson, 1978), but more often it has been found necessary to sow or bury replicated samples of viable seeds into small areas in which natural seed dispersal is prevented. Replicate samples are then retrieved at intervals and the number of emergent seedlings and buried seeds counted. The seed population in enforced dormancy (see Chapter 4) can be estimated by counting the number of seedlings that emerge from seeds maintained under favourable conditions. Those seeds which do not germinate (the innate and induced fraction of dormant seeds) can then be tested for viability using tetrazalium chloride (Smith & Thorneberry, 1951). The fates of the seeds of two species, determined using these methods, are shown in Fig. 5.6. For the winter annual *Vulpia fasciculata* (Watkinson, 1978) a small proportion of the seeds at the time of seed dissemination were dormant, but after a short period of after-ripening most of the seeds had lost their innate dormancy and were capable of germination. More than 99.5% of the seeds germinated in the year they were produced, and few seeds were eaten or died. In contrast, approximately 30% of the seeds of *Ranunculus repens* remained dormant after 14 months and approximately 50% were lost to seed predators (Sarukhán, 1974). Even fewer seeds (only about 7%) of the umbellifer *Lomatium dissectum* survived to germination after insect and mammalian seed predators had taken their toll (Thompson, 1985).

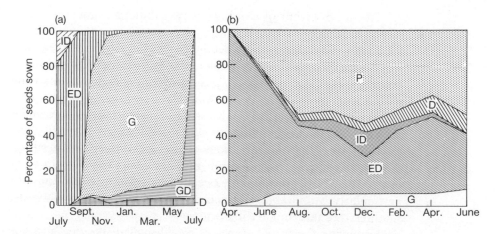

Fig. 5.6. The fates of seeds of (a) *Vulpia fasciculata* (Watkinson, 1978) and (b) *Ranunculus repens* (Sarukhán, 1974). G = seeds observed to germinate; GD = individuals that germinated and died; D = fraction of seeds that died without germinating; P = seeds lost to predation (this category is included in D for *Vulpia*); ED = seeds in enforced dormancy; ID = seeds in innate or induced dormancy.

The survival of weed seeds on agricultural land has been monitored over longer periods, and the number of viable seeds has been found to decline exponentially with time:

$$N_t = N_0 e^{-gt} \tag{6}$$

where N_t is the number of survivors at time t, N_0 is the initial seed density, and g is a constant that expresses the decay rate of the population (Roberts, 1972). The value of g appears to be specific for individual species (Fig. 5.7) and depends also upon the conditions of cultivation. In other words, for a given species at a particular site, the rate of loss of seeds from the seed bank is constant.

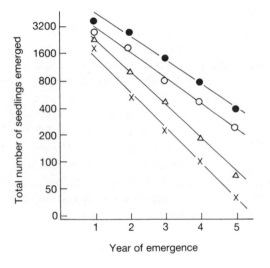

Fig. 5.7. The number of seedlings emerging in successive years from a horticultural soil where seed immigration was prevented. The decline in seedling emergence reflects a similar decline in the size of the seed bank. *Thlaspi arvense* (●); *Stellaria media* (△); *Capsella bursa-pastoris* (O); *Senecio vulgaris* (×). Combined data from experiments that began in 1953, 1954 and 1955. From Roberts (1964).

The survivorship of seedlings is very difficult to measure accurately, especially early in the life-cycle, unless frequent observations are made. Even then, those seedlings which die before emergence will not be recorded. For example, estimates of seedling mortality in populations of the annual *Vulpia fasciculata* obtained by mapping permanent plots at 2–3 week intervals, ranged from 1–10% (Watkinson & Harper, 1978), whereas estimates from an experiment in which seeds were radioactively labelled varied from 12–28% (Watkinson, 1978). Observations on a range of tree seedlings (Fig. 5.8) show that the risk of death declines continuously from birth (Sarukhán, 1980), presumably because the small tree seedlings are particularly vulnerable to such mortality factors as pathogens, shading, tree falls and herbivore activity.

Following the initial stages of establishment, the death rate is remarkably constant, at least for many perennial herbs. There is often no indication of good or bad years for survival, despite considerable variation in the weather from year to year. This point is

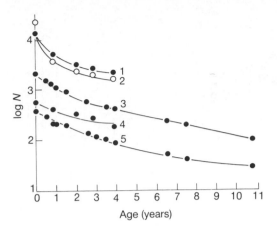

Fig. 5.8. Survivorship curves for cohorts of seedlings of 5 tropical trees: 1) *Parashorea tomentella;* 2) *Shorea gibbosa;* 3) *Shorea parvifolia;* 4) *Shorea ovalis;* 5) *Koompasia malaccensis.* From Sarukhan (1980).

illustrated for five cohorts of the man orchid, *Aceras anthropophorum* (Fig. 5.9a; Wells, 1981). Since the probability of death remains constant, it is possible to calculate a half-life for each of the cohorts as

$$\text{Half-life (years)} = \frac{t . \ln 2}{\ln N_x - \ln N_{x+t}} \qquad (7)$$

where N_x is the number of survivors at age x and N_{x+t} is the number surviving after a period of t years. For the five cohorts of *A. anthropophorum* the calculated half-life varied from 4.0–7.8 years. For a population of *Primula veris* studied by Tamm (1972) the half-life was 50 years, indicating that individual herbaceous plants (in this case probably genets) may be exceptionally long-lived. Where the fates of herbaceous perennials have been recorded at more regular intervals, it has generally been found that there is a marked seasonal rhythm to mortality, superimposed on the overall linear survivorship curves (e.g. Sarukhán & Harper, 1973). The risk of death is generally greatest in spring and early summer when the growth of plants is most rapid, and rather low during the winter. The major cause of death appears to be interference from actively growing neighbours, rather than from harsh weather conditions directly.

With the exception of short-lived plants, it is difficult to follow the survival of cohorts of individual genets through their entire life-span. Figure 5.10 illustrates the survivorship curves for ten species of winter annual from various habitats from seed maturation to flowering. The curves exhibit a wide range of shapes, from the extreme Deevey type I curve of *Vulpia fasciculata*, to the type III curve of *Spergula vernalis* (Watkinson, 1981a). Clearly there is a broad correlation between the fecundity of plants and the shape of their survivorship curves; the more fecund the plant, the more concave its survivorship curve. However, a single survivorship curve cannot be expected to characterize a species; different cohorts of the same

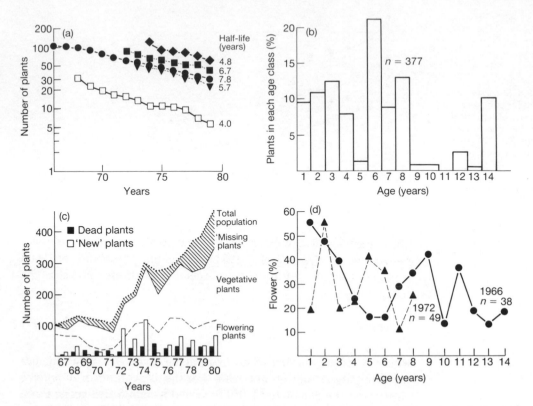

Fig. 5.9. Population dynamics of the man orchid (*Aceras anthropophorum*). (a) Survivorship curves for 5 cohorts of plants formed in 1966 (●), 1968 (□), 1972 (■), 1973 (▲) and 1974 (◆). (b) The age structure of a population of 377 plants in 1979. (c) The population dynamics of plants between 1966 and 1980. (d) The relationship between flowering and age of plant for the 1966 (●) and 1972 (▲) cohorts. From Wells (1981).

species may show markedly different survivorship curves in different places or at different times (Mack & Pyke, 1983). Nevertheless, these survivorship curves contain much useful information on the age-specific mortality of plants, and they are essential for an understanding of population dynamics.

For long-lived plants, information on age-specific survival must often be sought by indirect means. For example, it may be possible to age plants through some morphological marker such as annual growth rings or leaf and bud scars, from which the age structure of the population can be determined, and an estimate obtained of the probability of survival from one age class to the next (see Chapter 4). In making this estimate we must assume that the population has a stable age structure. This assumption involves two components. First, we must assume that the age-specific survival rates have remained the same from year to year, so that the probability of survival from one age class to the next is the same as would have been

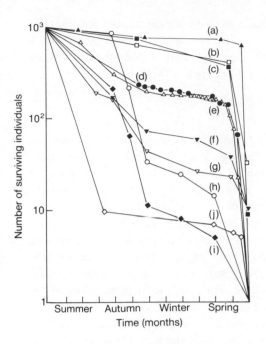

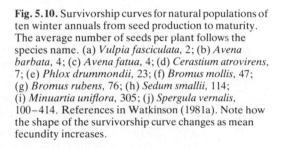

Fig. 5.10. Survivorship curves for natural populations of ten winter annuals from seed production to maturity. The average number of seeds per plant follows the species name. (a) *Vulpia fasciculata*, 2; (b) *Avena barbata*, 4; (c) *Avena fatua*, 4; (d) *Cerastium atrovirens*, 7; (e) *Phlox drummondii*, 23; (f) *Bromus mollis*, 47; (g) *Bromus rubens*, 76; (h) *Sedum smallii*, 114; (i) *Minuartia uniflora*, 305; (j) *Spergula vernalis*, 100–414. References in Watkinson (1981a). Note how the shape of the survivorship curve changes as mean fecundity increases.

obtained if a single cohort had been followed through time. Second, we make the extremely tenuous assumption that recruitment into the population is constant from year to year. These two assumptions hold true for very few plant populations, but in some cases, such as long-established, unmanaged forests, it may be possible to calculate approximate survivorship curves from data on age-structure (Chapter 4). Even so, the interpretation of the curves requires caution. It could be that a population with successively fewer individuals in each age class does *not*, in fact, have a stable age distribution, but is increasing in size; the predominance of young individuals in this case simply reflects an increase in recruitment into the population with time (Silvertown, 1982).

For a great many species' populations it is impossible to say anything about age-specific survival from the age structure of the population, since both recruitment and age-specific survival vary from year to year. This is particularly well illustrated for populations of the man orchid *Aceras anthropophorum* (Fig. 5.9b) in which the recruitment and half-life of plants varied considerably between cohorts (Fig. 5.9a). Moreover, the population was expanding (Fig. 5.9c), so any inferences about the population dynamics of this plant on the basis of age structure alone would be quite unwarranted.

Much of our discussion has been concerned with the survival of genets, but the fates of modules (whether they be the shoots of a tree, the ramets of a buttercup or the tillers of a grass) may be treated similarly. Survivorship curves for modules (e.g. Noble *et al.*, 1979;

Sydes, 1984) may vary as much as those for genets, but note that the death risk of modules early in life is less than that of genets since modules, unlike seedlings, develop with a continual supply of resources from the parent plant (Harper, 1977).

5.2.2   Patterns of fecundity

The age-specific fecundity of plants is calculated by monitoring the seed production of plants through time (Fig. 5.11), or, less precisely, by relating the fecundity of plants to the age structure of the population at a given time (Fig. 5.12). Typically there is a pre-reproductive period that may be of variable length. Whilst semelparous (monocarpic) plants die immediately following reproduction, the fecundity of iteroparous (polycarpic) plants generally increases with age to reach a maximum after which it levels off and perhaps declines. In some cases, however, the concept of age specific fecundity means almost nothing. For example, in populations of the man orchid (Fig. 5.9d) the age of plants had little influence on whether or not an individual flowered. Among 38 plants present in the population for 14 years, some individuals rarely flowered while one plant flowered *every* year (Wells, 1981). Moreover, in many herbaceous perennials there is no genuine ageing, since new tissues continually grow away from, and discard the ageing part of the soma (Harper & Bell, 1979). In this case, the concept of age-specific fecundity for the genet has little relevance, but this is not to deny that the flowering behaviour of individual modules may be age specific. Rather, the fecundity of genets may depend more on the number of modules than on the age of the genet (see Section 5.2.4).

5.2.3   Life-tables

Life-tables provide a convenient method of summarizing the data on age-specific mortality and fecundity in populations. Table 5.1 illus-

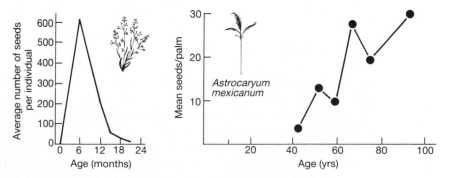

Fig. 5.11. Age-specific fecundity for (a) annual meadow grass, *Poa annua* (from Law, 1975); and (b) the tropical understorey palm, *Astrocaryum mexicanum*. From Sarukhán (1980).

trates the survivorship and age-specific fecundity of plants in a population of *Phlox drummondii* in Texas, USA (Leverich & Levin, 1979). All the seeds of this annual plant germinate more or less synchronously to form a single cohort of seedlings, and there is no carry over of seeds from one year to the next, so the plant has one discrete generation per year. Table 5.1 summarizes the data on survivorship and fecundity in the form of a standard life-table comprising 6 columns: 1) the census interval ($x-x'$) that relates to the age of the plants; 2) the length of that interval (D_x); 3) the number of individuals surviving to age x; 4) the probability (l_x) that an individual will survive from age 0 to age x; 5) the average number of seeds produced by an individual (b_x) in the interval $x-x'$; and 6) the product $l_x b_x$, the expected seed production (at birth) of a plant during the interval $x-x'$, (for further details see Begon & Mortimer, 1981).

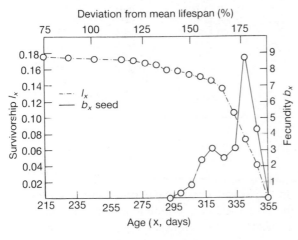

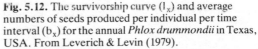

Fig. 5.12. The survivorship curve (l_x) and average numbers of seeds produced per individual per time interval (b_x) for the annual *Phlox drummondii* in Texas, USA. From Leverich & Levin (1979).

The formation of a standard life-table for *Phlox drummondii* (of the sort that would be used for human or red deer populations), is only possible because growth is determinate, and there is no persistent bank of seeds. It would make no sense to equate the survival and fecundity of a plant that germinated in the year of its production, with one that had lain dormant in the soil for ten years. Separate life-tables need to be constructed for cohorts of seeds, and for seedlings produced at different times (Schaal & Leverich, 1981).

While several authors have extolled the virtues of producing life-tables for cohorts of both genets and modules, very few have actually been produced. Even where they have been produced, little use has been made of them—the *Phlox* example provides a refreshing exception. Far too often, the production of survivorship curves and fecundity schedules are seen as ends in themselves.

Table 5.1. Fecundity and survivorship schedules for the annual *Phlox drummondii* at Nixon, Texas (from Leverich & Levin, 1979).

Age interval (days) $x-x'$	Length of interval (days) D_x	No. surviving to day x N_x	Survivorship l_x	Fecundity	
				b_x	$l_x b_x$
0–63	63	996	1.0000	0.0000	0.0000
63–124	61	668	0.6707	0.0000	0.0000
124–184	60	295	0.2962	0.0000	0.0000
184–215	31	190	0.1908	0.0000	0.0000
215–231	16	176	0.1767	0.0000	0.0000
231–247	16	174	0.1747	0.0000	0.0000
247–264	17	173	0.1737	0.0000	0.0000
264–271	7	172	0.1727	0.0000	0.0000
271–278	7	170	0.1707	0.0000	0.0000
278–285	7	167	0.1677	0.0000	0.0000
285–292	7	165	0.1657	0.0000	0.0000
292–299	7	159	0.1596	0.3394	0.0532
299–306	7	158	0.1586	0.7963	0.1231
306–313	7	154	0.1546	2.3995	0.3638
313–320	7	151	0.1516	3.1904	0.4589
320–327	7	147	0.1476	2.5411	0.3470
327–334	7	136	0.1365	3.1589	0.3330
334–341	7	105	0.1054	8.6625	0.6436
341–348	7	74	0.0743	4.3072	0.0951
348–355	7	22	0.0221	0.0000	0.0000
355–362	7	0	0.0000		

$$\Sigma = 2.4177$$

5.2.4   Age, stage or size

Population biologists in the Soviet Union have long recognized the need to classify and monitor the fates of individuals in relation to various 'age-states' as well as chronological age. For example, Gatsuk *et al.* (1980) employed a 10-state classification in their analysis of the ontogeny of woody plants (seed, seedling, juvenile, immature and virginile pre-reproductive plants; young, mature and old reproductive plants; and sub-senile and senile post-reproductive plants). Whilst there is a broad correlation between chronological age and age-state, plants may remain in a particular state for rather variable periods. For example, juvenile plants of ash *Fraxinus excelsior* may typically have a calendar age somewhere between 2–15 years, and immature plants a calendar age of 10–20 years. Herbaceous plants can be classified in a similar way (e.g. Harper & White, 1971; Sarukhán & Harper, 1973); again there is only a broad correlation between chronological age and age-state. A number of factors such as grazing may even result in a plant moving to a 'younger' age-state.

Studies on a range of herbaceous and woody perennials have shown differential mortality among young individuals attributable to

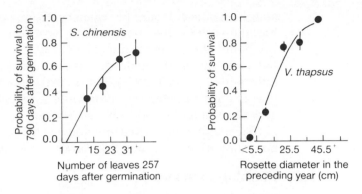

Fig. 5.13. Survival probabilities as a function of size for seedlings of (a) the desert shrub, *Simmondsia chinensis;* and (b) the biennial, *Verbascum thapsus*. From Sarukhán *et al.* (1984).

plant size or biomass (Fig. 5.13) rather than age; see Sarukhán *et al.* (1984) for a review. They have also shown that plant size as measured by height is an extremely good predictor of reproductive performance in *Astrocaryum mexicanum* (see Chapter 4). In populations of semelparous perennials, Gross (1981) has shown that the probability that a rosette will flower is strongly correlated with its size, but independent of its age. Rosettes of *Verbascum thapsus* less than 9 cm in diameter did not flower at all, but the probability of flowering increased steadily with size thereafter (Table 5.2); *all* rosettes with a diameter greater than 41 cm flowered in the subsequent year.

Table 5.2. Size-specific demography for the biennial *Verbascum thapsus*. Probability of dying, remaining vegetative or flowering, as a function of rosette diameter in the previous October. (– signifies no plants of this size and age.) (From Gross, 1981.)

Probability of	Dying		Remaining vegetative		Flowering	
Rosette diameter (cm)	Year 2	Year 3	Year 2	Year 3	Year 2	Year 3
$\leq$ 5.4	0.97	1.00	0.03	0	0	0
5.5–15.4	0.63	0.71	0.26	0.13	0.11	0.17
15.5–25.4	0.23	0.42	0.09	0.05	0.68	0.53
25.5–35.4	0.21	0.20	0	0	0.79	0.80
35.5–45.4	0	1.00	0	0	1.00	0
$\geq$ 45.5	0	–	0	–	1.00	–

There can be little doubt that both state/size and age are important determinants of survival and reproduction for many higher plants. Ideally it would always be preferable to monitor the fate of individual plants according to both their age and their size (Law, 1983), but this has seldom been attempted (see Werner & Caswell, 1977). Such large samples and such detailed monitoring are required to calculate all the necessary transition probabilities between the various age and size classes, that the labour involved is daunting. Most studies have analysed population behaviour in terms of either state/size or age. It must be remembered, however, that any analysis based on age or size alone, can only provide a partial description of the population. A population model based on age states provided a better predictor of population change in teasel, *Dipsacus fullonum* (Werner & Caswell, 1977), but it gave no indication of the flux of genets through time. This information is essential for analysing the genetic structure of populations. Similarly, population analyses of herbaceous perennials that monitor only the flux of modules, provide no information on the number of genets in a population.

5.3 Population models

5.3.1 Matrix models

We are now in a position to ask how the various patterns of fecundity and mortality affect the growth of populations. This can be done by using simple matrix models (those unfamiliar with matrices and matrix manipulation should consult Begon & Mortimer, 1981). P.H. Leslie (1945, 1948) was largely instrumental in promoting the use of matrices in ecology, and the matrix which summarizes the age-specific mortality and fecundity schedules of a population is often referred to as the Leslie matrix (Box 5.1). The value of λ calculated from the Leslie matrix is referred to as the 'finite rate of increase of the population', and is the most important single parameter in plant population dynamics. It represents the actual rate of population growth for a population characterized by a given set of survival and fecundity schedules. The value of λ determines whether the population will increase ($\lambda > 1$), decrease ($\lambda < 1$) or remain constant ($\lambda = 0$) in size.

In the best of all possible worlds, where recruitment was un-limited and microsites were freely available, the population could increase at its greatest rate. The value of λ determined under these conditions is used to compute another important parameter of population dynamics, namely the 'maximal intrinsic rate of increase'. This is calculated as:

$$r_{max} = \log_e \lambda$$

and its importance lies in the fact that it summarizes the rate at which the population can recover following disturbance (see May, 1982). The basic matrix model can also be used to describe the population dynamics of plants in which the fecundity and mortality schedules have been calculated in relation to size (Usher, 1966) or life-state (Sarukhán & Gadgil, 1974) rather than age (see Box 5.2).

A knowledge of the rate at which a population increases in size allows us to estimate the level of harvest that can be maintained without depleting the population (Usher, 1972b). This is particularly useful in planning the exploitation of naturally regenerating forest trees. The harvest that can be taken over one period of time if the population size increases from N to λN over that period is given by

$$H = 100\frac{\lambda-1}{\lambda} \qquad (16)$$

where H is expressed as a percentage of the total population. For populations of Scots pine, *Pinus sylvestris*, in Scotland where $\lambda = 1.204$ over a six year period, equation (16) implies that 17% of the individuals can be harvested during every six year period. But while this model allows us to calculate what proportion of individuals can be harvested, it tells us nothing about the size at which the population should be held. This would entail a much greater level of understanding of the dynamics of a species' population, including the degree of density dependence with different size distributions and at different densities.

Estimates of λ for other tree species indicate finite rates of increase very close to unity (Hartshorn, 1975; Pinero *et al.*, 1984) which allow only very low rates of exploitation. The average value of λ for *Astrocaryum mexicanum* calculated from six plots in a tropical rain forest in Veracruz, Mexico was 1.0046. This indicates a relatively stable population size with a doubling time of approximately 150 years. The sensitivity of the finite rate of increase to changes in the life history parameters allows an investigation into which elements of the life table play a key role in determining the numbers in a population (Pinero *et al.*, 1984). In a comparison of the relative importance of seed production and clonal growth in determining λ for three species of buttercup Sarukhán & Gadgil (1974) found that altering the seed production of *Ranunculus bulbosus* (its only means of population growth) had a far greater effect on λ than in *R. acris*, and that the population growth of *R. repens* was much more affected by changes in ramet production. This analysis refers to the growth rate of the total number of modules in the population, and should not detract from the importance of the occasional seedling input into the population in determining the number of genets (Soane & Watkinson, 1979).

Particular emphasis has been placed in this section on the dynamics of populations classified by age or stage because data on the fecundity and mortality of plants have usually been collected in this form. We await the collection of data appropriate for the application of population models classified by both age *and* size/stage (Law 1983).

5.3.2   Difference equations

For species which breed at discrete intervals and have non-overlapping generations (e.g. many annuals), it is possible to relate the

Box 5.1 The Leslie matrix

This is a square matrix defined by the expression

$$
L = \begin{bmatrix}
b_0 & b_1 & b_2 & \ldots & b_{k-1} & b_k \\
P_0 & 0 & 0 & \ldots & 0 & 0 \\
0 & P_1 & 0 & \ldots & 0 & 0 \\
\cdot\cdot & \cdot\cdot & \cdot\cdot & \ldots & \cdot\cdot & \cdot\cdot \\
0 & 0 & 0 & \ldots & P_{k-1} & 0
\end{bmatrix}
\tag{8}
$$

All the elements of the matrix are zero except for the subdiagonal and the first row. The subdiagonal elements (P's) define the pattern of age-specific survival and the elements in the top row (b's) define the age-specific fecundity. P_i is the probability that an individual aged i will survive to age $i+1$; b_i is the number of seeds produced per plant by a surviving individual in the ith age group during the interval t to $t+1$. We assume that the plants are hermaphrodite, but matrix procedures for dioecious species are described by Usher (1972).

If the population is divided into $k+1$ (0, 1, 2, . . . k) equal age groups corresponding to the transition probabilities in the Leslie matrix, then the number of individuals in each age class can be written in the form of a column vector.

$$
n = \begin{bmatrix}
n_0 \\
n_1 \\
n_2 \\
\cdot \\
\cdot \\
\cdot \\
\cdot \\
n_k
\end{bmatrix}
\tag{9}
$$

This vector summarizes the age distribution of the population at a given time and $\sum_{i=0}^{k} n_i$ gives the total size of the population. In order to calculate the state of the population after one time interval it is necessary only to multiply the column vector by the Leslie matrix.

$$
\begin{bmatrix}
b_0 & \ldots & & b_{k-1} & b_k \\
P_0 & \ldots & & 0 & 0 \\
0 & P_1 & . & 0 & 0 \\
\cdot & \cdot & \cdot & \cdot & \cdot \\
\cdot & \cdot & \cdot & \cdot & \cdot \\
\cdot & \cdot & \cdot & \cdot & \cdot \\
0 & 0 & . & P_{k-1} & 0
\end{bmatrix}
\times
\begin{bmatrix}
n_{0,t} \\
n_{1,t} \\
n_{2,t} \\
\cdot \\
\cdot \\
\cdot \\
n_{k,t}
\end{bmatrix}
=
\begin{bmatrix}
\sum_{i=0}^{k} b_i n_i t \\
P_0.n_{0t} \\
P_i.n_{it} \\
\cdot \\
\cdot \\
\cdot \\
P_{k-1}.n_{k-1,t}
\end{bmatrix}
=
\begin{bmatrix}
n_{0,t+1} \\
n_{1,t+1} \\
n_{2,t+1} \\
\cdot \\
\cdot \\
\cdot \\
n_{k,t+1}
\end{bmatrix}
\tag{10}
$$

If the matrices are represented by symbols then:

$$\mathbf{Ln}_t = \mathbf{n}_{t+1} \tag{11}$$

and

$$\mathbf{n}_t = \mathbf{L}^t \mathbf{n}_0 \tag{12}$$

To give a specific theoretical example (Williamson, 1972) for an iteroparous perennial with three age classes

Leslie matrix × population vector   new population vector

$$\begin{bmatrix} 0 & 9 & 12 \\ \frac{1}{3} & 0 & 0 \\ 0 & \frac{1}{2} & 0 \end{bmatrix} \times \begin{bmatrix} 12 \\ 8 \\ 2 \end{bmatrix} = \begin{bmatrix} (0 \times 12)+(9 \times 8)+(12 \times 2) \\ \frac{1}{3} \times 12 \\ \frac{1}{2} \times 8 \end{bmatrix} = \begin{bmatrix} 96 \\ 4 \\ 4 \end{bmatrix}$$

After repeated multiplication we find that oscillations in the relative numbers of individuals in each age class dampen down, and that the *proportion* of individuals in each age class remains the same. This 'stable age distribution' occurs in a population which is either *increasing or decreasing exponentially* (depending upon whether recruitment exceeds mortality), and is a characteristic of the transition matrix. It arises whatever the initial age distribution, n_0. Note also that after repeated multiplication, each age class is increasing at a constant rate, and that multiplying the population vector by the transition matrix is equivalent to multiplying it by a single number (λ). That is where λ is called the 'finite rate of increase' of the population.

$$\mathbf{Ln}_t = \lambda \cdot \mathbf{n}_t \tag{14}$$

number of individuals at time $t+1$ to those at time t by a difference equation of the form:

$$N_{t+1} = \lambda \cdot N_t \tag{17}$$

where λ is again the finite rate of increase of the population and is a measure of the balance between survival and reproduction. If N is the number of reproducing plants then λ takes into account both the number of seeds produced per plant and the survival of those off-spring to maturity. The inclusion of a seed bank into difference equation models is considered by MacDonald & Watkinson (1981). For *Phlox drummondii* (see Table 5.1) Leverich & Levin (1979) estimated λ as 2.42 (i.e. the population will multiply approximately two and a half times per year). Over a number of years population growth will be described by the equation

$$N_t = N_0 \lambda^t \tag{18}$$

or

$$N_t = N_0 \, 2.42^t$$

Box 5.2 The Lefkovitch matrix

A matrix based on stage rather than age is named after Lefkovitch (1965), who first used such matrices to investigate the population dynamics of insect pests classified according to stage. Depending on the time interval, it is possible for an individual to remain in a particular stage category, move up at least one stage class, or even revert to an earlier life-state (Hughes, 1984). In order to predict the number of individuals in each life state, however, it is necessary to assume the existence of a stable age distribution (Vandermeer, 1975).

In the example outlined below, the population vector lists the number of plants in each *state* or *size* category. For the creeping buttercup *Ranunculus repens* Sarukhán & Gadgil (1974) divided the population into five life states.

$$
\begin{bmatrix}
a_{ss} & 0 & 0 & a_{fs} & a_{gs} \\
0 & 0 & 0 & 0 & a_{gr} \\
a_{sa} & 0 & 0 & 0 & 0 \\
0 & a_{rf} & a_{af} & 0 & 0 \\
0 & 0 & 0 & a_{fg} & a_{gg}
\end{bmatrix}
\begin{bmatrix}
s \\ r \\ a \\ f \\ g
\end{bmatrix}
\tag{15}
$$

where s refers to seeds, r to ramets, a to non-flowering adults, f to flowering adults and g to flower and ramet producing adults. In the stage matrix the diagonal and subdiagonal elements define the probabilities of transition from one stage to another: those in the leading diagonal (a_{ss}, a_{gg}) define the probability of remaining within a category, those in the first subdiagonal (a_{af}, a_{fg}) refer to the probability of moving from one life state to the next, whilst those in the second subdiagonal (a_{sa}, a_{rf}) define the probabilities of transition to the next category but one in the population vector. The three top right hand elements define the number of seeds produced by a flowering adult (a_{fs}) and a flower and ramet producing adult (a_{gs}) together with the number of ramets produced by clonal growth (a_{gr}).

The Lefkovitch matrix can be used to predict the numbers in each state/size class at any subsequent census (as long as the conditions for the model hold). It may also be used to calculate the population's finite rate of increase and stable size distribution. For populations of *Ranunculus repens* in a North Wales pasture Sarukhán & Gadgil (1974) estimated λ to have a value of 1.04, indicating a relatively stable population size or one that is only just increasing. In contrast populations of *R. bulbosus* had a much greater capacity for increase ($\lambda = 1.603$). The application of the Lefkovitch matrix to the growth of individual plants in populations of *Dryas octopetala* is discussed by McGraw & Antonovics (1983).

It is plain to see that this equation leads to a population which, like the matrix models of Section 5.3.1, grows exponentially for ever!

There are two important flaws in all the models discussed so far. First, population size cannot increase indefinitely, and some or all of the parameters in the life table must be density-dependent. This is the subject of the next section. Second, the models assume that survival and reproduction are constant from year to year, and are not affected by changing physical conditions. Detailed study of the winter annual *Bromus tectorum* by Mack and Pyke (1983) has shown convincingly that the survival of seedlings varies greatly both between years and between different cohorts of the same generation, depending on year to year variation in the environment.

5.4 Population regulation

It was emphasized in Section 5.1.2 that in order to understand what determines the number of individuals in a population, it is not sufficient just to record survivorship and fecundity. It is also necessary to find out how the various elements in the life table depend upon the number and sizes of individuals in the population. Ideally, this should be carried out by experimental manipulation of natural populations

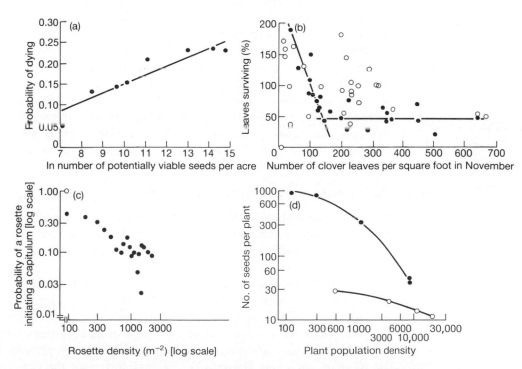

Fig. 5.14. Density-dependent processes in four plant populations. (a) Mortality in a population of sugar maple establishing from seed (Hett, 1971). (b) The effect of defoliation by wood pigeons on the survival of clover leaves in mid winter; 1962 (●) and 1963.(○) There is significant density dependence in 1962 but not in 1963 (Murton *et al.*, 1966). (c) The relationship between rosette density and the probability of flower-head initiation in *Hieracium pilosella* (Bishop & Davy, 1984). (d) The influence of flowering plant density on seed production per plant by *Salicornia europaea* on a low (●) and high (○) marsh (Watkinson & Davy, 1985).

to densities above and below their existing levels. If this is not practicable, it may be possible to measure plant performance at a range of naturally occurring densities. However, interpretation of this type of study requires caution because differences in plant performance due to density will be confounded by microenvironmental differences between sites (Antonovics & Levin, 1980).

Examples of density dependence have been recorded in all aspects of plant demography, including seed dispersal (Baker & O'Dowd, 1982), germination (Inouye, 1980), seedling establishment (Hett, 1971), survival during the vegetative phase of the life cycle (Symonides, 1983), leaf survivorship (Murton *et al.*, 1966), flower formation (Bishop & Davy, 1984) and the number of seeds set per plant (Watkinson, 1985a); see Fig. 5.14. Variations in plant performance with the level of crowding may result from factors both internal and external to the population. For example, an increase in population density may lead to a larger number of individuals competing for limited resources than at low densities. As a consequence fewer resources will be available to individual plants, fewer modules will be iterated and individual plant size and fecundity will be lower. At extremely high densities the death rate of modules on some plants may exceed the birth rate, and whole genets will die, the number increasing with density. Thus intraspecific competition for limiting resources could lead to both reduced survivorship and fecundity. The level of crowding in a population can also affect the level of herbivore and pathogen activity. For example, Augspurger & Kelly (1984) found that mortality due to damping-off disease increased with seedling density in populations of a tropical tree, *Platypodium elegans*. Likewise Murton *et al.* (1966) found that the grazing of clover by pigeons increased at high densities of leaves.

5.4.1 Plant competition and its course through time

The more closely plants are crowded together the more they interfere with each other's acquisition of the same essential resources. As a consequence, individual plants tend to be smaller in high density populations and to have a greater risk of death.

The various factors influencing the yield of individual plants are discussed in Chapters 11 and 12, and this section is restricted to a consideration of two important generalizations that have emerged from the study of yield-density relationships in monocultures. The first relates to the negatively density-dependent relationship between mean plant size and density. The second relates to the density-dependent mortality of plants that results from self-thinning.

Consider a set of plant monocultures, each grown under the same habitat conditions, but sown at different densities (Fig. 5.15). As crowding increases, the same amount of resources have to be shared

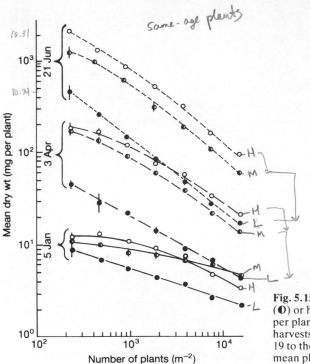

same-age plants

Disproportionate decline between High/Low

clonal?

Fig. 5.15. The influence of density and low (●), medium (◑) or high (○) nutrient regime on the shoot dry weight per plant of *Vulpia fasciculata* at three successive harvests. The fitted curves represent the best fits of Eqn. 19 to the data. Note how the lines become steeper as mean plant size increases. From Watkinson (1984).

between a larger and larger number of plants, so that after a given period of time plant size decreases with planting density. There is a reciprocal relationship between mean plant dry weight (w) and the density of surviving plants (N) of the form

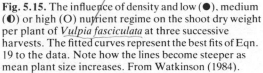

$0 < b \le 1$

$$w = w_m (1+aN)^{-b} \qquad = \frac{w_m}{(1 + aN)^b} \qquad (19)$$

The parameter w_m can be interpreted as the mean yield of an isolated plant, a as the area required to achieve a yield of w_m and b as the efficiency of resource utilization (Watkinson, 1980, 1984). Note that the equation is of the same form as that used by Hassell (1975) to describe the density dependent net growth rate from generation to generation in populations of insects, and to distinguish contest competition (b = 1) from scramble (b ≃ ∞). All three parameters increase with time, the value of b increasing from 0 to 1. It is only when the value of b = 1 that the total yield per unit area (B) becomes independent of plant density and the commonly observed 'law of constant yield' is found (Shinozaki & Kira, 1956). That is

$$B = wN = w_m (1+aN)^{-1} \qquad (20)$$

For large values of N

$$B = w_m a^{-1} \qquad (21)$$

per unit area

← Values of b > 1 lead to a decline in total yield at high densities.

Although equation 19 refers to the mean dry weight of plants, this

w decreases faster than N increases

 yield

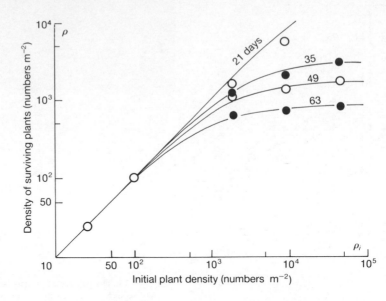

Fig. 5.16. Density-dependent mortality in buckwheat, *Fagopyrum esculentum*. The relationship between initial plant density and the density of survivors after various periods of time, showing the onset of density dependent mortality between 21 and 35 days. From Yoda *et al.* (1963).

is not to imply that there is an approximately equal partitioning of resources between individuals (scramble competition). Rather a hierarchy of exploitation develops (contest competition) that results in the unequal partitioning of resources between individuals and the development of size hierarchies with a few large individuals and numerous small ones (see Chapter 4). Such inequalities develop more quickly in high density populations, and as the individuals continue to grow, the capacity of some individuals to absorb competition by plastic responses is exceeded. Once this point has been reached the plastic response of plants may be augmented by self-thinning and density-dependent mortality. The highest density populations begin to suffer mortality first, with the smallest individuals in the size hierarchy being the first to die (see Chapter 4). As growth proceeds, mortality affects those populations sown at progressively lower densities. The relationship between the initial density of plants (N_i) and the density of surviving plants (N) in a series of populations sown over a wide range of densities (Fig. 5.16) can generally be described by the equation

$$N = N_i(1+mN_i)^{-1} \tag{22}$$

where m^{-1} is a parameter that represents the maximum density of plants that a habitat can support (Yoda *et al.*, 1963). This density decreases with time.

If a population of plants that is undergoing density-dependent mortality (self-thinning) is monitored through time, it is found that the relationship between the number of surviving plants (N) and the mean dry weight of the survivors can be expressed by the empirical

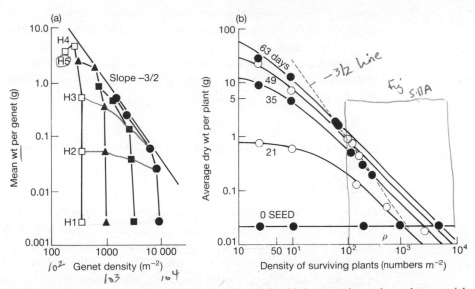

Fig. 5.17. Self-thinning in plant populations. (a) The changing relationship between the number and mean weight of surviving plants in populations of the grass *Lolium perenne* sown at four different densities, over four harvests (H1–H4). A thinning line of slope $-3/2$ is drawn to show the limiting condition which all the populations approach over time (Kays & Harper, 1974). (b) The relationship between weight and density of surviving plants at successive harvests in buckwheat, *Fagopyrum esculentum*, showing the interrelationship between the reciprocal equation curves, as described by Eqn. 19 (solid lines), and the $-3/2$ power law as described by Eqn. 24 (dotted line). From Yoda *et al.*, (1963).

equation (Yoda *et al.*, 1963):

$$w = cN^{-k} \tag{23}$$

or

$$\log w = \log c - k \log N$$

The meanings of the parameters c and k are described below. It can be seen that populations sown at very high densities and monitored through time will follow a trajectory with a straight line of gradient $-k$ when weight is plotted against density on logarithmic scales (Fig. 5.17a). Conventionally mean plant weight is plotted against density, but since it is an increase in weight that results in the death of individuals it might be more appropriate to plot log N against log w or log N against log B (where biomass B = N×w).

Note that equation (23) describes how plant weight varies with density in self-thinning populations that are monitored through time, whereas equation (19) describes the relationship between plant weight and the density of surviving plants for populations of the same age at a given time. The relationship between the two equations can be clearly seen in Fig. 5.17b where the thinning line (equation (23)) truncates a series of lines created by the reciprocal equation (19). Combinations of weight and density higher than the points of intersection cannot be realized because of the effects of self-thinning. The thinning line essentially separates permissible combinations of numbers and biomass, from combinations which cannot be realized.

As k in equation (23) typically has a value of approximately 1.5 (ranging from 1.3–1.8 for different species; White, 1980), it is often referred to as the −3/2 power law (Yoda *et al.*, 1963) or self-thinning rule. Since

$$w = cN^{-3/2} \tag{24}$$

it follows that the total yield per unit area (B = wN) will follow the trajectory

$$B = cN^{-1/2} \tag{25}$$

$B = \dfrac{c}{N^{1/2}}$ at the ⁻³/₂ trajectory

This means that in thinning populations total yield per unit area will increase as the number of individuals declines. It also follows that small individuals may occur at higher densities than large ones. Yield cannot, however, increase indefinitely. There is a limited amount of evidence (Lonsdale & Watkinson, 1982; Watkinson, 1984) that once the maximum standing crop of the population has been reached and there is no further increase in plant height, the thinning line changes from a slope of −3/2 to a slope of −1. This is the stage at which the further growth of survivors is only made possible by the death of others and the growth exactly compensates for the loss.

The −3/2 power law has frequently been used to describe the time trajectory of a population undergoing density dependent mortality, but it more generally defines the boundary line for those combinations of weight and density that are possible in plant populations. For a wide variety of species (White, 1980) the value of $\log_{10} c$ lies between 3.5 and 4.4 (where weight is measured in gram per plant and N in plants per square metre) although the value may be higher for some grass species (Lonsdale & Watkinson, 1983). This narrow range holds true over an impressive seven orders of magnitude of plant density and eleven orders of magnitude of mean biomass per plant. All combinations of weight and density are potentially possible to the left of the lines in Fig. 5.17 but none are possible to the right.

The wide applicability of the −3/2 power law to plant populations has led to studies on how shading (e.g. Kays & Harper, 1974, Lonsdale & Watkinson, 1982; Westoby & Howell, 1981), fertilizers (Yoda *et al.*, 1963) and plant architecture (Lonsdale & Watkinson, 1983) affect both the gradient and intercept of the thinning line. Other studies have shown that the −3/2 power law applies to mixtures of plants, but only when the component species are considered collectively as a single population (Bazzaz & Harper, 1976; Malmberg & Smith, 1982; Juloka-Sulonen, 1983). The applicability of the 'law' to modules has also been considered (Hutchings, 1979; Lonsdale & Watkinson, 1982; Pitelka, 1984).

Numerous attempts have been made to explain why thinning lines have a slope of $-3/2$ (Yoda *et al.*, 1963; White, 1980; Pickard, 1983; Charles-Edwards, 1984; Westoby, 1985). None of them is entirely satisfactory, but it appears that the $-3/2$ exponent relates to the way that three dimensional objects can be packed onto a two dimensional surface.

It is generally argued that the ground area or space (s) occupied by a plant is related to a square of the linear dimension of the plant (s α l^2) whilst canopy volume or mass relates to a cube of some linear dimension (w α l^3) thus leading to the observed $-3/2$ relationship between mass and density (w α $N^{-3/2}$), since the space available to each plant is inversely related to density (s α N^{-1}). This approach has been criticized (White, 1981; Westoby, 1984) but it would appear to be correct if height is taken as the characteristic linear dimension (Givnish, 1986). The relatively restricted range of values for c, which has the dimensions of $g/m^2/m$, can be interpreted in terms of the rate at which the amount of support tissue per unit canopy area, required to ensure mechanical stability, increases with plant and canopy height (Givnish, 1986).

5.5 Population models incorporating density dependence

5.5.1 Difference equations

A model to describe the population growth of annual plants with discrete generations needs to take account of both the plastic reduction in plant size that occurs with density (equation (19)) and self-thinning (equation (22)). The model proposed by Watkinson (1980) incorporates these two elements as follows:

$$N_{t+1} = \frac{\lambda' N_t}{(1+aN_t)^b + m\lambda' N_t} \tag{26}$$

where λ' is the finite rate of increase of the population and is a product of the mean number of seeds produced by plants in isolation (λ) and the probability of surviving density independent mortality (p). The parameters a, b and m are as previously defined in equations (19) and (22). Although this equation might appear complicated at first sight it is essentially a simple extension of the exponential growth equation

$$N_{t+1} = \lambda' N_t \tag{27}$$

that incorporates two density dependent elements. There is a term $(1+a\,N)^{-b}$ to describe the plastic reduction in plant size that occurs with density, and a term for self-thinning (m). In reality the plastic reduction in plant size and increased mortality in high density

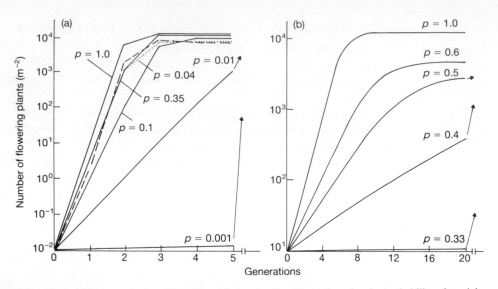

Fig. 5.18. Population changes calculated from Eqn. 26 showing the effects of varying the probability of surviving density-independent mortality, p, on the rate of colonization and equilibrium population size in (a) *Salicornia europaea;* and (b) *Vulpia fasciculata*. The actual colonization curve for *S. europaea* on the Lauwerszeepolder, Netherlands, is shown as a dashed line in (a). If it is assumed that p declines from 0.35 to 0.04 after 2 generations, then the model closely mimics the growth of the natural population. These predicted values are very close to those actually observed in the field. From Watkinson & Davy (1985).

populations both result from interference between neighbours, but it is convenient here to describe the two processes separately.

How much does this equation tell us about the population dynamics of plants in the field? Estimates of the parameters in equation (26) have been made for populations of three annuals: *Vulpia fasciculata* on fixed dunes in North Wales; *Salicornia europaea* on a low marsh in Eastern England; and *Cakile edentula* on dunes in Nova Scotia, Canada (Watkinson & Davy, 1985). Figure 5.18 shows the population growth curves for *Salicornia* and *Vulpia*. It can be seen that *Vulpia* takes a considerable number of years to reach equilibrium and that populations are extremely sensitive to changes in the level of density independent mortality. If the probability of survival is less than one third the population goes extinct. In contrast *Salicornia europaea* reaches an equilibrium density within three years if density independent mortality is low. Comparison of the theoretical growth curves for *Salicornia* in England with the observed colonization curve given by Joenje (1978) for the Lauwerszeepolder in the Netherlands indicates that the model mimics the growth of natural populations rather closely.

In describing the population dynamics of *Salicornia* and *Vulpia* by a simple equation we have assumed that fecundity and survival vary only with intraspecific competition and not with other factors such as grazing or physical conditions which may vary from year to year. Nevertheless it is possible to predict approximately and explain the densities that occur in natural populations of *Salicornia* (Jefferies *et*

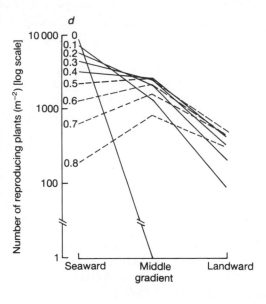

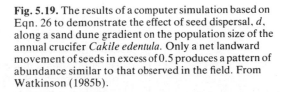

Fig. 5.19. The results of a computer simulation based on Eqn. 26 to demonstrate the effect of seed dispersal, d, along a sand dune gradient on the population size of the annual crucifer *Cakile edentula*. Only a net landward movement of seeds in excess of 0.5 produces a pattern of abundance similar to that observed in the field. From Watkinson (1985b).

al., 1981) and *Vulpia* (Watkinson & Harper, 1978) on the basis of the interaction between the density dependent control of fecundity and the level of density independent mortality. It is variations in the level of density independent mortality over space and time that result in variations in the abundance of both species.

This is not the case, however, for *Cakile edentula*. Keddy (1981) found that the abundance of the plant on sand dunes was typically greatest in the middle of a dune ridge and lower towards the seaward and landward ends. Along this gradient the levels of fecundity and mortality were such that the populations could be maintained only at the seaward end of the gradient. On the middle and landward sites the level of mortality exceeded reproductive output, so the populations would not have been able to persist in isolation. A computer simulation of the populations (Watkinson, 1985b) shows that the plants at the middle and landward ends of the gradients exist only because of immigration from the seaward end. There is a high annual dispersal of seeds (> 50%) landward from the highly fecund plants on the beach (Keddy 1982). This example (Fig. 5.19) neatly illustrates the complex interactions between various density dependent and density independent processes that affect the abundance and performance of plants. It is one of the few studies to examine dispersal explicitly as an agent determining population abundance. In another study, Burdon and Chilvers (1977) have shown that the exponential increase of pines in an area of eucalypt forest resulted almost entirely from a progressive increase in the dispersal of seed from a pine plantation.

Box 5.3 Dynamics of a model population of *Poa annua*

Law (1975) modified the basic Leslie matrix to describe the population dynamics of *Poa annua* by making seed production and seedling survival density-dependent, and by allowing for the existence of a seed bank.

$$\begin{bmatrix} p_s & 0 & b_1(N) & b_2(N) & b_3(N) \\ g_s & 0 & 0 & 0 & 0 \\ 0 & p_0(N) & 0 & 0 & 0 \\ 0 & 0 & p_1 & 0 & 0 \\ 0 & 0 & 0 & p_2 & 0 \end{bmatrix} \qquad (27)$$

where p_s is the probability of a seed surviving in the seed bank, g_s is the probability of germination, $p_0(N)$ is the probability of seedlings surviving to become young adults, p_1 is the probability of young adults becoming mature adults and p_2 is the probability of mature adults becoming old adults; $b_1(N)$, $b_2(N)$ and $b_3(N)$ are the number of seeds produced by young, medium and old adults respectively. The parameters with subscripts (N) are assumed to be density dependent. In each case, the functions are assumed to decrease monotonically with increasing density. The dynamics of this model are depicted in Fig. 5.1b.

5.5.2   Matrix models

Our understanding of how intraspecific competition affects plants of different ages and sizes is unfortunately limited. Yet this type of information is essential if we are to understand the dynamics of plants with overlapping generations. The responses to crowding described in the previous section were concerned entirely with mean plant performance. Certainly we have some understanding of size hierarchy development in monocultures (see Chapter 4) but for natural populations we have little information on how plants of different ages and sizes respond to crowding.

Small plants presumably perceive the level of crowding to be much higher than their larger neighbours. Indeed Law (1975) found that the survival of seedlings to young adults in populations of *Poa annua* was density dependent whilst the survival of older plants was not. In contrast, the age specific fecundity of three adult categories (young, medium, old) of plants was density dependent in each case, although the medium plants were the most fecund. An appropriate matrix model to describe the population dynamics of *Poa annua* (Law, 1975 in Begon & Mortimer, 1981) categorized into four age groups of approximately eight weeks duration (seedling, young adult, medium adult, old adult) together with a seed bank is described in Box 5.3. Iteration of the model with appropriate parameter values gives an indication of how a population establishing with 100 seeds grows towards an equilibrium population density with a stable age structure after eighteen time periods (Fig. 5.1b). Compare this simulated colonization curve with the actual colonization curve (Fig. 5.1a) observed by Law (1981).

5.6   Interactions in mixtures of species

So far we have considered the population dynamics of single species as though they occurred in an ecological vacuum. Natural communities, however, contain an assemblage of species that may interact in a great many ways. The number is almost endless but if the interactions are classified on the basis of the effect that the population of one species has on another then the number is greatly reduced. Essentially the population of one species may cause an increase ($+$), decrease ($-$) or have no effect (0) on another species, resulting in five different types of two-way interaction (Williamson, 1972), the most important of which are $-$ $+$ (plant/herbivore or plant/pathogen), $-$ $-$ (competition) and $+$ $+$ (mutualism). Commensalism ($+$ 0) and amensalism ($-$ 0) have received far less consideration than the other three types of interaction and are perhaps better considered as highly asymmetric forms of mutualism and competition respectively (see Chapter 4).

5.6.1   Interspecific competition

Where interspecific competition occurs for a resource which is in limited supply, the above definition leads us to expect that there will be a reciprocal reduction in the fitness of individuals of both species. Nevertheless it is possible that the individuals of one species may be very much more affected than those of the other; competition in this case is said to be one-sided or asymmetrical. Many studies have focused on the growth and survival of only one of the species in a mixture, so it is impossible to define the nature of the interactions between them. Moreover, few studies have investigated how the survival and fecundity of plants are affected by the presence of a second species. Thus despite the vast literature from both experimental and field studies on how plants interact in mixtures to affect yield or final biomass (see Harper, 1977), there is remarkably little that is relevant to understanding how interspecific competition affects the abundance and dynamics of plant populations.

Our knowledge of how interactions between plants may affect abundance comes from perturbation experiments in which species have been selectively removed from natural communities (e.g. Putwain & Harper, 1970; Abul–Faith & Bazzaz, 1979; Fowler, 1981). The removal of *Spartina patens* from a high marsh in North

Carolina by Silander and Antonovics (1982) produced a large and significant increase in the abundance of *Fimbristylis spadiceae* alone among six potential competitors. Similarly the removal of *F. spadiceae* produced a large increase in the percentage cover of *Spartina patens*, indicating reciprocal, symmetric competition. In contrast, the removal of *S. patens* on the low marsh had much less effect on the abundance of *F. spadiceae,* and no effect on the dominant *Spartina alterniflora.* Four of the subdominant species (including *S. patens*) increased by almost equal amounts when *S. alterniflora* was removed. The interaction between *S. alterniflora* and *S. patens* therefore appears to be highly asymmetric.

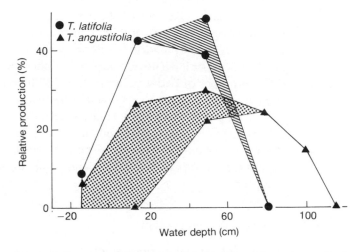

Fig. 5.20. The potential and realized distributions of the reedmaces *Typha angustifolia* (▲) and *T. latifolia* (●) expressed as relative production along a gradient of increasing water depth. The shaded areas show the loss of production suffered in the presence of the other species. From Grace & Wetzel (1981).

Evidence for the importance of competition in determining plant abundance also comes from the study of zonation along environmental gradients. Transplantation of two species of reedmace along a gradient of increasing water depth showed that *Typha latifolia* could grow only at the shallower end of the gradient in the absence of competition, whereas *T. angustifolia* was capable of growing over the entire gradient (Fig. 5.20). Strongly asymmetric competition, however, restricted *T. angustifolia* to the deeper areas (Grace & Wetzel, 1981).

Strong evidence that the results of interspecific competition may be influenced by abiotic conditions, and that these are instrumental in determining the zonₐtion of species along environmental gradients, also comes from a study by Sharitz & McCormick (1973) on the population dynamics of *Minuartia uniflora* and *Sedum smallii,* two annual plants evident on granite outcrops in the south-eastern

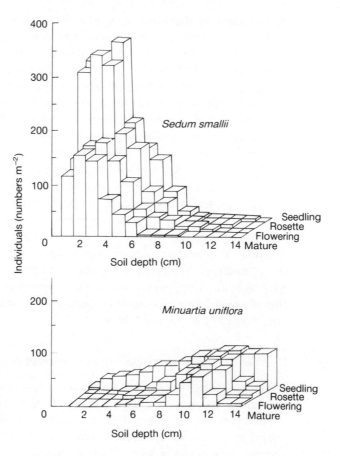

Fig. 5.21. The zonation of *Minuartia uniflora* and *Sedum smallii* on granite outcrops in Georgia, USA, according to soil depth at four stages of the life cycle. From Sharitz & McCormick (1973).

U.S.A. Figure 5.21 shows that the abundance and distribution of adult plants is strongly influenced by soil depth with *Minuartia* restricted to the deeper and wetter soils. Seedlings of the two species germinate over a much wider range of conditions, but competition results in a much smaller proportion of the seedlings of *Minuartia* surviving on the shallower soils and *Sedum* on the deeper soils. Experimental studies show that both species can tolerate a wide range of soil conditions and that it is interspecific competition that results in the observed zonation in the field.

All of these studies indicate that interspecific competition is important in determining plant abundance. The study by Silander and Antonovics (1982) also shows that the intensity of competition may vary greatly among pairs of species. In a species-rich grassland in North Carolina, Fowler (1981) found that the community was characterized by relatively weak but approximately equal competitive interactions amongst all of its component species. Goldberg & Werner (1983) argue that competitive interactions are often likely to be non-species specific and that there will be a large degree of equivalence in the effect of species of similar growth form on the ability of any particular species to establish within a community. Thus Peart (1982) found that grass seedling establishment and growth in monocultures of different grass species was dependent more on the biomass in the patches than upon which species dominated the patch. Similarly both Werner (1977) and Gross (1980) found that it was the abundance and biomass of particular growth forms (e.g. grasses, perennial dicots, shrubs etc.) they removed, rather than the identity of individual species, that determined seedling success in populations of *Dipsacus sylvestris* and *Verbascum thapsus* respectively.

A final note of caution is necessary. Where the removal of one species results in an increase in the abundance of another, it is tempting to assume that it is competition which determines the relative abundance of species. It is possible, however, that the removal of one species from a community leads to a reduction in the abundance of shared herbivores, and that it is the reduction in herbivore feeding, rather than reduced competition, which leads to the observed increase in the abundance of other species in the community (e.g. Parker & Root, 1983).

5.6.2   Competition between species: an experimental approach

Most experiments on plant competition have involved two species mixtures. In such a mixture competition between neighbouring plants may occur either between plants of the same species or members of different species. However, although intra- and inter-specific competition are different aspects of the same general phenomenon of interplant competition, they have generally been treated in isolation. The study of intraspecific competition has primarily involved an examination of the response of plants to varying density, whilst the study of competition in mixtures has generally involved growing pairs of species at a range of relative frequencies but at a constant overall density (so called 'replacement series' or 'de Wit series' after de Wit, 1960). Such substitutive experiments have led to considerable insights into the nature of plant interactions but their usefulness is considerably diminished by the fact that the outcome of competition can be predicted only in situations where total density is held constant.

An alternative, additive design where the density of one species is held constant while the density of the other is varied, has been used to mimic situations where a crop is subjected to different levels of weed infestation. In the past this experimental design has been criticized because the effects of frequency and density are confounded so that it is impossible to separate the effects of intraspecific from interspecific competition (e.g. Harper, 1977). More recently, however, a number of models have been proposed that allow us to analyse experiments where the densities of both species are varied (e.g. Watkinson, 1981b;Wright, 1981; Spitters, 1983). A consideration of equation (19) shows that the relationship between the mean yield per plant (w) and the density of surviving plants (N) in monoculture (Equation 19), can easily be extended to a two species mixture (N_1 and N_2) so that the effect of a second species on the yield of the other can be modelled by the equations

$$w_1 = w_{m1}(1+a_1(N_1+\alpha_{12}N_2))^{-b_1}$$

$$w_2 = w_{m2}(1+a_2(N_2+\alpha_{21}N_1))^{-b_2}$$

(28)

where α_{12} and α_{21} are competition coefficients that define the equivalence between species (Watkinson, 1981). Figure 5.22 shows the density response of two varieties of pea sown over a wide range of densities in pure populations, and also in the presence of a constant density of wild oats, *Avena fatua* (Butcher, 1983). It can be seen that the model provides a very good fit to the data and that the wild oats have a greater impact on the yield of the leafless pea variety Filby, than on the conventional pea variety Birte.

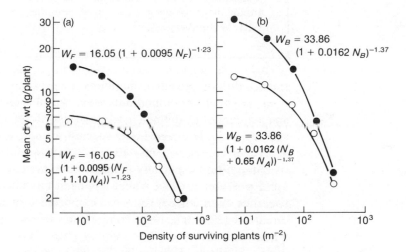

Fig. 5.22. The relationship between shoot dry weight and density of surviving plants in (a) leafless pea, var. Filby (F); and (b) conventional pea, var. Birte (B), in pure stand (●), and in the presence of a fixed density N_A of the weed *Avena fatua* (O). The curves represent the best fit of Eqns. 19 and 28. From Butcher (1983).

The competition coefficient for wild oats grown with the leafless pea indicates that a single wild oat plant is equivalent to 1.10 pea plants. To a certain extent this competition coefficient is illusory and it may even obscure the real mechanisms of competition (Pianka, 1981). It is certainly not a constant between a given pair of species, and may vary with the relative germination time of the two species or

Box 5.4 Multispecies competition models

If it is assumed that the competition coefficients between species are independent of one another, then equation (28) can be extended to predict the yield of species i in a community of n species

$$w_i = w_{mi}(1+a_i(N_i + \sum_{j \neq i}^{n} \alpha_{ij}N_j))^{b_i} \tag{30}$$

Rearranging gives

$$w_i = W_i(1+A_iN_i)^{-b_i} \tag{31}$$

where

$$W_i = w_{mi}(1+a_i \sum_{j \neq i}^{n} \alpha_{ij}N_j)^{-b_i}$$

and

$$A_i = a_i(1+a_i \sum_{j \neq i}^{n} \alpha_{ij}N_j)^{-1}$$

Equation (30) is identical in form to equation (19) which is the model that has been applied to yield density data for a range of annual plants growing in natural communities (Watkinson & Davy, 1985; Watkinson, 1985a).

the conditions in which the plants are growing (Chapter 2). Moreover, it should be abundantly clear that a single individual of one species cannot be defined in terms of so many equivalents of a second species, as individuals of both species vary so greatly in size (Chapter 4). Nevertheless, such models provide a powerful tool for the analysis of competition especially in agricultural systems. For example, the yield per unit area of wheat has been modelled by equation (28) together with an expression for mortality in mixtures based on equation (22) for a large combination of sowing densities of both wheat and the weed *Agrostemma githago* (Firbank & Watkinson, 1985). Similarly, Firbank *et al.* (1984) have modelled the impact of various levels of weed infestation by *Bromus sterilis* on the yield of winter wheat.

Such models can be extended to investigate the population dynamics of plants in multi-species communities (Box 5.4). For example, the best fit of equation (30) to data on the number of seed produced per plant (S) and the surviving density (N) of *Vulpia fasciculata* on sand dunes in North Wales is provided by the equation

$$S = 3.05(1+1.93 \ 10^{-4} N)^{-0.90} \tag{29}$$

This relationship describes the density response of *Vulpia* growing in a matrix of other annuals and prostrate perennials. The fact that this negatively density-dependent relationship between fecundity and density varies little over time or space despite considerable changes in the associated vegetation perhaps indicates that there is little competition between *Vulpia* and its associated species. Alternatively it might indicate the degree of competitive equivalence between species of similar growth form as suggested by Goldberg and Werner (1983).

5.7 The influence of herbivores

Evidence that herbivores (animals that eat living plant tissues) have a considerable impact on the population dynamics of plants comes from monitoring the fates of individuals, and from experimental comparisons of the birth, death and dispersal rates of plant populations vulnerable to, or protected from herbivores. Early in the nineteenth century Darwin (1859) followed the fate of 357 marked seedlings on a freshly dug piece of ground and found that 295 were killed, chiefly by slugs and snails. More recently it has been shown that mortality due to seed-feeding animals may be anything between 0–100% in a wide range of habitats (e.g. Janzen 1969; Sagar & Mortimer 1976). Animals that feed on seeds and seedlings have the most obvious effect on the survival of plant genets although occasionally herbivores may cause high mortality to adults. Laws *et al.* (1975) found that elephants had killed almost 96% of the mature trees in a *Terminalia glaucescens* woodland, for example. In general, however, mature plants are much more resistant to herbivory because most herbivores remove only parts of the plant or tap resources, leaving other plant parts that are capable of regeneration through the iteration of new modules. The impact of herbivory on the death rate of mature plants alternatively may be indirect. For example, defoliation of ragwort *Senecio jacobaea* by cinnabar moth *Tyria jacobaeae* greatly increases the mortality of plants from drought and frost (Harris, 1974), and defoliation of dock *Rumex crispus* by the beetle *Gastrophysa viridula* reduces root growth and so increases the probability that the plants will be washed away during storms (Whittaker, 1982).

Animals which remove plant parts externally by biting or chewing (e.g. rabbits, snails), suck resources from individual cells or from the plant's vascular system (e.g jassids, aphids), mine into their hosts (e.g. leaf mining larvae) or form galls (e.g. Cynipid wasps), have all been shown to influence fruiting, growth and form. The primary impact of herbivores on the growth of mature plants is through their effects on the birth and death rates of modules (Chapter 9). Comparisons of the fate of damaged and undamaged leaves on seedlings of five tropical tree species shows that the mortality of leaves is significantly increased if the leaf has previously been damaged by herbivores (Dirzo, 1984). Altering the birth and death rates of modules has a profound influence on the number of modules iterated and the age structure of these modules (e.g. Parker, 1985). As a consequence plant growth, form and fecundity are generally much more dramatically affected by herbivores than is the survivorship of individual genets. An example of the direct impact of herbivores on plant growth can be seen in Fig. 5.23. Here the wood growth of *Eucalyptus stellulata* is seen to increase 2–3 times after the exclusion of defoliating sawflies with insecticide (Morrow & La Marche, 1978). Similarly, bushes of scotch broom, *Cytisus scoparius*, sprayed regularly with insecticides, survived better and produced more seeds (Fig. 5.24) than the unsprayed bushes (Waloff & Richards, 1977).

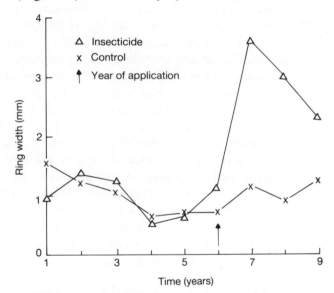

Fig. 5.23. The effect of defoliation by sawflies on wood growth in *Eucalyptus stellulata*, as shown by a comparison of insecticide sprayed (△) and control (×) trees. The year in which the insecticide was applied is indicated by an arrow. From Morrow & LaMarche (1978).

There is a considerable literature (see Crawley, 1983, for a review) that demonstrates the effects of herbivores on the survival, growth and reproduction of plants. It is one matter, however, to show that herbivores may have these effects, but quite another to show how important they are in determining plant abundance and dynamics. Some have argued that herbivores may regulate the abundance of plants (e.g. Brues, 1946) and that herbivores are in turn food limited. Others, in contrast, have argued (e.g. Hairston *et al.*, 1960) that herbivore abundance is determined by predators and parasites, and that herbivores are typically too scarce in relation to their food supply to have any impact on plant abundance. In fact there is probably a continuum between these two extremes, with the bulk of examples lying somewhere in the middle (Crawley, 1983).

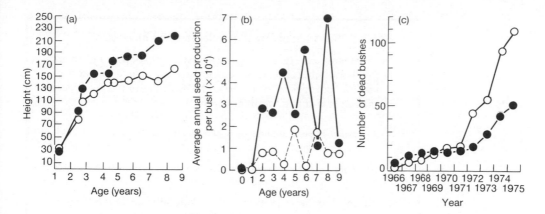

Fig. 5.24. The effect of removing plant-feeding insects by regular spraying with insecticides on (a) height; (b) fecundity; and (c) death rate of broom, *Cytisus scoparius*. Sprayed (●) and unsprayed (O) bushes. From Waloff & Richards (1977).

Simple monitoring of populations is not sufficient to demonstrate the role that animals play in determining plant abundance. High levels of seed predation may have no influence on the abundance of plants if seed densities are not reduced to levels below those which the population would have been reduced later as a result of competition, or when recruitment is microsite-limited. Likewise, the *absence* of obvious signs of herbivore damage does not mean that herbivores are unimportant in the dynamics of a plant population (Harper, 1977). It is only by perturbing natural populations that the role of herbivores in determining vegetation structure can be understood. Both plant and animal numbers may be altered, but the most common perturbation experiments involve the exclusion or introduction of herbivores.

The best documented cases of plant-herbivore dynamics following introductions come from the biological control literature where animals that feed on one species of plant have been introduced deliberately to control weeds (the target weeds, too, are generally introduced). One of the earliest successes of biological weed control involved the virtual elimination of prickly pear, *Opuntia vulgaris*, from Ceylon in the mid 19th century, following the introduction of a cochineal insect, *Dactylopius ceylonicus*. This was followed by an equally spectacular success in the late 1920s, in the eradication of *O. inermis* and *O. stricta*, from many parts of Australia after the introduction of the moth *Cactoblastis cactorum* from South America (Dodd, 1940). Today the moth still maintains the cactus at a low, stable equilibrium density and much of the area previously occupied by *Opuntia* is now in agricultural use (Monro, 1967).

The weed biocontrol systems provide examples of cases where a herbivore regulates the abundance of a plant and is itself food limited (Box 5.5). In contrast, the feeding of the cinnabar moth on ragwort

Box 5.5 Plant-herbivore dynamics in weed biocontrol

In cases where there is a reciprocal relationship between plant and herbivore numbers (i.e. the interaction is more or less symmetrical), it may be appropriate to represent the dynamics of the two species by differential equations of the form:

$$\frac{dV}{dt} = \text{change in plant abundance} = \text{plant gains} - \text{plant losses}$$

$$\frac{dN}{dt} = \text{change in herbivore abundance} = \text{herbivore gains} - \text{herbivore losses}$$

The most familiar equations of this kind are the Lotka/Volterra predator-prey equations, and numerous modifications of these can be employed to describe different kinds of plant-herbivore interactions (see Crawley, 1983).

The nodding thistle *Carduus nutans* for example, infests thousands of hectares of grazing land in Canada and the United States. A weevil *Rhinocyllus conicus* which feeds in the seed-head of the thistle, was imported from France, and released against the weed in 1969. The insect gradually built up in numbers and, after three years began to reduce thistle abundance. Weevil densities peaked six years after release and then fell off to what appears to be a stable equilibrium (Fig. 5.25). Thirteen years after release of the control agent, plant abundance had been reduced to less than 1/30th of its former level on one site and 1/450th on another (Harris, 1984).

The dynamics of these changes in plant and animal numbers can be described by making the following assumptions about gains and losses:

$$\frac{dV}{dt} = 0.5V \, \frac{100-V}{100} - 0.7N \, \frac{V}{V+8}$$

$$\frac{dN}{dt} = 0.5N - 0.25 \, \frac{N^2}{V}$$

The biology of the model is as follows. A scarce population of thistles would increase by a factor of 1.65 per year ($\lambda = \exp 0.5$), and would come to an equilibrium of 100 thistle heads/m^2 in the absence of weevils. The second term describes the impact of weevil feeding on plant dynamics; there is a functional response ($V/(V+8)$ see Hassell, 1978) when thistle heads are scarce, but when they are abundant, each weevil reduces seed heads at a rate of 0.7/m^2/year. The second equation describes the dynamics of the herbivore. It also increases by a factor of 1.65 per year, but density dependence operates on the insect through its mortality rate, which increases linearly with insect numbers *per plant* (i.e. with N/V).

A comparison of the two graphs shows that the model describes the pattern of change in reasonable qualitative terms. This does not mean, however, that the processes in the field are necessarily the same as those assumed in the model. Many different kinds of model could fit the data equally well. The value of these mathematical

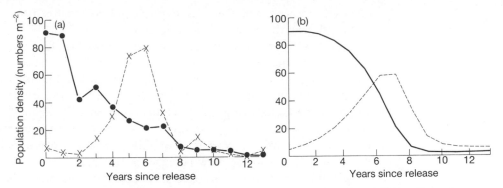

Fig. 5.25. Biological control of nodding thistle, *Carduus nutans* (solid line), by the weevil *Rhinocyllus conicus* (dashed line), in Canada. (a) Field data showing the build-up in beetle numbers to a peak 6 years after release, and a sustained reduction in thistle density over the first 8 years. (b) A simple model describing the increase of the insect and the decline of the weed. The encouraging fit between the model and the data does not mean that the model is necessarily a correct *explanation* of the dynamics.

models is not that they *explain* an observed pattern of dynamics, but that they are able to demonstrate the consequences of theoretical assumptions about population dynamics quite unambiguously. Understanding the *causes* of population dynamics demands manipulative experiments carried out under field conditions. What we learn from the model is that *if* the observed dynamics are a 'pure' plant-herbivore interaction, then: 1) the plant must experience density dependent limitation in the absence of the weevil; and 2) the weevil must experience density dependent mortality (or reduced fecundity) when herbivore numbers per plant are high in order for this pattern of numerical change to occur.

The model only tells part of the story, of course. For example, there are interesting changes in the size structure of thistle populations following control. Large, biennial rosettes predominated in the dense pre-release populations, whereas more plants grow as summer annuals in sparse populations following weevil attack. Again, the maintenance of low thistle numbers is closely linked to the establishment of a grass sward, and competition with pasture grasses is one of the main factors in the decline of the thistles (Harris, 1984).

provides one of the best examples of an asymmetric plant-herbivore interaction. On dry, sandy soils the numbers of caterpillars (Fig. 5.26) are food-limited, but the number of plants is determined largely by the weather conditions after seedling establishment (Dempster, 1971, 1975; Dempster & Lakhani, 1979; Meijden, 1979). Certainly the caterpillars regularly defoliate ragwort plants, but the plants can compensate for defoliation by regrowth and by the production of a second crop of flowers (Islam & Crawley, 1983; Crawley & Nachapong, 1985). If anything, defoliation tends to increase the number of rosettes, because damaged plants produce new rosettes from root buds and from the crown of the rootstock.

In many cases the plants and the herbivores which feed on them may have rather little impact on each other. For instance, most plant-eating insects are scarce relative to the availability of potentially limiting resources, due to the combined impact of density independent factors (e.g. bad weather) and natural enemies (e.g. insect parasitoids, insect predators, birds, pathogens; Strong *et al.*, 1984). For example, the exclusion of birds from understory shrubs of *Acer pennsylvanicum* in New Hampshire (Holmes *et al.*, 1979) has shown that the densities of externally feeding Lepidopteran larvae are significantly decreased by avian predation. More dramatically, plant-eating insect pests have been depressed to about one hundredth of their former abundance following successful biological control programmes (Beddington *et al.*, 1978).

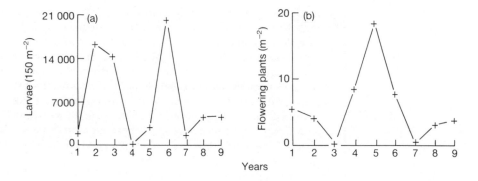

Fig. 5.26. Changes in (a) cinnabar moth, *Tyria jacobaeae*, density; and (b) numbers of flowering plants of ragwort, *Senecio jacobaea*, at Weeting Heath in eastern England, over a nine year period (data from Dempster & Lakhani, 1979). The number of larvae emerging each year is determined primarily by the abundance of flowering plants in the previous year. The number of flowering plants is determined largely by weather conditions at the time of germination and seedling establishment.

The exclusion of large herbivores from grasslands vividly demonstrates the effects that non-specialist grazing animals may have on the abundance of individual species (Mack & Thompson, 1982, and see Chapter 1). For example, Watson set up exclosures in the Serengeti in 1963. Four years later the grasses *Andropogon greenwayi* and *Sporobolus marginatus* which made up 56 and 20% of the standing crop respectively in the grazed areas were not found at all in the exclosures. Two tall grasses of the genus *Pennisetum* formed 98% of the standing crop in comparison with 8% outside the exclosure (McNaughton, 1979).

Unfortunately the dynamics of plants are rarely monitored following the exclusion of herbivores from an area. An exception is provided in a study on nutrient poor grassland by Bishop and Davy (1984). Rabbit exclusion had little impact on the rosette density of *Hieracium pilosella*, but significantly reduced the rate of turnover of rosettes resulting from clonal growth. The death of rosettes was related to herbivore feeding in a rather complex way, since only 11% of losses could be attributed directly to rabbit activity. Rosettes of *Hieracium* are semelparous, and senescence following flower initiation is the single most important cause of death in both grazed and ungrazed areas. However, the probability of flower initiation is higher in the grazed areas and, as a consequence, mortality is greater too. The higher mortality in the grazed areas was thus inextricably linked to the higher rate of flower initiation, as was the recruitment of rosettes, since clonal growth is also coupled to flower initiation. In other cases it is likely that the activity of herbivores may considerably alter the competitive relations between species (e.g. Bentley & Whittaker, 1979). The influence of herbivores on plant abundance within a community will then clearly depend on the selectivity of the grazer, the pattern of grazing and the competitive interactions between plants (Harper, 1977).

5.8   The influence of pathogens

Plant pathogens can be considered as tiny herbivores that leave their waste products inside the plant. Indeed if a plant is considered as a complex of food niches (Harper, 1977) it is clear that these have been exploited with remarkable parallelism by fungal pathogens and plant eating insects. It might be expected, therefore, that pathogens would have rather similar effects on the population dynamics of plants to those of herbivores. There are numerous reports in the literature of the effects of pathogens on the growth, survival and reproduction of agricultural and forestry crops (Klinkowski, 1970, Dinoor & Eshed, 1984), but there is surprisingly little information for other species. White rust (*Albugo candida*) and downy mildew (*Peronospera parasitica*) have been shown by Alexander and Burdon (1984) to have significant effects on the survival and reproductive output of affected individuals of *Capsella bursa-pastoris*. The impact of the diseases was dependent on the age of infection. Systemic infection of young seedlings from seed-borne or soil-borne spores resulted in the death of approximately 90% of plants while secondary infections later in development had little effect on mortality. However, fruit production of the infected plants was negatively correlated with disease severity.

In a study of tree seedling survival in *Platypodium elegans* on Barro Colorado Island, Panama, Augspurger (1983) found that damping-off disease was the major cause of death. The effects of the disease were again age-specific since the vulnerability of seedlings to damping-off disease decreased with age. During the first three months after emergence damping-off disease accounted for 64–95% of deaths. Subsequent experiments (Augspurger & Kelly, 1984) demonstrated that both an increase in dispersal distance from the parent tree and a decrease in seedling density reduced levels of the disease (Chapter 4). Distance between plants will generally be critical in determining infection since the spread of pathogens is usually passive. As a consequence the rate of pathogen spread is slower in low density stands of plants (Burdon & Chilvers, 1975) and the chance of seedling establishment will be greatest some distance away from mature plants where seedling densities are low (see Chapter 4). This is in marked contrast to many herbivores which can search out their food plants with uncanny ability (Williams, 1981).

Like herbivores, however, the degree of host-specificity (selectivity) varies enormously from those that will infect a wide variety of hosts with varying susceptibilities to those that are host-specific. It is the introduction of host-specific pathogens, either accidentally or deliberately, rather than the monitoring of natural populations that illustrates the enormous impact that pathogens can have on the abundance of plants. The accidental introduction of *Endothia parasitica* from south-east Asia to North America has resulted in the virtual elimination of chestnuts *Castanea dentata* from North American forests (Burdon & Shattock, 1980). Similarly the accidental introduction of dutch elm disease *Ceratostemella ulmi* into Britain has destroyed large numbers of elm trees throughout the country.

The use of host-specific pathogens in biological weed-control programmes provides some of the best evidence for the potential impact of pathogens on plant populations. The best documented case is provided by the introduction of the rust fungus *Puccinia chondrillina* to control skeleton weed (*Chondrilla juncea*) an important weed of arable land in much of south-eastern Australia (Burdon *et al.*, 1981). Skeleton weed is an apomictic composite with numerous different forms in its native Mediterranean region where typical stand densities are less than 10 plants/m² (Wapshere *et al.*, 1974). Only three types of the weed (Types A, B, C) have been found in Australia, and of these Type A had the most widespread distribution prior to 1971. In 1971 a single race of the rust was introduced from the Mediterranean to a number of sites in south-eastern Australia in an attempt to control the weed. This race had no effect at all on Types B and C but it rapidly spread and reduced densities of Type A from 200 to 10 plants/m² (Cullen & Groves, 1977). The rust destroys seedlings and rosettes, heavily damages flowering shoots and reduces seed output. Following the decline of Type A, however, the abundance and distribution of Types B and C have increased. It can be assumed, therefore, that competition from plants of Type A previously restricted the abundance of Types B and C prior to 1971. Certainly the development of competitive interactions between Types A and C is greatly influenced by the occurrence of the rust

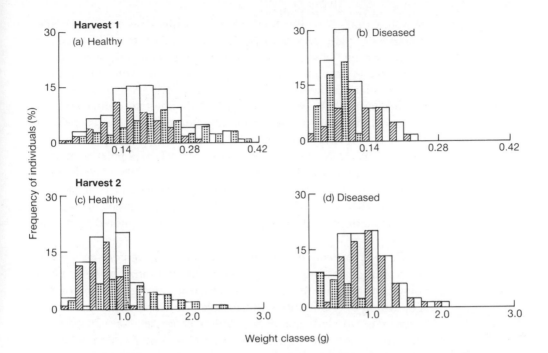

Fig. 5.27. The influence of pathogens on plant performance. Frequency distributions of individual plant weights during vegetative growth (Harvest 1) and at flowering (Harvest 2), from a competition experiment in which susceptible (dotted) and resistant (hatched) genotypes of skeleton weed, *Chondrilla juncea*, were grown in 50:50 mixtures. In (a) and (c) the plants were grown without the pathogen, while (b) and (d) received regular inoculations of the rust *Puccinia chondrillina*. Frequency distributions of individual plant weights for the population as a whole are also shown (open bars). In the presence of disease, all the largest plants were of the resistant genotype, while in the absence of the pathogen, the largest plants were of the susceptible genotype. From Burdon *et al.*, (1984).

under experimental conditions (Fig. 5.27; Burdon *et al.*, 1984). This study illustrates beautifully how an interaction between pathogen activity and a plant's competitive ability can potentially determine the abundance and population dynamics of a plant species.

5.9   Concluding remarks

For a long time, studies on the population dynamics of plants lagged behind those on animals, but an explosion in plant population studies over the last two decades has resulted in a considerable closing of the gap (although this is not yet reflected in general ecology texts). An enormous literature has emerged on the patterns of survival and reproduction in plant populations through a simple monitoring of the fate of individuals. But this straight-forward demographic approach is not enough. It is also necessary to take an experimental approach that involves deliberate perturbation of populations and study of their response. Only then can causal relationships be firmly established. The number of experimental studies is still too meagre for us to claim that we understand what determines the abundance and dynamics of plant populations. This is particularly true when it comes to elucidating the significance of mutualistic, competitive and plant-herbivore interactions. Our understanding of the role of mutualism in particular is almost non-existent (except in the cases of pollination studies and plant-ant interactions). It is not that mutualisms are unimportant or uncommon in nature. Rather it is the inconspicuous nature of many mutualisms (e.g. plant/mycorrhizal fungi and legume/*Rhizobium*), the inadequacy of standard population models in describing mutualistic interactions (see May, 1981; Vandermeer 1984) and perhaps an inappropriate socio-political climate in the past (Risch & Boucher, 1976), that have led to scant attention being given to the role of mutualistic interactions in the population dynamics of plants.

Central to an understanding of population dynamics is an appreciation of the importance of the spatial distribution of plants, both in relation to one another and also to their herbivores and pathogens. Yet attempts to quantify pattern have proved remarkably sterile in helping us to understand population processes (see Chapter 1). Despite the pioneering work of Watt (1947), population biologists are only just beginning to collect significant data on the spatial and temporal dynamics of plant populations. The study of Symonides (1983) is exceptional in providing data on both the spatial and temporal flux of the annual crucifer *Erophila verna*. Studies on neighbour relations (Chapter 4) and dynamic morphology (Chapter 9) undoubtedly have a major role to play in furthering our understanding of the factors that determine the patterns of abundance and dynamics which we are only now beginning to explore.

Chapter 6   Ecology of Pollination and Seed Dispersal

HENRY F. HOWE & LYNN C. WESTLEY

6.1   Introduction

It is easy for us to understand how animals like ourselves compete for mates, care for young, and generally enhance their contribution of genes to future generations (Hamilton, 1964). It is difficult for us to 'think like a plant'. Yet, if we are to understand plants, we need to shift our frame of reference to that of a sedentary and mindless organism which cannot choose where it lives, know its mating success, or assist its offspring for more than a minute fraction of their potentially long lives. The rules of the game of life are very different for plants than they are for those animals which are capable of decision and movement.

6.2   Challenges of a sedentary existence

Consider the reproductive problems of organisms as different as a squirrel and an oak. The squirrel actively searches for food and for a place to live, while the oak must grow where an acorn happens to land. A squirrel looks for a mate, but an oak casts millions of pollen grains to the wind, and attempts to snare the pollen of other individuals on its stigmas. A squirrel nourishes and protects its young until they stand a reasonable chance of survival, and perhaps 1 in 15 juvenile squirrels survives to reproductive age. Acorns are doomed unless a squirrel removes them from beneath the oak *and* forgets to eat them; perhaps 1 in 100,000 acorns survives to reproductive age. If a squirrel fails to breed in its second year, only one or two additional chances remain before its genetic inheritance is lost forever, while an oak may have hundreds of years in which to leave its offspring.

   This comparison underscores both the constraints, and in a sense the potential, of reproduction among stationary and uncomprehending organisms. High fecundity and long life give plants a hedge against disastrous losses of vulnerable seeds and seedlings. Yet the sheer scale of the reproductive potential of many plants opens up opportunities which are not available to many animals. For example, an extraordinarily successful mammal might raise several healthy youngsters in a lifetime, but a single tropical fig, capable of producing 5,000,000 seeds in a year, could leave tens of thousands of vigorous saplings if it grew at the edge of a landslip where conditions favoured its rapidly growing saplings (Garwood *et al.*, 1979). Even an oak, with its much lower fecundity, could leave hundreds of reproductive off-spring if it happened to be in the right place at the right time. Put another way, the behavioural characteristics that allow a squirrel to find a mate, raise her young, and minimize the losses to her repro-ductive efforts can only 'do her so much good'. The extravagance that allows a plant to hedge against huge losses of seeds and seedlings, carries with it enormous potential for parental fitness (Williams, 1975).

In short, the rhythm of plant reproduction may seem both un-hurried and profligate to impatient and exacting animals like our-selves. But it leads to fascinating questions. We want to explore the surprisingly varied ways in which plants mate and produce vigorous offspring, against apparently enormous odds. Why, for instance, do some plants rely on wind or water for the dissemination of pollen or seeds? Why do others rely upon animals for the critical processes of pollination and dispersal, when those animals are intent only on a meal from flowers or fruits? The reproductive interests of the plant would appear to be at odds with the interests of the hungry animal. The aim of this chapter is to explore the dynamics of pollination and seed dispersal in order to bring these ecological questions into bold relief, and suggest some routes towards answers.

6.3   Adaptive trends

Flowers disperse pollen, while fruits are dispersed and also provide for the embryos; however, the parallels between pollen and seed dispersal can be deceptive. A plant may lure pollinating insects with nectar and a showy flower, or may attract a fruit-eating bird with a bright and nutritious berry, but the pollen and the berry have quite different 'targets' (Table 6.1). The target for pollen from one flower is the receptive stigma of another (Chapter 7), while the target for a fruit is an establishment microsite (Chapter 5). Whereas seeds rarely

Table 6.1. Contrasts between pollen and seed dispersal, from the plant's perspective, inspired by Wheelwright & Orians (1982).

	Pollen dispersal	Seed dispersal
Target	Stigma of conspecific flower	Any site suitable for establishment
Animal motivation to target	Collect fragrances, nectar, and/or pollen	None: seeds are ballast to be discarded
Cues to target	Flower colour, shape, fragrance	None
Advantage to visitor constancy	High: ensures correct pollen transfer	High or low: depends on availability of effective or destructive agents

have a specific target, they gain some probablistic advantage by leaving the vicinity of the parent (Section 6.5.2 and Chapter 4). Most pollen grains fail because they never find their target, while many seeds fail *after* they reach a potentially suitable, if ill-defined, target.

Animals *exploit* flowers and fruits. Strictly speaking, no animals are adapted *for* pollination or seed dispersal. They are adapted for feeding on nectar or pollen, or, in odd cases like the orchid bees (*Euglossinae*), for collecting fragrances. Some animals possess morphological, behavioural or physiological traits which both help them to feed and, coincidentally, promote pollination or seed dissemination. Other animals carry pollen or seed without delivering either to the appropriate targets. A discussion of pollination or seed dispersal from the plant's perspective is a discussion of a reproductive process. A parallel discussion of animal traits which influence pollination and seed dispersal is a discussion of foraging which is only coincidentally associated with plant reproduction.

6.3.1   Flowers and pollinators

In the best of all plant worlds, an individual would disperse all its pollen to receptive stigmas, and would capture genetically superior pollen on each of its own stigmas. The myriad adaptations of flower form and function attest to the evolutionary attempts to approach this theoretical maximum. In the best of all animal worlds, a pollinator uses pollen or other floral rewards to further its own fitness. Its adaptations 'for pollination' are really adaptations for perceiving, foraging for, and eating either the male gametes (pollen) or other bribes that the plant may offer.

The ultimate goal of angiosperm pollination is fertilization (the union of a haploid sperm nucleus of a generative cell with a haploid egg cell). Pollination occurs when: 1) a pollen grain lands on a conspecific stigma; 2) the tube cell extends as a pollen tube to the nucellus; 3) one generative cell fertilizes the egg cell to form a diploid zygote; and 4) the second generative cell unites with the polar nuclei to form a triploid endosperm nucleus. A mature ovule is a seed, consisting of a diploid zygote or embryo, a triploid endosperm invested with nutrients from the parent, and diploid cells derived from those surroundings the original nucellus (see Raven *et al.*, 1981, for a full description). This is the fundamental process, assisted by petals, sepals and nectaries to attract, guide and reward pollinators. The pollinator environment has forged some order out of the vast range of potential adaptations for pollen transfer (Table 6.2). At least in a general way, flower colour, shape, size and reward are adapted to broad taxonomic associations of pollinating agents. Similarly, the pollinators' characteristics match the challenges of a huge variety of 'flower types'.

Table 6.2. Pollination syndromes. Derived from Faegri & van der Pijl (1971), Proctor & Yeo (1972), Baker & Hurd (1968). Nectar characteristics from Baker & Baker (1983a, 1983b).

Agent	Anthesis	Colour	Odour	Flower shape	Nectar
Insects the primary agents of pollination					
Beetles	Day and night	Usually dull	Fruity or aminoid	Flat or bowl-shaped; radial symmetry	Undistinguished if present
Carrion and dung flies	Day and night	Purple-brown or greenish	Decaying protein	Flat or deep; radial symmetry; often traps	If present, rich in amino acids
Syrphid and bee flies	Day and night	Variable	Variable	Moderately deep; usually radial symmetry	Hexose-rich
Bees	Day and night or diurnal	Variable but not pure red	Usually sweet	Flat to broad tube; bilateral or radial symmetry; may be closed	Sucrose-rich for long-tongued bees; hexose-rich for short-tongued bees
Hawkmoths	Crepuscular or nocturnal	White, pale or green	Sweet	Deep, often with spur; usually radial symmetry	Ample and sucrose-rich
Settling moths	Day and night or diurnal	Variable but not pure red	Sweet	Flat or moderately deep; bilateral or radial symmetry	Sucrose-rich
Butterflies	Day and night or diurnal	Variable; pink very common	Sweet	Upright; radial symmetry; deep or with spur	Variable, often sucrose-rich
Vertebrates the primary agents of pollination					
Bats	Night	Drab, pale, often green	Musty	Flat 'shaving brush' or deep tube; radial symmetry; much pollen; often upright, hanging outside foliage, or borne on branch or trunk	Ample and hexose-rich
Birds	Day	Vivid, often red	None	Tubular, sometimes curved; radial or bilateral symmetry; robust corolla; often hanging	Ample and sucrose-rich
Primarily abiotic pollination					
Wind	Day or night	Drab, green	None	Small; sepals and petals absent or much reduced; large stigmata; much pollen; often catkins	None or vestigeal
Water	Variable	Variable	None	Minute; sepals and petals absent or much reduced; entire male flower may be released	None

Abiotic gamete dissemination is the ancestral condition in vascular plants (Sporne, 1965), although the forms present in living angiosperms are almost certainly derived. Flagellated gametes of primitive tree ferns, for instance, swam through a film of water to get to the egg in the female archegonium. In comparison, water pollination in modern angiosperms is specialized; for example, in the waterweed *Vallisneria*, the entire male flower is released into the water (Proctor & Yeo, 1972). Likewise, wind pollination (anemophily) is common and perhaps ancestral in gymnosperms, but in angiosperms anemophily is derived. Common wind-pollinated grasses and sedges have greatly reduced floral parts, rather than primitive architecture (Stebbins, 1974); familiar wind-pollinated trees of temperate climates have insect pollinated relatives in the tropics.

Ancient angiosperms were almost certainly insect-pollinated (entomophilous), and beetle pollination is probably the primitive condition in flowering plants (Kevan & Baker, 1983). Contemporary beetle flowers (like the primitive Magnoliaceae) characteristically have little or no nectar, but produce enormous quantities of pollen, much of which is consumed by their visitors. Beetles show little modification for flower visiting beyond occasional elongation of the prothorax and neck; they eat pollen and frequently chew on other floral parts. Indeed, the angiosperm ovary may have evolved as a protection against 'ovule predation' by Jurassic beetles.

Other primitive pollination systems evolved rapidly. The truly repulsive scents of the orchid *Listera cordata* attract fungus gnats, and exemplify an early modification for pollination by flies lured to rotting fungi (Ackerman & Mesler, 1979). In such cases, the only really important attribute of the flies (apart from their capacity to be tricked) is a hairy body which traps pollen. Tubular flower construction, brightly coloured display, copious nectar, and viscous pollen attract long-tongued butterflies which sip nectar rather than eat pollen or flower parts (e.g. *Caesalpinia pulcherrima*; Cruden & Hermann-Parker, 1979). The strongly scented, night-blooming *Silene noctiflora* is a good example of a moth flower, designed to make use of the nocturnal equivalents of delicate butterflies (Proctor & Yeo, 1972). These primitive sorts of pollination would have been commonplace 100 million years ago.

Many familiar adaptations for pollination are rather more recent. Explosive anthers in 'buzz-pollinated' flowers like many *Solanum* species (Buchmann, 1983), and fragrant but nectarless orchids like *Cypripedium* (Dressler, 1981), epitomize specializations among very different plants for bee pollination. Bees, for their part, recognize and forage for distinctive flower shapes and colour patterns, and collect quantities of pollen on bristles, or in basket-like tarsal corbiculae (see von Frisch, 1967). Flower-visiting vertebrates are also relatively recent in the fossil record (Kevan & Baker, 1983). The conspicuous shaving brushes and tough, tubular bat-flowers found in the canopy trees of many tropical forests (e.g. *Pseudobombax* and *Ochroma*; Heithaus, 1982), probably evolved during the last few tens of millions of years. New World fruit bats become thoroughly dusted with pollen while visiting flowers. They only carry pollen from one tree to another accidentally. The understories of those same forests are often decorated by red bird-flowers like the many species of *Heliconia* (Dobkin, 1984), visited by voracious hummingbirds, honeycreepers or sunbirds (Carpenter, 1983). A modern forest or meadow is a living museum of many threads of evolutionary descent.

These are just a few of the thousands of different 'syndromes' which fit important groups of pollinators. Syndromes tell us that some plant groups have faced rather uniform pollination pressures throughout their histories. All magnolias, for instance, have primitive, generalized flowers. Other lineages have faced and responded to a remarkable variety of selection pressures from pollinators. For example Grant and Grant (1965) found that different species of *Phlox* were adapted to attract bees, flies, butterflies, moths, beetles, birds or even bats. As general organizing principles pollination syndromes are very useful indeed.

However, pollination syndromes are fallible for several reasons, and much of the exciting current work in pollination ecology explores these failures. First, the match between pollinator and syndrome is often loose. Even casual inspection of a meadow of flowers indicates that insects often visit the 'wrong' flower types, and that most flowers are visited by many kinds of pollinators (Sections 6.4 and 6.5). For example, it is not unusual for a 'bumble-bee flower' like *Delphinium nelsonii* to be pollinated by hummingbirds (Waser, 1983). Accessibility may be at least as important to foraging animals as the attributes of the flowers themselves. Second, some floral characters are inconstant, and draw different pollinators to different individuals of the same species. Colour or shape may vary within a species, and even within a patch of flowers. For instance, bees favour white morphs of *Succisa pratensis*, while butterflies prefer purple morphs in the same field (Kay, 1982). In at least one montane population of scarlet gilia *Ipomopsis aggregata* in the United States, individual plants shift flower colour from deep red to light pink as the season progresses. This allows them to take advantage of the high availability of hummingbird pollinators early in the season, and the high abundance of hawkmoths late in the season after the hummingbirds have migrated (Paige & Whitham, 1985). In this case, an adaptive adjustment by individual plants takes advantage of *two* syndromes.

More commonly, traits of interest to pollinators are intrinsically variable. Mass-flowering tropical trees bear small flowers in potentially huge numbers (Frankie *et al.*, 1983). In these plants, tree size and the proximity of flowering neighbours, rather than the characteristics of individual flowers, determine which pollinators are attracted. Similarly, nectars vary in the ratio of sucrose to hexoses (glucose and fructose); these sugar ratios show modes for different sorts of syndromes, but there is substantial variation about each mode (Baker & Baker, 1983a). Certain nectars also contain amino acids (Baker & Baker, 1983b) and lipids (Simpson & Neff, 1983) in concentrations sufficient to be nutritious. Variations in these substances are potentially important in attracting different pollinators, but these inherently variable characters also show that plants are not 'locked into' particular syndromes.

Further syndromes are defined as *we* perceive the flowers, not as animals see them. Birds see red but do not smell; bats smell but barely see at all; many insects see ultraviolet light reflected by flowers which to the human eye appear as unpatterned white or yellow (Kevan, 1983). Bees may visit flowers of the wrong apparent colour simply because *we* do not see the colour that matters to the foraging insect. Similarly, it is difficult for us to imagine the landscape of smells that guides a nectivorous bat through the darkness of an African rainforest. Finally, pollination of a species by markedly different agents may be adaptive. Grasses and oaks, for example, are normally wind-pollinated, but some of these species also attract insects to their flowers, indicating at least the potential for insect-mediated pollen dispersal. In evolutionary terms, labile syndromes may be a hedge against the failure of any particular agent of pollen dispersal.

6.3.2 Fruits and frugivores

In the best of all possible plant worlds, an individual would ensure the efficient dispersal of its embryos and provision them with enough lipid, protein and starch to give each offspring a start in life. In the best of all frugivore worlds, an animal would gorge itself on a single plant, feeding on pulp, but discarding the inedible seeds as useless ballast directly below the parent. In the real world, therefore, dispersal and provisioning conflict. For instance, a fruit with a heavy seed ballast is less likely to be eaten by a fruit-eating bird than a lighter fruit on the same or a neighbouring tree. Even if animals are not involved in dispersal, a heavy seed falls closer to the parent than a light one. Yet the larger the seed, the more vigorous the seedling. In such a trade-off of vagility and seed provisioning, there will be no such thing as the 'perfect seed' from the vantage of *both* seed dispersal and seedling vigour, simply because a seed cannot be large and small at the same time. Provisioning of a seed with a large, nutritious endosperm also carries the risk that an animal will consume and digest the seed itself before, during or after dissemination.

Plants meet the challenges of dispersal, provisioning and protection in a variety of ways (Table 6.3). Frugivores, however, do *not* have extensive modifications for dispersing seeds, presumably because the animal's interests are served only by eating fruits, and not by increasing the fitness of the plants which produced them.

Most fruits are adapted for dispersal by animals (Howe & Smallwood, 1982). Widespread though insect pollination is, regular seed dispersal by insects like ants (myrmecochory) predominates only in rather special habitats (Section 6.5.3). Most shrubs and trees, and many herbs, bear fruits which attract or stick to vertebrates. Seeds regularly hoarded by rodents or birds often have no dispersal adaptations other than a thickened seed coat and a rounded form.

Table 6.3. Dispersal syndromes. From van der Pijl (1972), Jansen (1983), Wheelwright *et al.* (1984), Gautier-Hion *et al.* (1985), and Howe (1986).

Agent	Colour	Odour	Form	Reward
		Primarily self-dispersed		
Gravity	Various	None	Undistinguished	None
Explosive dehiscence	Various	None	Explosive capsules or pods	None
Bristle contraction	Various	None	Hydroscopic bristles in varying humidity	None
		Primarily abiotic dispersal		
Water	Various, usually green or brown	None	Hairs, slime, small size, or corky tissue resists sinking or imparts low specific gravity	None
Wind	Various, usually green or brown	None	Minute size, wings, plumes, or balloons impart high surface to volume ratio	None
		Primarily vertebrate dispersal		
Hoarding mammals	Brown	Weak or aromatic	Tough thick-walled nuts; indehiscent	Seed itself
Hoarding birds	Green or brown	None	Rounded wingless seeds or nuts	Seed itself
Arboreal frugivorous mammals	Brown, green, white, orange, yellow	Aromatic	Often arillate seeds or drupes; often compound; often dehiscent	Aril or pulp rich in protein, sugar, or starch
Bats	Green, white, or pale yellow	Aromatic or musty	Various; often pendant	Pulp rich in lipid or starch
Terrestrial frugivorous mammals	Often green or brown	None	Tough, indehiscent often > 50 mm long	Pulp rich in lipid
Highly frugivorous birds	Black, blue, red, green or purple	None	Large arillate seeds or drupes; often dehiscent; seeds > 10 mm long	Pulp rich in lipid or protein
Any frugivorous birds	Black, blue, red, orange or white	None	Small or medium-sized arillate seeds, berries or drupes; seeds < 10 mm long	Various; often only sugar or starch
Animal fur or feathers	Undistinguished	None	Barbs, hooks, or sticky hairs	None
		Primarily insect dispersal		
Ants	Undistinguished	None to humans	Elaiosome attached to seed coat	Oil of starch body with chemical attractant

For instance, pinon pines *Pinus edulis* in the western United States lure seed-eating jays *Gymnorhinus* and nutcrackers *Nucifraga*, to their open cones and large, round, wingless seeds. Nutcrackers collect seeds in specialized throat pouches, and bury them in small caches (Bock *et al.*, 1973). Seeds dropped in transit or forgotten in the caches survive. Some tropical nuts have a pulp which partially satiates a rodent, and perhaps encourages burial rather than consumption (Bradford & Smith, 1977). This pulp represents a step in the evolution of dispersal adaptations because it offers a seed-eating animal something besides the seed itself, thereby reducing the conflict of interest between plant and dispersal agent.

Most spectacular are fruits adapted for consumption by frugivorous birds and mammals (van der Pijl, 1972). Beyond the familiar blue, black and red berries of temperate wayside shrubs, tropical fruits range from the brilliant red, arillate seeds of *Virola* in the nutmeg family (Myristicaceae) to the deep purple-black drupes of *Beilschmiedia* (Lauraceae). Dispersal agents include toucans in tropical America, and huge, flightless cassowaries in the rainforests of northern Australia, respectively (Howe, 1983; Stocker & Irvine, 1983).

Mammal fruits make up for their generally dull colouration with strong scents and often bizarre presentation. The nondescript capsule of *Tetragastris panamensis* (Burseraceae) dehisces to expose a bright purple core, which offsets highly aromatic, arillate seeds, much sought after by monkeys (Howe, 1980). Even more bizarre are the beans of *Swartzia prouacensis* (Leguminosae) which dangle on long threads as an enticement to fruit-eating bats (van der Pijl, 1972). Whatever the particular evolutionary route for such elaborate adaptations, their existence highlights the importance of dispersal to plants.

As with abiotic pollination, abiotic dispersal is a derived condition in angiosperms. Seeds without any evident means of dispersal are widespread, but only in the most severe desert habitats is the majority of species devoid of wings, plumes, holdfasts, or fleshy pulp (Ellner & Shmida, 1981). Water is probably an ancient mode of seed dispersal, but the corky husk, watertight shell and massive endosperm of the coconut *Cocos nucifera* show remarkable specialization for long exposure to salt water. Similarly, in contemporary angiosperms, winged samaras are secondarily derived in both ancient and much more recent families (e.g. Magnoliaceae and Leguminosae). Other exceptions prove the rule. Orchids have minute, dust-like seeds which are dispersed by air currents (Dressler, 1981). Wind dispersal might appear to be a primitive condition, but orchids are amongst the most highly specialized of plants!

Dispersal syndromes suffer many of the same shortcomings as pollination syndromes (Howe, 1985). Plants seem far more creative of structure and ornament than biologists are of categories, and much of the time, fruit-feeding animals seem to eat what they please. This is not surprising, if only because specialization for eating fruits does not greatly restrict the *kinds* of fruits which can be eaten. Frugivorous birds often have wide gapes, relatively short intestines, and sometimes lack a grinding gizzard (Moermond & Denslow, 1985). Highly frugivorous monkeys also have rather shortened guts compared with their leaf-eating relatives (Hladik, 1967). These adaptations for processing large quantities of fruit often provide some margin of safety for the seeds, because they increase the chances of their being discarded unharmed. But these gross physiological and morpho-logical adaptations do not greatly restrict fruit choice. For instance, classic 'bird fruits' such as the New World nutmegs *Virola*, also attract monkeys (Howe, 1983). Many fruits are eaten not only by specialized birds, bats or monkeys, which rarely digest the seeds, but also by less specialized herbivores like cattle and horses, which kill most of the seeds they eat, and leave the survivors in vulnerable heaps of dung (Janzen, 1982; Howe, 1985). The important point is that the animals which we observe eating a particular fruit may or may not disperse the seeds, and may or may not reflect an important evolutionary interaction in the history of the plant.

As with pollination syndromes, however, fruit syndromes have their uses. Syndromes suggest the selective forces with which a group of plants has had to contend. For example, diverse adaptations for dispersal can exist in a single plant family; van der Pijl (1972) has documented adaptations for seed dispersal by birds, bats, wind, water, adhesion to fur or feather, ballistics and simple gravity amongst the 17,000 species of the Leguminosae. Fruit syndromes also provide insights into the ecology of plant and animal interactions in whole communities. For instance, most fruits in the rain forest of Barro Colorado Island, Panama, appear to be adapted for dispersal by birds, bats or hoarding rodents, and these seem to be the important dispersal agents in this forest (Leigh *et al.*, 1982). In contrast, a similar forest in Amazonian Peru has eleven species of primates, and here there is a far higher proportion of yellow 'monkey fruits' (Terborgh, 1983). If nothing else, syndromes signal the kind of evolutionary processes that shaped the history of a group, or the ecological structure of the community in which a plant occurs.

6.4   Tactics and individual success

Pollination and dispersal syndromes illustrate the end results of selection upon the attributes of flowers and fruits. Comparisons between species of higher taxa can only tell part of the story, how-ever. Factors affecting the success or failure of individual plants within a species will often give the clearest impression of the sources of selection that act on flowering and fruiting. These include physical aspects of the environment, animal mutualists that benefit the plants to varying degrees, and other herbivorous animals that interfere with pollen and seed dispersal.

6.4.1   A matter of timing

There are several reasons why individual plants may flower or fruit in or out of synchrony with the other members of their species. For instance, the gloom of a tropical forest understorey probably favours an extended flowering season, rather than an energy-consuming burst of flowering, simply because the plants are severely light-limited (Opler *et al.*, 1979; and see Chapter 12).

A contrasting example is found in the highly seasonal tropical forest of Barro Colorado Island, Panama, where drought-deciduous canopy trees admit light to the understory during a severe dry season from December through April. Rare dry season rains tend to be torrential. A common understory shrub *Hybanthus prunifolius* takes advantage of these storms by bursting into flower within a week after a heavy rain. Flowers last only a day, and the entire flowering season is over in seven days. The seeds are dispersed explosively about one month later. The flowering season of *Hybanthus* is clearly keyed to climate; heavy rains provide both the cue for synchronous flowering and some assurance that the ground will be moist during fruit ripening and seed dispersal.

Augspurger (1980, 1981) found that many bees, butterflies, and hummingbirds visited *Hybanthus* flowers, but that virtually all effective pollinations were carried out by a small solitary bee *Melipona interrupta*. Furthermore, she discovered that fruits were often infested with microlepidopteran larvae which killed the seeds. One might therefore expect that pollination would be most successful when masses of flowers could attract the attention of *Melipona*, and that fruit ripening would proceed successfully only if huge crops of fruit could satiate the fruit-feeding moths. Augspurger carried out an experiment to examine the question of what would happen to shrubs that flowered out of synchrony with their neighbours. She ran a hose through the rain forest to simulate torrential rains, and induced some shrubs to flower well before the rest of the population. These early, experimental shrubs had only a 58% fruit set, as compared with 86% for the naturally synchronous controls; the key pollinating agent apparently did respond to massive displays of *Hybanthus* flowers, and paid less attention to stragglers. Asynchronous fruits did not escape the caterpillars, however. Fruits formed outside the normal fruiting period suffered twice the rate of seed predation (11 versus 5%) of the others (i.e. there was evidence of 'predator satiation' in the synchronous controls). Outliers, therefore, suffered both a reduced capacity to set fruits, and a reduced ability to mature them. Stabilizing selection on pollination and fruiting appears to maintain a strong advantage for flowering synchrony in *Hybanthus prunifolius*.

The *Hybanthus* example indicates that far more than individual reward is involved in flower function. Individual reproductive success is profoundly influenced by the presence of conspecifics in space and time, and individual plants succeed best when they synchronize with others. Of course, different species might find an advantage in *less* synchronous flowering, if, for example, the individuals compete for a rare pollinator. Just this kind of competition, but with respect to the dispersal of seeds by toucans, may account for the higher rates of fruit removal from *Virola* trees fruiting earlier or later than the majority (Manasse & Howe, 1983).

6.4.2   Seed provisioning versus dispersal

By comparison with pollination biology, little is known about intra-specific competition in fruit dispersal, but another example from the Barro Colorado forest does at least suggest some variables which are important to plants.

    Virola surinamensis is an enormous tree of the forest canopy. It is a member of the nutmeg family, and bears capsules the size of limes, each of which dehisces to expose a seed partially covered by a brilliant red aril (Howe, 1983). Large toucans (*Ramphastos swainsonii* and *R. sulfuratus*), guans (*Penelope purpurascens*), and several smaller birds and mammals, eat the arillate seeds, digest the aril, and regurgitate or defecate the seeds intact. This dispersal system is too unwieldy to contemplate the kind of direct experi-mentation that was so successful with *Hybanthus*, but dramatic differences in the extent to which individual trees succeed in dispers-ing their fruit crops does suggest that trees compete for dispersal agents. With *Virola*, an observational approach is required to tease apart the reasons why some trees are more successful in dispersing their seeds than others.

    Assuming that there is some direct advantage to seed dispersal (Section 6.5), the lifetime fitness of a perennial plant is determined (in part) by the sum of the proportion of its seeds dispersed annually, multiplied by the size of each annual crop (see Chapter 5). It is useful to ask whether variation in dispersal success is due to plant charac-teristics which might be moulded by natural selection, or is due simply to population or community characteristics that are beyond the scope of natural selection. In central Panama, *Virola surin-amensis* produces an average 5000 fruits per year. About half of the seeds are removed by birds and half fall directly beneath the parent plant (Howe, 1983). Most importantly, there is tremendous variation from tree to tree in removal success; as few as 13% or as many as 91% of the fruits may be taken away by animals. It turns out that neither variation in crop size, nutrient content of arils, nor different schedules of fruit production influence variation in the proportion of seeds taken from different trees by birds. However, birds are sensi-tive to the amount of ballast that they carry, and tend to prefer trees with small fruits (Fig. 6.1). Not surprisingly, removal success is greatest in trees with high ratios of edible aril to indigestible seed. The tree must trade off investment in edible pulp that will attract birds, against investment in the seed itself. Highly dispersible, small seeds do not produce seedlings as vigorous as those from the less dispersible, larger seeds (Howe & Richter, 1982). A seed cannot maximize both dispersibility and seedling vigour. Presumably, the distribution of seed sizes on trees in the Barro Colorado forest is a compromise reflecting the foraging preferences of the local bird assemblage and the need to provision the seeds so that they have some chance of establishing successfully.

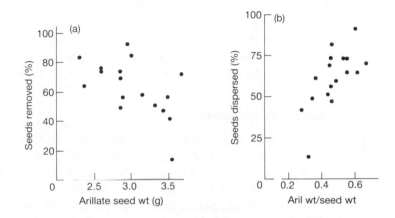

Fig. 6.1. (a) Negative correlation between the percentage of seeds removed and average weight of the arillate seed on each tree of *Virola surinamensis* (r = −0.67; p < 0.002). (b) Positive correlation between the percentage of seeds removed and the ratio of aril weight to seed weight on the same trees (r = 0.72; p < 0.001). From Howe & Vande Kerckhove (1980, 1981).

The trade-off between seed dispersibility and the need to provision seedlings is not unique to bird-dispersed plants (e.g. large winged seeds are less likely to be dispersed by wind than smaller ones), but the foraging habits of birds do add some interesting variables. The relationship between dispersal success and the aril:seed ratio in *Virola surinamensis* is much more pronounced during years (or shorter periods) of fruit scarcity than during fruit abundance in the forest (Howe, 1983). As with the smaller manakins and tanagers (Levey *et al.*, 1984), accessibility of fruits influences toucan foraging. The big birds simply do not bother to search out 'the best' *Virola* trees when food is abundant, but they compete actively for them when fruits are scarce. Competition between neighbouring trees determines the advantage of a particular dispersal 'tactic', with the result that the advantage of producing fruits with large arils and small seeds is only periodic; during plentiful years, plants that cheat the birds do just as well as those that do not.

6.5   Consequences of failure

Perhaps the cardinal lesson from field biology is that the reproductive efforts of plants almost always fail. If an oak and each of its descendents were to produce 100 reproductive offspring (a tiny fraction of lifetime acorn production), we would be blessed with 10^{19} oaks in 10 generations (there are only 10^{22} stars in the known universe)! Obviously, virtually all pollen grains and ovules fail in their ultimate mission, and the virtual certainty of failure encourages the versatility of plant reproductive strategies which we find so fascinating. Some individual plants produce more pollen than others, just as some produce more seeds than others. Plants may trade off the advantage of male reproductive function against the advantages of female function (Chapters 7 and 10). Either genetic predisposition or ecological circumstance can influence the likely success of either tactic.

6.5.1   Pollen failure

A pollen grain may fail to disperse to stigmas, may fail in competition with other pollen grains after reaching a stigma, or it may be rejected by the maternal plant (Willson & Burley, 1983). Each aspect of 'failure' has evolutionary implications.

Some pollen simply fails to leave the male parent; this happens when pollinator activity is low or when the weather is inclement for wind-pollination. Pollen that does leave the parent tends to be distributed in a highly peaked (leptokurtic) distribution, with the mode close to the parent, whether the pollinating agent is wind or animal (Chapter 7). It is easy to understand how this happens when we consider the fate of a pollen grain picked up by an insect from a flower. Assuming that the insect neither eats nor loses its load, a steadily decreasing number of pollen grains from each load is left on each successive flower that the insect visits. If a constant proportion is deposited at each flower visited, there will be an exponential decay in the number of grains left on each successive stigma. If the proportion of grains deposited per visit is high, then most pollen will be left close to the parent. If a plant can somehow ensure a low rate of deposition (e.g. with viscous pollen), then the pollen load is spread out over a greater number of stigmas and a much wider area (Fig. 6.2). Of course a pollinator may leave some, or even most, of the pollen grains on stigmas of the wrong plant species.

The simple failure of pollen dispersal is almost universal, but its magnitude varies between species. A wind pollinated flower may produce 1,000,000 pollen grains for every ovule, while an obligately self-fertilized species may produce as few as three pollen grains per ovule, reflecting extreme differences in the probability of successful pollen donation (Cruden, 1976). In general, species that differ in

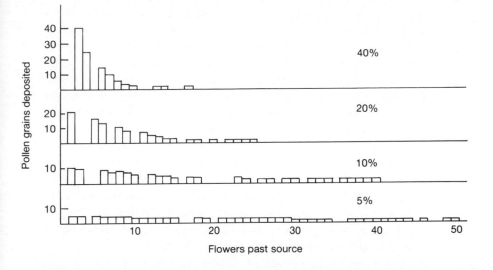

Fig. 6.2. Hypothetical pollen deposition curves, assuming an exponential decay and a constant proportion deposited on each stigma, ranging between 40% (top) to 5% (bottom). The model assumes that some flowers are 'blanks', neither giving nor receiving pollen. After Lertzman & Gass (1983).

opportunities for outcrossing produce predictable, and substantially different, numbers of pollen grains for each ovule (Table 6.4).

Pollen : ovule ratios also vary dramatically within species. Populations of *Caesalpinia* with largely hermaphroditic flowers (88% bisexual versus 12% male) have pollen : ovule ratios of 827, while outbreeding populations (11% hermaphroditic versus 89% male flowers) of the same species have ratios of 9,416 (Cruden, 1976). High levels of insect activity favour maleness and high pollen : ovule ratios, while ineffective pollen dispersal strongly favours plants which can self-pollinate. Quantitative studies of pollinator activity could contribute greatly to our understanding of male function in these plants.

Once the pollen grain reaches a receptive stigma, it must germinate and grow to the ovule. If pollinators are scarce, competition between pollen tubes is relatively light. But the race to the ovary can be extremely competitive. Mulcahy and his colleagues (1983) document pollen competition in *Geranium maculatum*; it takes over an

Table 6.4. Pollen-ovule ratios of plants with different breeding systems. Adapted from Cruden (1977).

Breeding system	Species (N)	Pollen/ovule	($\bar{x} \pm$ S.E.)
Cleistogamy (flower fails to open)	6	5 ± 1	
Obligate selfing (flower open)	7	28 ± 3	
Facultative selfing	20	168 ± 22	
Facultative outcrossing	38	797 ± 88	
Obligate outcrossing	25	5859 ± 936	

hour for an average pollen tube to reach the ovary in this species, so the order of pollen deposition and the timing of pollen germination, are critical. Pollen deposited by the first pollinator visit has an enormous advantage if visits are few and far between. But if pollinations occur in quick succession, markedly superior pollen deposited late may reach the ovule before inferior pollen deposited earlier. Pollen success on the stigma may be due to its own vigour or may be determined by female 'mate choice' (Willson & Burley, 1983). These alternatives cannot be easily separated by field observations, but their interaction certainly produces variation in male success. Bertin (1982b), for instance, found that some trumpet creeper vines *Campsis radicans* were more successful as males than as females, but those with high male success in contributing to the fruit crop had low female success, suggesting a trade off in sexual strategy. Since Bertin pollinated the flowers by hand, the data do not reflect contributions to plant maleness by pollinator movements, but rather they suggest differences in pollen vigour, or choice by the stigma and style of distinctive pollen grains (Chapter 7). It will be interesting to see whether natural pollination accentuates variance in male and female success in *Campsis*. In nature, hummingbirds *Archilochus colubris* deposit ten times as much pollen on the stigmas of *Campsis* as do insects (Bertin, 1982a). Even if all flowers attracted pollinators, those flowers especially capable of attracting wide-ranging hummingbirds would probably function better as males than those flowers dependent on sedentary bees.

6.5.2   Unfertilized ovules

Ovules often fail to yield seeds. Plants like milkweeds (*Asclepias*) may mature as many as 70% or as few as 5% of the fruits they initiate, depending upon the species, year and locale. Cruden & Westley (1986) found an average of 55% fruit set in outcrossing herbs and 11% fruit set in outcrossing trees. This may indicate that the gynoecium rejects 45–89% of its fertilized ovules, or simply that most stigmas fail to catch suitable pollen.

Fertilized ovules may fail to develop for several reasons including: 1) various herbivores may eat them; 2) abortion of damaged fruits is common; 3) the plants may be resource limited, and consistently abort fruits in such a way as to optimize the number of healthy seeds matured (Stephenson, 1981). In such cases the abortion may be quite selective; early or late pollinations may be favoured, or some pollen sources favoured and others excluded.

Optimal mate choice by selective abortion of fertilized ovules is only relevant in plants where several pollen grains reach each stigma, so that the gynoecium has at least a theoretical choice. Jack-in-the-pulpit *Arisaema triphyllum* is a common American aroid, pollinated

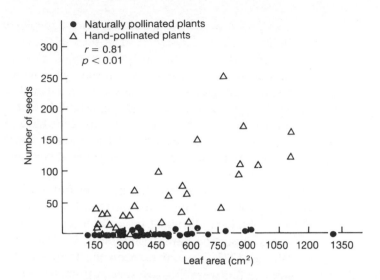

Fig. 6.3. The relationship between seed number and plant size in Jack-in-the-Pulpit (*Arisaema triphyllum*) near Brooktondale, New York, USA. Naturally pollinated individuals (●); hand pollinated plants (△). Only in the hand pollinated treatment did seed production increase with plant size. After Bierzychudek (1982).

by fungus gnats. The size of individual fruit crops is curiously independent of plant size. Bierzychudek (1982) demonstrated pollinator limitation by hand pollinating flowers on plants of different sizes, and comparing the percentages of seeds set with those of similar sized, control plants (Fig. 6.3). The number of seeds that *can* mature may be several times that observed in nature, suggesting severe pollinator limitation. The literature on pollinator limitation is inconclusive, however; some authors argue that pollen-limitation is common (Bierzychudek, 1981) while others contend that reproduction in most species is resource-limited (Stephenson, 1981). Discussion of either optimal mate selection or optimal life history tactics in plants must distinguish ovule failures due to the genetic attributes of pollen from failure due simply to pollen scarcity.

6.5.3 Undispersed seeds

Pollination is certainly a major bottleneck in plant reproduction, but mature seeds are by no means free of danger. While it is true that the vast majority of pollen grains is lost, it is equally important to appreciate that the vast majority of seeds and seedlings is virtually certain to be destroyed by insects, vertebrates or pathogens, or to succumb to desiccation, frost-heave, shade or competition (Chapter 5). The probability of survival can, however, be increased by adaptations for dispersal (Janzen, 1970; Howe & Smallwood, 1982). A seed removed from the vicinity of a parent tree might escape the very high mortality which afflicts dense aggregates of siblings beneath their parents (Chapter 4). The seed might encounter a gap in the canopy where establishment is possible, or perhaps even establish and persist as a stunted juvenile until a break in the canopy allows it to grow. In some species, a dispersal agent might carry the seed to the kind of microsite required for establishment and growth. These alternatives are not exclusive; what they have in common is a view that the consequences of seed dispersal are probabilistic, and that the probabilities may be altered by parental investment in fruit and seed characteristics.

Most seeds are not dispersed far from the parent. The dispersal curve for wind- or animal-dispersed seeds tends to be leptokurtic, while ballistic dispersal typically leads to a normal distribution of seeds about the parent (Salisbury, 1942; Wilson, 1983). Platt and Weiss (1977) present data on the terminal velocity and seed dispersion of several fugitive species from the same prairie habitat (Fig. 6.4); there are pronounced interspecific differences, with mean dispersal distance ranging from 0.4 m in *Mirabilis hirsuta* to 25.7 m in *Apocynum sibiricum*. As terminal velocity declines, so propagule dispersal increases.

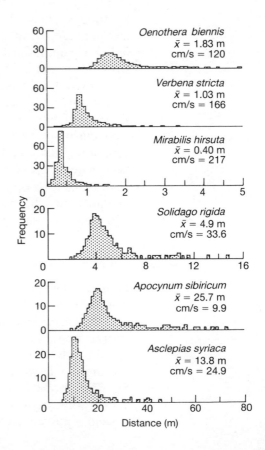

Fig. 6.4. Propagule dispersion of 6 species in 10–15 km/hr winds in the Caylor prairie. Mean dispersal distances $(\bar{x})$ and terminal velocities (cm/s) are given for each species. After Platt & Weiss (1977).

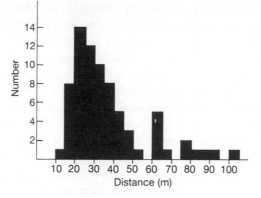

Fig. 6.5. The distribution of walnuts buried by squirrels in relation to the position of the walnut source in a cemetry. After Stapanian & Smith (1978).

Kohlermann (in Geiger, 1950) observed the distribution of 127,000 wind-dispersed propagules in 10 m wide concentric rings around a *Fraxinus* tree; the densities of seeds/m² dropped from 109 closest to the tree, through 30, 11, 7, 5, 2, 1.2, and 0.8 to 0.47 in the ring between 80 and 90 m; only 15% of the seeds fell more than 50 m from the parent. The modal dispersal distances on six species of herbs with ballistic dispersal ranged between 0.2–3.4 m, and few seeds from any species fell more than 4 m from the parent (Stamp & Lucas, 1983). For animal dispersed seeds, Stapanian and Smith provide data on the dispersal of black walnut *Juglans nigra* seeds by 6 squirrels from an experimental pile of 375 nuts (Fig. 6.5); the average walnut was moved 38.1 m and no seed was moved less than 15 m. Nuts were buried by the squirrels between 8 and 16 m from one another.

Water dispersal presents a rather different picture. Waser *et al.* (1982) intercepted seeds downstream from a patch of *Mimulus guttatus*, using soil-filled trays spaced at intervals of 25 m over a distance of 425 m. The trays were brought back to a greenhouse where the emerging seedlings were scored. There was no prolonged dormancy, so the number of seedlings provides a good estimate of the number of propagules in each trap. The number of seedlings averaged one per trap for the entire 425 m length of the sample, and it is clear that many seeds must have been transported beyond this distance. Platykurtic (flattened) seed distributions produced by water dispersal appear to represent an extreme contrast to the highly leptokurtic distributions of wind-dispersed seeds.

Seeds may be moved in a second bout of dispersal, once they have hit the ground. Westoby and Rice (1981) estimate that the median net gain in distance from the parent during the second episode is about 40% of the net distance covered in the second episode (since the seed is as likely as not to be moved *back* towards the parent in this second stage). Radioactively labelled seeds of the dune annual grass *Vulpia fasciculata* placed on the ground, were moved a mean of 3 cm in one site and 10 cm in another over a period of four months. Distance moved was negatively correlated with the amount of ground cover. In most populations, dispersal on the ground was less than that from the plant (this averaged 5 cm over both sites; Watkinson, 1978b).

Although biologists usually measure seed dispersal in absolute distances, it is clear that a distance of 10 m might have different implications for a tree and a herb. It is useful, therefore to express dispersal distances in units of 'plant diameters'; non weedy herbs disperse their seeds about 5 diameters, while weedy herbs and wind-dispersed trees disperse their seed an average of 10–15 diameters (Levin, 1981). In a sense, despite their vast differences in size, the weeds and the trees disperse their seeds approximately equal 'functional distances'. The degree to which such distances are 'functional' assumes, of course, that pathogens and seed predators adjust their depredations on seeds to something like crown diameters of plants. One could imagine a seed-eating weevil or deer searching for food more easily over 1–2 m than 100–200 m, regardless of the numbers of crown diameters involved (see the 'distance versus density' debate on the intensity of seed predation in Chapter 4).

The factors influencing seed and seedling survival are numerous (Chapter 5). We placed cohorts of *Virola* seeds and seedlings in concentric rings about fruiting *Virola* trees in Panama, and monitored their survival (Howe *et al.*, 1985). More than 99.96% of the seeds beneath the crown died within 12 weeks, while those 45 m away were more than 40 times as likely to survive the first three months (Fig. 6.6). Although rodents eat the seeds prior to germination, weevils (*Conotrachelus* spp.) which oviposit on the germinating seeds are responsible for the disproportionately high mortality underneath the parent trees. By 12 weeks of age, the weevil is no longer a threat, but rodents, tapirs, deer and other herbivorous mammals continue to kill the seedlings. *Virola* could not produce recruits without the services of the large toucans and guans which consistently drop seeds well beyond the edge of the crown (Section

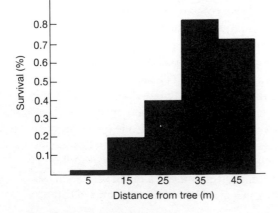

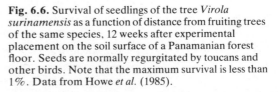

Fig. 6.6. Survival of seedlings of the tree *Virola surinamensis* as a function of distance from fruiting trees of the same species, 12 weeks after experimental placement on the soil surface of a Panamanian forest floor. Seeds are normally regurgitated by toucans and other birds. Note that the maximum survival is less than 1%. Data from Howe *et al.* (1985).

6.4.2). Smaller birds do take some fruits, but they are ineffective dispersal agents because they drop the seeds within 15 m of the fruiting *Virola*, where seed and seedling mortality is all but total.

A single example could be misleading. *Virola* seedlings and saplings are unusually shade tolerant, and its saplings are notably uncommon in light gaps in the forest (Hubbell, in prep). Seedlings and juveniles persist and even grow with 97% of the canopy overhead closed to light penetration (Howe *et al.*, 1985). Seedlings of other species need light. Augspurger (1983a,b) found that the wind-dispersed legumes of *Platypodium elegans* may be found up to 100 m from the parent trees. Most seedlings close to the parent die of fungal infections soon after germination. Within one year, virtually all seedlings are dead, regardless of their distance from the parent, unless they happen to occupy a light-gap. *Platypodium* recruitment is influenced by an interaction between dispersal distance, seedling density and light conditions, but the critical condition is a broken canopy that admits light (Augspurger & Kelly, 1984). In the *Virola* example, seed escape from the parent is the basic advantage to dispersal. In *Platypodium,* seed escape is secondary to the occupation of a large, sunlit gap in the forest (see Chapters 3 and 4).

Some seeds have a rather specific target for dispersal. In a study of seed dispersal by ants in the nutrient-poor, arid bushlands of Australia, Davidson and Morton (1981a,b) investigated recruitment of the shrub *Dissocarpus biflorus,* which bears abundant fruits with ant-attractive food bodies. In this system, the foraging ants *Rhytidponera* carry the seeds back to their mounds, where the seeds are discarded after the food bodies have been stripped off. Compared with the surrounding soil, the ant mounds are rich in nitrates and phosphorus; *Dissocarpus* grows vigorously on the mounds, but is absent or stunted elsewhere (Fig. 6.7). Unlike other plants which grow equally well on or off the mounds, dispersal of *Dissocarpus* seeds must be directed to mounds to be successful. Far from being an undefined place away from the parent (as with *Virola* and *Platypodium*), the targets for *Dissocarpus* fruits are special, discrete habitats that form a rather small proportion of the total habitat. Seed dispersal by ants is common among understory herbs in mesic

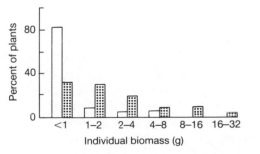

Fig. 6.7. Size distribution of *Dissocarpus biflorus* shrubs growing on and off ant mounds in the saltbush of arid central Australia. Plants off the mounds (open bars) are more skewed towards small size classes than those on the mounds (shaded bars). The largest plants were all found on ant mounds. From Davidson & Morton (1981a).

forests (Handel *et al.*, 1981), but reaches its greatest degree of specialization in the nutrient-poor, arid habitats of Australia and South Africa.

These examples highlight the importance of seed dispersal for a few species of plants. They do not come close to documenting the variety of ecological factors affecting the success or failure of seed dispersal. Many seeds, for instance, are capable of extended dormancy (Chapter 4). For some colonists of forest clearings, like pin cherry *Prunus pensylvanica* in the eastern United States, seed germination occurs years or even decades after seeds leave the trees. A large, clear-cut area may produce a solid stand of pin cherry from the dormant seed bank (Marks, 1974). For such a plant, the concepts of seed escape and directed dispersal are meaningless. Other trees lack seed dormancy and vagility, but various (and as yet unknown) defences give them the luxury of patience. For example, nearly 80% of the large, mammal-scattered seeds of *Gustavia superba* survive the seed and seedling stages through the first six months of life in the Panamanian rain forest (Sork, 1985). This is 2000 times the survival rate of *Virola* seeds beneath the parental crown in the same forest! Clearly, the consequences of dispersal failure vary dramatically between different species.

6.6   Unfeeling elements or selfish mutualists

Both physical and biotic factors influence the success or failure of pollination and seed dispersal. Abiotic dispersal is immensely wasteful; once released, wind- or water-carried pollen grains or seeds rarely find their targets. Plants can evolve means of ensuring visits by particular animals like bees or bats that are likely to carry pollen to neighbouring plants of the same species. Yet animals, too, can be unreliable: 1) they may be scarce, or vary unpredictably in abundance; 2) they may destroy pollen or seeds; and 3) they may simply fail to disperse whatever they collect. Whether unfeeling elements are more or less reliable than animal mutualists is determined by local conditions, just as local ecology determines the outcome of the conflict between plants attempting to establish their progeny, and animals which want only to eat the plants' reproductive parts.

6.6.1   Physical reality: oppression and response

Physical realities restrict plant options. Climate can be so severe, or so uncertain, that potential pollinators or dispersers are either scarce or unreliably variable in abundance. Weather also has a direct influence on flowering, fruiting and seed germination. Perhaps it is easiest to see the link between weather and pollination in severe and variable climates where both wind and animal pollination co-occur.

On the cold and wind-swept Faeroe Islands, for example, there are flies, beetles and moths, but no bees or butterflies. Most of the island's plants either self-pollinate, rely on the few available insects, are pollinated abiotically, or show exclusively vegetative reproduction (Hagerup, 1951). This depauperate sub-set of the British flora seems to be adjusting itself to novel conditions of insect scarcity.

Continental floras of temperate climates experience similarly capricious weather during early spring. Spring beauty *Claytonia virginica* is an abundant spring ephemeral herb of eastern North America. Flowers emerge just after snow-melt, before leaf flush of the trees overhead has reduced photosynthesis in the understory. Schemske (1977) found that *Claytonia* opens only a few flowers per day over several weeks. This makes the best use of andreniid bees and syrphid flies, the primary pollinators, since these insects are active only on warm days at a time of year when weather is unpredictable and often inhospitable. Careful observations show that seed set increases steadily through early spring, but that fruit abortion rises dramatically as the canopy closes overhead (Fig. 6.8). For *Claytonia*, flowering phenology is a race between plants that flower early and risk failure during poor weather (pollination limitation), and those that flower later and risk maturing in a shade so deep that their seeds cannot mature (resource limitation).

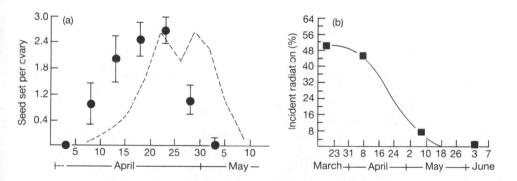

Fig. 6.8. (a) Seed set as a function of flowering time for *Claytonia virginica* in a deciduous woodland near Champaign, Illinois, USA. Solid circles and vertical bars represent means and twice standard errors for flowers which were first pistillate during a given time interval. The dashed line shows the total number of inflorescences per unit area (arbitrary units). (b) Canopy development as estimated by the percent of maximum incident radiation reaching the forest understory. After Schemske (1977).

In temperate climates, the success of seed dispersal is often unrelated to weather. Many species have seeds that overwinter before germination and are capable of remaining dormant for years or decades (Cook, 1980). Others, however, germinate in autumn and are vulnerable to frost-heaving or desiccation unless they are dispersed to microsites secure from these extremes of microclimate.

Perhaps the abundance of seed-eating insects and pathogens in the tropics changes the rules. Most tropical trees germinate within days or a very few weeks of being shed (Ng, 1978; Garwood, 1982). In seasonal forests like Barro Colorado Island, an eight month wet season is followed by four months with little or no rain. The woody plants of this forest vary tremendously in the details of their flowering schedules, but most fruits are shed in two rather brief periods, either at the very end of the wet season, or at the beginning of the next wet season (Fig. 6.9), regardless of flowering schedule. Foster (1982) found that flowering amongst insect-pollinated plants peaks early in the wet season when insects are most common, while canopy trees pollinated by birds or bats flower during the dry season when fewer insects are available. Garwood (1982) found that most seeds germinate at the beginning of the wet season. This means they germinate either directly after they are shed, or after a brief dormancy during the dry period. This allows seedlings to establish firmly before they have to endure the severe desiccation of the next dry season. For instance, *Virola surinamensis* seeds, like seeds of most species on Barro Colorado Island, germinate soon after they fall during the wet season (Section 6.5.2). But *Virola sebifera*, one of the unusual trees that fruits well before the dry season begins, has dormant seeds that germinate at the same time as *V. surinamensis*.

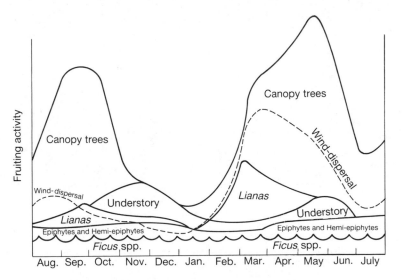

Fig. 6.9. Schematic diagram of the seasonality of fruiting activity in a 'normal' year on Barro Colorado Island, Panama. Activity represents the proportion of the forest vegetation that is fruiting. After Foster (1982).

In some ways it is artificial to distinguish between physical and biotic constraints on plant reproduction because climate influences the biotic environment by favouring some modes of reproduction over others, and by moulding the competitive environment in which plants and animals live. For instance, unpredictable and severe weather favours plants that are wind-pollinated over those that are animal-pollinated, with the result that the proportion of wind-pollinated plants increases with latitude and altitude (Regal, 1982).

6.6.2   Competition and facilitation

Flowering and fruiting plants live in communities occupied by dozens to hundreds of other plant species. Plants may compete with one another for pollinators or dispersal agents, or they may actually facilitate pollination or seed dispersal of neighbouring plants (Rathke, 1983). When flowers are generally scarce, pollinators are attracted to groups of flowers, and isolated plants may be ignored. At higher flower densities, inter-plant competition ensues because pollinators are satiated, and with vigorous competition for polli-nators, only the most attractive flowers set seeds. In principle, the same sort of effects might be expected among animal-dispersed plants (Howe & Estabrook, 1977), but the problem has not received detailed attention.

Competition for pollinators and dispersal agents might be *inferred* from the simple observation that many flowers are barren, or that many fruits are either not dispersed, or are eaten by animals that kill the seeds. It is far more difficult to show that two or more plant species really *do* compete for visitors. It is not enough to note that different species of plants flower at different times, because their apparent displacement may be indistinguishable from a random dis-tribution of flowering times (Pool & Rathke, 1979). In prairies, for instance, wind-pollinated plants show neither more nor less overlap in their flowering times than insect-pollinated species at the same site (Rabinowitz et al., 1981). Apparent competition between animals for plant resources is also difficult to quantify (Lawton & Hassell, 1984). For instance, bumble-bees (*Bombus*) visit different flower species where corolla length closely matches the tongue length of the bees (Heinrich, 1979). It is tempting to hypothesize that the bees have evolved different tongue lengths as a result of competition for these flowers, but it is also possible that plants and bees evolved elsewhere. For instance, the bees may have invaded New England and found flowers which they could visit, without their having had any role in the evolution of these flowers (Janzen, 1980). Even though it is plausible that competition for animal mutualists should occur at some level, direct evidence is hard to come by, and is best obtained by manipulative experiments (Waser, 1983).

Far from being displaced, many apparent competitors flower or fruit at the same times. Sometimes synchrony is a response to strong selection by weather, in which case physical factors override selection by competing pollinators or dispersal agents. However, plants some-times facilitate each other's pollination. Brown and Kodric-Brown (1979) found that eight coexisting hummingbird-pollinated plants in

Arizona shared pollinators, and Schemske (1981) found remarkable convergence among two bee-pollinated *Costus* species in Panama. In both systems, floral characteristics, phenology and presentation were more similar than might be expected by chance. What remains to be shown is whether the *absence* of one species reduces the seed set of the other.

One further source of confusion about competition should be mentioned. Animals may compete vigorously for flowers or fruits without helping the plants at all. Stingless *Trigona* bees fight furiously for both artificial and natural nectar sources in a Costa Rican dry forest, but the winners are not necessarily effective pollinators (Johnson & Hubbell, 1974). Similarly, territorial defence of fruit-laden shrubs of holly *Ilex aquifolium* by aggressive mistle thrushes *Turdus viscivorus* decreases, rather than increases, the rate of seed dispersal (Snow & Snow, 1984). The birds chase away many potential holly dispersers, but do not themselves eat all fruits on the bushes that they defend. Both the bees and the thrushes compete for food in ways which adversely affect the reproductive success of their host plants.

A direct demonstration that plants compete for pollinators or frugivores must show that the presence of one plant species has a negative impact on the fitness of the other. Waser (1978) provided such evidence for *Delphinium nelsonii* and *Ipomopsis aggregata* in the Rocky Mountains of the United States. Both the early flowering *Delphinium* and the later flowering *Ipomopsis* are pollinated by hummingbirds, and both suffer reduced seed set during the brief period when their flowering times overlap. Waser suggests that this competition has been sufficient to maintain the species' distinct flowering times.

6.6.3   Diversity and its consequences

Casual observations might suggest that only a few animals visit a flowering or fruiting plant. For some special species like tropical orchids (Dressler, 1981), figs (Wiebes, 1979) and certain desert wild-flowers (Linsley *et al.*, 1973) this impression may be correct. Most plants, however, are visited by a wide variety of potential pollinators and agents of dispersal. The insect records from 55 species of Asteraceae show that most have 26–75 species of visitors, while some plants are visited by over 150 kinds of insects (Schemske, 1983). A more detailed study of pollinator activity on *Calathea ovadensis*, a highland herb in southern Mexico, showed that both bees and butterflies visited the flowers, but that butterflies were poor polli-nators (0–8 fruits set per visit), that some bees were nearly as bad (2–8 fruits per visit), but that other bees were far more effective (30–34 fruits set per visit; Schemske & Horvitz, 1984). In this case, two genera of bees, *Bombus* and *Ryathymus,* were far more effective per visit than others, yet pollinated only 23% of the flowers. Abundant *Euglossa* bees were responsible for most seed set, even though they were less effective pollinators per visit. Specialization of *Calathea* to *Bombus* would be possible, but is unlikely because of the high frequency of visits by more common insects.

A much less complete literature on seed dispersal by ants and vertebrates suggests a similar generalized use of fruits (Howe & Smallwood, 1982). In ant–herb systems, several species of ants take the seeds of a given species of plant, and ants collect whatever suitably sized seeds they find. Similarly, fruiting trees and shrubs attract 10–30 species of frugivores in varying abundances (Sections 6.4.2 and 6.5.2), and there is little evidence of specialization beyond a rather broad set of dispersal agents.

Differences in species associations may or may not influence the efficiency of pollination and seed dispersal. Linhart & Feinsinger (1980) provide an example of differential pollen dispersal by hummingbirds on the large and species-rich island of Trinidad (4520 km^2), and on the much smaller, species-poor island of Tobago (295 km^2). *Justicia secunda* and *Mandevilla hirsuta* occur on both islands. The corolla of *Justicia* is easily accessible to most hummingbirds and the plant is effectively pollinated by both long- and short-billed species. *Mandevilla,* on the other hand, has a long, thin corolla accessible only to long-billed hummingbirds that are absent from Tobago. There was substantial pollinator dispersal of florescent dye dusted on *Mandevilla* on Trinidad (distances up to 180 m), but there was negligible dispersal on Tobago (only about 10 m). Seed set of *Mandevilla* was significantly higher on the larger island, but *Justicia* did equally well on Trinidad and Tobago. The specialized plant could not reproduce as effectively in the depauperate habitat.

6.7   Co-evolution or co-occurrence

Many modern plant communities would be unrecognizable if the plant species that rely upon animals for pollination and seed dispersal were removed. The fit between the size, colour, and shape of a flower and the size, tongue length and foraging preferences of a moth that visits it, is often so apparent that naturalists are tempted to assume that the flower and the moth are co-evolved. In its extreme form, this co-evolution would imply that the species of flower evolved in response to selection by the species of moth, and the moth evolved in response to selection imposed by the plant (for a detailed discussion see Feinsinger, 1983, and Schemske, 1983). Such reciprocal, symmetrical co-evolution of species pairs is certainly possible (Chapter 10), but is unlikely to be common for flower-pollinator or fruit-disperser systems. The same degree of fit could exist because related groups of plants adapt to a general body plan of a group of related animal species (e.g. different species of bees or of humming-birds), or to a similar body plan among unrelated animal taxa that resemble each other because of convergence (e.g. bumble-bees and butterflies both have long tongues). Likewise, the animals might adapt to eat nectar or fruits from one or more plant taxa which happen to have similar flowers or fruits. An apparent fit might well be fortuitous, involving animals and plants that co-occur in a contemporary community but do not share an evolutionary history.

The blurred distinctions between fruit syndromes suggest even less co-evolutionary precision than we see in pollination biology. For example, on the Australian sand plains, oppossums now scatter the seeds of the cycad *Macrozamia*, which might once have been dinosaur food (Burbidge & Whelan, 1982). Among the advanced seed plants, the enormous diversity of fruit types indicates plant adaptation to a wide variety of animals, but the degree of special-ization is probably rather loose. For example, Herrera (1984a) found that angiosperm shrub and tree species last about 30 million years in the fossil record, whereas mammal and bird species tend to be found for less than 1 million. If fossil plant and animal species are compar-able taxonomic entities, then Herrera's findings suggest that a plant adapts to a basic body plan of part of the animal community, rather than to particular, transient species.

6.7.1   Constraints on co-evolution

The historical record is far too fragmented to allow a definitive answer to the question of whether plant and animal species co-evolve. There are too few fossil insects even to address the question for pollinators. However, the suggestion that plant species last between 10 and 70 times as long as their potential avian and mam-malian dispersal agents tells us that adaptations for seed dispersal by birds and mammals must be rather general, and that the animals adjust to the plants, rather than vice versa (Herrera, 1984a). Recent ecological studies of pollination and seed dispersal have come a long way in suggesting why relationships between plants and the animals that visit them, usually involve many species, rather than a single, tightly co-evolved pair (Table 6.5).

Succession and plant community structure

Given the tremendous variability in plant and animal species compo-sition from place to place, it is virtually inconceivable that each plant population 'adjusts' its flower or fruit production to whatever poten-tial mutualists happen to exist in the vicinity (Howe, 1984). The wide variety of insects visiting populations of sunflowers in central Illinois

Table 6.5. Constraints on coevolution between species of mutualists which live apart, such as plants and their pollinators or dispersal agents. From Schemske (1983), Herrera (1984a), and Howe (1984).

1. Pollinators and dispersal agents vary in abundance, and therefore in effectiveness, in space and time.
2. Plant resources used by flower-visiting or fruit-eating animals vary dramatically in space and time.
3. Plants often have much longer generation times than animal mutualists, and consequently evolve more slowly.
4. Dependence of plants on animals and animals on plants is often asymmetric.
5. The genetic capacities for response to selection of plants or an animal mutualist may be different.

probably reflects their exposure to very different successional associations of both competitors and pollinators. The process of ecological succession undoubtedly favours generalist plants which can take advantage of a wide variety of pollinators or dispersal agents, and generalist animals, capable of visiting a wide variety of flowers or fruits.

Independent variations in the species composition of plant and animal communities make it unlikely that a plant species would have the opportunity to co-evolve with a single pollinator or dispersal agent. In southern Spain, for example, plant species growing in highland scrub encounter different abundances of dispersal agents and different arrays of competitors for those dispersal agents than the same plants growing a few km away in the lowlands (Herrera, 1984b). We find, as expected, that plant abundance, fruit quality, timing of fruit production, quantity of fruit production and the diets of the fruit-eating birds all differ between the two sites. Similarly, a given bird species will eat different fruits when it forages in the highlands than when it feeds in the coastal scrub.

Beyond local differences in successional stage, the composition of plant communities is invariably influenced by drainage, soil, altitude and climate (Chapter 1). For example, *Casearia corymbosa* trees growing in the canopy of the Costa Rican rain forest are dispersed largely by one tropical bird *Tityra semifasciata* which frequents the upper canopy, while stunted individuals of the same species, growing in dry forest 100 km away, are dispersed by the much smaller *Vireo flavoviridis*, a bird of the understory (Howe & Vande Kerckhove, 1979).

Demographic variation

Unless foraging behaviour is unaffected by the availability of other foods, an individual plant cannot have an 'optimal' level of flower or fruit production for particular pollinators or dispersal agents. What is best depends upon what the other plants are doing. While most insects, birds and mammals live for only one or a few years, plants commonly live for hundreds or even thousands of years. During their long lives, their flower production may vary by three or four orders of magnitude, with inevitable consequences for the animals that feed on their nectar or fruits. Ten young *Virola* trees, for instance, would produce only a small fraction of the fruits that ten large plants of the same species nearby would produce. If toucan population densities are more or less uniform, then most fruits may be taken from the small, young trees, but only a much smaller fraction from the larger, more fecund plants, with the consequence that the proportional dispersal success of the young plants is greater. Conversely, a mobile frugivore attracted to patches of high fruit density would disperse the fruits of large plants more effectively; small trees with low fruit crops might be completely ignored. Because animals are influenced by the accessibility of flower and fruit resources, in addition to their quality, stands of small adult plants are likely to attract a different guild of animals than mature stands of large, more fecund adults (Howe, 1983; Levey *et al.*, 1984).

Asymmetries

Finally, there are at least three kinds of asymmetries which influence the potential for plant-animal co-evolution. First, life history differences may cause plants and animals to evolve at different rates (although the advantage of rapid evolution in the short-lived animals may be more apparent than real; see Chapter 10). Second, the dependence of animal and plant species is usually asymmetrical. Schemske (1983) points out that some bee species visit only one or two species of flowers, while the flowers attract many species of pollinators. Similarly, a flower can require one kind of pollinator, while the insect visits many species of plants. For example, in Panama, *Hybanthus* is dependent on the single, local bee species *Melipona interrupta* for pollination during its four-day flowering season, but the bee visits flowers of many species in the course of a year (Augspurger, 1980). This kind of asymmetry is also common in seed dispersal. *Virola*, for instance, seems to be dependent on toucans in Panama, but the toucans eat fruits of many tree species throughout the year (Howe, 1983). In these two examples, the bees and the toucans impose stronger selection on the plants than the plants impose on the foraging behaviour of the animals. In brief, the animals are more important to the plants than these particular plants are to the animals. Third, the plants and their animal visitors may differ in their capacity to respond to natural selection (Howe, 1984). For example, the flower may be genetically 'stuck' with its corolla shape, whereas the tongue-length of the bee or moth which visits it may be flexible, and capable of responding to selection from other plants. There may be quantitative, rather than absolute, differences in heritability of traits influencing a mutualism, and this could result in differences in the abilities of each mutualist to respond to selection imposed by the other. It is of considerable interest to discover whether patterns of genetic variation constrain, or promote co-evolution.

6.7.2 Implications

A general implication of both the historical record and of contemporary ecological studies is that co-evolution usually amounts to a general adjustment of a group of plants to an array or morphologically, and often taxonomically, similar group of animal pollinators and dispersers. For instance, fruits of most New World *Casearia* trees are eaten by small birds, but the species of birds differ from one geographic locale to another. Similarly, insects, birds and mammals that forage for flowers or fruits select those kinds which are compatible with their nutritional needs and handling abilities.

Often the general body plan of a taxonomic group confers a degree of uniformity on structure, function and nutritional requirement. Hawkmoths are often nocturnal and hover while feeding, so it is not surprising that many flowers which they visit offer sugar-rich nectar and are pale enough to be visible at night. Given that one plant of a family (*Silene noctiflora*, Caryophyllaceae) has acquired hawkmoth pollination, it is not surprising to find that several other members of the family have similar traits and are also pollinated by hawkmoths. This does not mean that a particular pair of *Silene* and hawkmoth are co-evolved, but it does suggest that the *Silene* and hawkmoth taxa have a long, common history of compatibility.

No one knows for sure how the history of a plant-animal association begins. It may be that the plant species which founds a group is unusual because, for example in a depauperate island or desert community, it may co-evolve with an especially abundant pollinator or dispersal agent. Under changing circumstances, and a widening geographic range, the plant may come into contact with many equivalent pollinators or dispersal agents. Altered ecological circumstances make the initially 'specialist' plant appear to be 'generalist'. Eventually the taxon may splinter, producing descendent species with essentially similar flower and fruit structures, and attracting similar kinds of animals. For all the elegant natural history known about the processes of pollination and seed dissemination, we do not know precisely how they came about, nor what will provoke their further evolution. What we do know is that pollinators and seed dispersers play a vital, if often neglected, role in the population dynamics of many plants.

Chapter 7   Breeding Structure and Genetic Variation

DONALD A. LEVIN

7.1   Introduction

Plants grow, reproduce and die within an environment defined by both biotic and abiotic parameters. Their phenologies, forms, biomasses and seed yields are traditionally explained as responses to this environment. In habitats with certain prescribed features, plants of a given species may be robust and highly fecund, whereas in another habitat they may be depauperate and produce few seeds. Since favourable and unfavourable habitats are sometimes only a few centimetres apart (see Chapter 4), plants within the same population may differ in many respects by virtue of their divergent developmental responses. This phenotypic plasticity increases the probability that a plant will survive to reproductive maturity and produce at least a few seeds.

There is an abundant literature on plant responses to different culture conditions in the greenhouse and in the garden. Plants develop one phenotype in one culture and another phenotype in different culture; furthermore, these responses are predictable. With greenhouse studies in mind, we might conclude that differences in plant phenotypes in natural populations were entirely the result of phenotypic plasticity in heterogeneous environments. We would be mistaken, however, because the phenotype is the product of environmental *and* genetical influences. The relative importance of environment and genetics varies with the character; for example, differences in plant stature, leaf shape and flowering time are often the result of environmental factors, whereas differences in pathogen resistance, floral architecture and colour are largely the result of genetic factors. Comprehensive discussions of phenotypic determination from ecological and genetical perspectives are to be found in the reviews by Bradshaw (1965), Jain (1979) and Gottlieb (1984).

A number of fundamental relationships follow from the fact that the genotype governs, at least in part, the organism's phenotype: 1) the genotypes of organisms are important determinants of the amount and kind of phenotypic variation in populations; 2) through its effect on phenotype, the genotype is responsible for the ways in which a plant functions in, and interacts with, its environment; 3) the abilities and tolerances of individuals, and the ecological amplitudes and life histories of populations are functions of genetic variables;

217

4) as genotypes change in space and time, so the ecological properties of populations change; 5) as the amount of genetic variation changes in time and space, so the heterogeneity of response between individuals within population changes.

The amount, nature and organization of genetic variation within and between populations is governed by the pattern of mating, and by the spatial relationship between plants and their parents. Populations and species where mating and seed dispersal occur over substantial distances have very different genetical properties from those in which crossing and dispersal are restricted in space, and, in consequence, they have different ecological strategies and amplitudes. In this chapter I discuss the factors which determine crossing and dispersal spectra, the spectra themselves, and their genetic consequences at the population and species levels. I shall also demonstrate that mating and dispersal systems are not fixed attributes of species, but variables which respond to environmental factors changing in both space and time.

7.2   Breeding systems

Plants are the passive recipients of pollen brought from other plants by animals, wind or water. While they have various adaptations to attract pollinators and to facilitate pollen deposition by pollinators, they have no control over the pollen they receive from outside sources. The stigmas of most species receive some pollen from their own flower, or from nearby flowers on the same plant; indeed, most pollen on stigmas usually comes from the same individual. The proportion of self and cross pollen grains on stigmas does not dictate the level of self- and cross-fertilization, however, because most species are self-incompatible (i.e. they are unable to set seed following self-pollination). Before the male gametophyte (pollen grain) reaches the female gametophyte (embryo sac) with the egg cell and its polar nuclei inside, the pollen tube must grow through the pistil. It is in the pistil that the incompatibility reaction takes place.

7.2.1   Incompatibility systems

There are two kinds of incompatibility systems, known as the gametophytic and sporophytic incompatibility systems (de Nettancourt, 1977). Gametophytic incompatibility occurs in species belonging to 78 angiosperm families, whereas sporophytic incompatibility is largely restricted to species within the Compositae, Cruciferae and Convolvulaceae. The systems differ in their genetic mechanisms and in a number of structural and physiological features. Genetic differences are considered here, but a comprehensive treatment of compatibility systems and pollen-pistil interactions is to be found in Knox (1984).

The main features of *gametophytic* incompatibility are as follows: 1) incompatibility is controlled by one locus with a large number of alleles; 2) each plant has two different S alleles (e.g. S_1S_2 or S_3S_4, etc.) so that the pistil is heterozygous, and 50% of the pollen grains carry one S allele and 50% carry the other; 3) when pollen arrives on the stigma, it germinates and produces a pollen tube which grows through a special tract of tissue (or along a canal) in the style; 4) if pollen grains share on allele in the pistil, the pollen tube is arrested. In most species this occurs in the style, but in others it may occur on the surface of the stigma or in the ovary. If a pollen grain carries an allele not present in the pistil, the pollen tube proceeds normally and fertilization may occur. Therefore if an egg parent has a genotype S_1S_2 and a pollen parent has the genotype S_1S_3, 50% of the pollen (the S_3 grains) will be effective. With the same egg parent and an S_3S_4 pollen parent, all the pollen grains would be effective.

The main genetic features of *sporophytic* incompatibility are as follows: 1) incompatibility is controlled by one S locus having several alleles; 2) each plant has 2 different alleles, but the specificity of the *pollen* as well as that of the pistil is controlled by sporophytic tissue, so that each pollen grain behaves as though it had both S alleles, even though it only has one; 3) if pollen comes from a plant which has *either* allele in common with the egg parent, pollen tube growth is inhibited on the surface of the stigma, so that *all* self pollen is ineffective. If an egg parent were S_1S_2 and the pollen parent were S_1S_3, none of the grains will be effective, whereas if the pollen parent were S_3S_4, then all the pollen grains would be effective. In some species, interactions between S alleles may yield more complex crossing relationships (de Nettancourt, 1977).

In species with gametophytic incompatibility there are no structural differences associated with different S alleles; they are 'homomorphic'. In contrast, some species with sporophytic incompatibility are 'heteromorphic' (Ganders, 1979). In most species of this type there are two floral morphs, one with a long style and small pollen grains, and another with a short style and large pollen grains. The long style morph, known as 'pin', usually has the genotype Ss. The short style, or 'thrum' morph usually has the genotype ss. The system is controlled by two alleles, which govern both pollen size and style length, in addition to the incompatibility reaction. Compatible crosses occur when thrum pollen is deposited on pin stigmas, or when pin pollen is deposited on thrum stigmas. Self-pollinations, and cross-pollinations between like morphs, are ineffective. Heteromorphic incompatibility occurs in 24 families of flowering plants, although in most families it appears in only a few genera.

Plants which are self-incompatible can reduce the level of self-pollination through spatial separation of their anthers and stigmas (e.g. monoecy), and through temporal separation of pollen and stigma maturation (e.g. protandry). Nevertheless, self-pollination in hermaphroditic species is rarely avoided, because different flowers on the same plant tend to be at different levels of maturity (see Chapter 10).

7.2.2   Self-compatibility

Although the majority of angiosperms are self-incompatible, many species (especially annuals) are self-compatible. In self-compatible species, self-pollination is promoted by the close proximity of anthers and stigma, and by the synchrony of anther dehiscence and stigma maturity. These features also promote self-fertilization by ensuring a high ratio of self to cross pollen on stigmas. The bias towards self-fertilization is carried to the extreme in species which produce reduce flowers in which pollination occurs *before* opening (Lord, 1981; Uphof, 1938). These (cleistogamic) flowers may be produced on the same plant as normal (chasmogamic) flowers, with the proportion varying from plant to plant and population to population; this variation has both genetic and environmental bases (Section 7.7.3). Cleistogamy has been described in 287 species representing 56 plant families.

7.2.3   Breeding system balance

While self-compatibility is a necessary condition for self-fertilization, plants which are self-compatible are not necessarily attuned to high levels of self-fertilization. Consider the case of cultivated *Phlox drummondii* (Levin, 1975). Plants set approximately the same number of seeds per fruit with either cross or self pollen. However, most cross pollen germinates several hours earlier than self pollen, so that, if stigmas were pollinated with equal mixtures of self and outcross pollen, we would expect most progeny to be the products of cross fertilization. When cross pollen with a genetic marker allowing identification of cross progeny was mixed with an equal amount of self pollen, about 85% of the progeny were from cross fertilization. Clearly, then, the breeding system of this self-compatible plant favours cross-fertilization. Self pollen is like an insurance policy for seed production; if nothing better comes along, it is acceptable. The competitive disadvantage of self pollen is absent in the close relative *Phlox cuspidata*, where self pollen germinates synchronously with cross pollen, and pollination with 1:1 mixtures of cross and self pollen yields equal numbers of cross and self progeny.

In comparing congeneric species, it may be possible to judge the balance of the breeding system without experimentation, by noting the pollen:ovule ratio per flower. Cruden (1977) found that the greater the emphasis on self-fertilization within a species, the less pollen is produced per ovule (i.e. low P:O ratio). Species that are obligate out-crossers have an average P:O ratio of approximately 6000:1 against 800:1 for predominantly outcrossing species, 175:1 for predominantly selfing species and about 30:1 for obligate selfers. This pattern reflects the greater certainty and efficiency of pollination in predominately self-fertilizing species. In species with low P:O ratios, the relative performance of self pollen is likely to be better than in species with high P:O ratios.

It is important to realize that neither progeny ratios from dual pollinations in the greenhouse, nor P:O ratios demonstrate the level of self-fertilization which occurs in nature. They only speak of tendencies and potential. Such information can only be obtained through genetic analyses of progeny produced in the field.

7.3   Mating patterns

The plants (including the self) with which an individual plant crosses determine the mating pattern of a population. While the breeding system sets constraints on the mating pattern, the mating pattern is determined by the pollen population received by the stigmas. If all plants produced equal quantities of pollen of similar quality, and if pollen were deposited on stigmas from the whole population at random, then mating would be random. Indeed, random mating is a common assumption in many ecological and evolutionary models. However, it is very unlikely to occur in practice because: 1) pollen is not drawn from the pollen pool at random; 2) pollen quality may vary from plant to plant; and 3) the cross-compatibility of plants varies from one combination to another.

7.3.1   Pollen dispersion in animal-pollinated plants

The distances that pollen is transported can be determined if the donor plant has a dominant genetic marker not present in the recipients. A plant homozygous for this marker is planted within a population made up entirely of recessive homozygotes. Later in the season, the seeds are collected, and a record is made of the distance between the maternal plant and the pollen source. The progeny are then scored for presence of the marker.

Several experiments on wild and crop plants have employed this technique to demonstrate pollen flow. Two studies with contrasting pollen-source designs show that pollen dispersion is independent of the location of pollen sources, but does vary with the plant species. Datta *et al.* (1982) set up a paired experiment with jute *Corchorus*

Table 7.1. Mean hybridization (%) in *Hibiscus cannabinus* and *Corchorus olitorius* at different distances from the pollen source (after Datta *et al.*, 1982; and Maiti *et al.*, 1981).

Distance from marker plants (metre)	Hibiscus		Corchorus	
	Design I	Design II	Design I	Design II
0.5	2.41	3.38	2.47	2.99
1.0	1.91	2.13	1.10	1.48
1.5	1.06	2.24	0.56	0.61
2.0	1.30	1.69	0.28	0.29
2.5	0.58	1.13	0.24	0.22
3.0	0.43	0.97	0.24	0.16
3.5	1.28	0.68	0.12	0.13
4.0	1.32	0.95	0.12	0.11
4.5	0.79	1.27	0.13	0.05
5.0	0.34	1.15	0.08	0.05
5.5	0.91	0.36	0.09	0.03
6.0	0.49	0.60	0.09	0.03
6.5	0.53	0.85	0.05	0.29
7.0	0.63	0.44	0.04	0.05
7.5	0.51	0.51	0.06	0.06

olitorius which is pollinated by the bee *Apis dorsata*. In design I, a strain with reddish stems was sown in a central 5m square area. On each side of the square, an unpigmented strain was planted in 15 rows 0.5 m apart. In design II, the outer 2 rows were planted with the pigmented strain and the 15 inner rows with the unpigmented one (the central square was left unused); both designs employed 20×20 m plots. Red pigmentation is controlled by a dominant allele, and green by a recessive allele at the same locus. The percentage of hybrid plants declined sharply as the distance to the pollen source increased, and the crossing pattern was similar in the two experiments, despite the different location of the genetically marked pollen (Table 7.1).

Maiti *et al.* (1981) conducted a similar experiment on *Hibiscus cannabinus* which is pollinated by *Apis mellifica*. The pollen source was homozygous for a dominant allele which confers lobed-leaves, and the recipients were homozygous for a recessive allele which confers entire leaves (see Table 7.1). Crossing with the dominant strain declined with distance, but not as steeply as in jute. In *Hibiscus* about 0.5% of seeds from plants 7.5 m away involved crossing with the marker plants, as compared to 2.8% in the row adjacent to the marker. In *Corchorus*, 0.06% of the seeds from plants at 7.5 m were hybrids, compared with 2.75% in the row next to the marker. The patterns of effective pollen dispersion in these two species are similar to those of other bee-pollinated plants (Levin & Kerster, 1974). Although species differ in mean gene dispersal distances, the slopes of the curves are rather similar. Crossing is thus non-random in space, and most crosses are between plants and their closest neighbours.

Because the distance between mating plants is usually small, the spatial arrangement of two genotypes will strongly influence the incidence of crossing between them. If two genotypes are equally frequent, but occur in two juxtaposed patches (i.e. in a coarse grained mixture), most crosses will occur between plants of like genotypes. On the other hand, if the genotypes occur in a random, fine grained mixture, then about 50% of the crosses will be between genotypes. Nieuwhof (1963) demonstrated the effect of genotype arrangement on the frequency of crossing between genotypes in *Brassica oleracea*. He established a series of plots containing cabbage and kale in equal frequencies but interspersed to various degrees. The percentages of hybrids among cabbage progeny exceeded 38% when the two genotypes were planted in one- or two-row groups, but fell to 20% when the plants were grown in four-row blocks.

Pollinator foraging behaviour

The restricted spatial pattern of breeding is a consequence of pollinator foraging behaviour. Bees (and other pollinators to a somewhat lesser extent) tend to move from a plant to one of its near neighbours (Free, 1970; Levin & Kerster, 1974). The relationship between pollen dispersal and pollinator flight distance has been demonstrated in two Texas annuals, the Lepidopteran-pollinated *Phlox drummondii* and the bee-pollinated *Lupinus texensis*. In the *Phlox* study, a group of plants with a dominant flower marker was placed into a natural population lacking this marker (Levin, 1981). The *Lupinus* study involved the introduction of plants homozygous for an enzyme marker into the centre of a synthetic population lacking the marker (Schaal, 1980). Observations were made of pollinator flight distances between plants. The pollen flow and pollinator foraging distance distributions are compared for each species in Fig. 7.1, from which it is clear that both distances are relatively short. However, in both species, mean pollen flow distances were much greater than mean flight distances (2.6 m against 4.0 m in *Phlox*; 1.0 m against 1.8 m in *Lupinus*). This disparity is due to pollen carry-over; that is, pollen collected from one plant is deposited not only on the first plant visited, but also on some subsequent plants.

Where animals are the match-makers, anything which alters their behaviour will alter pollen dispersion in turn. The quantity and quality of resources for the pollinator are important variables in this respect. There is abundant evidence that pollinator flight distances are positively correlated with plant spacing. For example, Levin & Kerster (1969) studied flight distances of bees foraging in nine plant species, each represented by three populations with different densities. The correlation between mean flight distance and mean

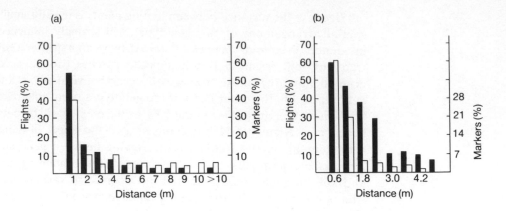

Fig. 7.1. Pollen dispersion against pollination flight distance in (a) *Phlox drummondii*; black bars denote flights, open bars pollen dispersion, (b) *Lupinus texersis*; black bars denote pollen dispersion, open bars denote flights. After Levin (1981) and Schaal (1980a).

spacing was compelling for each separate species, but even more striking was the fact that the correlation was still very strong when all nine species were considered collectively (r = 0.9; Fig. 7.2), indicating that bees respond similarly to plant spacing irrespective of species. Given the relationship between density and foraging, it follows that pollen flow will be more distant in sparse populations than in dense ones. That is to say, the average distance between mating plants is greater in sparse populations.

The distance between mating plants also depends upon the pollinator species, because different pollinators have different foraging behaviours. A population serviced predominantly by bees will tend to have a narrower mating structure than one serviced by butterflies or moths. This difference was very clear in species of *Senecio* studied by Schmitt (1980). She demonstrated that the mean

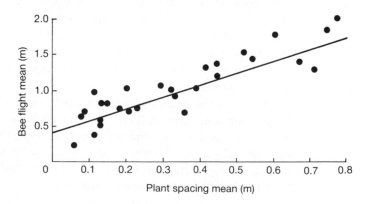

Fig. 7.2. Relationships between mean bee flight distance and mean plant spacing per population. Each dot represents one population. After Levin & Kerster (1969).

flight distance of butterflies is about nine times greater than that of bumble-bees working the same plant population (3.5 m versus 0.4 m). Different populations of the same species have different proportions of bee and butterfly pollinators, and presumably, different patterns of pollen dispersion; at one site there were more than three times as many bee visits as butterfly visits, but at another, the same *Senecio* was visited by roughly equal numbers of both pollinators.

Assortative mating

Crossing between plants may also be non-random with regard to phenotype. Pollinators like bees, lepidopterans and hummingbirds can discriminate between flower colours, odours, corolla sizes and petal configurations. They remain constant to a single phenotype even though other suitable flowers may be present (Grant, 1963; Levin, 1978; Faegri & van der Pijl, 1979). Constancy is short term, and may vary between individual pollinators. There is considerable variation in flower colour (and in scent and architecture to a lesser degree) in natural populations, so the potential for mating between like phenotypes ('assortative mating') is present.

In order to determine the extent to which differences in flower colour (red versus pink) may elicit assortative mating in the lepidopteran-pollinated *Phlox drummondii*, Levin & Watkins (1985) established synthetic populations in which the two colour variants were distributed in a checkerboard manner. Subsequently seeds were collected and the percentage hybrids (maroon flowers) was determined. 37% of the plants were hybrid, compared with the 50% expected from random mating, indicating that mating was partially assortative. They also studied the effect of plant stature differences on the level of assortment, by placing plants of one variant on raised platforms. A 20 cm platform gave 26% hybrid progeny, while raising the platform to 40 cm reduced hybridization to 16%. Greater assortment with increased stature difference is due to the tendency of pollinators to move horizontally between plants (see Chapter 6). Since neither plant variant was discriminated against, we may conclude that flower constancy was the basis for assortative mating.

Assortative mating can occur in three ways in natural populations (Kay, 1982): 1) it will arise from flower constancy in the absence of preference for one variant over the other; 2) it will arise if a population is serviced by different pollinators, some preferring one variant and some preferring another; and 3) assortative mating will occur if alternative variants are clumped in space, even if pollinators do not discriminate between them, since pollinator flights are principally between neighbouring plants.

7.3.2   Pollen dispersion in wind-pollinated plants

The pattern of effective pollen dispersal in wind-pollinated plants is like that of insect-pollinated plants, in that most pollen is distributed very close to the source, and the distribution is highly leptokurtic. This is well illustrated by the work of Paterniani & Short (1974) on maize. They planted maize homozygous for white endosperm (yy) and placed a single individual with the dominant yellow endosperm colour (YY) in the centre of the field. At harvest time the kernels developed after fertilization by pollen from the central plant were determined by their colour. Hybrid seed production was high only in the immediate vicinity of the pollen source (Fig. 7.3). The same authors considered the effect of plant density on effective pollen dispersion by planting two fields with row spacings of 40 cm and two with row spacings of 20 cm; on average, pollen flow in the sparsely planted plots occurred over greater distances than in the dense plots (Fig. 7.3). This is because plants in the sparse population have fewer potential mates, so the pollen from the central plant constitutes a higher proportion of the total pollen pool than in the dense population. The greater the spacing between plants, the greater will be the mean distances between mating plants.

The pattern of mating within a population of wind-pollinated plants is highly variable in time, simply because the atmospheric conditions which govern pollen dispersion (especially wind turbulence and speed) are so changeable. Pollen concentration at stigma height is related to the distance downwind from the source by the approximation:

$$\chi_{(x)} = \frac{Q}{\pi C^y C^z u x^m}$$

where χ is the pollen concentration, x is the distance downwind from the point source, Q is the number of grains released per second, C^y and C^z are diffusion coefficients for vertical and lateral directions, u is the mean wind velocity, and m is the degree of tubulence in the air (Sutton, 1947).

Bateman (1947) formulated a general mathematical expression for effective pollen dispersion which describes the patterns for animal and wind-pollinated plants rather well. If the level of crossing (F) is small, then:

$$F = \frac{y e^{-kx}}{x}$$

where y is the level of crossing at the point source itself, k is the rate of decrease of crossing with distance, and x, as before, is distance from the source. The model makes certain simple predictions: 1) successive increases in distance become less and less effective in retarding

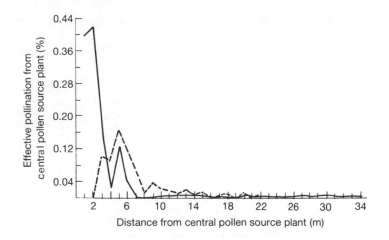

Fig. 7.3. Effective pollen dispersion in maize as measured by percentages of yellow kernels on plants according to their distance from the pollen source. Dispersion in the dense population is shown with a solid line, and in the sparse with a dashed line. After Paterniani & Short (1974).

crossing; 2) a few crosses may occur at distances two orders of magnitude greater than the mean distance; and 3) we can predict the level of crossing at any distance provided we know the level of crossing at two distances (although there are inevitable uncertainties in this kind of extrapolation).

From a knowledge of the pattern of effective pollinations for single plants, and assuming the pattern is similar, on average, for all plants, we can estimate the numbers of seeds that result from pollinations by plants located at different distances away. In Paterniani and Short's (1974) experiment on maize, the 10 plants within 1 m of the central plant produced 3691 kernels between them, of which only 1 was yellow! Each of the plants, therefore, received only 0.027% of its pollen, on average, from the central plant. In a 2 m radius around the central plant there were 33 plants producing 12,188 kernels of which 11 were yellow. Following the same logic as before, each plant received 2.97% of its pollen from plants within a radius of 2 m. Carrying this procedure to greater distances, they estimate that in sparse populations about 50% of the kernels of any individual plant result from pollen originating within a radius of 12 m. This involves over 800 plants as potential pollen parents.

7.4   Differential crossing success

Deviation from random mating may arise because some plants are more successful as pollen parents than others, or because some crossing combinations are more effective than others. Guttierrez & Sprague (1959) investigated genotype-dependent male success in a

Table 7.2. Percent of male functioning by the different stocks of maize in polycross plantings (after Guterrez and Sprague, 1959).

Stocks	Percent of male functioning by									Total no. F_1's
	sy	wt	gl_7v_{17}	lg_2gl_6	$wx\ su_{205}$	sh_1	su_1	bt_1	str	
sy	25.4	51.3	17.4	0.2	0.7	0.7	2.6	1.1	0.6	1405
wt	48.2	17.7	28.0	0.6	0.2	0.9	2.2	1.0	1.2	1238
gl_7v_{17}	6.4	7.8	58.1	1.0	2.7	7.2	10.2	5.9	0.7	2316
lg_2gl_6	1.6	1.3	9.1	12.4	3.5	12.5	19.2	20.2	20.2	1424
$wx\ su_{205}$	0.7	7.4	19.8	7.3	9.9	14.9	2.8	19.6	17.6	740
sh_1	11.3	4.4	37.8	2.6	3.8	5.9	19.8	8.1	6.3	1060
su_1	3.6	4.7	23.1	1.2	—	7.3	25.9	10.8	23.4	1582
bt_1	—	0.9	6.7	5.4	8.4	24.6	8.8	17.0	28.2	687
str	—	0.2	0.4	0.2	—	0.4	21.6	4.5	72.7	819
Total no. F_1's	1306	1323	3030	350	294	825	1451	1008	1684	11271
Total percent	11.6	11.8	26.9	3.1	2.6	7.3	12.9	8.9	14.9	—

population of maize. They used nine equally abundant stocks, each homozygous for a different, simply inherited recessive gene, and planted them in three contiguous latin squares. Plants were harvested individually, and male parentage determined. If mating had been at random, each of the nine stocks should have functioned as a pollinator 11.1% of the time. In fact, values ranged between 2.5 and 26.9%, and there was heterogeneity in paternity across single egg genotypes, as well as across all genotypes. For instance, two male stocks fathered 48% and 28% respectively, of all seed produced by a particular egg parent, whereas none of the other males (excluding the self) fathered more than 3% of the seed (see Table 7.2). Differences in overall male function were due to differences in pollen production, length of pollen shedding period, and plant height. Relatively tall plants, and strains producing more pollen and dispersing it over protracted periods, were the most successful.

Heterogeneity in crossing success was also demonstrated in a natural population of the trumpet-creeper *Campsis radicans*. Bertin

Table 7.3. Numbers of fruit resulting from pollinations using known pollen donors (after Bertin, 1982).

Recipient no. (female)	Donor number (male)								
	2	3	4	5	7	8	9	11	12
2	—	0	9	5	11	0	8	0	0
3	3	—	9	14	10	12	12	1	1
4	15	15	—	3	5	15	2	4	13
5	16	13	4	—	0	5	3	3	11
7	14	11	3	0	—	3	4	7	11
8	0	14	15	0	1	—	1	10	14
9	6	8	2	1	4	3	—	8	9
11	12	2	11	13	12	23	19	—	15
12	0	0	8	13	10	8	7	0	—

(1982) crossed nine plants in all possible combinations and recorded the number of fruits set per crossing combination. If all crosses were equally successful, fruit set would be the same, but this was not the case (Table 7.3); some crossing combinations yielded no or few fruits, while others yielded many. Fruit production was unevenly distributed between pollen donors in all of the plants examined. Surprisingly, the average male fruit success of a particular plant showed a significant negative correlation with the average female fruit success of that plant.

In many species, we find that the closer the relationship of predominantly cross-fertilizing individuals, the lower the level of cross-compatibility and the higher the proportions of seeds or fruits that abort (Willson & Burley, 1983).

7.5 Crossing between populations

Although pollen dispersion usually occurs over short distances, some pollen is transported beyond the bounds of the population. In many outcrossing crops, more than 5% of the seed produced in one population have their pollen parents in another population, when the distance between populations is less than 100 m. As distance increases, the level of interpopulation crossing declines to the point where less than 1% of the seeds are hybrid; and this distance is well known to plant breeders who are concerned with the maintenance of varietal purity (a range of isolation distances is shown in Table 7.4 for animal and wind-pollinated species). For some species, interpopulation distances of 300 m are sufficient, but in others the distance necessary to preclude cross-mating may be over 1000 m. It is clear from Table 7.4, however, that interpopulation crossing, where it occurs at all, tends to be restricted to populations in close proximity.

The percentage of seed produced from interpopulation crossing depends in part on the size of a population's pollen pool. Populations with few plants in flower (or few flowers per plant) will experience higher levels of interpopulation crossing, than those producing a profusion of flowers, all else being equal. In wind-pollinated plants, the pollen rain from adjacent populations may be independent of local pollen production, so that the lower the local production, the higher the proportion of all pollinations made by extraneous pollen, and vice versa.

Pollen is also transported into populations by insect-pollinators. The lower the number of pollinations accomplished before pollen vectors leave the population, the greater the importance of imported pollen. When their resources (pollen and nectar) are abundant, insects tend to be 'held' in populations, and carry out large numbers of pollinations. When populations are small, or have few flowers, then pollinators are less site constant. The effect of resource avail-

Table 7.4. Isolation requirements for seed crops (from Kernick, 1961).

	Breeding system[a]	Pollination agent[b]	Isolation requirement (m)
Gossypium spp.	S	I	200
Linum usitatissimum	S	I	100–300
Lupinus spp.	S	I	500
Camellia sinensis	S	I	800–3000
Lactuca sativa	S	I	30–60
Avena sativa	S	W	180
Hordeum vulgare	S	W	180
Oryza sativa	S	W	15–30
Sorghum vulgare	S	W	190–270
Papaver somniferum	SC	I	360
Phaseolus spp.	SC	I	45
Vicia faba	SC	I	90–180
Pastinaca sativa	SC	I	500
Nicotiana tabacum	SC	I	400
Coffea arabica	SC	IW	500
Hevea brasiliensis	C	I	2000
Helianthus annuus	C	I	800
Daucus carota	C	I	900
Lycopersicon esculentum	C	I	30–60
Brassica oleracea	C	I	600
Allium cepa	C	I	900
Raphanus sativus	C	I	270–300
Cucurbita spp.	C	I	400
Zea mays	C	W	180
Secale cereale	C	W	180

[a] S—Predominantly self-fertilizing; SC—self- and cross-fertilizations similarly important; C—Predominantly or exclusively cross-fertilizing.
[b] I—insect pollinated; W—wind-pollinated.

ability on the level of interpopulation crossing is seen in a red clover test garden studied by Williams & Evans (1935); in the first year when few plants flowered and individual inflorescences were small, 45% of the progeny had paternal parents outside the population, whereas when the population had 'an abundance of blooms' in the second year, less than 2% of the progeny were fathered from extraneous sources.

7.6   The proximity of parents and progeny

The breeding structure of populations depends upon the proximity of plants and their progeny, as well as on the spatial pattern of mating. If progeny are distributed at random throughout a population, the relatedness of plants will be independent of distance, and crosses between neighbouring plants will produce the same kinds of progeny as those between distant plants. On the other hand, if progeny are clumped about their parents, then crosses between neighbours will tend to involve relatives.

It would be well to document actual offspring distances using genetic markers, in much the same way as pollen dispersion has been documented. Unfortunately, this has never been done, so our discussion deals primarily with the dispersion of seeds, which provides a potential distribution of seed parent to offspring distances rather than a realized one (see Chapter 6).

7.6.1   An experiment on parent-progeny proximity

The demonstration of spatial relationships of parents and their reproductive progeny in a canopy tree would be almost a lifetime project, even if the genetic markers necessary for the unequivocal demonstration of kinship were available. The problem is much less formidable in annual species. Levin (1983) used flower colour variation in *Phlox drummondii*, an annual with ballistic seed dispersal, to investigate progeny-parent distances. He sowed a line of seed of a pink flowered variant into a field of red-flowered *Phlox*; and in the following spring there was a single line of 89 pink-flowered individuals. The pink plants crossed amongst each other and with the red phloxes, producing pink and magenta-flowered progeny, respectively. The distribution of pink-flowered progeny relative to the sowing line in the second year is shown in Fig. 7.4a. The mean distance of pink plants from the line was 0.83 m and the maximum distance was 2.18. Clearly, progeny *are* established near their parents.

The field also contained hybrids whose location was determined both by pollen and by seed dispersal. Pollen dispersal was from the sown line to various red-flowered plants, whereas hybrid seed dispersal from red-flowered plants was random in direction.

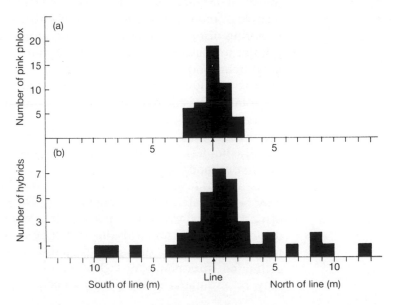

Fig. 7.4. The location of (a) pink flowered *Phlox*; and (b) hybrids, relative to the sowing line in a natural population. After Levin (1983).

Accordingly, the position of hybrids reflects the effective pollen dispersal distances (Fig. 7.4b). The mean distance from the line was 3.35 m and the maximum was 14.2 m, which means that the distance between mates is greater than the distance between seed plants and their progeny. However, in both instances the distances are small. Pollen and seed dispersion in *Phlox* is similar to that of many other animal-pollinated annuals, so the observations made in this study may be broadly representative of spatial relationships in annual plant populations.

7.7   Mating system models

Mating systems are considered within the context of neighbourhood and mixed mating models. The neighbourhood model uses data on pollen and seed dispersion, while the mixed mating model uses data on the genetic composition of progeny.

7.7.1   Neighbourhood model

The mating system of populations may be characterized by effective pollen and seed dispersal patterns; by the distances between mating plants and between seed-parents and their progeny. Wright's (1969) neighbourhood model is most useful for thinking about mating structure and its correlate, the genetic structure of populations. A neighbourhood is defined as the area from which the parents of some central individual may be treated as if they were drawn at random. If mating and seed dispersal distances were random, then the neighbourhood would encompass the entire population. With restricted dispersal distances, the area is very much less. The neighbourhood is a circle of radius 2σ with an area $A = 4\pi\sigma^2$, where σ^2 is the parent-offspring dispersal variance measured around a mean of zero and relative to a single reference axis passing through the population. The neighbourhood size is $N_e = A.d$ where d, the effective density, is approximately the density of flowering plants, if the population is constant in size and has a stable age structure, and if the distribution per parent of offspring reaching maturity is Poisson. Deviation from these conditions undoubtedly exists, and the effective density will usually be less than the number of flowering plants within a neighbourhood area.

Crawford (1984) has shown how the neighbourhood parameters may be estimated for plant populations. The total parent-offspring dispersal variance is $\sigma^2 = 0.5\ \sigma_p^2 + \sigma_s^2$, where σ_p^2 is the pollen dispersal variance and σ_s^2 is the seed dispersal variance. If there are no genetic markers, the pollen dispersal variance may be roughly estimated from pollinator flight distances. Seed dispersal variance may be estimated in the field, or under experimental conditions

Table 7.5. Neighbourhood estimates in herbaceous plants.

	Neighbourhood size	Neighbourhood area
Phlox pilosa	1409	108
Liatris cylindracea	1260	63
Liatris aspera	176	38
Viola rostrata	167	25
Viola pensylvanica	310	42
Primula vulgaris	175	30

where seeds are easier to follow. Neighbourhood size and area estimates for six animal-pollinated herbs are presented in Table 7.5. In this sample, neighbourhood area varied between 25 and 108 m^2, showing that the parents of an average individual would be drawn from within a relatively small radius. The number of individuals in that area (the neighbourhood size) varied from less than 200 to over 1000, indicating that the number of potential parent pairs for an individual is substantial (but, of course, these values are small compared to the case of random pollen and seed dispersal, where the neighbourhood would contain between 10^3 and 10^4 plants).

The neighbourhood properties of a species vary in space and time because pollen dispersion is not constant, so the values in Table 7.5 should only be viewed as representative. Levin & Kerster (1969) found substantial differences in the neighbourhood attributes of four populations of *Liatris aspera* which differed in plant density (Table 7.6). Neighbourhood size was positively correlated with density, varying from 45 in the sparse population (1 plant/m^2) to 363 in the dense population (11 plants/m^2). Neighbourhood area, on the other hand, was negatively correlated with density, varying from 45 m^2 in the sparse population to 33 m^2 in the dense population. Neighbourhood structure varies with density because pollen dispersion, as judged from inter-plant pollinator flight distances, was inversely related to density. In the sparse population, the axial variance was 2.30 compared with 0.36 in the dense population.

The neighbourhood diameter is useful in measuring the degree to which two subpopulations are isolated by distance. The neighbourhood concept is so formulated that gene migration over one neighbourhood diameter occurs at a given rate per generation, regardless of other neighbourhood parameters. Therefore, the greater the number of neighbourhood diameters separating two subpopulations, the greater their isolation, and the greater the potential for their genetic divergence. Looking back at the data for *Liatris aspera* (Table 7.6), we see that neighbourhood diameter is much greater in the sparse population than in the dense. This means that two clusters of plants, 100 m apart, in the dense population are isolated to a greater extent than those in the sparse population.

Neighbourhood size and area also provide insight into the extent

Table 7.6. Neighbourhood estimates in relation to plant density in *Liatris aspera* (after Levin & Kerster, 1969).

Population	I	II	III	IV
Density (plants m^{-2})	1	3.25	5	11
Pollinator flight σ^2 (axial)	2.30	1.24	0.67	0.36
Neighbourhood area (m^2)	45	38	35	33
Neighbourhood size	45	123.5	175	363

to which local populations may differentiate into distinctive subunits either in response to selection or genetic drift. Wright (1969) showed that even large populations distributed continuously over an area will differentiate if gene dispersal is restricted. This process is referred to as isolation by distance. The smaller the neighbourhood size, the greater the spatial differentiation of gene and genotype frequencies. A considerable amount of random subdivision would be expected if the neighbourhood size were less than 20, and a moderate amount would be expected if it were less than 200. With values as large as 1000, random differentiation would not occur. Thus, the values for the species in Table 7.5 suggest that their populations could undergo limited random local differentiation.

7.7.2   Mixed mating model

Another model used to analyse mating systems is referred to as the mixed mating model (see Clegg, 1980). The model divides the mating process into two components: 1) random mating; and 2) self-fertilization. The model rests on three fundamental assumptions: 1) mating is due to random outcrossing (with probability t) or self-fertilization (with probability $1-t$); 2) the same pollen population is available to each maternal plant; 3) the rate of outcrossing is independent of the maternal genotype. As we have already seen, these assumptions are not met in full. Nevertheless, the model provides considerable insight into the extent to which mating departs from random. The model differs from the neighbourhood model in specifically describing the *results* of mating.

The level of outcrossing is estimated from the gene frequencies in the pollen pool, and from the maternal parent genotype. In mixed mating models, deviations from random mating are reflected as self-fertilization, so that an obligatory out-crossing plant may have 'self-fertilization' rates $(1-t)$ of several percent, even though no self-fertilization actually occurred. Mating between relatives (in-breeding) is the prime contributor to this apparent selfing.

The level of random outcrossing as reviewed by Schemske and Lande (1985) ranges from less than 1% in some grasses (*Hordeum* and *Festuca*) to over 95% in some trees (*Pinus*) and herbs (*Oenothera*). The distribution of mean outcrossing rates for species is

Table 7.7. Interpopulational range of outcrossing
rates in flowering plants. Listed in order of increasing
magnitude of ranges (after Schoen, 1982).

Taxon	Range of outcrossing rates
Hordeum jubatum	0.01–0.03
Avena barbata	0.01–0.08
Trifolium hirtum	0.01–0.10
Hordeum spontaneum	0–0.10
Eucalyptus obliqua	0.64–0.84
Eucalyptus pauciflora	0.62–0.84
Lupinus affinis	0–0.29
Helianthus annuus	0.60–0.91
Plectritis congesta	0.48–0.80
Lupinus bicolor	0.13–0.50
Limnanthes alba	0.43–0.97
Clarkia exilis	0.43–0.89
Clarkia tembloriensis	0.08–0.83
Gilia achilleifolia	0.15–0.96
Lycopersicon pimpinellifolium	0–0.84
Lupinus nanus	0–1.00

bimodal, the extreme classes of primary selfing ($t \leq 0.2$) and primary outcrossing ($t \geq 0.81$) containing most of the species. Of particular interest is the interpopulation range of outcrossing rates in many species (Table 7.7). Populations of *Lupinus nanus* range from almost complete selfing to almost complete random outcrossing, and similarly wide ranges have been reported from *Lycopersicon pimpinellifolium* and *Gilia achilleifolia*. In some species, differences in outcrossing are related to a particular habitat variable so that, for example, outcrossing in *Avena barbata* is higher in mesic habitats than in xeric ones (Clegg, 1980). The reasons for this are yet to be determined, but it is possible that more pollen is produced per plant in moist than in dry conditions (Chapter 12).

Interpopulation differences in mating pattern have genetic as well as environmental causes. This is seen in the perennial, insect pollinated *Thymus vulgaris*. Domée (1981) planted several genotypes (clones) in sites containing natural populations of thyme. Genotypes differed in selfing rates within sites, and single genotypes differed in rates between sites; for example, genotype P_4 had an outcrossing rate of about 94%, whereas C_7 had an average rate of less than 60%, varying between sites from 20% up to 80%. The average outcrossing rate for all genotypes at a site ranged from 95% to less than 60%.

Intergenotypic differences in selfing may be due to differences in pollinator attraction, the proximity of stigmas and anthers, the synchrony of stigma and anther maturation, or the level of self-compatibility. An informative study on this subject was carried out on the bee pollinated *Nicotiana rustica* by Breese (1959), who found that populations with anther dehiscence shortly after flower opening had about 20% outcrossing, compared with 40% in those where anther dehiscence was delayed by two to three days. Higher out-crossing rates in the flowers with delayed anthesis are the result of lower rates of self-pollination relative to cross-pollination by visiting bees. The position of the anthers relative to the stigma is also important, and can affect the rate of self-pollination (both mechanically and by bees). Populations with the stigmas above the anthers had higher rates of outcrossing than the others (30% com-pared with 20% in populations where the anthers were level with, or lower than the stigma). However, differences in the timing of dehiscence and stigma receptivity were more important in affec-ting outcrossing than were the proximity of anthers and stigma (Breese, 1959).

Pollinators may discriminate between floral variants, giving the preferred ones more service, with the consequence that some plants will receive more cross-pollen than others, and phenotypic dif-ferences in outcrossing rates will arise. We understand this process best where it involves different petal colours. Consider the case of the morning glory *Ipomoea purpurea* which is pollinated primarily by bumble-bees. It is characterized by striking colour polymorphism, including white, pink and dark blue. Brown & Clegg (1984) found that there was a significant preference for blue or pink over white, but no preference between pink and blue, which leads us to predict that the white variant would have a lower outcrossing rate than the other two. This, indeed, was the trend observed; $t = 0.48$ in white, 0.73 in pink and 0.69 in blue.

The amount of pollinator service is also dependent upon the number of flowers per unit area. Plants in sparse populations may receive less service per flower than those in dense populations, so that the outcrossing rate declines with plant density. The effect of spacing on selfing is clear in *Thymus vulgaris*; when individuals are more than 1 m apart, the percentage of selfing reaches 80%, whereas when they are less than 1 m apart the percentage selfing varies between 5 and 40% (Valdeyron *et al.*, 1977). The effect of plant numbers on selfing is evident also in *Nicotiana rustica*, where populations of 1000 plants had an outcrossing average of 26%, compared with only 7% in populations of 6 plants (Breese, 1959).

The level of self-fertilization may be a function of the total number of flowers per plant. Crawford (1984) estimated that plants of *Malva moschata* with fewer than 100 flowers had selfing rates less than 0.4 whereas those with more than 500 flowers had rates exceed-ing 0.7. He also found negative correlations between the daily selfing rates of single plants and the numbers of flowers on them.

7.7.3   Cleistogamy versus chasmogamy

In most species with mixed mating systems, all flowers have the potential for both cross- or self-fertilization. However, some species have cleistogamic and chasmogamic flowers. The cleistogamic flowers are preordained for self-fertilization, while the chasmogamic flowers may undergo different levels of cross-fertilization, depending upon the compatibility system and the mating pattern. In *Lithospermum caroliniense*, cleistogamy is paired with heteromorphic incompatibility, so that all chasmogamic seeds are cross-fertilized. The rate of outcrossing is thus equivalent to the proportion of the total seed production which comes from chasmogamic flowers. In a survey of 14 populations, Levin (1972) found that outcrossing varied from 52–100%.

In most species with cleistogamic flowers, the chasmogamic flowers are self-compatible, so the level of cross-fertilization cannot be measured from flower-specific seed production. Nevertheless, the balance in the breeding system may be judged from the proportion of the two seed types; the greater the proportion of chasmogamic seeds, the greater the level of outcrossing. Consider the genus *Impatiens*; mean chasmogamic seed production per population varies from 0–90% in *I. capensis* (Waller, 1980), 54–100% in *I. pallida*, and 72–98% in *I. biflora* (Schemske, 1978).

Greenhouse experiments and measurements of *I. capensis* in the field revealed that the level of chasmogamic flower production was positively correlated with light and soil moisture, through their effects on plant growth (Waller, 1980). Larger plants had a higher percentage of chasmogamic flowers. In one population, the first chasmogamic flowers were predictably located in the axils of the 8th, 9th and 10th nodes. Smaller plants never produced this many nodes, so they never produced any chasmogamic flowers. There are several other examples, especially in grasses, of cleistogamy being more pronounced when site conditions are unfavourable (Schemske, 1978).

There have been few studies demonstrating genetic control of the degree of cleistogamy, and only one in which the genetic and environmental control was quantified in nature. Clay (1982) reported that a population of *Danthonia spicata* contained plants with 12–65% cleistogamic flowers. From an analysis of progeny cloned and planted back into the field, he estimated that 53% of the phenotypic variance in the field was attributable to genetic factors.

7.8   Consequences of self-fertilization

Self-fertilization brings about predictable changes in the genetic composition of populations, and consequently, in the characters of the plants within them. I shall consider the genetic issue first, and then its primary consequence, inbreeding depression.

7.8.1   Changes in the genetic composition of populations

The level of self-fertilization in populations is an important determinant of their genetic composition. In order to appreciate how this may be, it is necessary to understand the results of repeated self-fertilization. Consider a population in which all individuals are heterozygous (Aa) at a single locus, and which reproduces only through self-fertilization. After one generation of self-fertilization, heterozygosity declines to 50% and both homozygotes are present with equal frequency (Table 7.8). In the next generation, homozygotes have only homozygous offspring (like themselves), while only half the progeny of the heterozygotes are heterozygous. The level of heterozygosity is thus halved each generation so that after, say, ten generations, the population is almost completely homozygous at a given locus.

Table 7.8. Results of selfing in a population started entirely with heterozygotes, Aa.

Generation	Genotypic frequency			Heterozygote frequency
	AA	Aa	aa	
0	0	1	0	1
1	1/4	1/2	1/4	0.5
2	3/8	1/4	3/8	0.25
3	7/16	1/8	7/16	0.12
4	15/32	1/16	15/32	0.06
5	31/64	1/32	31/64	0.03
n	$1-(1/2)^n$	$(1/2)^n$	$1-(1/2)^n$	$(1/2)^n$
	1/2	0	1/2	0

The rate at which heterozygosity declines depends on the level of self-fertilization; the lower the level, the less the reduction in heterozygosity from one generation to the next. In the absence of other factors, populations would eventually become homozygous even if the level of self-fertilization was low, because the level of heterozygosity would decline in each generation.

Self-fertilization in any degree leads to a reduction in genetic variability within populations, measured in terms of genotype number, proportion of loci which are polymorphic (i.e. have more than one allele), and number of alleles per locus (Wright, 1969). Finally, the greater the level of self-fertilization, the less is the impact of pollen from distant sources, and thus the more isolated are populations or assemblages within populations. Accordingly, the greater the level of selfing, the greater the proportion of the total species variation that will reside between populations, and the greater the proportion of the total variation within populations that will reside between sub-populations. Selfing promotes divergence within and between populations, independent of the effects of natural selection.

7.8.2   Inbreeding depression

The mating pattern determines the genotypic structure of the next generation at the zygotic stage of the life cycle. If all zygotes had the same likelihood of developing into viable seeds regardless of genotype, and if germination, establishment and survivorship were also independent of the genotype, then the genotypic composition of zygotes and mature plants would be similar. However, genotype independence does not appear to be the rule. Most species display inbreeding depression, and the products of self-fertilization and matings between relatives have lower survivorship and vigour than the products of matings between unrelated plants.

Self-fertilization

In outcrossing species, the total inbreeding depression due to selfing is often considerable, frequently exceeding 50% (i.e. the products of selfing may, on average, leave less than half the number of progeny as the products of outcrossing; Lande & Schemske, 1985). The average individual is heterozygous for one or more recessive lethal factors, as well as for many deleterious genes with small, nearly additive effects on fitness. Habitually self-fertilizing species usually have much smaller inbreeding depression than outbreeders.

The effects of selfing are most pronounced during seed development. Seed abortion in conifers and angiosperms is usually 2–10 times greater following self-fertilization than cross-fertilization. Levin (1984) analysed the effect of self-fertilization on seed mortality in several populations of *Phlox drummondii*, an almost exclusively cross-fertilizing annual. All populations exhibited rather low levels of seed inviability with outcrossing (mean = 12.6%; range from 3–25%), but in every population the percentage seed-inviability was greater with selfing, and reached 41% in one population. For all populations, the average abortion rate following selfing was twice that with outcrossing, and the relative survivorship of self to outcrossed seed varied from 0.57–0.98 with an average of 0.83.

The germination and survivorship of juveniles of self relative to cross progeny are poorly understood for herbaceous plants under field conditions. One of the most informative studies was conducted by Schemske (1983) on species of the neotropical perennial *Costus*. He planted self and outcross seeds of *C. allenii* and *C. laevis* embedded in their natural vegetation along sun and shade transects, and subsequently monitored germination, seedling survivorship and biomass. The percentage germination of selfed seeds was less than that of outcross seed in both sun and shaded sites. Seedling survival did not vary in relation to parentage, but the biomass of outcross plants was greater than self plants in both sites.

A similar study was conducted by Schoen (1983) on the Californian annual *Gilia achilleifolia*. He planted cross and self seed in natural sites, monitoring germination, survivorship, and fecundity, from which he calculated net reproductive rates. The response to selfing was rather different than in *Costus*. Whereas self seed germinated as well as outcross seed, self seedling establishment was only 11% compared with 16% for outcross progeny. Seedling to adult survivorship was also less in the self progeny (Fig. 7.5), so that the net reproductive rate of the crossed plants was 19.9 compared with 11.1 for the selfed plants.

Germination and survivorship of self progeny are also relatively low in conifers and other tree species. Consider the case of *Eucalyptus regnans*; the average germination rate from self-pollination was only 63% of the level obtained from cross-pollination. Transplanted seedlings were followed over a 12 year period in the field; cross progeny had higher survivorship than self progeny. The difference between self and cross survivorship depended upon the maternal tree and the difference was exaggerated with time. Cross progeny also grew faster as reflected in their greater heights and diameters (Table 7.9).

In the redwood *Sequoia sempervirens* self seed had about 5% germination compared with 25% for cross seed (Libby *et al.*, 1981). Grown under conditions of high root-rot incidence, seedling survivorship was about 15% for self versus 60% for cross progeny, whereas with low levels of root-rot, self seedlings had 80% survivorship compared with 85% for the cross. These data demonstrate

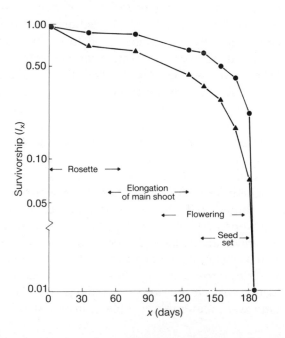

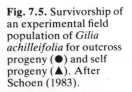

Fig. 7.5. Survivorship of an experimental field population of *Gilia achilleifolia* for outcross progeny (●) and self progeny (▲). After Schoen (1983).

Table 7.9. Comparison of selfed (S) and open pollinated (O) families of four *Eucalyptus regnans* trees (after Eldridge & Griffin, 1983).

Trait	Age from planting yr	Parent tree number 189		192		184		193	
		S	O	S	O	S	O	S	O
Survival %	1/2	100	100	91.6	95.8	95.8	95.8	100	100
Survival %	1–1/2	95.8	95.8	83.3	79.2	83.3	95.8	66.7	91.6
Survival %	5	95.8	95.8	70.8	70.8	41.7	91.6	62.5	87.5
Survival %	9	91.6	95.8	62.5	62.5	33.3	91.6	62.5	87.5
Survival %	12–1/2	62.5	83.3	25.0	45.8	20.8	70.8	16.7	75.0
Height m	1/2	0.55	0.59	0.56	0.48	0.59	0.67	0.49	0.75
Height m	1–1/2	2.06	2.07	1.93	1.90	1.81	2.10	1.68	2.46
Height m	5	9.75	9.88	8.14	8.44	7.86	9.27	8.23	9.97
[a] Height m	9–1/2	23.6	24.2	20.7	20.9	19.3	22.2	16.7	24.1
Diameter cm	5	11.4	12.7	9.2	9.7	10.4	11.2	8.4	12.4
Diameter cm	9	16.7	19.6	14.3	15.8	16.0	18.0	10.6	19.0
Diameter cm	12–1/2	21.1	23.9	21.9	22.5	2.29	25.2	15.6	25.1

that the 'cost' of self-fertilization depends upon the habitat; the more extreme the conditions, the greater the liability of self-fertilization.

Mating between relatives

The quality of outcross progeny may also vary as a function of the distance between parents, because parental relatedness is negatively correlated with distance (Section 7.6). The more restricted the dispersion of seeds and pollen, the stronger becomes the inverse relationship between distance and relatedness. One of the most interesting examples of distance-dependent progeny quality comes from the work of Coles & Fowler (1976) on white spruce *Picea glauca*. They crossed several mother trees with their own pollen and with pollen from other trees located up to 3200 m away, and scored

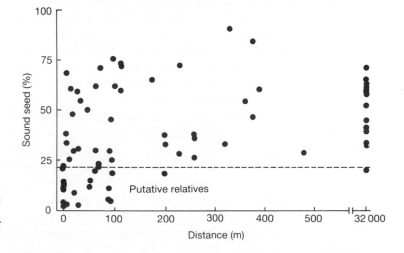

Fig. 7.6. Percent sound seed in relation to the distance between the mother tree and the pollen parent in a population of *Picea glauca*. Note that the average percentage of sound seed increases with separation between male and female parents. After Coles & Fowler (1976).

the percentages of viable seed (Fig. 7.6). Crosses between trees less than 100 m apart had a lower percentage viable seed set than crosses between parents more than 200 m apart. Many of the short-distance crosses gave viability percentages similar to those obtained by selfing, which suggests that the trees involved were probably close relatives.

The only study of seedling mortality as a function of crossing distance involved the herbaceous perennial *Delphinium nelsoni*. Waser & Price (1983) made a set of crosses with pollen and egg parents 1, 3, 10, 30, 100 or 1000 m apart. The resulting seeds were planted in a nearby plot and seedling survivorship was monitored for two years. Only 5% of the self seedlings survived to year 2, while survivorship varied between 8 and 12% for the crosses up to 100 m, but showed no consistent trend with distance. However, survivorship from crosses 1000 m away was significantly higher (16%), suggesting that plants within 1000 m of each other are more closely related than those 1000 m apart.

7.9   Asexual reproduction

The flowering plants employ a wide variety of asexual reproduction methods (apomixis) which complement or replace the sexual process (Heslop-Harrison, 1982). The most familiar processes are those which come under the heading of vegetative reproduction (clonal growth). These depend on the distribution and separate establishment of physiologically independent individuals from segments of the parent plant (Chapter 9). These segments may be apical meristems carried on stolons as in strawberry, fragments of the shoot as in jointed cactus, underground rhizomes as in many grasses, or small propagules, readily detachable from the shoot, such as the bulbils of onions. In a sense, sexual and asexual reproduction are competitive, because they depend upon the same limited resources (Chapter 12), and the balance between the two often depends upon soil conditions, light and temperature, and on the intensity of competition with neighbours (Abrahamson, 1979). In general, the balance shifts towards vegetative reproduction in unfavourable years or habitats, and towards sexual reproduction in better conditions or when population density increases.

In certain species, normal seed is produced but no sexual fusion occurs during its production; progeny have exactly the same genetic constitution as their parents. This process, known as agamospermy, has the advantages of the seed habit (dispersal and the potential for extended dormancy) added to the advantages of asexual reproduction (copying a successful genotype and no 'wastage' of resources on male function). In some species, pollination is required to stimulate seed development, even though fertilization does not occur. The most common form of agamospermy involves the formation of seed from a diploid embryo sac.

The level and nature of agamospermy may vary between populations. Four populations of the New Zealand grass *Agropyron scabrum* showed four different patterns; in one the plants were completely sexual, in another they were partially apomictic, in another predominantly apomictic, and in a fourth they were obligatorily apomictic (Hair, 1956). Apomictic and sexual populations of many species have different ecogeographical distributions. Consider *Antennaria parlinii*, which is a dioecious agamospermous perennial herb of the eastern United States. Its populations may consist of male and female plants which produce only sexual seed, or they may consist entirely of female plants producing agamospermous seed. Bayer & Stebbins (1983) found that populations in the unglaciated region of Ohio were sexual, whereas nearly half those in the glaciated region were exclusively, and others partially, agamospermous.

The genetic control of agamospermy has been comprehensively reviewed by Asker (1980). Agamospermy is most commonly controlled by two unlinked loci, between which there may be functional interactions. Typically, the alleles controlling this character are recessive.

7.10   The amount and distribution of genetic variation

7.10.1   Protein polymorphism

Geneticists and evolutionists have long sought to describe genetic variation in natural populations. The application of protein electrophoresis during the past two decades has revolutionized our understanding of the amount and distribution of genetic variation, and has permitted interspecific comparisons. However, the struggle to *understand* variation in morphological, physiological and biochemical characters still continues.

Electrophoresis

Protein electrophoresis has been used to determine the genetic correlates of different levels of selfing and outcrossing in populations. The primary evidence observed in studies of electrophoretic variation is a band of colour in a slab of starch or acrylamide gel. The coloured bands, which are produced by applying specific reagents to the gel, denote the position to which a protein or enzyme has migrated during electrophoresis. By analysing such gels, it is possible to determine whether a particular homologous enzyme extracted from different plants has the same, or different, mobility. Different molecular forms of an enzyme may be coded by different alleles at the same locus, and are called allozymes. Most allozymic differences are reflected in migration differences, and hence show up as different bands on the gel.

Electrophoresis has a number of advantages for describing variation: 1) inheritance of electrophoretically detectable protein variants can easily be demonstrated; 2) most protein variants at a single locus (allozymes) are codominant, and allele frequencies can be calculated without the need for genetic crosses; 3) genetic variation can be compared directly between populations or species. Before the advent of electrophoresis, the population geneticist studied morphological characters governed by many genes, whose effects were neither distinguishable from environmental influence, nor individually recognizable. The main drawback of electrophoresis, is that we do not know the effect of the observed genetic variation on the fitness of the phenotype, nor which phenotypic traits are altered by each allozyme. Electrophoretic techniques, the interpretation of electrophoretic data, and the application of such data to the solution of genetic and evolutionary problems, have been comprehensively reviewed by Gottlieb (1981); other reviews describe enzyme polymorphisms within and between plant species (Brown, 1979; Hamrick *et al.*, 1979; Loveless & Hamrick, 1984).

Amount of variation

Protein electrophoresis provides fundamental information on the amount of genetic variation in plant populations in the form of: 1) the proportion of protein loci that are polymorphic; 2) the number and relative frequency of alleles at these loci; and 3) the proportion of gene loci heterozygous per individual. Allozymic variation in outcrossing and selfing species is compared in Table 7.10; as predicted by theory, outcrossers are more variable than selfers. They have a higher proportion of polymorphic loci, a larger number of alleles at

Table 7.10. Allozyme variation in plants. Differences between selfers and outcrossers significant at 0.01 (**) or 0.001 (***). L = mean proportion of loci polymorphic; A = mean number of alleles at polymorphic loci; HET = observed mean heterozygosity in an average individual. Data based on an average of 16 loci, 9 enzymes and 11–14 populations (after Gottlieb, 1981).

Species	Category	L (species)	A	L (population)	HET
Selfers	Mean	0.183	2.260	0.044	0.001
	SE	0.032	0.099	0.014	0.0004
	n	28	19	28	28
Outcrossers	Mean	0.51***	2.90**	0.37***	0.086***
	SE	0.06	0.17	0.05	0.017
	n	21	20	21	21

each polymorphic locus, and greater mean heterozygosity than selfers.

Variation differences are apparent at many levels. For example, there ar two closely related annual species of the genus *Phlox* which are endemic to central Texas; *P. drummondii*, an outcrosser, and *P. cuspidata*, a selfer (Levin, 1978b). Whereas both were polymorphic at 20% of the 20 loci analysed, the mean percentages of loci which were polymorphic per population were 5.5% in the selfer and 14.5% in the outcrosser; the mean number of alleles per polymorphic locus per population was 1.2 for the selfer and 1.7 for the outcrosser; the mean level of heterozygosity was only 0.8% for the selfing *P. cuspidata* compared with 5.2% for the outcrossing *P. drummondii*.

Schoen (1982) investigated variation between conspecific populations, and showed that a group of predominantly outcrossing populations of *Gilia achilleifolia* have a higher percentage of polymorphic loci per population (52 versus 31), more alleles per locus (a mean of 1.6 versus 1.3) and greater heterozygosity (0.14 versus 0.06) than a group of predominantly inbreeding populations. The difference between outcrossing and selfing populations within a species is carried to the extreme in *Clarkia xantiana*, where outcrossing populations were polymorphic at 16 out of 32 loci, while the selfers showed no variation at all for the same loci (Gottlieb, 1984b).

Apportionment of variation

We may now ask how this variation is apportioned within and between populations. The total allelic diversity (H_T) for each polymorphic locus is calculated as follows.

$$H_T = 1 - \Sigma p_i^2$$

where p_i is the mean frequency of the ith allele at a locus. If the frequency of two allozymes at a locus were each 0.5, then for that locus $H_T = 1 - (0.25 + 0.25)$. This total allelic diversity can be partitioned into the allelic diversity within populations (H_S) and the allelic diversity between populations (D_{ST}), so that $H_T = H_S + D_{ST}$. The proportion of the allelic diversity due to the between population component is the ratio D_{ST}/H_T, and mean values for H_T, H_S and D_{ST} for a species are obtained by summing over all polymorphic loci.

The mating systems of species have a significant effect on the amount and distribution of genetic variation (Hamrick, 1983; Table 7.11). Outcrossed, animal-pollinated species and mixed mated species have the highest H_T values, while selfing species have the lowest. On average, selfers have a greater proportion of allelic variation distributed between populations than species with mixed mating, and these have a greater proportion distributed between populations than outcrossing species. The difference between selfers

Table 7.11. The relationship between reproductive traits and the distribution of allozyme variation within and between natural plant populations (after Hamrick, 1983).

Characteristic	Number of studies	H_T	H_S	D_{ST}/H_T
Mode of reproduction				
Asexual	1	0.172	0.159	0.080
Sexual	108	0.280	0.194	0.284
Both	13	0.325	0.257	0.209
Mating system				
Selfed	35	0.250	0.141	0.437
Mixed-animal	26	0.284	0.181	0.304
Mixed-wind	3	0.712	0.560	0.189
Outcrossed-animal	23	0.352	0.238	0.221
Outcrossed-wind	35	0.256	0.248	0.056
Seed dispersal mechanism				
Large	22	0.260	0.181	0.305
Animal-attached	13	0.262	0.155	0.425
Small	37	0.312	0.196	0.349
Winged or plumose	38	0.260	0.238	0.073
Animal-ingested	12	0.344	0.210	0.330

H_T = total allelic diversity; H_S = mean allelic diversity within populations; D_{ST}/H_T = the ratio of the allelic diversity among populations to the total allelic diversity.

and wind-pollinated outcrossers is quite striking; selfers have 44% of their variation between populations compared with only 6% for wind-pollinated species. It is noteworthy that wind-pollinated species have lower proportions of variation between populations than animal-pollinated species in both mixed mating and outcrossing categories, suggesting that the level of interpopulation crossing may be higher in wind-pollinated species.

We compared outcrossing and selfing *Phlox* and *Gilia* in terms of amounts of variation earlier in this section. Now let us see how the mating system affects the apportionment of this variation. In the selfing *P. cuspidata* 40% of the variation is between populations, whereas in the outcrossing *P. drummondii* only 25% is between populations (Levin, 1978b). In *Gilia achilleifolia*, the predominantly selfing populations had 39% of their variation distributed between populations, while the predominantly outcrossing populations had only 17% (Schoen, 1982).

As we have seen (Chapter 4), populations are defined by the boundaries we draw round them, and they can be arbitrarily subdivided into a series of quadrats of equal size. We expect selfers to show a greater proportion of the total allelic diversity distributed between populations than outcrosses (because self-fertilization is equivalent to zero distance gene dispersal). Thus subunits within a population are isolated by distance to a greater extent in selfers than in outcrossers; subpopulations of selfers may be rather different from

Table 7.12. Studies of microgeographic differentiation in plants (after Brown, 1979).

	Sub popns.	H	$D_{ST}/H(\%)$
Inbreeders			
Avena barbata	7	0.27	35
	17	0.27	50
	20	0.29	15
Avena fatua	16	0.15	59
Bromus mollis	10	0.21	3
Outbreeders			
Lolium multiflorum	4	0.26	1
	4	0.31	1
Silene maritima	6	0.32	4
Liatris cylindracea	66	0.09	7
Pinus ponderosa	8	0.33	3

one another, whereas those of outcrossers are very much alike (Brown, 1979; Table 7.12). For example, in two studies on predominantly selfing, annual wild oats *Avena*, about 50% of the variation was between subpopulations.

Variation in asexual species

We are only just beginning to understand the amount and distribution of genetic variation in asexual species. So far there does not appear to be a substantial difference at the population or species level, when all species are considered together (Table 7.11). As far as differences within a species are concerned, however, Usberti & Jain (1978) found that sexual populations of the grass *Panicum maximum* had higher total allelic diversity ($H_T = 0.40$) and lower between population variation ($D_{ST}/H_T = 0.03$) than had asexual populations ($H_T = 0.17$; $D_{ST}/H_T = 0.08$). The amount of within population diversity was also greater in the sexual populations ($H_S = 0.38$) than in the asexual ones ($H_S = 0.16$). Although variation in asexually reproducing populations of *P. maximum* was relatively low, it is important to recognize that local populations which are predominantly asexual may have very high levels of allelic diversity (see Chapters 4 and 10). For example, the trembling aspen *Populus tremuloides* is a deciduous tree whose primary mode of reproduction is asexual (by root suckers), yet Chelik & Dancik (1982) found that in seven populations the mean allelic diversity was 0.42.

Up to this point, we have measured variation in partially asexual species in terms of the number and frequencies of alleles at polymorphic loci. Now let us consider the number of genotypes per sampled individual. Lyman & Ellstand (1984) have compiled information (Table 7.13) which shows that clonal plant species have less than 0.1 genotypes per individual sampled. In other words, if 100

Table 7.13. Genotypic diversity among populations of clonal plant species (after Lyman & Ellstrand, 1984).

Taxon	Pops/N	Clones	Clones/Individuals
Erigeron annuus	3/300	17	0.056
Taraxacum officinale	3/284	4	0.014
Taraxacum officinale	22/518	47	0.091
Lycopodium lucidulum	16/242	19	0.078
Oenothera biennis	3/75	4	0.053
Oenothera biennis	44/2200	46	0.021
Oenothera laciniata	60/2400	108	0.045

plants were sampled from a study population, on average less than 10 different genotypes would be obtained. This contrasts with sexual species in which the number of genotypes per individual sampled approaches 1.0 in all but the most highly inbred species (see Chapter 4). Thus a sexual population will almost certainly be more genetically diverse than an asexual population of the same size.

Variation in relation to seed dispersal

The genetic structure of a species is a consequence of seed dispersal as well as of the mating system. The greater the mean and variance of dispersal distances within and between populations, the greater the allelic diversity within populations, and the lower the proportion of the total diversity residing between populations. Hamrick (1983) discusses genetic diversity in relation to seed dispersal mechanisms. He shows that there are no striking differences between species with large or small, winged or plumed, wind-dispersed or animal-ingested seeds, in total allelic diversity, or allelic diversity within populations (Table 7.11). However, there are large differences in the apportionment of diversity. Species with winged or plumed seeds have little variation between populations (7%), but species with other kinds of seeds have at least 30% of their variation between populations. It is tempting to conclude that species with airborne seeds experience greater interpopulation seed exchange, but this result may simply be an artefact. Most of the species which have wind-dispersed seeds happen to be wind-pollinated conifers, and, as a group, these plants show little between population variation. Thus it seems likely that interpopulation pollen exchange rather than interpopulation seed exchange is responsible for the small proportion of total allelic diversity between populations of species with airborne seeds.

Long distance dispersal may bring seeds into suitable sites unoccupied by the species, and a new population may be founded. Genetic differences between source and colonial populations are best known in plants which have been introduced into a new continent during the past 200 years. Clegg and Brown (1983) show how colonial populations have much less genetic variation than the source populations (although identifying the precise source of alien plants is often extremely difficult). For example, the Mediterranean source populations of the grass *Avena barbata* have an average allelic diversity of 0.43 compared with 0.18 for the alien populations in California. Similarly, English source populations of the grass *Bromus mollis* have an average allelic diversity of 0.23 compared with 0.18 in Australia where the grass is introduced. The loss of genetic variation associated with intercontinental colonization is expected because: 1) initial population sizes are small; 2) there is no migration to enrich or maintain genetic variation; and 3) the new environment may have novel, or unusually severe selective pressures.

Genetic differences between a disjunct population of *Phlox pilosa* ssp. *detonsa*, several hundred kilometres and probably a few thousand years removed, and the subspecies as a whole, have been studied using electrophoretic procedures (Levin, 1984). The disjunct form has fewer polymorphic loci and fewer alleles per polymorphic locus. It is missing alleles present in the progenitor, but has no new ones, and is less heterozygous.

7.10.2 Morphological and physiological variation

So far, our discussion of genetic variation has been confined to allozyme loci. This is not to suggest that loci controlling morphological, physiological, biochemical or phenological traits are invariant. Indeed, there is considerable genetic variation present for these characters, both within and between populations (Bradshaw, 1984a; Venable, 1984; Lawrence, 1984). The spatial pattern of variation in these characters is more likely to be shaped by natural selection than is the pattern of allozymes, and as such, the effect of the mating system alone is likely to be confounded. In a sense, the mating system is like the producer of a play who chooses the participants. Natural selection, on the other hand, is like the director, who determines the roles of each participant and the amount of time they spend on stage.

Because plants are sedentary, patterns of differentiation of population tend to follow very closely the underlying patterns of the environment (Chapter 4). The various characters of morphology and physiology are typically controlled by different gene systems, and, since all characters are not subjected to the same selective differentials in space, most characters will have unique patterns of variation in space.

Studies of adaptive differentiation sometimes include genetic investigations of character differences. One of the classic studies was conducted by Clausen and Hiesey (1958) on *Potentilla glandulosa*. They crossed plants from coastal and alpine populations and, from

Table 7.14. Estimated minimum number of genes controlling the differences between the coastal and alpine races of *Potentilla glandulosa* in fourteen characters (after Clausen & Hiesey, 1958).

Character	Genes	Character	Genes
Winter dormancy	3	Leaflet number	1
Flowering time	many	Seed weight	6
Density of inflorescence	1	Seed colour	4
Glandular pubescence	5	Sepal length	5
Anthocyanin	4	Petal width	2
Stem length	many	Petal length	4
Leaf length	many	Petal colour	1

patterns of variation in the F_2 generation, they estimated the minimum number of genes governing the inheritance of several characters. They showed that some characters, like leaflet number and petal dimension, are controlled by one or a few genes, while others, like flowering time and stem length, are controlled by many (Table 7.14).

Not all observed patterns of variation match obvious environmental gradients, and certain patterns may result from genetic drift rather than natural selection. Perhaps the best demonstration of this is in the diminutive, self-incompatible annual *Linanthus parryae*. The character in question is a flower colour, white or blue, which is under the control of a single gene. Epling and Dobzhansky (1942) described flower colour variation at 427 stations in the Mojave desert. The distribution pattern of the 10% blue-flowered plants was complex; there were enclaves of pure blue and pure white, as well as various mixtures of the two morphs. Populations of pure or predominantly white plants may occur within a few hundred metres of others of pure or predominantly blue plants. The frequencies of the alleles for flower colour become more divergent as the distances between stations increase. At distances of 250 m or less, the correlation in gene frequencies between samples is greater than 0.8, at 800 m it is 0.65 and declines further to 0.58 at 1.6 km and 0.17 at 13 km. This variation pattern is probably due to restricted pollen and seed dispersal, and a small neighbourhood size (of the order of 100).

If *Linanthus* populations have small neighbourhood sizes, some populations might be expected to undergo substantial change in genetic composition over time. This issue was addressed by Epling and his colleagues (1960) who studied morph proportions in fixed quadrats for eight years during the period 1944–1953. They found that the frequencies of blue and white morphs remained remarkably stable in spite of large fluctuations in density. This stability is attributed in part to restricted pollen and seed dispersal, but also to the long-lived seed bank which buffers the population against changes due to genetic drift, selection or immigration.

A long-lived seed bank from which germinating individuals are drawn causes an extensive overlapping of generations, which may have important genetic consequences (Templeton & Levin, 1979). For those loci that yield constant fitness effects, the seed bank retards the annual rate of gene frequency change. The buffering property of the seed bank can also greatly reduce the 'fitness uncertainty' generated by cyclic or random environments, and can free the population from genetic responses to the fitness conditions realized in every year.

7.11   The genetic variable in ecology

In closing, let us consider the role of the genetic variable in plant ecology. Bradshaw's recent observation is particularly apropos; he writes that 'until now ecologists . . . have, perhaps unintentionally, tended to envisage that the range of habitats which a species occupies is related to its innate physiological tolerance, properties which can be measured in single genotypes. It is imperative that we now look for evolutionary causes' (Bradshaw, 1984b). As he and others point out, the ability of species to invade different habitats, to develop and reproduce, and to persist in those habitats in spite of environmental change, depends upon the amount and organization of genetic variation within and between populations. The greater the amount of variation, and the more it is distributed between populations, the greater will be the species' ecological amplitude in time and space. Through its effect on variation, the mating system and parent-offspring proximity become important determinants of ecological amplitude both locally and across the range of the species.

Chapter 8 Life History and Environment

MICHAEL J. CRAWLEY

8.1 Introduction

A central theme running through this book is that fundamental processes like seed production, growth, mortality and dispersal occur in the context of very small numbers of individuals, and that the spatial arrangement of plants is of vital importance in determining their fitness. Unlike animal ecology, where the individuals are assumed to blunder about at random, like molecules obeying the ideal gas laws, each individual does *not* have the potential to meet and interact with every other. For example, plants compete strongly with only the six or so individuals in their immediate vicinity (their 'first order neighbours'). Furthermore, this competition is highly asymmetric, with the larger plants exerting a much greater influence on the fitness of the smaller individuals than vice versa (Chapter 4). Plants exchange pollen over very limited distances, even in wind-pollinated species (Chapter 7), and most seeds fall in the immediate vicinity of their maternal parents (Chapter 6). While the influence of herbivorous animals and plant pathogens may be felt over areas of several to many hectares (Chapter 5), it is clear that most of the important interactions between plants are extremely local. This means that theoretical plant ecology needs a different logical base than the 'gas laws' of theoretical animal ecology (Pacala & Silander, 1985).

The emergent properties exhibited by plant communities (diversity, biomass, evenness, productivity, and so on) result from the interactions of *individual* plants with their immediate physical environments, and with a very limited circle of neighbouring plants of the same or of different species (Chapter 1). Where there is large scale uniformity in plant communities, this speaks of large scale uniformity of substrate (e.g. soil nutrients or water conditions), or of uniformly severe disturbance (e.g. fire, or defoliation by large, mobile vertebrate herbivores).

Further, we expect that plants as different from one another as trees and annual herbs, will exhibit different patterns of demographic behaviour, and that different kinds of plants will flourish in different environments. The purpose of this chapter is to introduce the variety of plant growth-forms and life histories, and briefly to describe the

principle factors which determine the kinds of plants that are
successful under different environmental conditions.

8.2   Patterns of life history

Different plant species exhibit different patterns of development,
age-specific reproduction and mortality. These traits are conveniently summarized by the phrase 'life history'. The range of plant
life histories extends from unicellular algae which divide after only a
few hours, to trees which live for a thousand years. Amongst the
vascular plants there are some which set seed once then die (the
monocarpic or semelparous species), while others reproduce several
to many times after reaching maturity (the polycarpic or iteroparous
species). Other life history traits with implications in population
dynamics include the prevalence of vegetative reproduction, the
relationship between seed size and seed number, the mating system
and mode of seed dispersal, the kind and degree of seed dormancy,
and the size and periodicity of seed crops (e.g. whether or not the
species shows mast fruiting). Further, the ideal life history depends
upon what the other plants in the environment are doing (Chapter 9),
and upon the average frequency with which germination microsites
become available in the habitat (see below).

A good deal of attention has been directed by theorists to the
question of which patterns of life history are likely to prosper in
different kinds of environments (Cohen, 1966; Stearns, 1976, 1977).
While a number of important generalizations have emerged from
these studies, it is important to bear in mind that several different
kinds of life history are likely to be found in any given plant community, or at any particular stage in succession. There may be only
one mathematically optimum life history for a given environment,
but history and circumstance will have thrown together many species
which exhibit a variety of solutions to the problems posed by survival
and reproduction in any one place. The mathematical optimum may
also be quite different from the best *practical* solution.

One of the first distinctions in life history theory was between *r*-
and *K*-selection (MacArthur, 1972). If superior fitness at low population density is incompatible with superior fitness at high density,
then a range of different plant types may evolve. In Fig. 8.1 type A
increases faster than type B at low density, but type B increases faster
than type A at high density. Type A is *r*-selected and type B is
K-selected. From these traits, Pianka (1970) predicted that at high
densities (where competition is assumed to be more severe) the
optimum strategy is 1) to allocate fewer resources to reproduction
and more to adult maintenance; and 2) to produce fewer, larger
offspring (Gadgil & Solbrig, 1972; Schaffer & Gadgil, 1975). It soon
became clear, however, that this dichotomy of traits was very crude,

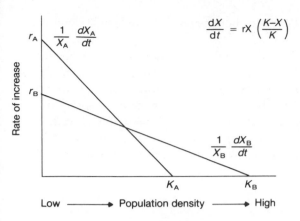

$$\frac{dX}{dt} = rX\left(\frac{K-X}{K}\right)$$

Fig. 8.1. The model behind the notion of r- and K-strategists. At low population densities, species A increases more rapidly than species B, but at high densities the position is reversed. Both species grow more slowly at high densities, but species B increases more rapidly than species A. The figure is drawn on the basis of the linear, logistic model (upper right), but the precise shape of the curves is not crucial. Because the intrinsic rate of increase (r_A) of species A is greatest it is called an r-strategist (relative to species B). Because the carrying capacity of the environment (K_B) is greater for species B, it is called a K-strategist (relative to species A). Of course, the two curves need not intersect. It is perfectly possible for one species to have higher values of both r and K; it would win at both high and low population densities. After MacArthur (1972).

because fitness is affected by factors influencing development rate, survivorship *and* fecundity, and these may be influenced in quite different ways by population density. Grime (1979) extended the r and K classification to account for what he called the intensity of 'stress' to which plants were subject in different habitats; he classified habitats in terms of the levels of disturbance, competition and stress experienced in them (see Chapter 12). The major problem with this approach is that 'stress' is not measurable. Conditions that would be extremely stressful for one species may be optimal for another.

Further sophistications to life history theory have developed from the 'principle of allocation' in which organisms with finite resources must allocate them between competing demands (Cody, 1966). Thus, for example, there must be a trade-off between traits promoting fecundity and those promoting survival (Williams, 1966; Stearns, 1976). The aim of life history theory is to compare the fitness of phenotypes which differ in their development rates or age-specific fecundity- or survivorship-schedules in specified ecological circumstances. Fitness is measured as the intrinsic rate of increase, r, of the phenotype, p, in environment, E according to the equation:

$$1 = \sum_{t=0}^{\infty} e^{-rt}\, 1_t(p,E)\, b_t(p,E) \tag{1}$$

where t is the age of the plant in years, $1_t(p,E)$ is the fraction of plants which survive from seed dispersal to age t, and $b_t(p,E)$ is the fecundity of phenotype p at age t in environment E. This is an adequate measure of fitness in constant, density independent environments, but also holds for density dependent populations (Charlesworth, 1980). Indeed, maximizing the intrinsic rate of increase in density dependent populations is equivalent to maximizing the carrying capacity of the environment. This equivalence is not widely appreciated by ecologists, many of whom consider maximizing the rate of increase and maximizing the carrying capacity as *alternatives*, as in the dichotomy between *r*- and *K*-selection (Sibly & Calow, 1983).

For clonal growth, Sibly and Calow (1982) show that more, smaller offspring should be produced in conditions which are good for individual growth. For sexually reproducing forms, if there are no interactions between variables, then the number of offspring per breeding should be maximized, survival until first (or next) breeding should be maximized, and time to first (or next) breeding should be minimized. If, as would seem likely, there are trade-offs between

Table 8.1. Trade-offs between different aspects of plant life history. Five life history attributes are considered: time to first breeding; time between subsequent breeding attempts; fecundity at each breeding attempt; survival from seed to first reproduction; and survival between successive breeding attempts. All ten pair-wise trade-offs are considered, but some are clearly of more general significance for plants than others. There are countless caveats to the interpretation of these trade-offs; for example, if young (small) plants suffer low rates of pollination or ineffective seed dispersal (low offspring survivorship), then selection for reduced age at first breeding may be weak. Again, there is no direct evidence that increased parental investment can hasten sexual maturity in perennial plants (see 4, 7 and 9).

1. Offspring per breeding attempt	*vs* Survival from seed to first reproduction	The classic seed-size/seed-number trade-off; small seeds give less competitive seedlings.
2. Offspring per breeding attempt	*vs* Survival between successive breedings	'Risky reproduction' depletes parental reserves, so increased seed production is bought at the price of reduced adult survivorship.
3. Survival between successive breedings	*vs* Survival from seed to first reproduction	By investing more in a given number of seeds (at the price of reduced adult survival), larger seeds (with higher survivorship) can be produced.
4. Survival between successive breedings	*vs* Age at first breeding	Increased parental investment (e.g. larger seeds) may reduce development time.
5. Survival to first reproduction	*vs* Age at first breeding	Increased risk by young plant (e.g. lower anti-herbivore defences) may reduce development time.
6. Survival between successive breedings	*vs* Time between successive breeding attempts	The plant could breed quickly, but only at the risk of depleting its reserves. By waiting to build up a store, the mature plant increases its chance of survival, but delays subsequent reproduction.
7. Offspring per breeding attempt	*vs* Age at first breeding	Increasing the number of seeds at the expense of individual seed size may mean that smaller seeds produce seedlings which achieve maturity later.
8. Offspring per breeding attempt	*vs* Time between successive breeding attempts	Increased delay allows a larger build-up of reserves, and hence more seeds.
9. Time between successive breedings	*vs* Age at first breeding	Fruiting every year may mean the production of smaller seeds, and hence delayed first breeding.
10. Time between successive breedings	*vs* Survival from seed to first reproduction	Longer gap allows larger seeds with higher competitive ability. Production of larger seed crops (masting) may increase seed and seedling survival by satiating seed predators.

variables, then the optimum trade-off will be that which maximizes fitness. In cases where juvenile mortality is low, there should be selection towards semelparity with many offspring per brood, while if juvenile mortality is high there should be selection towards itero-parity with few offspring per brood (Sibly & Calow, 1983). Further trade-offs between reproduction, survivorship and development time are sketched in Table 8.1.

One of the most severe practical problems in plant ecology derives from the fact that there is often pronounced polymorphism in life history traits within species, and even within a single clutch of offspring produced by an individual parent. At best, this means that very large samples of plants must be analysed in order to determine representative mean values for life history parameters. At worst, it means that average values of life history parameters are meaningless as characterizations. These difficulties are clearly illustrated by the polymorphism for cyanogenesis shown by the legume *Lotus corniculatus* (see Chapter 12); not only do individual plants differ from one another in their production of HCN, but the *same* plant may vary its phenotype, sometimes appearing cyanogenic, at other times acyanogenic (Ellis *et al.*, 1977). Also, what happens to an individual plant depends very much upon when, exactly, recruitment occurs. The fates of plants in different cohorts can be radically different depending upon the phenology of the species and the timing of recruitment, due to differences in weather conditions, natural enemy activity, and so on. From their detailed study of the grass *Bromus tectorum*, Mack and Pyke (1984) found that the probability of a seed germinating, or a seedling dying varied dramatically both within and between years, so that population dynamics were determined by 'the chronology of environmental events'. Clearly, then, generalizations about the life history attributes of 'the average individual' must be treated with great caution.

8.3 The growth forms of plants

Amongst the vascular plants there is a marvellous variety of growth forms, from tiny annual plants which produce ten seeds in only a few weeks, to gigantic coniferous trees which produce huge crops of seeds periodically over the course of many centuries. Several classification systems have been proposed to describe the range of plant growth forms, but easily the most valuable was devised by the Danish ecologist P. Raunkaier in the 1920s. His system is based on the precise position in which the plant maintains its perennating buds for the duration of the unfavourable season (see Fig. 1.9). If there is no unfavourable season (e.g. in the continually moist tropics), then most plant species of closed communities are tree-like; they maintain their perennating buds on aerial, usually woody, shoots. Raunkiaer called these plants 'phanerophytes', distinguishing between large ($>$ 30 m), medium (8–29 m), small (2–7 m) and tiny ($<$ 2 m) species as mega-, meso-, micro- and nano-phanerophytes. When the unfavourable season is extremely cold, plants with their perennating buds very close to the soil surface are favoured (chamaephytes), while in extremely arid conditions, perennials which die back to subterranean storage tissues may prevail (geophytes, and other cryptophytes). Despite its anachronistic terminology, the classification system has survived because it is thorough and is based on ecologically significant characteristics (see Chapter 1).

Leaf size and shape vary widely both within and between different life forms of plants. Different forms of venation, degrees of division, serration of the leaf edge, thickness and kinds of surface-covering, occur in countless combinations, each influencing light interception, heat balance, temperature regulation, water balance and CO_2 diffusion (Parkhurst & Loucks, 1972; Givnish, 1979). There is a broad correlation, for example, between water availability and leaf size; the largest leaves are found in tropical rain forest, medium-sized leaves in temperate forests, and small leaves in deserts, tundra or heathland communities which are dry or cold. The longevity of leaves also differs markedly both between species within a habitat, and between environments of different kinds (Chabot & Hicks, 1982, and see Chapters 11 and 12).

8.3.1   Annual plants

In terms of fitness, annual plants appear to have two great advantages: 1) they reproduce early, so they have the potential for very high intrinsic rates of increase; and 2) they can survive adverse conditions as dormant seeds in the soil. Weighed against these attributes, however, are two rather severe constraints: 1) it is difficult for annual plants to grow sufficiently large in one season that they can compete with tall, perennial plants; and 2) annual plants rely upon the availability of establishment microsites in *every* generation—if there are no microsites, there can be no recruitment. The balance between these opposing forces is determined by the severity of the unfavourable season, the density of competing vegetation, and the frequency of disturbance.

Put formally, if F is the average fecundity of plants surviving to reproduce, and S is the fraction of seeds which survives to reproduce, then the annual rate of increase, λ, is:

$$\lambda = S . F \qquad (2)$$

When λ is less than 1, the population declines; when it is larger than one, the population increases exponentially. Static populations would have λ exactly equal to 1 (Chapter 5). It is a mistake, however, to equate an annual life history with r-selected demographic behaviour (see Section 8.2), since some annual species have extremely low intrinsic rates of increase (e.g. *Vulpia fasciculata*; Watkinson, 1978).

If recruitment only occurs at intervals of more than one year because of intermittent microsite availability, then the mortality suffered by seeds in the bank must be taken into account, and the rate of increase must be discounted in calculating the annual average value. For example, if recruitment occurred every other year, and the survival of seeds in the bank was B per year, then S.B.F. seeds would be produced every two years, so the annual rate of increase would be much lower, namely:

$$\lambda = (S.B.F.)^{1/2} \tag{3}$$

Note that we take the square root of the two-yearly rate (*not* half). If gaps appear on average only every x years, then:

$$\lambda = (S.B.^{x-1}.F)^{1/x} \tag{4}$$

(for details, see Kelly, 1985). Clearly, the potential advantages of an annual life history are reduced considerably when recruitment is limited by the intermittent availability of microsites, especially if survival in the seed bank is low. Under these circumstances, a long-lived life history, with resources devoted to high adult survivorship, and with only intermittent reproduction (triggered, ideally, by the same factor which makes microsites available) might confer greater fitness.

In habitats characterized by moist winters and dry summers, many of the annual species germinate in autumn and produce a rosette of overwintering leaves (so called 'winter annuals'). Rapid growth begins very early in the year, flowering occurs in early spring, and seeds are ripened before the onset of the summer drought. This kind of life history is common in temperate sand dune communities (e.g. many small crucifers and grasses) and garden weeds (e.g. *Cardamine hirsuta, Lamium purpureum*). The phenology of winter annuals ensures that the vegetative stage escapes the attentions of most invertebrate herbivores. However, the plants are vulnerable to vertebrate herbivores which forage throughout the winter (e.g. rabbits and voles), because they offer young, green foliage at a time of year when many other species are leafless, or carry only old leaves of low nutritional value. Summer annuals overwinter as dormant seeds, then germinate in spring and grow rapidly to ripen their seeds in summer or early autumn. A few summer annuals are able to grow sufficiently large in a single season that they are capable of forming closed, monospecific stands (e.g. the alien *Impatiens glandulifera* on muddy river banks in England). Others, like *Galium aparine*, are able to persist in dense, mesic vegetation composed of woody and herbaceous perennials, by virtue of large seeds, early germination, rapid growth and a scrambling habit (Chapter 12).

Because recruitment from seed is so vital to the perpetuation of the genes of annual plants, it comes as no surprise to find that annuals display a variety of traits which ensure that all their eggs are not put in one basket (so called 'bet-hedging' tactics; Stearns & Crandall, 1981). It would be suicide, for example, for an annual plant to have synchronous germination of its entire seed crop in an environment, or at a time, where conditions were likely to deteriorate so badly that all the seedlings would die. Thus, for example, we find that individual plants produce seeds which, within a single clutch, display a wide range of dormancy-breaking and germination requirements (Leon, 1985).

By altering the relative costs and benefits of dispersal, environmental uncertainty also determines whether it is better for the plant to stay where it is, or to move on (De Angelis *et al.*, 1979; Venable & Lawlor, 1980). (Note, however, that selection is likely to favour dispersal even in stable environments; Hamilton & May, 1977). Many annual plants adopt a mixed strategy by equipping some seeds for long-range dispersal and others for staying put. This trait is particularly common amongst members of the Compositae, where fruits of the outermost flowers are large and have no parachute, while the more numerous, smaller fruits from the centre of the inflorescence have a pappus of hairs and are wind dispersed (Zohary, 1950; Baker & O'Dowd, 1982). The question of how many large, non-dispersing seeds, and how many small, far-dispersing seeds to produce, is a problem for analysis as an Evolutionary Stable Strategy (see Chapter 9), since the production of large numbers of non-dispersing seeds would lead to competition between the sibs. The ESS is to produce an increasing proportion of far-dispersed seeds as total seed production increases. Where plants have been found to show variable morph ratios with clutch size (e.g. plants producing both subterranean and aerial fruits), they conform to the ESS, producing larger, near-dispersed fruit first, and changing the ratio in favour of smaller, far-dispersed fruit as clutch size increases (Silvertown, 1985).

In variable environments, selection will tolerate some loss of average (arithmetic mean) fitness, if there is a corresponding reduction in the *variability* of fitness, and this, too, will lead to the evolution of bet-hedging, risk-avoiding strategies (Bulmer, 1985). Ritland and Jain (1984) found just such patterns in two contrasting annual species of *Limnanthes* which differed in their seed dormancy, vegetative growth rate, timing of reproduction, allocation to reproduction and seed size.

8.3.2   Monocarpic perennials

Plants which spend one or more years in a vegetative condition before flowering once then dying have been called 'big bang' strategists. At one end of the spectrum are strict biennial plants like *Melilotus alba* which flower and die in their second year of life. At the other extreme are long lived species like the sword-leaved *Lobelia wollastonii* from the mountains of Uganda, or the silversword *Argyroxiphium sandwicense* from Hawaii, which grows until it achieves a rosette diameter of about 60 cm, then flowers and dies; this takes about seven years. *Puya raimondii* from the Bolivian Andes does not flower until it is over 100 years old. Between these extremes are many kinds of plants which are more or less long-lived, depending upon the opportunities for growth, and upon their histories of defoliation and crowding.

There has been considerable debate about the relative fitness of annual, biennial and perennial monocarpic life histories (Hart, 1977; Silvertown, 1983; Kelly, 1985), but, as we have seen in Section 8.3.1, these comparisons are of limited value in the absence of considerations about the frequency and reliability of microsite availability. Indeed, for a great many so-called biennials, the populations themselves are ephemeral, and, as Harper (1977) wryly observes, most populations of biennials are on their way to extinction by the time an ecologist decides to study them! Be that as it may, if the plant spends x years in the vegetative stage prior to flowering, its average annual rate of increase will be:

$$\lambda = (S^x.F)^{1/x} \tag{5}$$

where S is the annual survival rate. For a given fecundity, F, the rate of increase falls rapidly as the delay in reproduction, x, is increased. Weighed against this are two potential advantages for big bang reproduction: 1) fecundity of the larger plants may be sufficiently high to compensate for the delay; or 2) the large seed crop may satiate local seed predators, so that seedling recruitment is assured. Furthermore, the long gap between seed crops may reduce the risk of mortality from specialist seed-feeding species because the herbivores cannot increase in abundance during the intervening period (Janzen, 1976).

8.3.3   Herbaceous perennial plants

Herbaceous perennial plants comprise an extremely diverse array of different growth-forms (Table 8.2); each type will have more or less unique aspects of their population dynamics and life history. For example, the main difference in life-style between trees and the renascent herbaceous perennials is in their occupancy of the aerial environment. Both maintain tenure of a piece of ground for many

Table 8.2. Growth forms of·herbaceous perennials.

Renascent herbs	Sendentary	Multicapital rhizomes Mat geophytes with	*Vincetoxicum*
		a) stem tubers	*Crocus*
		b) root tubers	*Ophrys*
		c) bulbs	*Lilium*
		d) tuberous stems	*Cyclamen*
	Mobile	Rhizome geophytes on	
		a) sand	*Ammophila*
		b) woodland	*Anemone*
		c) wet mud	*Phragmites*
		d) grassland	*Cirsium arvense*
Rosette plants	Sedentary	Short internodes & close-set leaves	
		a) alpine fell fields	*Draba*
		b) elongate leaves	*Taraxacum*
		c) succulent leaves	*Sedum*
		d) leathery leaves	*Saxifraga*
		e) *Musa*-form	
		f) branched tuft trees	*Yucca*
		g) unbranched tuft trees	palms
	Mobile	Stoloniferous plants	*Fragaria*
Creeping plants	Mobile	Prostrate, rooting, leafy shoots of	
		a) woods	*Lysimachia nummularia*
		b) grasslands	*Trifolium repens*
		c) wetlands	*Menyanthes trifoliata*
Cushion plants	Sedentary	Richly branched densly packed shoots	*Silene acaulis*
Undershrubs	Sedentary	Flowering shoots die after blooming	*Lavandula*
		Weakly lignified tropical forest spp	Acanthaceae
		Rhizomatous	*Vaccinium*
		Canes	*Rubus idaeus*
Succulent-stemmed plants	Sedentary	Cactus-form	*Opuntia*
Aquatic herbs	Free-floating	Surface leaves	*Lemna*
		Submerged leaves	*Utricularia*
	Rooted	Surface leaves	*Nuphar*
		Submerged leaves	*Litorella*

years, but the shoots of renascent herbs die back to ground level at the end of every growing season. The advantage of this growth form is that the shoots are not vulnerable to drought, cold or natural enemies during the unfavourable season. The main disadvantage, compared to a woody habit, is that the individual's position in the canopy of plants competing for light must be re-established every year. Weighed against this, however, is the fact that herbaceous perennials need not divert resources into the production or maintenance of permanent, woody, supporting structures.

A further advantage of the herbaceous perennial habit (shared by a few woody species) is the ability to 'move about' by virtue of rhizomes, stolons, or rooting shoots. These plants are capable of spreading radially to form large, clonal stands (e.g. *Holcus mollis*), of exploiting patchy habitats by 'jumping over' barriers of un-inhabitable substrate (e.g. the long, tough runners of *Potentilla reptans*), or of insinuating themselves through a dense matrix of taller plants (e.g. the 'guerilla' growth form of *Trifolium repens*; Chapters 4 and 9).

In a study of rhizome development in patchy environments differing in soil salinity, Salzman (1985) found that *Ambrosia psilostachya* showed a distinct tendency to spread into less saline (and presumably more favourable) soils. There was a strong correlation between the salt tolerance of a clone and its preference for non-saline patches; while the tolerant clones were almost indifferent, the intolerant clones put over 90% of their shoots from new rhizomes in the non-saline patches. This could have been due to: 1) direct suppression of growth or increased rhizome mortality in saline soils; 2) increased investment in rhizome growth in less saline patches; or 3) both of these. While there is some evidence to suggest selective investment (plants produced fewer shoots at a given salinity when given the 'choice' of a less saline patch in which to grow than when grown in a habitat of uniform salinity), it is not yet possible to distinguish critically between these processes under field conditions (Salzman, 1985).

The Achilles heel of an herbaceous species is its perennating organ (rhizome, corm, bulb, tuber or fleshy rootstock). If the plant is to survive for a protracted period, these structures must be resistant to herbivores and pathogens, as well as being capable of surviving the rigours of the unfavourable season. One of the reasons that we know so little about the ecology of most herbaceous perennials, is because it is so difficult to study this subterranean world. For example, we know virtually nothing about the extent and importance of feeding by herbivores on (or in) rhizomes, corms and tubers in the population ecology of wild plants.

Grasses and sedges are amongst the most important groups of herbaceous perennials. They display a remarkable range of form and life history in different environments: they may be evergreen or summer green, rhizomatous or tussock-forming, prostrate or erect, cryptophyte or chamaephyte, outbreeding, inbreeding or not-breeding-at-all! Perhaps their greatest claim to fame, however, lies in their ability to withstand grazing. Those species with prostrate growth form, buried meristems, rapid tillering and substantial, inaccessible reserves of carbohydrate, achieve dominance under continuously high levels of vertebrate grazing (Johnson & Parsons, 1985). Not all grasses are grazing-tolerant, however, and species with vulnerable meristems or storage organs are quickly eradicated under grazing (Holmes, 1980). Grasses also possess a range of chemical defences against defoliation (especially in the early seedling stages; Bernays & Chapman, 1978) and many species have sharp or coarse silica-rich tissues which are unpalatable to vertebrates (Section 8.4.15).

8.3.4   Trees and tree-like plants

In order to buy themselves a place in the sun, trees delay investment in reproduction until they have grown sufficiently large to ensure their survival over a protracted period. Then they can devote resources to reproduction as and when they are available (e.g. in years of exceptionally good weather for pollination, in times of unusually favourable net production, or in years when herbivores are scarce). Plants which are unable to obtain a place in the canopy will usually die without leaving any progeny at all.

With presently avilable data it is virtually impossible to calculate the annual rate of increase of tree species, because we have no accurate data on the fraction of seeds which survives to reproductive age (this is hardly surprising given the longevity of most ecologists)! We can make some rough estimates, however, by considering two extreme cases. First, we assume that only the first bout of seed production is important (i.e. the trees have approximately the same rate of increase as monocarpic perennials). Second we assume that the surviving plants reproduce at an average rate F each year, once they have begun to reproduce at age x (i.e. they are essentially immortal). In this case

$$(l_x . F)^{1/x} < \lambda < \exp(l_x . F) \tag{6}$$

In making calculations of the rate of increase of long-lived plants we must resist the temptation of estimating survivorship by dividing the numbers of mature plants found per unit area by the numbers of seeds falling on the same area, since this procedure implicitly assumes that the population is stable; i.e. we are trying to estimate λ, but assuming that $\lambda = 1$! For instance, if we found 100 mature oak trees/ha and 50 acorns/m^2 we could say that survivorship is 0.0002 *only if* the population is stable. However, we can *not* use this information to calculate the actual rate of increase (or the potential rate of increase) of the oak population (see Chapter 5).

The recent literature on the canopy architecture of trees (e.g. Hallé *et al.*, 1978) has drawn attention to the wide variety of different ways in which woody plants can exploit the aerial environment. For example, we can distinguish between 'monolayers' which have their leaves spread in a single, horizontal layer, and 'multilayers' which have leaves distributed in several, vertically separated layers. Other things being equal, monolayers should have higher rates of *net* photosynthesis at very low light intensities (because there is no self-shading), while multilayers should be more drought resistant than monolayers (because they experience reduced heat loads; see Horn (1971) for details). Clearly, there are problems with simple generalizations like these: 1) three-dimensional branching patterns are difficult to quantify; 2) it is difficult to separate genetic and micro-environmental determinants of branching; and 3) the selective advantage of different branching patterns between species and between habitats is by no means clear (Chapter 9). With field work and theoretical modelling, however, it may be possible to distinguish between alternative hypotheses of canopy architecture. We could ask, for example, whether plants are selected: 1) to maximize cover per unit biomass; 2) to overtop and shade-out other individuals; 3) to maximize photosynthetic area per unit volume; 4) to minimize self-shading of branches, and optimize leaf area index; 5) to optimize canopy microclimate; or 6) to maximize the efficiency of pollination and/or fruit dispersal (Cody, 1984). These themes are taken up in Chapter 9.

8.4   Environmental factors affecting plant performance

The life history patterns and growth forms discussed in Section 8.3 place constraints on the survival and reproduction of plants, which mean that they are competitive with other species only under a *limited* range of environmental conditions (Orians & Solbrig, 1977). Some species may be 'Jacks-of-all-trades; but *no* single species can be 'master of all' (Chapter 1). In this Section I shall describe some of the morphological, physiological and phenological traits of plants which appear to be associated with particular environmental factors. Remember, however, that without careful, long-term experimentation, it is all too easy to misinterpret the 'adaptive significance' of a plant's features (see Section 9.6.1).

Throughout this section, a distinction is made between the **avoidance** and **tolerance** of extreme conditions. Similarly, we should distinguish between those responses that reflect constraints imposed by the plant's genetic programme, and those that reflect phenotypic plasticity in response to local environmental conditions. Within a species there will be both genotypic variation and phenotypic plasticity. Indeed there may often be genotypic variation *for* phenotypic plasticity (Bradshaw, 1973). The ecological range occupied by a species is determined by the interplay between all these factors.

Finally, it must be stressed that single factor explanations will usually prove to be inadequate as descriptions of a plant's environment. In most communities, plant performance is affected by the *interaction* of several kinds of factors. So, for instance, many Australian ecosystems are extremely arid, extremely nutrient-poor, and very prone to fires. Similarly, high alpine plant communities suffer extremely short growing seasons, extremely intense sunlight and extreme exposure to high winds.

Whole books have been written on each of the topics discussed in the following sections, and the material presented here is intended as nothing more than a key to the modern literature.

8.4.1   Fire

Fire is one of the most important factors in many of the world's plant communities (Kozlowski & Ahlgren, 1974; Mooney *et al.*, 1981). The likelihood of fire is determined by the interplay between the weather (how dry it is) and the amount of fuel present. This, in turn, depends upon the time since the last fire, and on the inflammability of the plant material, both living and dead (Boerner, 1982). Communities differ in the intensity of fires, in their frequency and in the area burnt during a typical outbreak. For example, some communities may suffer frequent surface or 'grass fires', while others experience much less frequent (but more devastating) canopy fires. Peatlands may experience 'ground fires' in which the peat itself is burned, with severe consequences for the seed bank and for below-ground perennating organs (Mallik *et al.*, 1984).

The results of fire include: 1) release of nutrients from accumulated standing dead biomass; 2) breakdown of hydrophobic plant litter which leads to improved soil wettability and opens up establishment microsites; 3) breaking dormancy in many fire-adapted species (Went *et al.*, 1952); 4) removal of inhibiting chemicals including allelochemicals from the soil surface. The impact of a fire on the plant community is determined by the season in which it occurs (because species vary seasonally in their vulnerability) and in the area burnt (as this influences the proximity of surviving plants which can act as seed parents). The great majority of species are killed by fires of even moderate intensity, and frequent fires greatly restrict the pool of species capable of persisting in a community (Naveh, 1975; Kruger, 1982; Mooney & Conrad, 1977).

The four main kinds of adaptations to fire are: 1) resistance, where plants have thick, fire-proof bark, as in *Pinus resinosa*; 2) regeneration by sprouting from root stocks or surviving stems, common in many deciduous trees like *Populus* and *Betula*; 3) possession of specialized underground organs like the lignotubers of certain *Eucalyptus* species; 4) specialized, long lived fruits which accumulate on the plant over a number of years, only opening to release their seeds after the passage of a fire (as in the serotinous cones of *Pinus banksiana* and the woody follicles of Australian *Hakea*).

Frequent grass fires and infrequent canopy fires exert quite different selection pressures, a fact exemplified by the contrasting life histories of different species of pines. Seedlings of most pines are killed by grass fires, but species like *P. montezumae* have grass-like seedlings which do not produce a woody aerial shoot for several years after germination. Instead, they produce herbaceous 'grass foliage' which dies back to the ground each year, and the seedlings build up an underground reserve in a woody rootstock over a period of years. When the reserve is large enough, the plants suddenly produce a shoot which grows tall enough in a single season to take it out of the range of grass fires. In contrast, pines from communities dominated by canopy fires (like bishop pine *P. monticola*) are not severely harmed by grass fires. They grow for 60 years or so, accumulating serotinous cones on their branches and shedding no seed. After a canopy fire kills the mature trees, the accumulated seeds of 40 or more years of reproduction are released into the ash in a single burst, where they germinate free of interspecific competitors.

8.4.2   Drought

The availability of soil water is a prime determinant of plant community structure. Communities where water is in short supply (i.e. where potential evaporation greatly exceeds actual evapotranspiration) tend to be dominated by plants showing one or more adaptations to drought tolerance ('xeromorphic' features), or by annual plants which escape the drought as dormant seeds in the soil (the annuals only germinate after rainfall, and grow rapidly in conditions of relatively high water availability).

Drought tolerant plants improve their water relations both by increasing their efficiency in extracting and storing water, and by reducing the rate at which they lose water through evapotranspiration (Slatyer, 1967). Water extraction is improved by modifications to the root system, and where the water table lies far beneath the surface, long-lived plants may develop very deep root systems (e.g. mesquite, *Prosopis juliflora*, may have roots which reach down to 12 m). The young, shallow-rooted individuals of such species are particularly vulnerable to death through desiccation, and frequently exhibit novel means of water gathering (e.g. mesquite seedlings can absorb dew through their leaves). Where the water table is beyond the range of even the deepest-rooted plants, and where rainfall occurs as brief, light showers, then plants with rapidly growing, extensive, but shallow roots may be dominant (e.g. saguaro cactus, *Carnegia gigantea*).

Water loss through leaves is reduced in a wide variety of different ways including: 1) reduced leaf area; 2) abscission of leaves during the dry season (drought-deciduous plants); 3) modifications of the stomata (e.g. setting them in deep pits in the epidermis); 4) possession of a thick, waxy epidermis; 5) positioning of the stomata on the inner surface of tightly inrolled leaves; 6) orienting the leaves vertically to minimize the direct radiation they receive; or 7) investing the leaf surface with a clothing of hairs. It is well known, for example, that species which are normally glabrous, become hairy in dry places, and that hairy species become more hairy (e.g. *Ranunculus bulbosus*). Further improvements in the efficiency of water use are brought about by a range of physiological modifications to photosynthesis and respiration (Chapter 11), and by tolerance of high levels of tissue dehydration (Kozlowski, 1972).

The second problem faced by plants in sunny, arid environments is over-heating. Unable to cool their leaves by increased evapo-transpiration, they show a range of morphological features which reduce heat loads, and physiological adaptations which permit the tolerance of higher temperatures. Over-heating is avoided by: 1) increasing the surface area for heat dissipation; 2) protecting the plant surface from direct sunlight; and 3) increasing the reflectance of the surface. All of these can be accomplished by covering the surface with a dense felt of hairs. The hairs also produce a boundary layer of still air next to the stomata, which traps moisture and reduces evaporation (note that the hairs may also have a role in defence against herbivorous insects). Not all xerophytic plants are hairy, however, and some desert species have completely hairless, smooth surfaces. In these cases, water loss is restricted by a thick covering of wax, or by modifications to the walls of the epidermal cells (Levitt, 1972).

Water shortages are not restricted to plants for desert regions. Xeromorphic features are also exhibited by plants growing in habitats where water is unavailable because it is frozen or too salty to extract (this is sometimes called 'physiological drought'). Even temperate species from mesic environments may suffer water shortage during one or more of the summer months. In tropical rain forests, many of the canopy tree species show xeromorphic leaf features on the mature individuals which emerge into the intense sunlight above the surface of the leaf canopy, while saplings and shaded mature individuals of the same species have larger, thinner shade leaves lacking xeromorphic features (Section 8.4.4).

Indirect consequences of drought on plant performance include an increased susceptibility to damage by pathogens and insect herbivores (White, 1974). Many insects, for example, show a preference for wilted rather than turgid tissues (Lewis, 1979), and aphids feeding on irrigated crops do far less damage than on unwatered plots (Cammell & Way, 1983). Water shortage also leads to an increase in free amino acids in the foliage and in the phloem, which may increase the quality of these tissues as food for insect herbivores (see also Section 8.4.8).

8.4.3   Waterlogging

Waterlogging of the soil by flooding is common in many habitats following snow melt or after very heavy rains. This has both direct effects on plant root systems and indirect effects on soil structure and function. Physical attributes of the soil are altered, including restriction of gas exchange (depletion of molecular oxygen, and accumulation of soil gases like N, CO_2, CH_4 and H_2), thermal effects (altered radiation absorbtion and reflectance, modified heat flux, and so on), and alterations to soil structure (increased soil plasticity, breakdown of crumb structure, and swelling of soil colloids, especially in sodic soils with high clay content).

Within hours of flooding, normal soils are virtually devoid of O_2 and aerobic soil organisms are replaced by facultative anaerobes. Prolonged flooding leads to the replacement of these forms by obligate anaerobes. For example, fungi and actinomycetes are dominant in freely-drained, aerobic soils, but are largely replaced by bacteria under waterlogged conditions, leading to a reduction in the average rate of decomposition. Not all waterlogged soils are uniformly anaerobic, however, because the roots of many vascular marsh plants contain special air-conducting cells (aerenchyma) and give off oxygen, making their rhizospheres aerobic (Jackson & Drew, 1984). Flooding also has profound effects on the electrochemistry of the soil solution; there is an immediate dilution of the solution and a great reduction in redox potential, usually accompanied by an increase in pH (Armstrong, 1982).

Many plant species are killed by even brief exposure to waterlogging. Others are capable of a variety of responses which enable them to survive flooding of their root systems. The most frequent responses are swelling of the stem base (hypertrophy), wilting, reduced leaf growth, abscission, epinastis, root death and rotting, with adventitious rooting higher up the stem. As a consequence, waterlogged plants often show signs of nutrient or hormone deficiency. There is often an alteration in the pattern of hormone distribution within the plant, with increased levels of ethylene, auxin and abscissic acid in the shoot, and decreased levels of gibberellins and cytokinins. Flooding tends to increase the rate of flower and fruit abortion, and prolonged waterlogging leads to increased mortality, even amongst somewhat flood-tolerant species. Plants in water-logged soils are also more susceptible to disease (Kozlowski, 1984).

Certain species, of course, inhabit permanently wet soils. Special-ized 'flood trees' are capable of growing with their root systems permanently immersed in fresh water (*Taxodium distchum*, *Nyssa aquatica*) or salt water (*Rhizophora mangle*). In these plants there are specialized root morphologies which allow oxygen transportation to the growing tips. These include 'knee roots' (pneumatophores) and surface-sprouting adventitious roots, coupled with hyper-trophied lenticels which allow gas exchange and toxin release to the surface waters (Keeley, 1979). The seeds of flood trees are tolerant of long submersion, and often germinate immediately the floodwaters recede. Some, like *Populus deltoides*, can even germinate under-water, and their seedlings can survive considerable periods of submersion (Clark & Benforado, 1981).

Herbaceous vascular plants from permanently wet sites show a range of morphological and physiological traits including: 1) anatomical features allowing oxygen transportation to the roots; 2) the ability to exclude or tolerate soil toxins such as Fe^{2+} and H_2S; and 3) biochemical features which allow prolonged fermentation (anaerobic glycolysis) in the roots (Fitter & Hay, 1981).

8.4.4   Shade

The relationship between plant performance and light intensity is discussed in detail in Chapter 11, and here I shall simply describe some of the features exhibited by plants which live in more or less permanent shade (see Boardman, 1977, for a review). In temperate, deciduous forest, the most typical class of shade plants exhibit 'phenological escape'. They grow rapidly and flower in the spring before the forest floor is shaded (e.g. *Hyacinthoides non-scripta* and *Anemone nemorosa*). They show few morphological modifications for shade tolerance and their leaves die back to ground level when the canopy closes in summer. It is interesting that plants with this life history are absent from deciduous woodlands on very infertile soils, where the ground flora tends to consist of a rather stable community of slow-growing, evergreen herbs (Grime, 1979).

Plants living in permanent shade include many ferns, mosses and lichens, as well as the (usually evergreen) vascular plants of the floor of evergreen forests. These species must maximize their photo-synthetic gain from the low levels of energy they receive, by means of: 1) a reduced respiration rate—this lowers the compensation point so that the plant can maintain positive rates of net photosynthesis at lower light intensities; 2) increased unit leaf rate (the photosynthetic rate per unit energy per unit leaf area); 3) increased chlorophyll per unit leaf weight; 4) increased leaf area per unit weight invested in shoot biomass (by having thinner leaves, or arranging leaves in flat tiers to minimize self-shading; see Chapter 9). Interesting case histories are described by Bjorkman (1968) for sun and shade eco-types of *Solidago virgaurea,* and by Solbrig (1981) who compared the performance of three species of *Viola* in reciprocal transplant experiments in different degrees of shade.

Leaf morphology is highly plastic, and many tree species in tropical forests possess large, thin shade leaves as juveniles, but smaller, thicker sun leaves when they emerge into the full sunlight of the upper canopy. Other differences include: 1) a single-layered upper and lower epidermis in young shade leaves, but a multi-layered epidermis in sun leaves; 2) short, broad palisade cells in shade leaves, but long, thin cells in leaves of the adult tree; 3) fewer, larger stomata in the shade leaves of young trees; 4) almost hairless leaves in the shade but dense indumentum on sun leaves; and 5) low frequency of vascular bundles (each with weak bundle sheaths) in the shade leaves, but high frequency (with very strong bundle sheaths) in sun leaves of the adult tree (Roth, 1984). For other factors influencing leaf morphology see Section 8.3.4, and Chapters 11 and 12.

8.4.5 Disturbance

The word 'disturbance' is used in two different ways in the literature of plant ecology: 1) to cover various accidental injuries suffered by mature or seedling plants (e.g. landslip, burial by sand, fire, or storm damage), defoliation by herbivores, or intervention by man (e.g. ploughing arable land, road construction, etc.). Grime (1979) defines this usage of disturbance as 'the mechanisms which limit the plant biomass by causing its partial or total destruction'; 2) to cover processes leading to the creation of bare ground, loose soil, or light gaps, which constitute microsites in which recruitment can occur. Clearly, the two meanings overlap, but I shall concentrate on the second meaning in this section—disturbance as a creator of establishment microsites (Mooney & Godron, 1983; Pickett & White, 1985).

In most communities, individual plants have tenure over pieces of real estate for protracted periods. This 'site pre-emption' means that other plants are kept out, unless or until the owner of the site dies naturally, or succumbs to accident or attack by its natural enemies. The degree to which the occupant of a site must be debilitated before its space is invasible by other individuals is poorly understood, but is clearly of vital importance in studies of long-term vegetational dynamics (e.g. successional studies; see Chapter 2).

Specific adaptations related to disturbance and microsite creation can be summarized in two categories: 1) morphological, pheno-logical or physiological attributes which increase the probability of getting into (and staying in) a newly created microsite; and 2) traits which enable a plant to resist disturbance, and to keep other individuals out.

Seed and fruit morphology are amongst the most important determinants of microsite colonizing ability. In general, the larger the seed, the less the disturbance necessary for successful establishment. Thus oaks *Quercus robur* with their large acorns can produce vigorous seedlings in dense vegetation, whereas small-seeded trees like birch *Betula pendula* are light-demanding. In addition to their direct role in creating microsites (Crawley, 1983; Burggraaf-van Nierop & Meijden, 1984), herbivores may also influence plant recruitment when defoliation leads to the production of smaller seeds, which are capable of recruitment only under a more restricted range of conditions than can be exploited by the larger seeds from undefoliated individuals (Crawley & Nachapong, 1985). Species with wind-dispersed seeds (dust-like seeds of orchids, the pappus fruits of many Compositae, or the winged fruits of trees like *Acer* or *Fraxinus*) are especially prevalent in areas subject to erratic, large-scale disturbance (Grime, 1979). Similarly, in habitats where there is frequent, severe disturbance, many of the short-lived species retain a bank of dormant seeds in the soil (Hastings, 1980; Chapter 4).

In addition to attributes of seed dispersal, the probability of obtaining tenure of a microsite is greatly increased by such *seedling* traits as shade tolerance, increased resistance to fungal attack and low palatability to herbivores. Seedlings which are capable of prolonged survival can exploit ephemeral resources (e.g. temporarily improved light or temperature conditions), whereas seed germination is an all-or-nothing process, irreversible if conditions deteriorate.

The second set of disturbance related traits concern the ability of an individual plant to keep others out. Tolerance of factors such as fire, grazing, trampling, and drought all enable a plant to maintain its tenure of a site in the face of severe disturbance. Other traits may (perhaps fortuitously) enable an established individual to exclude potential invaders either by: 1) casting a dense shade; 2) allelo-chemical effects; 3) the production of deep, persistent leaf litter; 4) harbouring high densities of invertebrate herbivores which kill seedlings beneath the canopy; 5) desiccating the soil surface by virtue of a shallow, but highly competitive root system.

8.4.6   Low nutrient availability

The principles of nutrient uptake, translocation and use are described in Chapter 12, and the implications of the relative availability of different nutrients for the outcome of competition between plant species are discussed in Chapter 2. Here I shall list the features of plants found in environments where certain soil nutrients are available at very low rates of supply.

Perhaps the most conspicuous features of plants from low nutrient environments are their small size, their tendency to have small, leathery, long-lived leaves, and their high root:shoot ratios (Chapin, 1980; Vitousek, 1982). Since the individuals have relatively small shoot systems, plant communities on nutrient-poor soils tend to be open, with bare gaps between the plants. Nutrient use efficiency might be expected to be high in low nutrient environments, as indicated by: 1) the production of a given amount of plant tissue from a smaller amount of nutrients; 2) a lower rate of leaching of essential nutrients from live foliage, minimizing uptake needs; 3) reabsorbtion of a larger proportion of foliar nutrients prior to leaf-fall (Vitousek & Mattson, 1984; Boerner, 1984). The root systems of species tolerant of low soil nutrients often consist largely of storage tissues, which means that their typically high root:shoot ratios do *not* necessarily reflect extensive systems of nutrient-absorbing roots.

Physiological traits of plants from nutrient-poor soils include: i) slow growth rates; ii) high investment in anti-herbivore defences; iii) selectivity (e.g. ability to take up calcium in the presence of high concentrations of Mg on serpentine soils; see below); iv) low saturation rates of nutrient uptake coupled with failure to respond to fertilizer application; v) storage of nutrients for use when the supply rate drops ('luxury uptake'); vi) efficient nutrient utilization; vii) flexible allocation patterns (e.g. plasticity within the same root system, so that there is proliferation of lateral roots in localized zones of higher nutrient availability); viii) high investment in mycorrhizal development; ix) efficient mechanisms of internal nutrient recycling to ensure minimal losses through leaf fall, exudation or leaching (Clarkson & Hanson, 1980).

Perhaps the most fascinating adaptation to low-nutrient soils is found amongst the insectivorous plants. Some species exhibit movements in the capture of their prey, as in the Venus fly trap, *Dionaea*; others possess traps or pitfalls to ensnare unwary insects, like the pitcher plants, *Nepenthes*; a third group trap insects on sticky leaves (e.g. butterworts, *Pinguicula*) or with special, sticky glands (e.g. sundews, *Drosera*). A full description of these plants is given by Kerner (1894, and see Fig. 8.2). Recent experiments using radio-labelled amino-acids have shown that the plants can take up intact amino-acids from their prey, and also that valuable supplies of phosphorus are also obtained by this means (Chandler & Anderson, 1976).

The flora of serpentine soils has attracted a great deal of attention from plant ecologists chiefly because of the number of rare plant

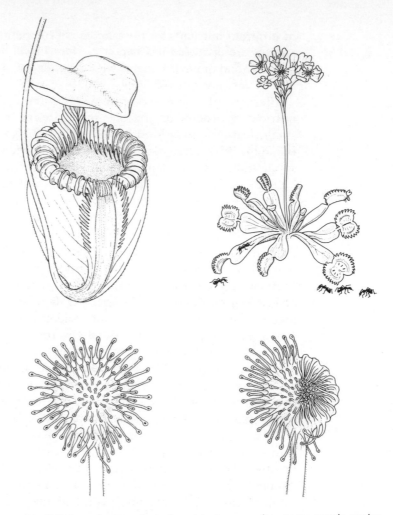

Fig. 8.2. Insectivorous plants. In nutrient-poor environments, certain species supplement their nitrogen and phosphorus requirements by trapping and digesting small, invertebrate animals. Top left, the pitcher plant, *Nepenthes*; top right, the Venus fly trap, *Dionaea*; below, the leaf tip of a sundew, *Drosera*, showing its open sticky tentacles (left), and when inflected over a captured insect (right). After Kerner (1894).

species which are found there. The communities are open, and support low densities of rather small plants. Many of the constituent species are found *only* on the highly nutrient-poor and magnesium-rich serpentine soils (Proctor & Woodell, 1975). The question which has occupied ecologists is whether these 'serpentine endemics' have a *requirement* for some specific factor provided only by serpentine soils (e.g. high magnesium), or whether the serpentine endemics are simply poor competitors in closed vegetation? Although there is some evidence for both processes (e.g. *Poa curtifolia* has a high Mg requirement), the weight of evidence points to the low competitive ability of the serpentine endemics on normal soils. When serpentine ecotypes of *Plantago erecta* were sown experimentally in cleared plots on normal soil they were healthy and produced good crops of seed. In competition with native vegetation on uncleared plots, however, they were suppressed and failed to set seed at all. Their inability to compete on normal soils is thought to be a consequence of their intrinsically slow growth rate (see Chapter 12).

8.4.7   Soil acidity

The hydrogen ion concentration of the soil (pH) depends upon the balance of basic and acid substances, and on the degree of dissociation of the acids. Several processes contribute to acidification of the soil, including: 1) mineral weathering which releases relatively large amounts of bases; 2) humification of plant litter which gives rise to organic acids and to acid stable humus; 3) respiration by roots and soil organisms producing carbon dioxide which acidifies the soil solution; 4) uptake of certain nutrients as cations which leads to an increase in hydrogen ions in the soil (e.g. ammonium nitrogen; see Chapter 12). Soil pH tends to change with depth, reflecting differences in the balance between leaching and accumulation (Etherington, 1982). The role of industrial pollutants in the acidification of soils and water is described by Paces (1985).

Because pH is so easy to measure, there has been a vast number of studies correlating plant distribution to soil acidity (e.g. Table 8.3). This ease of measurement is not matched, however, by ease of interpretation! While it is clear that some species have a very narrow range of pH tolerance, and are therefore good indicators of soil acidity (these tend to be species characteristic of either extremely acid or extremely basic soils), others, including most species found in the mid range of pH values, can grow under a rather broad span of soil acidity. Away from the extremes, plant stature and abundance tend to be determined by the interplay between competing plant species, nutrient availability, heavy metal concentration, and calcium availability, rather than by pH as such.

In general, we find that high pH tends to be associated with 'unfavourability' (like deficiencies of iron or phosphorus), whereas low pH is associated with toxicity (e.g. high concentrations of dissolved aluminium and iron; Fitter and Hay, 1981). Hydrogen ion concentration of the soil solution exerts its most important ecological effects by influencing nutrient availability and the concentration of potentially toxic ions. High pH may lead to a decrease in potassium, phosphorus and iron availability due at least in part to competition for sites on soil ion exchange complexes. However, increased pH may also lead to increased rates of nitrogen fixation by symbiotic bacteria (Andrew, 1978), and to a change in availability for predominantly ammonium–nitrogen at low pH to nitrate–nitrogen at high pH. Hydrogen ions have a twofold effect on nutrient uptake: 1) they have a direct effect by displacing cations; and 2) an indirect effect by favouring bicarbonate ions which compete with phosphate ions (Olsen, 1953).

The solubilities of many metallic cations vary markedly with pH, and aluminium, for example, may be released in soluble form at toxic concentrations at pH values of about 4 (Jones, 1961). Soluble aluminium interferes with nutrient uptake and can inhibit root growth, particularly that of seedlings (Clymo, 1962). Certain species which appear to perform best on acid soils (calcifuges) may, in fact, simply be more competitive than other species when soil levels of Al^{3+} are high (e.g. the grass *Deschampsia flexuosa,* which has its optimum pH for growth under experimental conditions at 5.5–6.0, yet is found in the field on much more acid soils; Hackett, 1965).

In addition, the structure of the soil itself is influenced by pH, and factors such as coherence, swelling, porosity, aeration and water-holding capacity may all be altered; the more a species is affected by any of these, the more it is affected by soil acidity (Stalfelt, 1972). Soils may also show a striking degree of small-scale heterogeneity in pH (e.g. from pH 4–7 over only a few metres of dune heath; Ranwell, 1959; Maarel & Leertouwer, 1967).

Where the role of pH has been investigated critically, it has been found that the *interaction* between acidity and calcium ion concentration can be the vital factor in determining plant performance. For example, many *Sphagnum* mosses suffer disproportionately severe depression of growth in the presence of high pH and high

Table 8.3. The relation between soil pH and distribution in Denmark of some meadow species. (After Olsen, 1923.) The figures represent the percentage distribution in the various pH ranges. Note the increase in species richness with increasing pH.

Species	pH range								
	3.5 to 3.9	4.0 to 4.4	4.5 to 4.9	5.0 to 5.4	5.5 to 5.9	6.0 to 6.4	6.5 to 6.9	7.0 to 7.4	7.5 to 7.9
Deschampsia flexuosa	54	31	15						
Calluna vulgaris	31	23	*31*	15					
Molinia coerulea	25	25	25	11		4	4	7	
Potentilla erecta	18	18	18	*21*	10	5	3	8	
Hieracium pilosella			20	*30*	30	20			
Anthoxanthum odoratum	7	7	15	17	*26*	17	7	2	2
Deschampsia caespitosa				3	*30*	27	12	15	12
Galium palustre				11	29	*32*	14	11	4
Veronica chlamaedrys				7	29	14	*36*	7	7
Scirpus sylvaticus							*63*	12	25
Succisa pratensis			8	17	8	8	17	*42*	
Tussilago farfara						11	11	*44*	33
Agrostis stolonifera							14	29	*57*
Total species	5	5	7	9	7	9	10	10	7

calcium availability together, but suffer rather little under conditions of high pH or high calcium availability alone (Clymo, 1973).

The biology of the so-called calcicole (lime-loving) and calcifuge (lime-hating) plants is closely linked with soil acidity. Botanists have long recognized that the floras of chalk and limestone regions are exceptionally rich in species, and have sought to understand the properties of those plants which appear to be restricted to calcareous or to very acid soils (Table 8.3 and Fig. 8.2). Away from chalk and limestone substrates, soils tend to be calcium-deficient because of weathering and leaching. However, calcium deficiency tends to be associated with low pH, simply because calcium is the most abundant base cation. Because of this correlation, it is difficult by field observation alone to distinguish between plants which actually *require* high calcium, and others which are simply not competitive on soils of low pH.

Most of the ecological effects of calcium on plant performance appear to be indirect: 1) calcium modifies soil nutrient availability, and excess calcium can bring about deficiencies of iron, magnesium or trace elements (e.g. 'lime-induced chlorosis'; Grime & Hodgson, 1969); 2) calcium may render phosphoric acid unavailable, so that plants with high phosphate requirements appear to be calcifuges; 3) calcium-rich soils tend to be porous and to have low soil water content, with the consequence that soil temperatures are relatively high and aeration is good (e.g. the grass *Bromus erectus* is restricted to calcareous soils in the moist, temperate parts of Europe, but occurs on non-calcareous soils in warm, dry Mediterranean regions; Salisbury, 1921); 4) calcium modifies the relative competitive abilities of plants for different nutrients (see Chapter 2). Calcicoles, therefore, tend not to be plants which require high calcium as such, but those which are sensitive to other environmental factors *influenced* by calcium, such as acidity, nutrient availability, aluminium toxicity, soil temperature or aeration.

A detailed study of calcicoles and calcifuges was carried out in 'chalk heath' by Grubb *et al.* (1969). This is an unusual plant community where calcifuges like heather, *Calluna vulgaris*, grow alongside chalk grassland species like *Asperula cynanchica*. The classical explanation of this phenomenon was based on an alleged stratification of the root systems; the calcifuges were supposed to be shallow rooted plants living in an acid, upper soil layer overlying the chalk, while the calcicoles were deep rooted species with their root systems in contact with the chalk. This hypothesis failed to explain how the seedlings of the lime-loving plants were supposed to survive in the acid, upper layer. It was also at odds with the observation that many of the calcicoles were actually *shallow* rooted! By concentrating on the dynamics of plant recruitment, Grubb *et al.* (1969) were able to show that both kinds of plants form healthy root systems

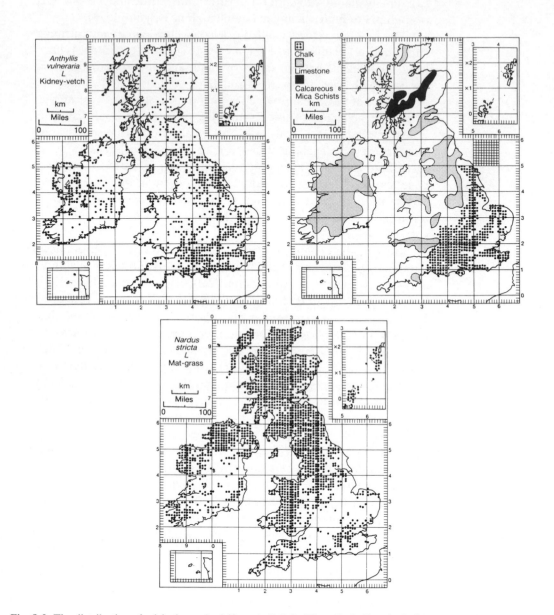

Fig. 8.3. The distribution of calcicoles and calcifuges in Britain. The calicole (lime-loving) legume *Anthyllis vulneraria* follows closely the distribution of the chalk in southern England. It is more scattered on the limestones of northern England and Ireland, and scarce on the high mountains of the Scottish Highlands. It is also found in coastal habitats where there are calcareous dunes of shell sand. The calcifuge (lime-avoiding) grass *Nardus stricta* is found on acid soils throughout upland Britain. It is absent from the chalk, but occurs in limestone regions where acid glacial drift overlies the calcareous rocks. Note that species distributions cannot be explained by reference to *single* environmental factors. For example, *Nardus* requires high rainfall and cannot tolerate heavy defoliation, while *Anthyllis* is absent from exposed, high altitude calcareous rocks, and from wet substrates generally. Also, different environmental factors tend to be correlated with one another, so that chalky soils, for example, are usually thin, relatively warm and freely drained, while acid soils are often peaty, cool and wet. Maps by kind permission of the Biological Records Centre, Monks Wood.

at pH 5–6, and that soils with this pH were relatively frequent amongst a mosaic of more acid and more calcareous patches. However, unless heavy grazing kept rank-growing grasses and acidifying shrubs at bay, the lime-loving species were unable to regenerate. In short, what set the calcicoles aside was: 1) their intolerance of soil pH < 5; 2) some resistance to drought; and 3) some degree of sensitivity to competition with tall grasses.

In this same area, relaxation of rabbit grazing after the introduction of myxomatosis, led to very rapid acidification of the soil, because of vigorous growth by *Calluna vulgaris* and *Ulex europaeus*, both of which produce strongly acidifying litter. The pH of the top 1 cm of mineral soil fell from about 5.5 to about 4.0 in only ten years. This example illustrates the important point that plants may affect pH just as much as pH affects plants (Grubb *et al.*, 1969).

8.4.8   Heavy metals in soil

A great deal of work has been carried out on the ecology of 'heavy metals' in soils, particularly in relation to the re-vegetation of mineral spoil heaps and the reclamation of derelict land (Bradshaw & Chadwick, 1980). The term heavy metals covers a diverse range of elements like lead, cadmium, zinc and copper which are classed as environmental pollutants, and a number of others like nickel, chromium or mercury which are less important pollutants but may exhibit important effects on the growth of certain plants. Because the range of elements is so broad, there are few generalizations about the responses of plants to heavy metals as a group, and detailed responses vary considerably from one plant species to another (Antonovics *et al.*, 1971; Hughes *et al.*, 1980).

The effects of heavy metals on plants are felt in three main ways: 1) through direct toxicity, leading to stunting and chlorosis; 2) through antagonism with other nutrients, often leading to symptoms of iron deficiency; and 3) through inhibition of root penetration and growth.

There is no doubt that high concentrations of heavy metals can act as a potent agent of natural selection, and where soil levels are high, a specialized, metal-tolerant flora may arise. Some species are so substrate-specific that they can be used in prospecting for commercial ore deposits (e.g. the 'Rhodesian copper plant' *Becium homblei*, a labiate used by copper and nickel prospectors in Central Africa; Howard-Williams, 1970). The metal-tolerant plants exhibit one or more processes for containing the otherwise toxic metals: 1) the metals may be excluded by modification to the root's uptake mechanisms; 2) the metals may be complexed internally to render them unchanged, but innocuous; 3) they may be degraded to innocuous by-products; or 4) the plant may possess resistant enzymes which can function in elevated levels of the pollutant (Bradshaw, 1976).

In species which exhibit polymorphism in heavy metal tolerance, the tolerant genotypes are usually little different in fitness from the non-tolerant genotypes when grown on normal soils (e.g. *Agrostis capillaris*). There are exceptions, however. Zinc-tolerant *Anthoxanthum odoratum* is distinctly less fit than normal genotypes, and zinc-tolerant *Armeria maritima* will not grow on normal soils without an application of zinc. These heavy metal tolerant species have contributed greatly to our understanding of the evolutionary ecology of plants (Chapter 7). In particular, the sharpness of the boundary between tolerant and normal populations has emphasized that the strength of natural selection under field conditions can be sufficiently great to maintain polymorphism even in the face of strong gene flow (e.g. in wind pollinated grasses; Bradshaw & McNeilly, 1981).

8.4.9 Salinity

Soils dominated by Na^+ and Cl^- ions are found in coastal salt marshes, and ionic dominance by Na^+ and SO_4^{2-} occurs in 'salt deserts' (arid, inland environments where freshwater drainage is impeded, and evaporation exceeds precipitation). Although plants growing in salty soils (halophytes) do not normally *require* high salinity, some, like *Halogeton glomeratus*, will not survive in non-saline culture solution, while others, like *Salicornia*, grow only poorly (Wainwright, 1984). Some halophytes can tolerate extraordinarily high salt concentrations, and *Suaeda maritima* can grow in $500 \ mol/m^3$ NaCl (Greenway & Munns, 1980).

The principal effects of salinity are felt via the plant's altered osmotic balance. The low external solute potential (e.g. -20 bar for sea water of roughly 3% NaCl) means that in order to take in water the plant must achieve even lower intracellular potential, which can lead to: 1) reduced growth; 2) depressed transpiration rate; 3) reduced water availability; and 4) an excessive accumulation of ions and reduced uptake of essential mineral nutrients (Slatyer, 1967; Ranwell, 1972).

Salt tolerant plants tend to display one or more of the following traits: 1) ion selection (e.g. the ability to absorb ions like potassium in the presence of high concentrations of sodium in the external medium); 2) ion extrusion (e.g. the possession of specialized 'salt glands' by species like *Spartina*); 3) ion accumulation in tissues away from active metabolic sites (e.g. *Agropyron elongatum* accumulates chlorides in roots which, as in most grasses, are shed annually along with their accumulated ions); 4) ion dilution (as in succulent halophytes like *Aster*, where succulence has the effect of increasing ion dilution by increasing the volume to surface-area ratio of the plant; see Chapman, 1977, for details).

8.4.10   Atmospheric pollutants

Atmospheric pollution with gases such as sulphur dioxide, ozone, hydrogen fluoride, peroxyacetyl-nitrate (PAN), or oxides of nitrogen can have dramatic effects on plant growth and community structure (Grace *et al.*, 1981; Koziol & Whatley, 1984). For example, along a 60 km transect downwind from a smelter in Ontario, Canada, there were no trees or shrubs at all in the first 8 km, and there was high mortality of mature trees as far as 25 km away. The species richness of the ground flora was reduced for up to 35 km downwind (Gordon & Gorham, 1963). Recently, a heated controversy has raged over the role of 'acid rain' in the destruction of coniferous forests in Germany and Scandinavia (see references in van Breemen, 1985, and Paces, 1985).

The gases responsible for air pollution are typically produced by large-scale industrial processes and from burning fossil fuels, although substantial air pollution can arise from volcanic eruptions and from natural forest fires. These pollutants affect plants both directly and indirectly. The direct, phytotoxic effects influence net photosynthesis, stomatal resistance, and metabolic and reproductive activity (Treshow, 1984). For example, 8 hours of exposure to SO_2 at 785 $\mu g/m^3$ produces symptoms such as chlorosis and bleached spotting of the leaves, and 4 hours of exposure to O_3 at 59 $\mu g/m^3$ causes flecking of the leaves and necrosis of conifer needle tips. Other pollutants take longer to act, but are effective at very low concentrations. Exposure to hydrogen fluoride at concentrations of only 0.08 $\mu g/m^3$ for five weeks causes tip- and margin-burn to leaves, dwarfing and leaf abscission (Stern *et al.*, 1984). This visible damage is due to the death of mesophyll and/or epidermal cells. Whether or not it is associated with yield reductions, or with impaired plant survival or fecundity, depends upon a host of other factors, including weather conditions, the intensity of plant competition and the abundance of herbivorous animals. Substantially impaired performance may result from air pollution levels far lower than those which produce visible damage symptoms (Ashenden & Mansfield, 1978).

Some species are so sensitive to air pollutants that their presence can be used to produce quite accurate maps of mean levels of air pollution. Lichens have been particularly useful in this kind of study since they demonstrate such a wide range of pollution sensitivity, from species like *Lobaria pulmonaria* which are exceptionally sensitive and can survive only in the cleanest air, to others like *Lecanora conizaeoides* which thrive even in the most polluted urban environments (Hawksworth, 1973; Seaward, 1977).

The indirect effects of air pollutants act via the alterations in plant biochemistry which they induce. One of the most interesting side effects involves the plant's invertebrate herbivores. It has been known for some time that plants which are exposed to extreme conditions (e.g. drought, pollutants, physical disturbance) exhibit altered nitrogen metabolism; they tend to show increased tissue concentrations of nitrogen and altered patterns of amino acid composition (White, 1974; Jager & Grill, 1975). It now appears that these increased levels of amino acid availability can represent substantial improvements in food quality for the insects feeding on these plants, with the result that air pollution may even lead to insect outbreaks (Port & Thompson, 1980; Edmunds & Alstad, 1982). In a detailed study of blackfly *Aphis fabae* on broad beans *Vicia faba*, Dohmen *et al.*, (1984) suggest that increased pest-status of blackfly in Essex, downwind from London, is due to alteration in the concentration and composition of amino acids in the bean plant, induced by SO_2 and NO_2 in the London air. Experiments with these gases, and with filtered and unfiltered London air, show that the gases have no effect on the insects directly, and that the increase in aphid growth which they observe in polluted air is mediated entirely through induced changes in host plant chemistry.

8.4.11   Exposure

In habitats like mountain ridges and sea cliffs exposed to strong, persistent winds, or in aquatic communities subject to strong wave action, the sheer mechanical battering to which plants are subject imposes severe limits to the growth forms which can survive there. Indeed, one of the arguments put forward to explain why vascular plants have been so conspicuously unsuccessful at colonizing the rocky intertidal zone is that the algal holdfast is greatly superior to the vascular plant root system as a means of anchorage under these conditions.

The most severely exposed terrestrial habitats (e.g. dry, vertical rock faces) are characterized by life forms such as endolithic cryptogams which actively bore into the surface of the rock itself (e.g. certain lichens, cyanobacteria and blue-green algae). Vascular plants of exposed places tend to be dwarfed ('nanism'), prostrate (e.g. *Salix herbacea*), or to form dense, hemispherical cushions which nestle in relatively sheltered crevices amongst the rocks (Bliss, 1956). Tall plants can only survive in exposed habitats if they are very securely anchored, have snap-resistant stems, and leaves which do not tatter. The plants of sea cliffs have to contend with both strong winds and salt spray (see Section 8.4.9), while mountain plants are exposed both to the wind and to prolonged coverage by snow.

The main effects of strong winds are felt through their drying action, and the direct damage they cause to leaves, shoots and buds (Grace, 1977). Many plants of exposed places show pronounced xeromorphic features (see Section 8.4.2). The wind sculptured canopies of krumholtz trees growing at the timberline on mountains, and the close-cropped, umbrella-like crowns of seaside coniferous trees, are vivid testimony to the pruning action of the wind. Over the course of many years, long-lived plants in exposed habitats may actually move downwind (e.g. 'islands' of *Picea engelmannii* and *Abies lasiocarpa* moved at an average rate of about 2 cm per year, due to rooting of low, horizontal branches on the sheltered side, while the exposed, windward edge died back through freezing and desiccation of the shoots; Benedict, 1984).

Some extraordinary plants are found in exposed habitats. For example, the Adriatic coast of Jugoslavia is one of the stormiest in the world, and is buffeted throughout the year by very strong winds known as the Bora. It has served as a refuge for numerous bizarre plants, which are relics of the Palaeo-Mediterranean flora. Many of the characteristic endemics belong to the Cruciferae (e.g. *Brassica cazzae*) and Umbelliferae (e.g. *Seseli palmoides*), and form columnar, palmoid treelets with huge, monocarpic inflorescences. Other endemics include shrubby, bottle-like succulents ('hyper-pachycauls'), while further species look like miniature baobabs (e.g. *Astragualus dalmaticus* (Leguminosae) and *Centaurea lungensis* (Compositae)). A common feature of these plants is that they have persistent petioles which form a collar around the shoot apex protect-ing it against desiccation and freezing (Lovric & Lovric, 1984).

Species from sheltered, lowland habitats are often less productive when grown in windy environments because of: 1) direct tattering and abrasion of their leaves; and 2) the diversion of limited resources into extra supporting tissue (see Chapter 12).

8.4.12   Trampling

The contrast between the vegetation of footpaths through grassland and that of the surroundings is often extremely marked, especially in winter when the green of the footpath stands out in vivid contrast to the dead remains of plant life on either side. It is also noticeable that livestock prefer to graze the footpath rather than the surrounding grassland (Bates, 1935). Trampling by man and domestic animals creates a characteristic plant community dominated by species whose morphology gives them a certain tolerance of bruising, compression and other physical abuse. For example, in mesic grasslands in Britain, the central part of the path is dominated by grasses like *Poa pratensis* and *Lolium perenne*, and by herbs such as *Trifolium repens* and *Plantago major*.

Trampling tolerance in *Poa* and *Lolium* derives from the unusual structure of their leaf sheaths, with their conduplicate stem and folded leaf section (compared with the normal rolling of the leaf in other grasses). The leaves thus offer a flat surface to the crushing action of the foot. Further, the grasses of the footpath are crypto-phytes, with their perennating buds buried just below soil level, whereas the grasses of the surrounding area tend to be hemicrypto-phytes or chamaephytes with their buds at, or above the surface. Trampling tolerance in the herbs derives from their prostrate growth habit. In *Plantago major,* the tough leaves are held in a ground-hugging rosette, with the upper leaves protecting the lower. On the path this species grows as a cryptophyte, whereas in untrodden areas nearby it is a hemicryptophyte with semi-erect leaves. The trampling tolerance of *Trifolium repens* derives from its tough, rather flattened, prostrate runners, but its leaves and perennating buds can be damaged by treading. It tends to be found on the edge of the path, and its presence may be due more to reduced competition for light, than to trampling tolerance as such.

Another characteristic plant community occurs in gateways and on farm tracks. Annual plant species like *Chamomilla suaveolens* and *Polygonum aviculare* and perennials like *Potentilla anserina* and *Plantago major* are particularly abundant in these places, owing their presence to an ability to withstand the pressure of wheels and the puddling action of animal hooves. The annual species germinate late in the year, and because of this, they may be unable to establish in closed vegetation (Bates, 1935); indeed, most of the species of tracks and gateways appear to be light-demanding. *Potentilla* persists by sending out copious runners over the disturbed ground during the summer, but the plant does not tolerate much direct treading. Treading resistance in *Polygonum aviculare* derives from its tough, wiry, prostrate stems, whereas the normally upright *Chamomilla suaveolens,* possesses an extremely pliable stem, which is tough, fibrous and does not snap if bent double.

8.4.13   Extremes of heat

Low temperatures

Since biochemical reaction rates are temperature-dependent, the most obvious effects of reduced temperature are in reduced net photosynthesis and reduced growth rates. The consequence of this is that plants take longer to reach their threshold size for flowering, or fail to accumulate sufficient resources during the growing season to flower at all (Went, 1953). Thus, in arctic and alpine environments where the growing season is very short, annual plants are extremely uncommon, and many of the perennial plants have foregone the production of seed in favour of entirely vegetative reproduction (Bliss, 1985).

The physiological effects of low temperature are complex, and still rather poorly understood (see Fitter & Hay, 1981). Some cold tolerant species, however, can be chilled to $-38°C$ without damage. Chilling of tropical plants brings about a phase change (liquid to solid) in their membrane lipids, and inactivation of membrane-bound enzymes in the mitochondria. Temperate plants, however, do not suffer until ice begins to form, leading to mechanical damage, progressive dehydration and eventually to cell death. Slow cooling is much less harmful than rapid cooling to the same temperature, and rapid thawing may also have lethal effects (Levitt, 1978).

Plants from extremely cold environments exhibit a number of common traits: 1) if any seeds are produced they show marked dormancy which can only be broken by cold treatment; 2) the plants often possess carbohydrate storage organs which allow them to grow very rapidly in the spring, and also to accumulate resources over several brief growing seasons before investing in a burst of seed production; 3) flower buds can be formed one or more years before flowering; 4) leaves tend to be small, long-lived (investment in leaves is paid off over a long period), and dark-coloured (an accumulation of anthocyanin pigments in the leaves allows them to heat up rapidly because dark-coloured leaves absorb more radiation); 5) the tissues can accommodate intercellular ice without damage; 6) they are tolerant of the cell dehydration caused by freezing; and 7) they are able to acquire freezing resistance by 'hardening' through exposure to temperatures below $5°C$ for several days once growth has stopped and dormancy has been established.

High temperatures

Plant temperatures in excess of $40°C$ are usually associated with failure of the cooling system, when stomatal closure in response to water shortage cuts off evaporation and its consequent transpirational cooling (see Section 8.4.2). The result of high tissue temperatures is that cell metabolism is severely disrupted, possibly by protein denaturation, membrane damage or the production of toxic substances (Levitt, 1972).

Plants from hot environments show one or more of the following traits: 1) small, dissected leaves which increase the rate of convective heat loss; 2) leaves which reflect a high proportion of the incident radiation (e.g. some plants possess a thick, white, reflective pubescence at times of the year when temperatures are highest, but less hairy, green leaves, in cooler seasons; Ehleringer & Mooney, 1978); 3) a C_4 photosynthetic system which continues to function up to $45–60°C$ rather than a C_3 system which operates only up to $35–45°C$ (see Chapter 11); 4) physiological tolerance of very high tissue temperatures, especially in those species like succulents and sclerophylls which cannot cool their tissues by high rates of transpiration during the daytime (the maximum recorded plant temperature is

65°C from an *Opuntia* cactus). As with frost and drought resistance, the mechanisms of heat resistance are not yet clear, but resistance can be induced by 'thermal hardening', which may be related to the stability of protein structure (Fitter & Hay, 1981).

8.4.14   Mutualists

There has been a veritable explosion of interest in the ecology of mutualism in recent years (references in Boucher, 1985). Almost all kinds of mutualisms are of concern to plant ecologists, from the intimate symbioses of algae and fungi living together as lichens (Seaward, 1977), to the loose, facultative mutualisms involved in the dispersal of fruit and the stimulation of germination (Krefting & Roe, 1949). Many aspects of mutualism are covered elsewhere in the book: pollination in Chapters 6 and 7; fruit dispersal and germination-enhancement by animals in Chapter 6; nodulation of legume roots in Chapter 2; mycorrhizae in Chapter 12; plant defence by ants in Section 8.4.15. Because nodulation and mycorrhizae occur under-ground, their study has suffered from the 'out of sight, out of mind' syndrome, and it is only recently that the study of these vital mutualisms has begun in earnest (Alexander, 1983; Harley & Smith, 1983; Sprent, 1983).

Root nodulation by *Rhizobium* bacteria occurs in most species of legumes, and nodulation has also been reported in 158 species from 14 genera of non-legumes, including *Alnus*, *Myrica*, *Dryas*, *Casuarina* and *Hippophae* (Bond, 1976). Not surprisingly, nodulation is particularly prevalent on nitrogen-deficient soils, and during the early stages of primary successions (Sprent, 1983). Furthermore, nitrogen fertilization of soils tends to reduce the incidence of nodule formation (Chapter 2).

Mycorrhizal development is prevalent in most habitats and most plant taxa, but can be classified into three main types: 1) many of the dominant tree species found in low-diversity or monospecific stands have sheathing (ectotrophic) mycorrhizae; 2) the great majority of herbaceous plants support vesicular-arbuscular (V–A, or endo-trophic) mycorrhizae; 3) heaths have their own, distinct 'ericaceous' mycorrhizae.

In a mycorrhizal association, the fungus obtains carbohydrate from the root system of the plant, and the plant obtains nutrients from the fungus (Bowen, 1980; Fogel, 1980; Malloch *et al.*, 1980). (The relationship between germinating orchid seeds and their fungi is untypical, in that the vascular plant parasitizes carbohydrate from the fungus; Summerhayes, 1951.) While the principle advantage of mycorrhizal association is usually said to be an enhanced ability to gather immobile soil nutrients like phosphates (Caldwell *et al.*, 1985), Alexander (1983) emphasizes its potential importance in the nitrogen economy of plants. Studies of nutrient transfer between plants whose root systems are bridged by mycorrhizae have demonstrated exchange of labelled phosphorus, but, as stressed by Ritz and Newman (1984) this does not imply *net* movement from one plant to another. Just as nitrogen fertilization reduces nodulation, so phosphate fertilization tends to reduce the development of mycorrhizal fungi.

The amounts of carbohydrate paid to the nodulating bacteria for the nitrogen they fix and to the mycorrhizal fungi for the phosphate they gather are unknown. Large amounts of carbohydrate 'leak' from the root system into the soil in nutrient-rich systems (e.g. agricultural soils; Coleman *et al.*, 1984; Whipps, 1984), but little of this material is payment to soil mutualists. The likely functions of secreted carbohydrates include lubrication of root tips penetrating between soil particles, defence of roots against soil pathogens, and the production of allelopathic compounds. In nutrient-poor soils, however, carbohydrates passing out of the roots are more likely to be payment for the services of mycorrhizal fungi or nodulating bacteria (Newman, 1985; Read *et al.*, 1985). These topics are expanded more fully in Chapter 12.

8.4.15   Enemies

Plants are protected against their enemies by a wide range of structural, biochemical, phenological and ecological features. A number of these are discussed elsewhere (Chapters 5, 9, 10 and 12), and I shall simply introduce some of the key references here (for details, see Crawley, 1983, Strong *et al.*, 1984, and Rhoades, 1985).

The protective function of many of the surface features of plants, such as hooked or glandular hairs, spines, thorns and thick waxes, can be seen by comparing the damage done to genetic strains of plants which lack these traits, when exposed to vertebrate or invertebrate herbivores (Maxwell & Jennings, 1980). There is also a bewildering variety of biochemicals ('secondary plant compounds') contained within tissues (Stumpf & Conn, 1980), or produced by tissues in response to pathogen attack (Green & Ryan, 1971) or herbivore damage (Haukioja, 1980). While it was originally thought that secondary plant compounds were simply 'metabolic waste products', it is now clear that they have a central role in the defence of plants against their pathogens and herbivores (Fraenkel, 1959). Biochemical features are frequently the basis of attempts by crop breeders to introduce pest resistance into strains of crop plants. The compounds may act directly as poisons, or indirectly as feeding deterrents. Conversely, the *lack* of a biochemical attractant or feeding stimulant may protect a plant from damage by its specialist herbivores. Excellent reviews of defensive plant biochemistry are to be found in Rosenthal & Janzen (1979) and Harborne (1982).

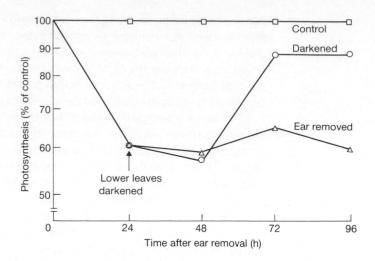

Fig. 8.4. The source-sink hypothesis states that the rate of photosynthesis in the leaves and other green parts (the 'sources') is controlled (at least in part) by the availability of 'sinks' for the carbohydrate produced (e.g. rapidly growing or respiring tissues). The effect is shown in this classic experiment by King *et al.* (1967). In wheat, the flag leaf exports most of its photosynthate to the developing ear. When the ear is experimentally removed at time 0, sink strength is reduced and photosynthetic rate falls to only 60% of control levels. However, if a new sink is created by shading the lower leaves, the photosynthetic rate of the flag leaf recovers to levels not significantly different from the controls. Feeding by herbivorous animals can lead to increases in the photosynthetic rate when they increase the sink strength (e.g. sucking insects like aphids), or when they reduce the size of the source (e.g. defoliating insects and grazing vertebrates).

Plants also possess many morphological and physiological features which increase their *tolerance* of defoliation: 1) inaccessible meristems and rapid refoliation (many pasture grasses); 2) reserves of dormant buds (many trees); 3) readily mobilized reserves of carbohydrate and proteins which allow regrowth; 4) plasticity in the distribution of their photosynthate between roots and shoots and between reproductive and vegetative functions within the shoot; and 5) the ability to increase the rate of photosynthesis per unit area of surviving leaf tissue (Fig. 8.4 and see Chapter 11). These and other aspects of plant compensation are reviewed by Crawley (1983).

Defence against insects through 'phenological escape' appears to be rather common, especially in seasonal environments. Individuals which leaf out very early will escape herbivore damage because their leaves are mature and unpalatable by the time the insect herbivores emerge from their overwintering stages. Weighed against this, early-leafing plants are extremely vulnerable to late frosts which could kill all their leaves. Individuals which produce their leaves very late may escape herbivore damage because the emerging insects starve in the absence of young foliage on which to feed. However, late flushing trees are prone to attack by larger caterpillars emigrating from nearby trees later in the season. Late flushing trees also 'waste' a certain amount of potentially interceptible sunlight by remaining leafless until relatively late in the season, and would be at a competitive disadvantage in years when herbivore attack was slight. Clearly, the precise timing of bud burst is a delicate compromise depending, at least in part, upon what other plants in the population are doing.

Another ecological means of escape from enemies involves life history characteristics. Very short lived plants may be able to escape the attentions of specialist herbivores by developing so quickly that their enemies are unlikely either to discover them or to build up in damaging numbers upon them. Such plants are said to be 'unapparent' to their enemies, in contrast to 'apparent' plants (such as trees) which are 'bound to be found' by their herbivores during the course of their long lives (Feeny, 1976; Rhoades & Cates, 1976). A second kind of ecological defence is 'associational resistance' (Root, 1973). This is not an evolved characteristic of individual plants, but occurs when the rate of attack suffered by an individual plant is influenced by the structure and species composition of the surrounding vegetation. Plants may simply be hidden from their enemies by taller vegetation, or the surrounding plants may mask the chemical cues which the herbivores use in locating their hosts. Other protective effects of complex vegetation may include alterations to microclimate which favour the plant more than its enemies, and increased predator or parasite densities in the mixed community (Bach, 1980; Risch, 1980).

An important consideration in the evolution of plant defence involves the trade-off between investment in growth and reproduction, and the production of defensive structures and compounds. In a number of tropical and temperate trees (Coley, 1983; Crawley, 1985) there is a consistent and rather unexpected correlation between herbivore damage and plant growth. It appears that the plants with the highest levels of leaf damage are those which grow most rapidly! Species which have high maximum relative growth rates (e.g. those growing where water, light and nutrients are freely available), behave in a way which suggests that leaves are cheap to produce, or low in value (or both). Presumably, this correlation between productivity and defence reflects the costs of herbivore damage to plant fitness. In productive environments the plant is able to grow so rapidly that it can compensate for a good deal of herbivore damage, and there is no net gain in fitness from the production of *extra* defensive chemicals. In unproductive environments, however, it may be so difficult for the plant to make good the losses it suffers to herbivores, that there *is* a net gain in fitness from an investment in costly defensive traits. This, presumably, is one of the reasons why plants of low-nutrient environments have such heavily defended, long-lived leaves (Coley *et al.*, 1985, and see Section 8.4.6 and Chapter 12).

Finally, some plants pay 'protection money' to ants in order to secure defence from their herbivorous enemies or from neighbouring plants attempting to invade their air space. Ants may be attracted to the plants to feed from specialized extrafloral nectaries, or to collect honeydew from sucking insects. Once on the plant, the ants kill or expel foliage feeding caterpillars, or nip off the shoot tips of invading plants. In other, more specialized cases (e.g. 'ant acacias'), the plants harbour their protectors within special hollow (or hollowable) stems where the ants culture colonies of scale insects or other Homoptera (Janzen, 1985). Plant species defended by ants do not invest in such costly chemical defences as their related non-ant species (Chapter 12).

8.5   Conclusions

Much of the interest in the factors described in Section 8.4 arises in circumstances when their prevailing levels *change*, and particularly when they change as a direct or indirect result of the actions of the plants themselves. For example, whilst it is vital that we understand the effect of fire on the survival rate of mature individuals, if we are to understand community dynamics, we need to know how the plants influence the likelihood of a fire occurring at all. Similarly, we know that soil chemistry and biochemistry are vital to root function, but we know rather little as yet about the extent to which prolonged site tenure by an individual plant changes the chemistry of the soil beneath it.

Perhaps the principle message of this chapter is that the 'ecological optimum' and the 'physiological optimum' of plants are not one and the same. The conditions under which plants occur at their maximum abundance in the field (the so-called ecological optimum) are often quite different from the levels of those same environmental conditions under which the plants perform best in single-factor, single-species experiments (the physiological optimum). The reasons for this discrepancy are many, complex and interacting, but they can be summarized as follows. Plant abundance under a given set of field conditions is determined by both its *relative competitive ability* with other plants, and its *relative susceptibility* to herbivores and pathogens. As often as not, therefore, plant distributions in the field are refuges from competitors or enemies, rather than places which present the plant with 'ideal environmental conditions'.

Chapter 9   The Dynamics of Growth and Form

DONALD M. WALLER

9.1   Introduction

Plant forms, often diverse and striking, are objects of both aesthetic wonder and scientific curiosity. No less intriguing are the elaborate, yet inherently simple processes of growth by which these forms are generated. Historically, ecology has provided the context for interpreting variation in form, and this tradition continues today. A plant's growth responds sensitively to the environment, so that its form at least partially reflects its circumstances of growth. In a real sense, the plant's development and its form are physical embodiments of its behaviour; they represent its response to variation in its environment. As with animal behaviour, plant form is primarily directed towards the acquisition of resources, be they light, water, nutrients, or pollinating and dispersal agents. For plants, most of these resources are sparsely distributed through space, often favouring sparse and efficient support structures. Unlike animals, however, most of these behavioural decisions are developmental and therefore irreversible. A plant's genetically determined pattern of growth represents its evolved programme for responding to an unpredictable environment. Its size reflects its immediate success in capturing resources, whilst its form (if persistent) records a history of past environments and its behavioural responses.

Plant cells are encased in rigid cell walls. This simple fact has profound implications for how plants grow and for how we study them. Although they make plants immobile, these rigid cells do allow plants to grow, compete with each other, defend themselves from herbivores, mate, and reproduce. Growth of plants is peripheral, confined to localized areas of cell division and expansion known as meristems. The organs that result are standardized, disposable, and often mass-produced.

In this Chapter, I shall show how plant growth constrains form, and how growth and form affect our approach to studying the ecology of plants. First, I shall describe the essentially modular nature of plant design, and the mechanisms of plant growth. After reviewing some of the methods used to describe and analyse form, I shall examine how plant growth and form respond to variations in growing conditions. This leads us to consider the question of how often plants are limited in their responses by structural or architectural constraints. I shall then describe the growing body of work that attempts

291

to interpret plant form in adaptive terms, and assess how well these efforts are succeeding. Finally, I shall describe how plant growth and form may be modified by the influence of pathogens and herbivores to their own ends.

9.1.1   Adaptive arguments

Ever since the great period of botanical exploration in the 19th century (Schimper, 1903), botanists have struggled to interpret the bewildering diversity of plant growth forms that populate the globe. Plant geographers soon progressed beyond simple descriptions of the overall appearance (the physiognomy) of plant communities to speculate on the functional, or 'epharmonic' relationship between plant form and environment. The first step in such an analysis was a systematic description and classification of plant forms, so that they could be related to climate (e.g. Raunkiaer's system based upon the position of the plant's perennating buds; see Chapter 8). The field of plant ecology traces its early 20th century roots in North America to attempts by physiologists to account for plant 'tolerances' and their responses to variable environmental conditions (Cittadino, 1980).

Ecologists have continued to interpret plant forms as adjusted in an appropriate way to their environment. Indeed, the conclusion seems inescapable with many aspects of form, like the production of sun leaves in areas of high light. The ultimate criterion of suitability is evolutionary fitness, but adaptive arguments and analyses are usually made with reference to maximizing some component of fitness, like competitive ability, herbivore avoidance, or seed set. Physiological ecology, plant demography and morphology all have a role in illuminating the ecological significance of various forms and patterns of growth.

Physiological ecology has now matured to the point where it makes detailed statements, sometimes based on elaborate models, of how physical and physiological characters affect photosynthetic performance (Chapter 11). Plant population biology has approached variability in growth and form from the contrasting perspectives of concern with the relative performance of different species and individuals within species. This approach frequently boils down to asking how components of fitness depend upon the response of the subunits, like buds, leaves and branches, to their biological and physical environment. In addition, population analyses provide the tool for translating these physical aspects of performance into the currency of individual fitness.

There are, of course, several dangers with the adaptational, or optimality approach. Although few dispute the idea that a plant's form has profound implications for its performance in a given environment, the ease with which plausible 'just-so' stories can be concocted, makes them seductive. Gould and Lewontin (1979), worried by the plethora of adaptive accounts, accused ecologists of adopting a 'panglossian paradigm' in which plants and animals are broken down into individual parts, each of which is then interpreted as the adaptive solution to some functional problem. They point out that many structures may simply be historical 'left overs', adjusted to a species' past environment, or the incidental result of selection for some other trait (Harper, 1982). This begs the question of the apparently straightforward, but actually very difficult problem of defining a trait in the first place. Selection on one trait also produces correlated changes in other traits, as a result of genetic or developmental linkages. The nature of these genetic and developmental correlations is an area of considerable current interest, especially in life history evolution (Dingle & Hegmann, 1982).

We need some notion of adaptation, however, to guide our interpretation of form; to ignore it would rob us of many insights. Horn (1979) has reasoned that adaptive arguments can be made more realistic and testable if they are cast as a comparison of *relative* performance. He also argues that because plants use carbohydrates to pay for their structure, defence and reproduction, adaptive arguments based on this economy may be safer than similar arguments for animals (but see Section 9.5 and Chapter 12). Suggested functions should always be backed up by some kind of comparative data or a plausible model which demonstrates the efficiency of the proposed function in an appropriate context. Even then, the usefulness of the function is only *supported*, not proved.

An essential tool in the synthesis of adaptive arguments is the notion of a strategy that, if common in the population, could not be invaded by any alternative type. This unbeatable strategy is called an ESS (evolutionarily stable strategy), and embodies the idea that the particular structure or pattern of behaviour which succeeds best *depends upon what the other individuals in the vicinity are doing* (Maynard Smith & Price, 1973). ESS models are based on some clear criterion of performance (ultimately fitness) and often contain explicit trade-offs, or cost-benefit relations between competing activities (although these rarely incorporate developmental constraints explicitly). Such models are now widely used in evolutionary and behavioural ecology (Maynard Smith, 1984), and have proved useful in illuminating such general patterns as those of sexual allocation (e.g. Charnov, 1982).

9.2   The nature of plant growth and form

9.2.1   Differentiation for efficiency

The emergence of vascular plants like *Cooksonia* and *Rhynia* onto land during the Silurian Period, over 400 million years ago, opened vast ecological opportunities, but also posed severe structural and physiological problems. Their unknown ancestors were essentially two-dimensional sheets, or simple, sprawling cylinders, restricted to wet surfaces. As many of these primitive plants struggled with each other for access to light, those with rudimentary vascular tissue for support and supply were favoured. Simultaneously, competition for below-ground resources favoured differentiation into true roots, specialized for penetration and absorbtion. Simple, upright photosynthetic stems proved to be no match for plants with true leaves mounted in orderly arrays on petioles along branches. Further specializations included lignified fibres which gave improved physical support, and a cutin-laden epidermis punctuated with stomata which gave greater control over water loss. Thus the evolutionary radiation of the vascular plants reflects the ecological advantages of functional specialization.

Selection for efficiency in physical support and leaf display has continued unabated, so we should not be surprised that many aspects of plant form appear highly adaptive (see Section 9.6). However, the mechanical and physiological bases for these forms also pose many constraints on the way plants grow (Section 9.5). The fact that no one design dominates in all, or even many, environments implies that the specialization which has accompanied selection for efficiency in any given environment involves trade-offs which make that same plant less competitive in other environments. There is tremendous diversity among all types of organs (see Chapter 8); amongst leaves, for example, there are large and small, thick and thin, broad and narrow, simple and compound, smooth-edged and serrate-edged, each, presumably, reaching competitive dominance in some native environment. It is a mark of ecology's increasing maturity that we are beginning to be able to interpret the adaptive basis for this variation.

9.2.2   Mechanisms of growth

Plant form clearly cannot be divorced from the dynamic processes of growth that generated it. All plant growth derives from highly localized meristems, where there is rapid cell division and tissue differentiation. Primary growth results from the activity of terminal meristems which extend the root and shoot axes (Fig. 9.1). In primary growth, the root and shoot meristems give rise to an axis and an array of lateral primordia. In higher plants these lateral primordia differentiate into leaves and their associated axillary meristems, arranged in a characteristic, geometric pattern known as 'phyllotaxy'. If these structures are laid down continuously by the growing terminal meristem, they give rise to a pattern of alternate leaves (often arranged radially at an angle of 137.5°). If they are produced in

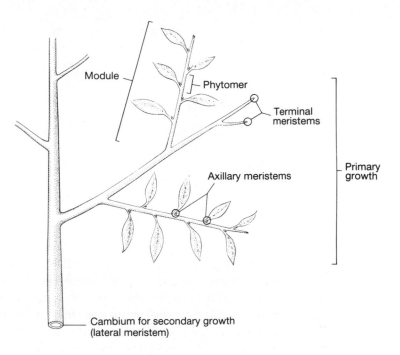

Fig. 9.1. Diagram illustrating the meristems that contribute to primary and secondary growth, and two of the finer sub-units recognized in plants: the module and the phytomer (or metamer).

rhythmic pulses, they produce opposite or whorled leaves. The axillary meristems either begin creating new shoots immediately (as in many tropical trees), or they first lie dormant within scale leaves, forming a bud (as in temperate trees). The points along the stem where leaves and their buds are located are termed nodes, and the gaps between them are known as internodes.

The elongating terminal meristem of a woody dicot shoot leaves behind it a ring of dividing cells which allows the twig to increase in girth. These lateral meristems form two cylindrical zones of cell division termed the vascular cambium and the bark cambium, which produce tissues known collectively as secondary growth (Fig. 9.1; and see Esau, 1977). The mechanism of stem thickening depends upon the formation of reaction wood in response to physical stress; this wood thickens branches and trunks in order to ensure their stability and to allow limbs to grow away from the stem at their characteristic angles. Curiously, gymnosperms thicken their limbs on the underside (with 'compression' wood) whereas angiosperms reinforce the tops of their limbs (with 'tension' wood). Further information on the details of plant growth and anatomy can be found in Raven *et al.* (1981) and Wilson (1984).

Because meristems are capable of continuing their growth indefinitely, they give plants the potential for immortal growth. Why, then, do plants not live forever? Arguments from the evolutionary study of life histories (Williams, 1957; Hamilton, 1966) suggest that selection favours the ability to survive and reproduce at earlier ages over traits which contribute to fitness only at a later age. Thus, selection does not favour immortality directly, and the potential for unending cell division is simply a by-product of meristematic growth.

9.2.3   Mechanisms that control growth

Growth and form are controlled on three levels: genetic, developmental and physiological. The patterns of developmental and physiological response themselves have, of course, an ultimate genetic basis, but this Chapter is concerned only with the proximate factors influencing growth. Each level differs in its time-scale and its reversibility. Thus, genetic changes accumulate gradually over many generations in response to changes in the selective forces affecting the population, whilst each individual plant develops uniquely in the particular spot where it is rooted. Both of these responses to the environment are irreversible, at least in the short term. However plants also respond tactically to short-term changes in their environment, making physiological adjustments (as when leaves wilt in response to temporary water stress). This section considers the first two categories of response, and describes the genetic and developmental bases for growth. Physiological responses to environmental changes are discussed in Chapter 8.

Evolutionary changes in plant form have occurred frequently, causing even closely related taxa to diverge in architecture or floral morphology. Many of these differences in structure could be due to changes in just one or a few genes of large effect (Gottlieb, 1984). Other characters, particularly those that display continuous variation like height or weight, are controlled by many genes of small effect. Gottlieb suggested that discrete character differences evolve frequently in plants because their open, plastic morphogenesis is not seriously disrupted by such changes. While the patterns of variation are suggestive, we are still some way from elucidating the genetic bases of morphological change in most species.

Despite the uniformity of their appearance and design, the subunits of which the plant is constructed are not always on an equal footing with one another. There is a clear need for the plant to organize growth in such a way that plant parts do not interfere with one another later in life, and to ensure a balance between leaves, supporting shoots and roots. This is especially true in trees since they must grow tall to avoid being over-topped by competitors, and in the temperate zone where plants must synchronize their growth with the seasons.

To ensure a reliable hierarchy of investment, plants have evolved mechanisms to limit the growth of certain subunits, and coordinate their development. This control is frequently local rather than extended (Section 9.2.5); branches and twigs that grow vigorously this year, tend to produce more and bigger twigs next year, while stressed subunits frequently die back or remain dormant. Such auto-correlation of growth is clearly advantageous since it allows a plant to respond opportunistically to favourable environmental patches, while minimizing investment in areas that are shaded or otherwise inappropriate for growth. These simple processes are sufficient to account for much of the variation in plant form that we observe between individuals of the same species (e.g. in open-grown versus closed-grown trees, in the 'flag' and krumholtz trees that grow exposed to the wind at the timberline, or in the lateral growth of tree crowns in response to neighbouring gaps). These sculpturing forces of nature have been imitated by human pruning in the Japanese art of Bonsai.

Plant hormones are the control substances implicated in mediating this control, and ultimately in determining plant form. Leader shoots, for example, inhibit the growth of buds below them on a shoot by producing auxin (indoleacetic acid) and thereby enforce apical dominance. Damage to the terminal meristem cuts auxin production and releases lower buds which may then grow into replacement shoots. Alternatively, a lower module may respond by accelerating its growth and re-orientating to replace the damaged leader. Another auxin-mediated response works in relation to asymmetric illumination of shoots; light hitting just one side of a shoot affects the interconversion of the two forms of phytochrome in a way which cuts auxin production on the shadier side. This stimulates elongation of the shady side and causes the shoot to bend towards the light. Gibberellin can greatly enhance stem elongation, initiate flowering in certain long-day plants, and break dormancy in some seeds. Other important hormones are cytokinins, abscissic acid and ethylene. For a full discussion of plant hormones and their effects, see Salisbury & Ross (1985) and Galston & Davies (1970).

Despite our knowledge of the effects and modes of action of plant hormones, we are still some way from understanding how they act in concert to coordinate growth and reproduction. Many hormones have complex and subtle effects, especially when they act in combination with other hormones and with nutritional levels within the plant. Our failure to discover some kinds of hormone which we might expect to find (e.g. a single 'florigen' or flowering hormone) probably reflects this complexity. Although physiologists frequently assume that their findings with a particular plant are general, species could differ in their responses to the various hormones in ecologically interesting ways. Unfortunately, only a few (e.g. *Coleus* and tobacco) have been studied in detail.

The vascular constrictions in trees studied in Zimmermann (1978) represent another structural means for favouring certain modules over others, similar in effect to (and perhaps due to) apical dominance. This work has shown that lower limbs develop vascular constrictions which equalize water potential across a tree's crown, or perhaps favour high leader shoots. Such a mechanism is clearly necessary if lower branches are not to intercept and transpire all the available water before it reaches the higher limbs of the tree.

9.2.4   Modular construction of plants

In most vascular plants, growth is indeterminate, with meristems continuing to produce new roots and shoots until the opportunities for further local growth are exhausted. This is as true for a stoloniferous, clonal herb snaking through a meadow, as it is for the branches within the crown of a tree. It reflects the essential similarity of these two means of vegetative growth.

Most plants are composed of a hierarchy of subunits. To describe and analyse plant growth and form we need simple and consistent terms to represent these subunits. A growing apical meristem produces, by elongation, a shoot with its leaves, buds and associated internodes (e.g. a twig). When such structures are determinate (i.e. those whose apical meristem dies or produces a terminal inflorescence), morphologists define them as 'articles' (a French word first used by Hallé & Oldeman, 1972). This term has generally been translated as 'module' and it seems useful to generalize its ecological definition to include both determinate and indeterminate shoot axes (Fig. 9.1). Modules are usually made up of smaller structural subunits, termed 'phytomers' (Gray, 1874) or 'metamers' (White, 1984), consisting of a leaf (or whorl of leaves), their axillary buds, and the associated internode. Thus phytomers connect linearly to form a module, and modules, in turn, may be arranged to form larger branch systems. It is the integrated growth of these subunits that gives rise to the hierarchy of phytomers, modules and branches in plants.

The iterated, modular nature of plants and their meristems gives them the potential for both indefinite growth and relative autonomy among plant parts. Despite the fact that continued growth produces great plasticity in final size, most plants have a recognizable form, implying that they follow characteristic, but probably rather simple, rules of growth. Even the two-dimensional rhizome patterns of some species may be distinctive (Fig. 9.3). These features make plants more than superficially different from multi-celled animals (excluding modular, branched animals like ascidians and bryozoans) which often undergo complex cellular gymnastics during development and are, of necessity, more fully integrated. Growth is often decentralized in modular organisms, with any segment capable of regenerating the pattern of the whole (Section 9.3.3).

9.2.5   How integrated are plants?

The presence of hundreds or thousands of subunits within a single plant raises the question: to what degree are these subunits acting independently in their response to the environment? This is an important question, both in understanding the mechanism of plant growth, and in our ability to abstract plant behaviour into simple and local 'rules of growth' (Bell *et al.*, 1979) or mathematical models.

Roots and shoots clearly are integrated in the sense that they depend on one another via the phloem and xylem, but the question remains as to the degree that different modules or metamers interact with each other. Structures like leaves act primarily as sources of photosynthate, while developing meristems, roots, and flowering shoots act mainly as sinks (Chapter 11). In general, the degree to which a given sink draws upon a particular source, depends on both its size and the proximity of alternative sources. Instead of all sources being equally shared between all sinks, it appears that the cost of mobilization and active transport favour local sinks over more distant ones. Watson and Casper (1984) review evidence for the existence of such 'independent physiological units' in plants (see Chapter 10).

Pitelka and Ashmun (1985) reviewed studies on the degree of vascular integration found in clonal herbs and trees, and concluded that local integration tends to be the rule, with sinks such as young or flowering shoots being subsidized from nearby modules. For example, the flowers, fruits and developing leaves of the clonal herb *Maianthemum canadense*, draw most of their resources from adjacent mature leaves. Interestingly, *Aster acuminatus*, a gap-colonizing herb, has ramets that quickly gain physiological independence, while *Clintonia borealis*, a long-lived, shade-adapted forest herb, is more integrated, with strong sinks drawing reserves from distant ramets. Such integration may allow *Clintonia* ramets to average soil and light resources over a wider area, in order to maintain their dominance at a site and to respond to disturbance.

The mechanisms of integration depend on both the vascular connections and communication via hormones. Plant hormones often control the pattern of growth, and allow some species to co-ordinate their responses over substantial distances within large individuals. Flowering, in particular, is often coordinated in this way. Grafting a flowering soybean plant to a non-flowering plant will usually induce the latter to flower, even when it has not experienced the long days necessary to induce flowering. Hormonal communication may even occur between individuals, as suggested by some recent work on herbivory where attacks on one tree apparently induce defensive reactions in nearby trees (Baldwin & Schultz, 1983). Hormonal communication between individual plants could also occur through mycorrhizal root connections (which do transfer nutrients), (Chiariello *et al.*, 1982).

There may be certain advantages for a plant to be broken up into more or less autonomous pieces; for example, a compartmentalized plant may be less vulnerable to sucking insects like aphids or whitefly, or to defoliation by large, vertebrate herbivores like the giraffe or bison, because damage is more localized than in an integrated plant. Likewise, a tree's response to trunk injury consists mostly of walling off the injured section with a proliferation of parenchyma cells well-stocked with defensive phenolics and resins. The pattern of vascular connections and hormonal effects often connects metamers on the same side of large plants. Such 'sectorial transport' suggests that there may be an advantage to correlating growth between roots and branches on the same side of a plant. Again, the degree to which plants are integrated, and how this is achieved, probably depends upon ecological circumstances in ways that have not yet been explored.

9.3   Methods to describe and analyse growth and form

9.3.1   Demography and growth analysis

Traditional 'plant growth analysis' is concerned with aggregate quantities like biomass, duration of growth, and relative growth rate, and with how these differ between sites and genotypes (Evans, 1972; Hunt, 1978a,b). Since it requires destructive sampling, this method is more appropriate in replicated experiments than in the monitoring of individual plants in the field. By ignoring the often complex details of spatial geometry and the internal turnover of parts, plant growth analysis summarizes the overall performance of a set of plants growing under a particular set of conditions; it has the advantage of simplicity.

Agronomists and evolutionists interested primarily in final seed yield from annual plants have applied a second type of growth analysis to their plants (Chapter 12). 'Yield component analysis' partitions total yield into a series of multiplicative factors like branches per plant, flowers per branch, and seeds per flower. On a log scale, these components are additive, allowing comparisons between them and facilitating statistical analysis. Such an approach clearly reflects the hierarchical nature of plant form, but because it is an essentially descriptive technique, not tied to the mechanisms of growth (e.g. that branch or flower performance often varies with position or timing of development), the results may not generalize across environments or across plants of different sizes.

The demographic tracking of individual plant parts has now emerged as a third basic tool in plant population ecology. Because plants are composed of populations of replicated subunits, it is natural to estimate birth, growth and death rates of the subunits to analyse plant growth (Sarukhán & Harper, 1973; Bazzaz & Harper 1977). This approach has the advantage of accounting rather thoroughly for the internal dynamics of growth, and of illuminating differences in initiation and turnover in plants of different sizes or in different environments. It has the disadvantage of being labour intensive, but there may be ways to economize on effort, either by exhaustively censusing only a few plants, or by sampling portions of larger plants, using a stratified random sampling technique. Although most often applied to herbs (e.g. Lovett Doust, 1981; Newell *et al.*, 1981; Smith, 1983), the technique has recently been extended to trees (Maillette, in Harper, 1980; Waller, 1986).

9.3.2   Developmental and architectural descriptions

Whilst counting leaves, tillers, or branches provides numerical insight into a plant's growth and allocation, we need further tools to describe the developmental and geometric relations between plant parts. Such relations are often crucial for understanding constraints on growth and how the parts interact as the plant grows. A branch's neighbours, especially those overhead, are of great significance to its ecological performance.

In the 1970s, morphologists studying the ontogeny of tropical tree form, identified 23 developmentally distinct 'architectural models' of plants (Hallé & Oldemann, 1970; Hallé *et al.*, 1978). These models are based on characters like the periodicity of growth flushes, the orientation of primary shoots, and the position of inflorescences (whether terminal or axillary). These are qualitative descriptions of the genetically fixed pattern of shoot development and do not normally specify the number of modules or branches that a given plant will produce. Nevertheless, such developmental models often help to specify the form that a plant will assume. When a tree loses a limb through injury, or regenerates a branch from epicormic sprouts, the modules that regrow usually mirror the characteristic architecture of the crown as a whole in a process termed 'reiteration' (Hallé *et al.*, 1978).

Whilst models can reveal the underlying pattern of development, it is difficult to make ecological generalizations about any but the simplest models. This is because several distinct developmental models may give rise to similar final crown shapes (an example of ecological convergence; see Fig. 9.2), while a single model shared by several species can produce ecologically quite divergent crowns, depending on factors like the relative elongation of axes, and the exact arrangement of leaves (Fisher & Hibbs, 1982). Also, we find many growth forms in a single plant community (Chapters 1 and 8).

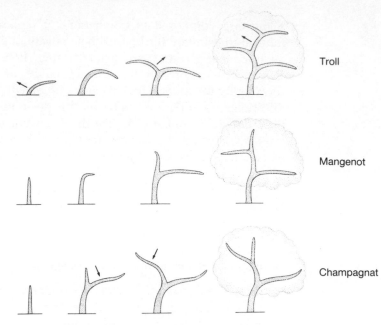

Troll

Mangenot

Champagnat

Fig. 9.2. Three developmentally different architectural models that can produce ecologically similar crown shapes. After Hallé *et al.* (1978).

Thus, whilst they are useful for understanding how a given form may develop, these models lead to unambiguous ecological predictions only for the very simplest systems.

Analyses of stream networks by hydrologists has inspired ecological morphologists to attempt to interpret branching patterns in

(a)   (b)

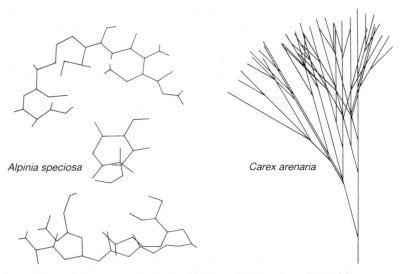

Alpinia speciosa   *Carex arenaria*

Fig. 9.3. Two contrasting rhizome morphologies. (a) View from above of the arrangement of the actual rhizome systems in three plants of *Alpinia speciosa* (Zingiberaceae). (b) Computer projection of the rhizome system of one *Carex arenaria* plant after eight years of growth; branch angles, lengths and probabilities based on observed values. After Bell & Tomlinson (1980).

trees by examining the relative number of branches in different branch 'orders'. These are usually defined according to the 'Strahler ordering technique' developed for streams, that begin at the edge of the canopy (first order branches) and work their way towards the trunk, incrementing the order of a branch each time it intersects the junction of two, similarly-ordered branches. Whilst the ratio between the number of terminal and subterminal branches may be of direct ecological interest and relevance, higher order 'bifurcation ratios' are harder to interpret, since the centripetal ordering scheme bears no clear relation to the *centrifugal* pattern of tree development (Steingraeber *et al.*, 1979).

9.3.3   Mathematical models of growth

A full description of a plant's growth would include not only information on the number, size and timing of plant parts, but also information on their three-dimensional arrangement. Such complete data would allow one to simulate exactly how a plant's form develops, and could eventually permit analyses of how parts of a plant interfere with each other and with parts of neighbouring plants. The magnitude of such an enterprise has discouraged most demographers from attempting it, but the obvious value of such data has encouraged some early work. Waller & Steingraeber (1986) recently reviewed models of branching and they provide a taxonomy for existing quantitative models based upon whether they are spatial or non-spatial, deterministic or stochastic, and whether they incorporate changes in the pattern of growth in response to surrounding conditions.

The problem of describing and analysing form is greatly simplified in two dimensions, and it is not surprising that work on spatial form has concentrated on rhizome geometry, or on the arrangement of branches within essentially two-dimensional tiers. Bell and his colleagues have analysed the rhizome branching patterns of several herbs and trees in detail, and identified two major patterns: 1) near-hexagonal forms, well suited to filling and occupying space efficiently; and 2) linear forms, capable of rapid expansion into new areas (Fig. 9.3; Bell *et al.*, 1979; Bell & Tomlinson, 1980). These groupings match well with Lovett Doust's (1981) labels for describing extreme clonal growth tactics as either 'phalanx' or 'guerilla' (Chapter 8).

Several researchers are currently building computer models based on observed data to represent the geometry of growth and incorporate interactions between plant subunits. It is relatively easy to develop models that produce graphic output which is immediately recognizable as the species from which the data were obtained (Fig. 9.3). The hope is that such models, when extended to incorporate interactions between divergent growth forms, could be used to predict the outcome of competition in a given type of environment. This is an exciting and potentially powerful approach, but it could be misleading unless the assumed mechanisms of interaction are rigorously tested in real plants.

Several ecologists are attempting to extend such analyses to three-dimensional tree crowns. Working initially with *Terminalia catappa* (Combretaceae), a tree whose branches are arranged into discrete tiers, Honda and Fisher (1978, 1979) progressed from a simple model that represented branching deterministically without interactions between the branches (Honda, 1971), to explore the effects of varying branch angle and branch length on branch overlap and effective leaf area (Fig. 9.8). They have now extended the models further to include interactions between the branches (Honda *et al.*, 1981) and decussate branching (opposite branches, alternately perpendicular) with geotropic effects that represent a full three-dimenional crown as in *Cameraria* or *Tabernaemontana* (Apocynaceae) (Honda *et al.*, 1982). We can expect more complex and realistic models of this sort which eventually could be used to test the adaptiveness of observed plant forms (Section 9.6).

9.4   The flexibility of growth and form

Plants respond sensitively to the environment, not by moving their bodies, but by varying their physiology and growth. In this section, I shall show how plants respond to variable environments by changing both the number of their subunits, and by altering their shape or arrangement. Some circumstances favour plasticity more than others, and the responsiveness of a plant depends critically on both its physiology and its plan of development.

9.4.1   Plasticity in size

A plant's size directly reflects the conditions under which it is growing. Branches and parts of branches that receive more light grow and divide more prolifically than shaded branches. Differential proliferation and directed growth lead eventually to a crown that occupies areas with more light. For example, trees planted in a free-standing group only have branches on the outward-facing sides of their trunks. Roots are correspondingly opportunistic in discovering, and growing towards, locally rich patches of water or nutrients (termed 'foraging' by Cook, 1983; Salzman, 1985). Such opportunistic growth can occur rapidly, changing the form of the plant, as when dormant epicormic buds are released to grow into adventitious trunk or root sprouts. The reproductive shoots of long-lived perennials, in particular, are frequently initiated only in sunnier microsites (Chapter 10).

Plants also respond to stimuli in more subtle ways. The orientation of their axes depends on the responses they make to stimuli like gravity, light, or obstacles (collectively called 'tropisms'). The hormones and control mechanisms which bring about these responses have been studied for many years by physiologists. Thus, limbs that normally grow out at a characteristic low angle from a tree, often reorient to become more vertical by producing reaction wood, when they are released from apical control by the loss of the leader shoot and its production of auxin. Whilst it seems natural for roots to grow downwards and for shoots to grow up towards the light, these generalizations do not always apply, reflecting the ability of selection to modify these basic tropisms. Thus, while the primary root usually grows downwards, most roots spread laterally and are insensitive to gravity. Even more surprising are certain lianes whose apices actually orient towards darkness initially, a response which often leads them towards the trunk of a suitable host tree.

9.4.2   Developmental plasticity

The response of a plant to its environment often includes changes in its morphology or physiology in addition to changes in its size or the number of its parts. A familiar example is the ability of many species to produce thin 'shade leaves' when growing in a forest, and smaller, thicker 'sun leaves' when growing in the open (Chapter 12). This morphological response is usually matched by increased levels of carboxylating enzymes and decreased concentrations of chlorophyll in sun leaves.

Sugar maple *Acer saccharum* is a typical late successional tree of north-eastern North America that displays plasticity on several levels (Steingraeber *et al.*, 1979; Steingraeber, 1982a,b). When a horizontal branch is growing under the low light typical of forest understories, the petioles twist around so that the opposite, lobed leaves are displayed in a characteristic and efficient planar array. In contrast, a branch growing in a sunny environment will usually be more erect, with the petioles extending directly outwards. The leader shoots in such environments are also capable of differentiating several pairs of leaves in addition to the four pairs pre-packaged in the bud, and these smaller, less lobed, later leaves have shorter petioles, allowing them to fit within the earlier leaves, and avoid self-shading (Fig. 9.4). Furthermore, branching also varies in response to light, both within and between trees. Open grown trees and the tops of forest canopy trees have more terminal twigs per subterminal branch (the $R_{1:2}$ branching ratio changes from 4.06 to 6.47). Such plasticity clearly adds to the leaf area, and the number of leaf layers, under sunny conditions in a way that enhances photosynthesis (compare with

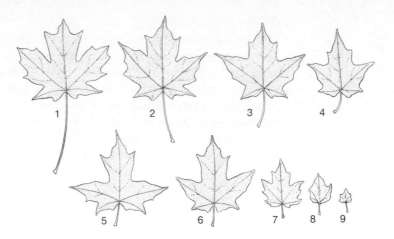

Fig. 9.4. Leaves from a vigorous terminal shoot on sugar maple, numbered in order of their appearance. Leaves on nodes 1–4 were preformed in the bud. Note the smaller size, shallower lobes, and shorter petioles of later (neo-) formed leaves. After Steingraeber (1982a).

Horn, 1971). Bell and Tomlinson (1980) list many other plants which display morphological flexibility in the face of changed conditions or age, but they also found many species which appear to be unable to respond with such flexibility.

9.4.3   When to be, and when not to be, responsive

A key problem in assessing plant plasticity is to recognize when plasticity reflects a genetically programmed (and presumably adaptive) response, and when it only reflects a lack of developmental control. This problem is obviously related to the general question of the genetic basis of plant form. If we assume for the moment that some genetic component of this plasticity is itself adaptive (see Section 9.1.1), then we can ask whether some ecological circumstances might favour such plasticity more than others.

We should expect adaptively flexible growth in those plants that occupy ecological conditions which favour opportunistic growth. These plants include short-lived weeds, all our cultivated annuals, and other ruderals that originally occupied unpredictable habitats where such responsiveness allows them to grow quickly to whatever size the local conditions will permit. In contrast, longer-lived plants that grow in close competition with other individuals benefit from types of tactical plasticity in morphology and/or physiology that allow them to invade and dominate nearby spaces that open up temporarily (Chapter 3). Any plant that does not quickly take advantage of all the resources available to it, will be more likely to be overtopped by a neighbour and competitively suppressed. It is worth noting that competition for light may be rather different from competition for water or nutrients (Chapter 12); plants 'scramble' for soil resources, and success is probably proportional to size, whilst plants 'contest' for light, so that a slightly larger individual can win disproportionate success by interfering with the growth of subordinate plants (Turner & Rabinowitz, 1983; Weiner, 1985).

Many angiosperm trees, like the sugar maple referred to earlier, display considerable flexibility in branching and overall crown shape, giving them the ability to exploit nearby gaps in the forest canopy opportunistically. In contrast, most gymnosperms, with their single trunk, strict whorls of horizontal branches, and overall conical shape, are less capable of such opportunism (hemlock, *Tsuga* is an exception; Hallé *et al.*, 1978). It is interesting to reflect that these pines, firs and spruces often occupy either marginal land with fewer competitors, or recently burned (or clear cut) areas with uniformly high light intensities, that allow even-aged stands to develop; such stands contain relatively few gaps.

There are probably other circumstances where flexibility is not favoured, and could even be disadvantageous. Plants that habitually occupy very cold, dry or nutrient-poor soils are not usually faced with competition from close neighbours. Although these plants grow under extreme physiological conditions, the conditions may be rather stable or fairly predictable (what Grime, 1979, termed 'stress-tolerant'). Such plants often show a conservative pattern involving slow, steady growth, even when conditions are temporarily favourable. Under ideal growing conditions, the plants are more likely to hoard nutrients than to accelerate their growth (Chapin, 1980). This pattern might seem to be maladaptive, but it could be favoured when temporarily accelerated growth produces a body too large to be sustained once conditions deteriorate. More likely, however, it reflects a physiological constraint which developed under selection for growth or nutrient efficiency in the face of prevailing environmental scarcity (see Chapter 12).

9.5   Constraints on growth and form

Despite their flexibility in coping with variable environments, plants are clearly also constrained in their growth, both by their genetic programme, and by their physical structure. The genetically fixed basis for form is captured in the expressive German term 'bauplan' (literally 'building plan'), and the importance of bauplan for channelling the evolution of form has recently been re-emphasized (Gould, 1977; Gould & Lewontin, 1979). We are not used to thinking in terms of what is *not* possible, and this makes it hard to elucidate these growth constraints and their implications. Nevertheless, their importance makes it imperative that we incorporate them into our evolutionary and ecological reasoning. This section provides some examples of these structural and developmental constraints, but, undoubtedly, many further instances await discovery.

9.5.1   Structural constraints

Aerial shoots face the obvious forces of gravity, wind, and in certain areas, epiphytes, all tending to pull their branches down. This means that a plant body must devote an increasing fraction of its total biomass to support as it grows larger, since the strength of a column or beam scales with the square of its diameter, while its mass increases with diameter squared times length. This is an example of an allometric relationship; that is, one in which an organism changes shape or some other feature at a rate that is not simply proportional to its size (Calder, 1984). Engineering analyses reveal that allo-metries are the rule when comparing organisms of different sizes, and often serve to preserve some other property. The ubiquity of allo-metric relationships indicates how often constraints are imposed on structures by the nature of the materials of which they are composed. Adaptive patterns (Section 9.6) often emerge, therefore, as rela-tively minor wrinkles on overriding allometric relationships.

The scaling of tree trunks and limbs provides a good example of such an allometric structural constraint. The diameter of a trunk or limb tends to enlarge faster than its length, generally scaling with length to the 3/2 power (McMahon, 1975; McMahon & Kronauer, 1976). This particular allometry preserves dynamic, or elastic simi-larity, making a limb of any size bend to a constant deflection per unit length, and suggesting the importance of dynamic stresses in trees. If small trunks and limbs had the same ratio of diameter to length as large ones, the allometric exponent would be 1.0, while if they grew so as to preserve equal static (weight bearing) stress across all sizes of limbs, the exponent would be 2.0. This explains why the largest trees are stouter than the shorter denizens of old fields and forest under-stories. It also implies that static load stresses increase faster than a trunk's strength with increasing size. This may be why huge trees sometimes seem to drop limbs or topple over spontaneously, and perhaps ultimately why trees do not grow larger than they do (along with the other obvious constraint of requiring a transpirational stream under great stress in order to lift water to such great heights).

McMahon (1975) found that trees which hold records for their size appear to be designed quite conservatively, being only about one-quarter of their theoretical buckling height. These trees, how-ever, were generally open-grown, and the formula used to choose them gave disproportionate weighting to diameter, and so would be expected to pinpoint 'obese' trees with fat trunks. Trees growing in close competition in a forest are unlikely to tolerate such large safety margins, at least not until they reach the canopy. In fact, data from self-thinning stands of birch and aspen reveal trunks near, or even beyond, their theoretical buckling height (where individuals are apparently propped up by their neighbours).

I have already mentioned the notion that many plants are composed of more or less autonomous subunits (Section 9.2.5), and raised the question of how integrated their growth might be. One reason they may have this design is that it minimizes the difficulty of 'programming' development. Complex rules of growth would require elaborate coordination during development, requiring further genetic information, control substances, and the presence of all necessary subunits. Most theoretical models of plant growth and allocation implicitly assume that the total pool of photosynthate is available for use at any time, and that transport and storage are free. While this greatly simplifies the models, these assumptions are often far from true, and so far, adaptive models have failed to incorporate these kinds of constraints.

Plant anatomy and morphology can also present structural constraints on the timing of leaf activity. In eastern North American deciduous forest trees, the earliest leafing species (like aspens *Populus*, birches *Betula*, and maples *Acer*) all have xylem vessels of narrow diameter, while later leafing species (like ash *Fraxinus*, oak *Quercus*, elm *Ulmus* and walnut *Juglans*) have large xylem vessels (ring-porous anatomy) with more vessels per unit wood volume (Lechowicz, 1984). This is apparently because large-diameter vessels, whilst very efficient for water transport, tend to cavitate in winter and become unusable. Thus, trees with large vessels must build new xylem each spring before leafing out fully. The reverse may also be true; plants adapted for growing at certain times of year appear to be constrained in their morphology and physiology. A study of over 70 herbaceous plants in a Virginia forest revealed that these plants fell into a small number of distinct phenological guilds. Species within a guild tended to resemble each other in leaf size, shape, thickness, height and arrangement far more than plants from different guilds (Givnish *et al.*, 1986).

9.5.2 Architectural and developmental constraints

The nature of three-dimensional space itself imposes some geometrical constraints on the way in which surfaces may be combined (Stevens, 1974). For example, larger organisms have a larger surface area to volume ratio due to allometry, requiring them to tap specialized organs of high surface area (like root hairs or mycorrhizae) to compensate. In addition, trees and herbs grow from a seed through a sequence of stages, each of which must be physiologically and ecologically competent, besides leading to some ecologically appropriate final form. This implies further constraints, especially for a tree faced with different ecological roles during each of the stages of its development. For example, while it might seem that a geodesic dome could serve as a structurally efficient lattice for supporting a photosynthetic crown, there is no such plant because there is no way for a small geodesic dome to grow into a large one.

Besides these constraints on form, plants may also be limited in their growth by a shortage of the appropriate meristems (Harper, 1977). Water hyacinth plants that start to flower experience a pronounced drop in their vegetative growth rate as meristems switch roles, long before they have allocated appreciable biomass to flowers (Watson, 1984). We do not yet know if this is a general phenomenon, but it does raise questions about using biomass alone to measure resource-allocation patterns (Chapter 12).

9.6   Adaptive aspects of plant form

9.6.1   Plant height

The height to which a plant should grow is an obvious example of the kind of context-sensitive behaviour which is amenable to an ESS analysis (see Section 9.1.1). In any densely vegetated environment, leaves serve both as the plant's solar collectors (its 'jaws') and as its weapons for interfering with the growth of neighbouring plants (its 'claws'). In the absence of neighbours, a plant is free to grow sideways, investing minimally in structural tissue for support and supply. However, once plants begin to encounter neighbours, they gain a distinct competitive advantage by growing taller, so that a small size difference can be amplified into a continuing competitive advantage (c.f. competition for soil resources: Section 9.4.3).

Because light increases more or less exponentially as distance beneath the surface of the canopy is reduced, while the costs of structural tissue escalate less rapidly with height, increasing competition favours increased investment in height growth (Givnish, 1982). For example, in a diverse community of herbs growing in the Virginia Piedmont, height responds sensitively to cover (Fig. 9.5). Taller plants are forced to allocate proportionally more resources into supportive stems and so usually grow only as tall as necessary to compete. This idea receives further support from a study of the wood nettle *Laportea canadensis* which displays as much height variation between environments as the different herb species showed within the Virginia herb community (Menges, 1985). Interestingly, height levelled off in the densest monospecific stands, hinting at some between-stem 'altruism' in this clonal plant (see Chapter 4).

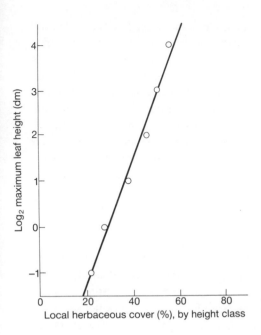

Fig. 9.5. How a diverse set of herbaceous plants responds to increased competition by growing taller. Maximum leaf height plotted against average ambient cover experienced by that height class. Leaf height roughly doubles for each 7% increase in cover. Note the $\log_2$ scale. After Givnish (1982).

9.6.2   Leaf arrangement

The arrangement of leaves into a crown obviously affects their efficiency, and we can expect different arrangements to be favoured in different environments. Because photosynthesis is essentially saturated at about one-quarter of full sunlight, a plant growing in the open can achieve higher total productivity by producing several diffuse layers of leaves, each of which is photosynthesising at capacity, than by producing one continuous layer. In contrast, a plant growing in the shade (below one-quarter sunlight) would do best to avoid self-shading by supporting a single, continuous layer of shade-adapted leaves (Horn, 1971). There are therefore two extreme hypothetical canopy arrangements; *multilayered* crowns with leaves scattered throughout the interior of the canopy, and *monolayered* crowns with all their leaves concentrated in a thin shell at the edge of the crown. Horn suggested that leaf arrangements in early versus late successional trees matched the prediction that multilayers would do best in the open, but monolayers in the shade. The leaf arrangements found on many open versus shade grown herbs follow the same broad pattern.

   A tree developing its canopy in the open is on the horns of a dilemma; it may either branch prolifically to produce a wide, low crown with immediate high productivity, or it may restrict its branching, channeling its resources into a taller leader axis. While a tall, narrow crown is not immediately as productive as a low, wide crown, it has the capacity to overtop and eventually to shade-out such prodigal branchers. Many early successional trees like sumac (*Rhus*) strike a medium course, by branching moderately, but building compound leaves whose long rachises act as 'throw-away branches' (Givnish, 1978). This permits them to occupy a wide and diffuse crown without investing in permanent support tissue. Devil's walking stick *Aralia spinosa* provides an extreme example of this phenomenon; its two to three-pinnate leaves are 1 m or more in length and allow the trunk to grow without any branching at a rate of 70 cm per year for the first few years of life. By contrast, flowering dogwood invests 7–15 times as much wood to support the same leaf area (P. White, 1984).

9.6.3   Developmental decisions

Because of their plasticity, plants face innumerable developmental decisions. How many branches and buds should be released? Which meristems should receive priority? How should resources be allocated between vegetative growth, defence and reproduction? This last question forms the basis for a large body of literature on life histories and resource allocation patterns (Chapter 12).

Annuals face the particular decision of when to switch from vegetative growth to the production of flowers and fruit. If the end of the growing season is predictable, and resources devoted to each function bring returns in direct proportion to investment, then the

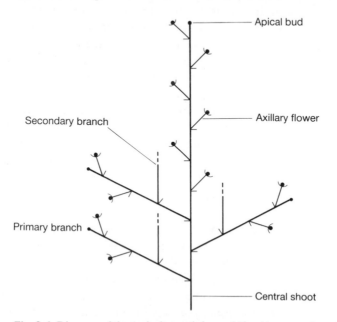

Fig. 9.6. Diagram of the typical growth form of *Floerkia proserpinacoides*, a forest annual, growing under non-competitive conditions. Whether the plant develops a branch or a flower at node 3 varies with density in an apparently adaptive fashion. After B.H. Smith (1984).

best solution is to switch completely to reproductive investment at a fixed time prior to the end of the growing season (Paltridge & Denholm, 1974; Macevicz & Oster, 1976). Many annuals display this 'bang-bang' strategy, but others show a period of *partial* allocation to both vegetative growth and reproduction before their final, all-out commitment to seed production. This strategy is favoured either when the end of the growing season is unpredictable (Cohen, 1971), or when the plants face close competition that favours increased height growth.

A special case of this general problem which directly involves plant form is provided by a study of the small woodland annual *Floerkea proserpinacoides* (Smith, 1984). Using demographic data on the survival and fecundity of flowers and branches produced at different times, Smith calculated the expected return from producing either a branch or a flower at each of several nodes (Fig. 9.6). The calculations estimated that the best anatomical switch from producing branches to producing flowers would be at node number three; this was indeed shown to be the case. The exact decision to switch however depends upon growth rate as influenced by competition from neighbours.

9.6.4   Patterns of branching

A further aspect of plant form which has been analysed in adaptive terms is the arrangement of branches within the crown. Because the number, length and angle of divergence of branches may all vary, and because the performance of a given crown should be averaged over the daily and seasonal course of the sun, the investigation of optimal branching can be complex. Efforts so far have concentrated either on patterns of rhizome branching or essentially two-dimensional branch tiers for simplicity, but have still had to rely extensively on computer simulation. These models are now being extended to simulate the support and photosynthetic efficiency of simple, but fully three-dimensional crowns (Fig. 9.7; Niklas & Kerchner, 1984).

Using field data from the tropical tree *Terminalia catappa* whose branches are conveniently arranged into distinct tiers, Honda and his colleagues (see Section 9.3.3) developed a computer model to explore the effects of changing branch angle and branch length on the non-overlapping area of leaves within a tier, and used this as a simple measure of ecological performance. They found that observed branch angles were very close to those predicted as being optimal on the basis of this criterion (Fig. 9.8). The observed ratios of branch lengths, however, were significantly lower than those predicted by the model, perhaps due to the omission from the model of any allometric increase in cost per unit length of longer branches.

Because interactions are so important in tree crowns, models which hope to portray their structure accurately must incorporate *non-stationary* effects; those effects which change with time, with plant size, or with the precise position of the module in question. Ideally, the model's reactions should mimic those of the real plant, and it is important to test the assumed interactions of the model before any trust is placed in it. Agreement between a model's predictions and the growth pattern of a real plant does not constitute a proof of the validity of the model, since many models might describe a particular pattern equally well. Constructing a model does, however,

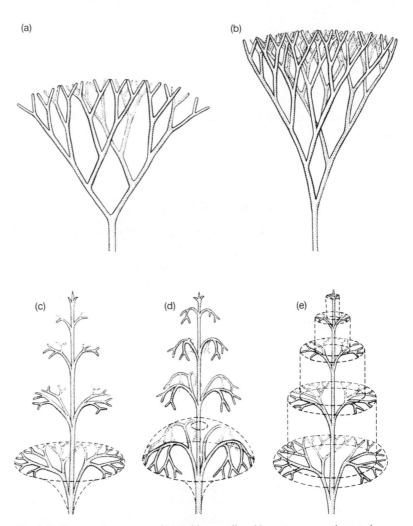

Fig. 9.7. Geometric patterns of branching predicted by computer analyses to increase photosynthetic efficiency and reduce mechanical (bending) stress. (a) Dichotomous branching at equal angles of primitive photosynthetic stems. (b) Dichotomous branching at unequal angles to reduce self-shading. (c) and (d) Branches overtop one another and merge into planes to minimize self-shading. (e) Differentiation into mechanical support tissue and photosynthetic (leaf) tissue, favoured for increasing photosynthesis while reducing total bending stress. After Niklas & Kerchner (1984).

demonstrate clearly what consequences emerge from a given set of assumptions, and this is a vital function. Conversely, model predictions that do not match an observed pattern of branching do not necessarily invalidate the model's original assumptions, since there may just be additional interactions that have not yet been included. Finally, a model that is just sufficiently complex in order to mimic plant behaviour accurately is more useful than one which is so complex that the consequences of particular assumptions cannot be readily seen.

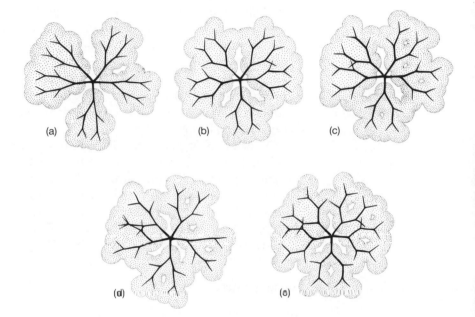

Fig. 9.8. Computer-generated predictions of the effect of branch angles on effective leaf area within a branch tier of a tree with *Terminalia*-style branching. After Honda & Fisher (1978).

9.7 Herbivores, pathogens and plant growth

A vast range of species of vertebrate and invertebrate herbivores, fungal pathogens, bacteria and viruses influence the growth and form of plants. We are most familiar with the impact of crop pests, but the role of herbivores and pathogens is no less important in the dynamics of natural plant populations. Feeding by herbivorous animals affects both the growth rate of the plant, and influences plant shape directly (by destroying meristems) and indirectly (by altering the pattern of resource allocation within the plant). This topic is reviewed by Crawley (1983); I shall restrict discussion to the major classes of impact.

9.7.1   Gall formers

Perhaps the most spectacular modifications of plant growth are caused by gall-forming animals. Figure 9.9 shows some of the variety of gall forms which are produced on a single plant species *Quercus robur* by different species from a single genus of gall-forming wasps in the family Cynipidae. The precise mechanism by which these galls are formed is not yet fully understood, but the intriguing possibility exists that these insects are expert 'genetic engineers'. The hypothesis is that the female insect injects a fragment of genetic material (probably in the form of viral DNA) into the plant when she lays her egg. This is incorporated into the plant's genome so that subsequent mitotic cell divisions, incorporating the new genetic programme, produce the unique form of the resulting gall (Cornell, 1982). The viral DNA is probably inherited via the maternal cytoplasm of the wasps. One of the fascinating twists to this story is that many Cynipids produce two distinct forms of galls each year (for the sexual and agamic generations, respectively), so that two different genomes seem likely to be involved.

Other kinds of galls formed by sucking insects or by fungi (e.g. witches' brooms) may be produced by the plant in response to the organism producing plant hormone mimics or otherwise interfering with the normal mechanisms of the plant for restraining mitotic cell division. The spruce gall adelgid *Adelges abietis* produces pineapple-like galls on young shoots which check growth and cause disfigured canopy development (Metcalf & Flint, 1951).

9.7.2   Sucking insects

Insects like aphids, frog hoppers and whitefly tap the plant's phloem or xylem or remove the fluid contents of individual cells. Their impact on growth is caused both by the carbohydrate and protein loss that they cause by removal from the phloem thus reducing the growth rate of sinks nearby (Chapter 12), and by the saliva which they inject when inserting their needle-like mouthparts. Salivary secretions can lead to blocking of the phloem which causes death of the upper parts of the shoot (e.g. 'silver top' in flowering grasses fed upon by bugs). Severe aphid attack can so reduce carbohydrate availability that seed production is completely prevented. Perhaps the greatest effect of sucking insects on plant growth and form is through their acting as vectors of viral diseases which may kill or severely stunt the plant (Thresh, 1981).

9.7.3   Defoliating herbivores

A vast army of caterpillars and vertebrate herbivores make a living by eating plant leaves. Their impact on plant growth is determined by

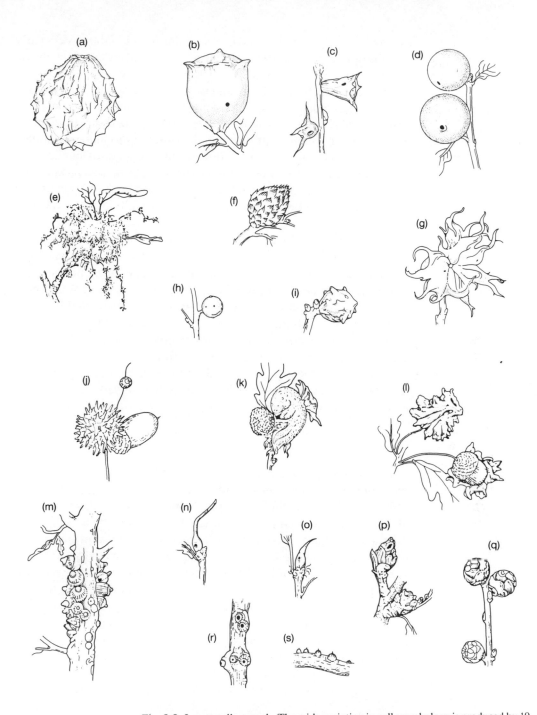

Fig. 9.9. Insect galls on oak. The wide variation in gall morphology is produced by 19 species of cynipid wasps all belonging to the genus *Andricus*, attacking a single species of host plant, *Quercus robur*, in Europe. (a) *Andricus hungaricus*; (b) *quercustozae*; (c) *polycerus*; (d) *kollari*; (e) *quercusramuli*; (f) *fecundator*; (g) *coriarius*; (h) *gallaetinctoriae*; (i) *tinctoriusnostrus*; (j) *sekendorffi*; (k) *dentimitratus*; (l) *quercuscalicis*; (m) *testaceipes*; (n) *aries*; (o) *solitarius*; (p) *inflator*; (q) *lignicola*; (r) *rhyzomae*; (s) *quercuscorticis*. Figure by M.J. Crawley.

the amount of leaf taken and by the timing of defoliation. Very early defoliation may be readily compensated in trees, for example, by the production of new leaves either from existing unopened buds, or from the production of new buds (Heichel & Turner, 1976), resulting in little impact on plant survival. Similarly, late defoliation may occur after the leaves have completed their export of carbohydrate to the plant, and may mean little more than the loss of nutrients which might otherwise have been remobilized prior to natural leaf fall. The most damaging defoliation is likely to be that which occurs on middle-aged leaves; i.e. at a time after refoliation could profitably restore the lost leaf area, but before the leaf has repaid its full carbohydrate debt to the plant.

Most plants subject to periodic defoliation have a reserve of unopened buds which can rapidly restore leaf area in case of severe early attack. They also have the ability to produce new buds on older tissues to compensate for defoliation later in the season. While most trees can compensate sufficiently that they are not killed by occasional heavy defoliation, even low levels of herbivore feeding may reduce their seed production. For example, Crawley (1985) found that oaks (*Quercus robur*) protected from insect herbivores with insecticide produced between 2.5 and 4.5 times as many seeds as unsprayed trees, despite the fact that the unsprayed trees lost only 8–12% of their leaf area.

9.7.4 Browsing herbivores

Like the sculpturing effects of high wind, browsing animals like elephants and deer remove buds and twigs and so can dramatically affect the shapes of the woody plants they browse. Removing twigs also alters apical dominance (Section 9.2.3), so that heavily browsed plants take on a stunted, highly branched and bushy appearance. On certain free-grown plants this may increase productivity (just as a gardener's pruning of fruit trees encourages the development of the most productive shoots). In competitive environments like forests, however, browsing will almost always reduce competitive ability and increase the likelihood of death. The tolerance of many woody plants to such pruning is evidenced by the example of garden topiaries which show how greatly clipping can modify plant shape.

9.7.5 Pathogens

The catalogue of grotesque plant horrors found in manuals on pest damage is testimony to the potential of fungi, bacteria and viruses to reduce plant growth and alter plant form. Any of a number of organs may experience accentuated growth while others may be reduced or wither away. For example, the fungus *Puccinia fusca* attacks the European wood anemone *Anemone nemorosa*, causing the petioles to elongate about 30% more than usual, while the leaf blades shorten by a similar amount and become less divided.

Beyond changing the relative size of organs, some parasites modify the growth and form of the plants they infect even more directly. Witches brooms on silver fir (*Abies alba*) cause the normally horizontal branch sections they infect to curve upwards in a dense growth that resembles an epiphyte. Normal twigs are 2-ranked in this species, but become whorled with infection and develop precociously (prolepsis). Finally, the evergreen leaves that normally last for six to eight years in this species, yellow and drop off within a year. In general, it appears that pathogens are adept at stimulating growth in those organs or tissues that they depend upon for their own growth.

9.8 Conclusions

In this Chapter we explored how meristematic, modular growth has given plants a wide range of form, and the ability to respond flexibly to their variable environments. Ruderal and competitive plants (Grime, 1979), in particular, need this flexibility to respond to unpredictable conditions and local opportunities.

Despite this plasticity in size and sometimes in shape, the growth of all plants is confined in ways that are often difficult to appreciate. Plants grow by extension, expansion of girth, and branching, all of which must be balanced within the constraints imposed by their structural material, and the need for an ecologically efficient architecture for intercepting light and capturing nutrients. Their modular nature often implies that the individual subunits of plants, such as ramets, branches and modules respond semi-autonomously to their environment. Such independence reflects the decentralized nature of plant growth responses (and incidentally simplifies the modelling of plant growth). Whether this independence represents the constraints imposed by a simple but sufficient evolutionary design, or is in some way adaptive, remains to be demonstrated.

Models of plant growth are increasing in complexity and detail, benefitting from the increase in data available on the demography of plant parts. By incorporating interactions between the parts, like shading and competition for resources, they are gradually becoming more realistic, and their predictions are beginning to be tested against empirical patterns. Soon, they may be used to derive patterns like the $-3/2$ thinning rule (Chapter 5), or extended to include interactions between species (Chapter 2). Because the models are so complex, however, it will be difficult to distinguish between plausible alternatives unless direct tests are also made of the mechanisms of growth on which they are based.

It should be clear from this Chapter that plant ecology is currently

experiencing a renaissance of interest in the dynamics of morphology, and in elucidating the relations between plant form and
ecological performance. Physiological ecology and plant demography have 'discovered' classical morphology, which is itself
emerging from a preoccupation with static form, to pose questions
about the functional roles played by various structures. Thus, ecological plant morphology may serve as a link that will enrich all these
disciplines.

Chapter 10 Individual Plants as Genetic Mosaics: Ecological Organisms versus Evolutionary Individuals

DOUGLAS E. GILL

10.1 Introduction

The principle of natural selection emphasizes the unequal contributions of individuals to future generations through differential reproduction and survival. Although selection at other levels of biological organization, like cells or groups of individuals, might result in unequal representation of genes in descendants, the fundamental unit of natural selection is assumed to be the *individual* organism (Williams, 1966; Harper, 1967; Lewontin, 1970; but see Dawkins, 1976). The plants and animals which botanists and zoologists regularly observe functioning as independent individuals, such as plantains and beetles, are indeed the units of natural selection upon which classic theory rests. But the equivalency breaks down in clonal species such as cattails and corals, and the problem of defining evolutionary individuals becomes extremely difficult when organisms such as viruses, aphids or dandelions are considered (Addicott, 1979; Horn, 1940; Janzen, 1977, 1979).

The difficulty of recognizing evolutionary individuals within cloning species of plants is illustrated in swards of pasture grasses and forbs. Harberd (1961) found single clones of *Festuca rubra* as wide as 200 m and perhaps 400 years old. On the other hand, one clone of *Trifolium repens* existed as five discrete patches (Harberd, 1963). There were four distinct types among 58 isolates of *Holcus mollis* and one genotype covered an area of 800 m in diameter (Harberd, 1967). The conclusion drawn from these observations is that distinct evolutionary individuals are not easy to identify on the basis of either discrete spatial boundaries, size or morphology.

A classic example of the evolutionary individual being confused with the ecological organism is the case of sterile worker castes in eusocial insects. Darwin (1859) himself was puzzled by the problem of how sterile individuals could evolve, and pondered the possibility that the colony as a whole was the actual unit of selection. Wheeler (1911) supported that interpretation and argued that queen ants were functionally the gonads of a colony and the sterile workers were analogous to appendages, such as nutrient-gathering roots or stinging nematocysts. In striking contrast, Hamilton (1964) proposed the now popular theory of kin selection, whereby natural selection can favour sacrifices in reproductive effort, even to the point of sterility, if the celibate individuals assist sufficiently the reproduction or survival of their close relatives. By considering the inclusive fitness of the worker castes in eusocial insects, Hamilton (1964) and Wilson (1971) conferred an evolutionary legitimacy to sterile individuals that Darwin and Wheeler had denied.

Many plants are, in a sense, the opposite of the eusocial insects. Whereas an ant colony is best thought of as a single evolutionary individual (the queen) made up of many ecologically distinct organisms (the workers), a large tree can be viewed as a single ecological organism made up of many evolutionarily independent units (branches). Although a tree functions as a single ecological unit by virtue of its unified root and vascular systems, it could also function as a set of *many* evolutionary individuals, if each branch were a unique genotype that generated an independent path of inheritance.

My thesis in this chapter is that the concept of individual plants as genetic mosaics has profound implications for the ecology of plant-enemy interactions, the evolution of resistance in long-lived plants, the rates of evolution in higher plants in general, and the genetic consequences of mass-flowering as a common plant breeding system. I frame the hypothesis in the context of large canopy trees, but it applies equally well to planar clones of grasses and other highly branched plants.

10.2   The hypothesis of genetic mosaicism in plants

My central hypothesis is that large plants with extensively branching architecture are actually colonies of many heritable genotypes. The genetic diversity within their structure is generated by the accumulation of developmental mutations that arise spontaneously in the meristems of the proliferating modular parts. Depending on the species, the number of modules expands sigmoidally over the course of development and approaches an asymptote of between ten thousand and one million per plant. Because each module is produced by an apical meristem capable of producing every kind of plant tissue, and because meristematic tissue is highly mutagenic, the number of developmental mutations should also expand logistically as the tree grows. Since meristems are totipotent and can give rise to both vegetative and reproductive organs, the developmental mutations have the dual potential of being expressed somatically in the vegetative tissue *and* inherited through the gametophytes.

The hypothesis makes several assumptions and several predictions. First, it assumes that plant architecture is modular and that the modules gain developmental independence with time. Second, it assumes that significant phenotypic variation exists between modules, and predicts that some of the between-module variation is genetic in origin. Third, it assumes that the phenotypic variation confers fitness differentials between the plant parts, and predicts that natural selection *within* the genetically diverse population of modules results in net differential reproduction of genotypes. And fourth, it assumes that intraplant pollination (geitonogamy) can effectively recombine genetic variants within self-compatible plants, and predicts that mass-flowering is an adaptive phenology that promotes significant genotypic variability in seed crops.

The principal contribution that the hypothesis makes to basic theory concerns the interaction of plants with their enemies. There are three important implications. First, genetic mosaicism is expected to generate phenotypic variation in susceptibility and resistance to pests between the many branches of a tree. As a tree is attacked each growing season, susceptible branches should experience greater damage than resistant plant parts. Consequently, growth will be greater by the resistant than by susceptible modules, and a greater proportion of the tree will be resistant in succeeding growing seasons. In effect, the hypothesis predicts that the genotype of a tree changes (evolves) over time as resistant modules expand most rapidly each growing season.

Second, the excessive damage done to susceptible modules by pests not only retards their growth but also curtails their reproduction. Whilst undamaged (resistant) branches produce abundant fruit, heavily defoliated or diseased (susceptible) branches produce less or none at all. As a consequence, the fruit drop from mosaic trees will be biased towards genetically resistant modules. Thus, the interaction between modules and pests naturally selects for genotypes resistant to current enemies through differential reproduction of genotypes *within* a tree. Operating each year, this process of intraplant natural selection accelerates the rate of plant evolution by changing the time scale from the slow, demographically defined generation time (perhaps scores to hundreds of years) to an annual time scale.

Third, the rates of evolution in the pests are expected to *decelerate* in response to mosaicism. As Whitham (1981, 1983) has developed so clearly, mosaicism within large plants breaks the macrohabitats of insects into myriad small patches that can set strong limitations to their abundance, dispersal and distributions. The action of highly localized selection on particular ramets can inhibit the capabilities of the pests to colonize other modules on the same tree or other trees.

A second contribution that the hypothesis of intraplant genetic mosaicism makes to basic theory relates to breeding systems in plants. In the traditional concept of plants as monogenotypic individuals, the genetic effect of depositing self pollen on flowers elsewhere in the tree (geitonogamy) would be no different from returning it to its own flower (autogamy); both are acts of self-fertilization and potentially would lead to damaging inbreeding. Cross-pollination from separate individuals (xenogamy) is assumed to be the principal mechanism whereby recombinants are produced. Thus, mass-flowering in animal-pollinated species with bisexual flowers has puzzled evolutionary botanists because of the high cost (litres of nectar and grams of pollen lost in the 99% self-fertilization) for the rare instances of out-crossing (Frankie, 1976; Augspurger, 1980).

The proposition of genetic mosaicism predicts that mass-flowering adaptively encourages pollination among flowers on different branches of the same plant (geitonogamy) and thereby effectively generates recombinants from the putatively rich genetic variation available within the plant. Because seedlings that are genetically different from parent plants may escape pathogens (Rice, 1983; Augspurger, 1984; Augspurger & Kelly, 1984), breeding mechanisms that promote genotypic diversity in seeds have high selective value for individual trees. The hypothesis of genetic mosaicism suggests that mass-flowering is one such mechanism in large, self-compatible plants. The developmental and genetical consequences of genetic mosaicism within arborescent plants was developed in some detail by Grant (1975). The ecological and evolutionary consequences of this hypothesis were first elaborated by Whitham & Slobodchikoff (1981) and Whitham et al. (1984). Here I present my own predictions of the hypothesis and the evidence that supports their underlying assumptions.

10.3   Modular construction of plants

The empirical models of tree architecture proposed by Hallé et al. (1978), and Shinozaki et al. (1964a,b) all have as their basic premise that plant development proceeds by the repetitious multiplication of structurally equivalent modules. In fact, simple computer algorithms that assemble modules using very few variables can generate a wide diversity of plant architectures (see Chapter 9). Each module develops independently and divergently from the others, and older modules give rise to new ones by lateral budding. In a very real sense, each module undergoes a complete life cycle of birth, growth, maturation, senescence, and finally death (Harper, 1977; Buss, 1983). As a consequence, a large, mature tree with its myriad branches, is really a population of modules with a distinct age structure, that lends itself to rigorous 'demographic' analysis (Chapter 9).

Several lines of evidence support the notion that developmental modules of plants are functionally independent of one another: 1) independent patterns of phenology between branches; 2) competitive interactions between modules for limited resources; and 3) highly localized effects of defoliation.

10.3.1   Independent phenology of branches

The functional independence of modules within plants is well illustrated by the unique patterns of flowering and fruiting of individual branches. The consistent between-year variation in timing and duration of flowering by individual branches of Australian *Eucalyptus regnans* suggests a genetic influence (Griffin, 1980). Alternate bearing (the alternation of heavy and sparse flowering or fruiting in successive years) is common among trees in an orchard, but has also been recorded for branches within trees in the wild (Davis, 1957). One branch may carry a good fruit crop while the remainder of the tree is bare; in the following year the reverse pattern is found. Sex expression may differ from branch to branch within the same monoecious tree, and the sexuality of different branches may alter from year to year with changes in environmental conditions (MacArthur, 1977; Sakai & Oden, 1983).

10.3.2   Competition between modules

Competition between parts of the same plant for limited resources is well known to foresters and gardeners. After new shoots spring from a cut stump in a forest, or epicormic buds flourish on a *Eucalyptus* following a fire, the number of sprouts declines steadily with time (Kozlowski, 1971). A single sprout often assumes dominance and finally redefines the trunk of the tree. The heavy mortality of the shoots is probably due to competition for water or critical nutrients, rather than to shading or herbivory (Kozlowski, 1971).

There is extensive evidence of intense competition between reproductive and vegetative parts of plants for limited resources (see Chapter 12). Reproductive activity is capable of suppressing vegetative growth (Gourley & Howlett, 1941; Kozlowski, 1971; Kramer & Kozlowski, 1960; Lenz, 1967), and extensive fruit abortion (even in the absence of pest damage) is indicative of competitive interactions between reproductive parts (Stephenson, 1982). Of course, some low seed sets and poor fruit productions may also result from other causes (e.g. poor pollinator service; see Chapter 5).

10.3.3   Defoliation of branches

The functional independence of modules is most clearly shown by the results of experimental defoliations of branches within trees. The negative effects invariably remain highly localized and are not integrated across the entire plant. Heichel & Turner (1984) subjected single, large 'representative' branches of three red oaks *Quercus borealis* and three red maples *Acer rubrum* to experimental hand-defoliations of 0, 50, 75 and 100% intensities for three successive years. Leaf number and growth were significantly reduced in the defoliated branches compared to undefoliated controls. Rockwood (1973) performed hand-defoliation experiments on six species of native Costa Rican trees in which he removed the leaves from small twigs, larger branches and entire sectors of many experimental trees. The defoliated tree parts, irrespective of size, produced no fruit, while the paired controls on the same tree produced normal quantities of fruit. Janzen (1976) performed a similar experiment on a single female Kentucky coffee tree (*Gymnocladus dioicus*) with three matched pairs of defoliated and control branches. The fruits on the defoliated branches had significantly lighter pods, and fewer, lighter seeds than those on the control branches.

The localized effects of defoliation are also thoroughly documented in the horticultural literature. Reduced fruit crowding on vines and fruit-trees leads to improved quality and volume of yields. Pruning fruit and flower buds in the immediate vicinity of developing fruits increased yields (Abbott, 1959, 1960, 1965; Barlow, 1964, 1966; Maggs, 1963; Quinlan & Preston, 1968, 1971), whilst removing those leaves closest to the developing fruits reduced both the quality of fruit and the rate of fruit growth (Haller & Magness, 1925; Magness, 1928; Kozlowski & Keller, 1966). Between 30 and 40 neighbouring leaves are necessary for optimal fruit development (Weinberger, 1931; Weinberger & Cullinan, 1932; Winkler, 1930); Haller & Magness (1925) demonstrated that fruits draw on the photosynthates of 'feeder' leaves up to but not beyond 1 m away.

In summary, theoretical models of plant architecture, intraplant variation in phenology, competitive interactions among plant parts, and highly localized consequences of experimental defoliations, all support the concept that plants are constructed from iterated modular parts. Each module attains a developmental, physiological, reproductive and ecological identity that imbues it with the character of an independent individual.

10.4   Phenotypic variation between plant parts

There is abundant evidence that extensive phenotypic variation in morphology, biochemistry and phenology exists among branches of large plants. However, phenotypic variation can derive from both genetic and environmental sources, and it is often difficult to separate the principal factors responsible for particular cases of observed variation (Smith-Gill, 1983; see Chapter 8). I shall restrict discussion to: 1) chemical variation between branches; 2) general plant responses to pest damage; and 3) immunity and gall formation by plants.

10.4.1   Chemical variation within trees

That trees are chemically tessellated is well known. The variability within individual trees in the quantity of secondary compounds such as cuticular waxes in cypress, leaf phenolics in oak, and volatile oils in spruce is frequently as great as the variability from one tree to another (Dyson & Herbin, 1968; Parker, 1977; von Rudloff, 1969, 1972b). This variation presents serious difficulties for sampling and statistical analysis (Dyson & Herbin, 1968, 1970; von Rudloff, 1968); for instance, samples of leaves taken from different heights (Roberts, 1970; Ogilvie & von Rudloff, 1968) and from north versus south-facing sides of a tree (Hanover, 1966c) differed significantly in their monoterpene composition.

Different tissues have unique biochemical profiles as a consequence of ageing and developmental differentiation (Upadhya & Yee, 1968; von Rudloff, 1962b), and seasonal climatic changes are another prominent source of within-tree variation in chemical characteristics (Zavarin *et al.*, 1971; Dement & Mooney, 1974).

10.4.2   Responses of pests to plants

There is abundant evidence in the literature of genetically-based chemical variation between plants that relates directly or indirectly to resistance to pests (reviewed by Maxwell & Jennings, 1980). The pioneering insect-transfer experiments of Edmunds and Alstad (1978, 1982) demonstrate highly specific interactions between colonies of black leaf pine scale *Nuculaspis californica* (Homoptera: Diaspididae) and individual trees of *Pinus ponderosa* in the western United States. Colonies of the insects typically fail to spread onto neighbouring trees because they are genetically specialized to the unique chemical properties of each tree (Alstad & Edmunds, 1983).

Similar experiments were performed by Service & Lenski (1982) and Service (1984) with the aphid *Uroleucon rudbeckiae* on the composite herb *Rudbeckia lacinata*. Clones of aphids were cultured on several clones of the composite under laboratory conditions. Age of first reproduction, daily reproductive rate, age-specific fecundity and age of death of aphids were all significantly affected by the host plant clone.

Whilst these examples deal with between-individual variation, some chemical variation between branches *within* plants has been demonstrated. Differences associated with nutrient composition, defensive chemistry and tissue longevity can have profound effects on the performance of the plant's enemies. The most persuasive documentation of within-plant phenotypic variation is found in the aphid-cottonwood system studied by Whitham (1981, 1983). The aphids *Pemphigus betae* show consistent patterns of infestation between branches of individual cottonwoods *Populus angustifolia*. Zucker (1982) provides evidence that leaf phenolics play a significant role in host selection by the aphids. These studies extend the important conclusions of Edmunds and Alstad (1982) to variation *between* branches with respect to pest resistance.

Experimental exchanges of aphids between branches demonstrate both differential phenotypic and genetic properties of resistance and susceptibility between the branches (Whitham *et al.*, 1984; Whitham & Slobodchikoff, 1981). These experimental results allow several important conclusions: 1) individual trees are heterogeneous environments or mosaics of suitability for herbivores; 2) within-tree variation can be as great as the differences between trees; and 3) within-tree mosaicism can slow the evolutionary responses of herbivorous insects to large plants by making it difficult for the herbivores to track their food plants in space and time. Whitham proposes that much of the within-plant variability may have a genetic basis and may be critically important in the evolution of resistance to pests by plants.

10.4.3   Leaf galls on witch-hazel

In a long-term study of gall-formation by the aphid *Hormaphis hamamelidis* on leaves of witch-hazel *Hamamelis virginiana*, I have found significant variation in attack rates between years, regions, genets, ramets within genets, and branches within ramets. For two separate bushes (about 4 km apart), variation in the levels of galling between ramets within each bush was highly significant, despite the fact that the ramets intertwine with one another through the same microclimate over a height range of 1–3 m. The difference in galling between the bushes was ten-fold, but there was a twelve-fold difference in infestations between the ramets in one bush and a 58-fold difference in the other. The witch-hazel sheds all its leaves each year, and the aphids are reported to leave the tree for their alternate host plant *Betula nigra* between one leaf crop and the next (Pergande, 1901). If the aphids settled at random and the branches were phenotypically uniform within bushes, then the pattern of infestation of individual branches should be Poisson distributed and uncorrelated between successive years. However, Gill and Halverson (1984) reported significant autocorrelations in levels of infestation on a branch-by-branch and ramet-by-ramet basis between two successive peak years. They interpreted this as consistent with (but not proof of) the hypothesis of genetic mosaicism within genets of witch-hazel in traits of susceptibility to aphid infestation.

10.4.4   Plant immunity

One important kind of plant plasticity that generates phenotypic variation is the active induction of defensive structures and chemistry in response to attack by pests ('induced defences'). Since the possibility of acquired immunity was first suggested by Chester (1933), several reviews of the topic have appeared (Price, 1940; Painter, 1958). Clarke and Knox (1979) summarize the reactions of plants to parasite attack into two categories: 1) the initial recognition of foreign assault; and 2) the response. The ability of plants to recognize foreign versus self is well known from graft rejection (Libby, 1974; Campbell & Jetzer, 1972) and fungal parasite-plant interactions (Ellingboe, 1978). The response may consist of two processes; 1) 'walling off' infected tissue from healthy host tissue; and 2) production of secondary compounds at the site of injury. Interferons are induced in plants by virtually every pest from viruses to parasitic higher plants (Atanasoff, 1964). Equally widespread among higher plants are phytoalexins, which are produced in the few cell layers immediately adjacent to the site of injury (Hare, 1966; Yarwood, 1967).

Certain plant responses to injury are delayed and occur at sites some distance from the point of assault. Damage to potatoes and tomatoes by Colorado potato beetles *Leptinotarsa decemlineata* causes release of a proteinase inhibitor that spreads rapidly throughout the plant. Within 48 hours the inhibiting factor can account for 10% of the total soluble protein in the entire plant (Green & Ryan, 1972). Similarly, birch trees in Finland may actively defend themselves against herbivores for as long as two years, by elevating the concentration of secondary compounds in leaves adjacent to those damaged by the caterpillars of moths and sawflies (Haukioja & Niemela, 1976, 1977, 1979). However, in both these examples, the responses are clearly non-specific, and can be mimicked by mechanical tearing of leaves. Despite the extraordinary reports by Rhoades (1982) and Baldwin and Schultz (1983) of rapid changes in tree leaf chemistry induced by damage in *neighbouring* trees (the 'talking-trees' effect; see Fowler & Lawton, 1985), it is clear that higher plants generally show highly localized responses, and not the combined properties of recognition of non-self, phagocytosis, memory, and systemic transport of antibodies that characterize animal immunities. Thus, it is best not to use the phrase 'plant immunity' in the context of mechanisms of plant resistance to pests.

In summary, there is abundant documentation of phenotypic variation between branches within trees with respect to morphology, composition of defensive secondary compounds, and the distribution of damage or injury by herbivores. While it is clear that most of the variation is the direct consequence of external forces and developmental processes, some of the variation may be genetic in origin and heritable through gametophytes.

10.5   Genetic variation within plants

Several lines of evidence indicate that there are substantial quantities of both allelic and chromosomal variation between modular parts of large plants. For example, heterophylly of the aquatic *Ranunculus flammula* is controlled not only by a dominant inhibitor gene for lamina expansion, but also by the immediate environment of the leaf buds (Cook & Johnson, 1968). Libby *et al.* (1969) recommended that bristle-cone pines, *Pinus aristata*, some individuals of which started as single genotypes nearly 4000 years ago, should be examined for the acquistion of spontaneous genetic variants among their branches.

Cytogenic variation is well documented between plant parts. The classic example is *Claytonia virginica*, where nearly 68% of the individuals in wild populations are chromosomally chimeric (Lewis, 1970; Lewis *et al.*, 1971). Intraplant variation in the B-chromosomes is common in herbaceous species, especially in the genus *Crepis* (Stebbins, 1974), and in certain trees (Jones, 1975). Evidence for significant genetic variation within plants includes: 1) vegetative propagation of bud and branch sports; 2) analysis of the histogenic origins of plant variegation; and 3) inheritance of variation through gametophytes (pollen and seed).

10.5.1   Vegetative propagation of bud sports

Spontaneous bud sports are well-known in horticultural history and many have been exploited for commercial purposes (Bartram, 1955; Dermen, 1960; Grant, 1971). Clones of almond and walnut trees in Turkey have diverged genetically with respect to fruit characteristics and viral disease resistance after centuries of mutation and selection during vegetative propagation (Sykes, 1975). Clones of nectarines derived from a single ancestral parent vary genetically in surface ultrastructure (Fogle & Faust, 1975). Albino suckers have been recorded in aspens, and somatic and bud mutations occur in many aspen shoots (Green *et al.*, 1971).

Bud sports have been used extensively in the development of commercially desirable fruits and crops. Seedless grapes, navel oranges, and other seedless fruits are all cloned cultivars derived from spontaneous bud variants. Pink grapefruit arose from a single branch bud sport in 1906 (Hartmann & Kester, 1975). Of the thousands of cultivars of apples, most are propagations of sport branches found in orchards (A. Thompson, pers. comm.). During the last century, bud sport variations accounted for the origins of over 70% of all colour variants in commercial flowers (East, 1908).

Evidence of genetic variation also emerges from selective culture of clones from such asexual species as garden strawberries (Comstock *et al.*, 1958) and dandelions (Gadgil & Solbrig, 1972).

Of great interest to the hypothesis of genetic mosaicism is the burgeoning research on somaclonal and gametoclonal variants in cell cultures (Evans *et al.*, 1984; Williams, 1984). Most extensively developed in potatoes and tomatoes (Sheppard *et al.*, 1980; Evans & Sharp, 1983), cloning research has uncovered a diverse array of mutations, including single gene changes at both new and previously mapped loci, large and small chromosomal aberrations (polyploidy, deletions, rearrangements, etc.), and cytoplasmic variants (Evans *et al.*, 1984). Astonishingly, the frequency of single gene mutations in tomato cultures, where the genetic analysis has been most detailed, is as high as one mutant out of every 20–25 regenerated plants! First viewed as hampering the technology of propagation of valuable cultivars from cells and tissues, the abundance of spontaneous mutants is now enthusiastically embraced as a rich source of heritable variation in morphology, yield, and pest resistance (Evans *et al.*, 1984).

In programmes for breeding pest-resistance in forest trees, Painter (1966) advocated investigation of within-tree genetic variation as a potentially rich, untapped source; 'Bud mutations are probably as common in field crops as [they are in] woody plants, but a forest tree has so many more branches. Only a single insect resistant branch is necessary for a resistance program' (Painter, 1966). I am unaware of any major commercial efforts to follow up this intriguing advice.

Cloning tillers from pasture grasses has revealed unexpected quantities of genetic variation within individual genets. The within-clone variance of such traits as date of ear emergence, length of stem, length of panicle and length of leaf of the pasture grass *Festuca ovina* was as great as the between-clone variance. Breese *et al.* (1965) found that the expression of tiller height and rate of growth in *Lolium perenne* could be altered by selecting for high and low lines of tillers cloned from the same parent plant. Clones of tillers from plants that had a long history of asexual propagation failed to respond to the selection regimen, but clones which were derived from seed-grown plants were very responsive.

Graft-incompatibility between species and between clones within species of fruit trees is a common and serious problem in programmes of vegetative propagation (Libby, 1974). It is used as one criterion of clonal identity (Sykes, 1975; Posnette & Cropley, 1962), but within-clone variability in graft-compatibility has been observed in apples (Campbell & Jetzer, 1972) and pears (Coles & Campbell, 1965). These intraclonal differences in graft-compatibility may indicate genetic differences that have arisen within clones (Sykes, 1975).

10.5.2   Cytogenetic and histogenetic origins

Although cells in apical meristems are totipotent and are capable of giving rise to an entire plant (Clowes, 1961; Zimmerman & Brown, 1971; Shepard *et al.*, 1980), specific tissues differentiate from distinctive germinal layers in the apex. Thus, not all mutants that appear spontaneously in cells of shoot meristems will necessarily appear in all tissues in all species. We need to know the likelihood that a developmental mutation will be inherited through pollen or seed in order to appreciate the evolutionary importance of intraplant mosaicism.

The histogenesis of floral and vegetative parts has a long history of controversy (Klekowski & Karinova-Fukshansky, 1984a). Out of the historical disagreements between the two-layer (tunica and corpus) and the three-layer hypotheses of apical meristematic anatomy emerges the conclusion that there exist three or four basic histogenic organizations of apical meristems among major taxa of vascular plants (but see Graham & Wareing, 1984): 1) bryophytes and pteridophytes have a single apical cell that gives rise to all tissues (Bierhorst, 1977), so spontaneous mutations in that cell will be present in *all* the vegetative and reproductive tissues derived from it (e.g. Klekowski, 1984); 2) some gymnosperms maybe monohistogenic (Imai, 1935) while others are multi-layered (Clowes, 1961); 3) most dicots are trihistogenic with multiple layers parallel to the shoot-tip surface (Stewart, 1978); and 4) most monocots and some dicots are dihistogenic (Clowes, 1961).

Whether a developmental mutation will appear in the gametophytes of a dicot therefore depends on the germinal layer in which the mutant initial cell occurs. Only if the mutation occurs in an initial of the second layer (L–2), will it be transmitted through the gametophytes. The L–2 layer produces a diverse array of tissues including parenchyma, lamina palisades and spongy mesophyll, but also pollen grains and embryo sacs. If the mutation originates in an initial of the outermost (L–1) layer of a meristem, then it will be present only in the thin (1 cell deep) epidermis which covers all parts of the plant. Mutations in the third germinal layer (L–3) are expected to be present primarily in the vascular systems of xylem and phloem. However, both L–3 and L–2 are multilayered, because of abundant periclinal divisions (parallel to the meristem surface), and the two layers often exchange cells with one another. Therefore some mutations that originate in the initials of L–3 can shift into the embryonic mass of cells of the L–2 layer, and hence into pollen and ovules.

The critical point is that some traits for resistance to enemies, were they to arise spontaneously in development of shoots, are expected to be inherited through pollen and seed, while other traits probably will not. Those morphological, physiological and bio-chemical traits in the spongy mesophyll and palisade layers of leaves that confer resistance to leaf-mining and leaf-chewing insects, are likely to be inherited via pollen and seed. Also, some L-3 traits, such as structures of vascular walls that might prevent entry by fungal hyphae or bacteria, or block stylets of sucking insects like aphids and leafhoppers, have a small, but finite chance of transmission through sexual reproduction. However few, if any, developmental mutations in L-1 that change anti-herbivore traits in the epidermis, such as the production of hairs, cuticular waxes, etc., will be inherited sexually.

In summary, developmental mutations are routinely discovered among modular parts within plants under natural conditions, and bud sports have been exploited extensively as a source of new cultivars in commercially valuable fruits. That some distinctive branches are unique genetic entities is shown in the persistence of the traits in vegetative propagation of cuttings. But whether developmental mutations are inherited through the gametophytes depends on their occurrence in the germinal layer of apical meristems that gives rise to pollen grains and ovules.

10.6 Fitness differentials between modules

The central theses of this Section are that every individual tree functions as a colony of evolutionary individuals, and that significant differentials in survival and reproduction between branches can be demonstrated. Many of the preceding examples, including compe-tition between modules and the results of natural defoliation, con-tribute towards differential fitness of plant parts. The following sections discuss additional evidence for differential mortality and reproduction between modules within plants.

10.6.1 Shedding of plant parts

Shedding of portions or whole parts of branches, stems, roots, and reproductive structures is a common and regular process in forest trees. Approximately 13% of the annual increment in stems and branches is shed, and discarded branches compose as much as 30% of the forest litter (Kozlowski, 1973a). Both positive and negative con-sequences result from the process of shedding. As a natural process, shedding can rid a tree of injured, diseased or senescent parts, as well as those less vigorous modules that are outcompeted by other modules. On the other hand, induced or premature shedding can result in nutrient and water losses, which may render the plant less competitive with its neighbours and more susceptible to attack by pathogens and herbivores.

Shoot tip abortion is common in temperate canopy trees at the beginning of the growing season, and has puzzled developmental biologists for years (Millington & Chaney, 1973). The apparent waste of photosynthetic potential is not readily explained by hypotheses of competition or self-pruning of diseased parts. Unlike the phenomena of fruit and seed abortion which have plausible adaptive explanations (Sweet, 1973; Stephenson, 1980), the significance of shoot tip abortion from non-pathogenic causes has remained obscure. The hypothesis of genetic mosaicism offers an explanation. If mutations occur spontaneously in apical meristems at the conventional rate of 10^{-8} per locus per cell division (Vogel, 1972), and if most of the developmental mutations are deleterious (Veleminsky & Gichner, 1978), one would expect to find numerous dying buds and shoots throughout a tree. The causes are expected to be subtle, internal, non-obvious, and not identifiable to any external factor of climate, pathogen, or pest.

10.6.2 Differential reproduction

It is both a common observation in natural history and frequent horticultural experience, that fruit production is variable between years, between trees, and between branches within a tree (Goodrum *et al.*, 1971; Downs & McQuilkin, 1944; Sykes, 1975). Few data, however, compare the variation in fruit production between branches on the same, individual tree with the between-individual tree variation. I have gathered data on fruiting in Costa Rican caper trees *Capparis odoratissima* and Virginian witch-hazels *Hamamelis virginiana*, which show that modular parts of large trees can differ significantly in a number of important reproductive traits.

Fruit was collected from several branchlets on major branches on two caper trees in 1979 and 1981 (Gill, 1983). The results showed that in 1979 (but not in 1981) variation in fruit production between major branches was greater than the variation between trees. In both years, variation between major branches in the number of seeds per pod was greater than the variation between trees.

Data have also been gathered over the five year period 1980–1984 on witch-hazels in the Shenandoah Mountains of western Virginia (Gill & Halverson, 1984; see Table 10.1). In three of the five years, variation in fruit production between ramets exceeded variation between the whole individual bushes. Seed production, especially as it relates to attack by the seed-feeding weevil *Pseudanthonomus hamamelides* (De Steven, 1981, 1982), was not evaluated, but it is clear that differences in reproduction between modules can be as great or greater than differences between individual plants.

Possible causes of variation in fruit production between branches include differences in the number of female flowers (Saki & Oden, 1983), hormonal control of fruit development (Wright, 1956), rate of

Table 10.1. Fruit production in two genets of witch-hazel, *Hamamelis virginiana*, over the five years 1980–1984. Mean ($\bar{x}$) and standard error (S.E.) refer to the actual number of fruits per ramet. Data were transformed to ln (number of fruits per ramet + 1) prior to statistical analysis. t^2 = Student's t statistic squared (Sokal & Rohlf, 1981, p. 220). DTMS = difference between transformed means of LSM and DRI squared. ARVP = among ramet variance within bushes pooled between the two genets. Only in 1984 was the difference between the mean fruit production from the two bushes significant ($P < 0.05$).

Year	Bush	No. stems	$\bar{x}$	S.E.	t^2	DTMS	ARVP
1980	LSM	14	132.07	52.52	2.27	3.20	4.98
	DRI	5	30.40	21.34			
1981	LSM	14	0.86	0.78	5.29	1.93	1.35
	DRI	5	9.20	4.66			
1982	LSM	14	6.50	0.52	1.39	0.66	1.74
	DRI	5	27.20	21.39			
1983	LSM	14	31.50	8.74	0.00	0.00	3.09
	DRI	5	22.20	11.15			
1984	LSM	12	1.42	0.56	14.90	5.57	1.32
	DRI	5	42.00	15.96			

fruit abortion (Stephenson, 1980), competition among fruit buds (Quinlan & Preston, 1968, 1971), shading (May & Antcliffe, 1963), leaf area and position relative to fruit buds (Haller, 1933), alternate bearing (Davis, 1957), defoliation (Crawley, 1983), and fruit or flower-bud predation (McKey, 1975). These data simply document the kind of phenotypic variation in fruit production between plant parts which is a necessary precondition for natural selection to operate within individual plants. Whether the combination of phenotypic variation and differential reproduction between ramets actually produces differential inheritance of genotypes, remains to be shown in this witch-hazel system.

10.7   The evolution of plant resistance to pests

Evidence in the literature demonstrates overwhelmingly the negative impact of defoliation on plant growth, reproduction and survival, all of which are components of plant fitness (Kulman, 1971; Crawley, 1983). Thus, in principle, herbivory must be a powerful force of natural selection in plants for mechanisms of resistance, and a vast array of morphological, biochemical and phenological traits in plants have been interpreted as having their primary function in defensive roles against phytopathogens and insects (Fraenkel, 1959; Ehrlich & Raven, 1964; Whittaker & Feeny, 1971; Hanover, 1975). Whilst many evolutionary ecologists no longer question the view that phytophagous insects and other herbivores and microbes have been a major evolutionary force shaping many characteristics of plants, the issue is still controversial (see Jermy, 1984).

It is not necessary that plant numbers be regulated by herbivory

(Hairston *et al.*, 1960) in order that resistant plant phenotypes have a selective advantage over susceptible plants. Phytophagous enemies which consume only a small portion of the foliage can still have a significant selective impact. 1% differentials in fitness can have major evolutionary significance but would be undetectable in the demography of a species. In general, it is not necessary for factors of evolutionary importance to have measurable ecological (demographic) effects; conversely, factors causing major reproductive deficits or imposing significant mortality are bound to have evolutionary consequences when there is genotypic variation in their effects.

A major unsolved question in coevolutionary theory is how long-lived organisms, such as large canopy trees, evolve resistance against short-lived enemies like microbes or insects. If a plant individual replaces itself in a population once every 100 years, and an enemy replaces itself once every year, then 100 multiplicative episodes of selection and potential change in allelic frequency occur in the enemy population for every one in the plant population. Obviously, the potential rate of spread of a favourable gene in the enemy population is over 100-fold faster than the annual rate of increase of a comparably favourable gene in the plant population. Consider the difficulties faced by the coast redwood, *Sequoia semprevirens*, which may have a longevity of several thousand years, in evolving resistance to insects and bacterial pathogens which may have many generations each year (Molisch, 1938). The ability of short-lived organisms to evolve rapidly is illustrated by the capabilities of many herbivorous insects to detoxify the most extraordinary and novel chemicals (Smith, 1962), and by the rapid evolution of resistance to pesticides such as DDT.

One form of the theory of coevolution (Dawkins & Krebs, 1979; Jermy, 1984) asserts that there is a perpetual arms race between herbivores and plants (or predator and prey, or parasite and host) so that for every adaptive advance made by the plant host, there is a countering adaptive advance made by the pest. For such an interaction to persist, it is necessary that the rates of evolution of the two participants be commensurate in time and equivalent in intensity. If the rates are unequal, the interaction is unstable. Clearly, if new mutant genes for resistance become fixed in a plant population at a rate greater than the rate of fixation of genes for pathogenicity in the enemies, then the host will be continuously refractory to infection or defoliation. On the other hand, if the rate of fixation of pathogenic genes in the parasite is faster than the rate of evolution of resistance in the host, then repeated outbreaks of highly damaging parasites, and frequent local extinctions of susceptible host populations should prevail in nature. In both cases, asymmetry of evolutionary rates produces instability and the genetic interactions between plant and parasite would be brief over evolutionary time.

Two lines of evidence encourage the view that the behaviour of plant-herbivore interactions is dynamic and has the features of an arms race. They are: 1) the gene-for-gene theory of the evolution of resistance, and 2) the disparity of generation times between pests and trees.

10.7.1   The gene-for-gene theory

The best evidence for rapid genetic interactions between plants and their pests comes from the research on cereal grasses and their fungal pathogens (Anikster & Wahl, 1980; Ellingboe, 1968, 1972, 1978; Flor, 1971; Person, 1967; Person & Sidhu, 1971). Decades of agronomic research have led to the formulation of the 'gene-for-gene' hypothesis, which states that for every gene specifying resistance in the host plant there exists a complementary gene for pathogenicity in the parasite (Person et al., 1962; but see Barrett, 1985). The pattern most commonly found is a series of dominant (or co-dominant) alleles for resistance at very few loci in the plant, and numerous, non-allelic, recessive genes for pathogenicity at many loci in the parasite (Flor, 1971; Hooker & Saxena, 1971; Jewell, 1966; Person, 1966, 1967). Somatic mutation and recombination have also proved to be sources of genetic variation in the pathogens (Ellingboe, 1961; Ellingboe & Raper, 1962; Anagnostakis, 1982).

The gene-for-gene theory predicts instability in the frequency of resistant and susceptible genes in host plants, and complementary instability in the frequency of pathogenic and avirulent genes in a plant parasite. Coupled oscillatory behaviour is predicted in both sets of genes because of the frequency-dependence that is inherent in the evolutionary interaction of hosts and parasites (Anderson & May, 1982). Indeed, the agronomic history of cereal grasses is characterized by repeated appearances of new virulent strains of fungal pests (Day, 1973; Person, 1967; Johnson, 1961) that shorten the useful field life of each new resistant cultivar of cereal to roughly five to seven years (Frey & Browning, 1971). The more genetically homogeneous the planting of crops (Adams et al., 1971) and stands of timber (Hartley, 1939; Gerhold, 1966; Stern, 1974), the more rapid the appearance of new virulent strains of pests (Person, 1967). Were it not for the rapid production of new resistant varieties of the cereals in breeding programmes, the pathogens would periodically dominate the arms race, with disastrous agricultural consequences.

10.7.2   Disparity of generation times

A principal reason for expecting undamped oscillations and eventual collapse of arms races between plants and their enemies is the frequent disparity in generation times between the two participants, and the unequal evolutionary rates they generate. Only if some compensatory mechanisms oppose the apparently great advantage to the short-lived enemy can the rates of evolution in the two interactions be made equal. Possible compensatory mechanisms include: 1) readjustment of the time scale from the usual generations of the organism to either absolute time or to the number of cell divisions; 2) mechanisms (e.g. higher chromosome numbers or polyploidy, or higher rates of mutation and recombination) in long-lived plants that generate large quantities of genetic variation upon which natural selection can more forcefully operate; 3) conflicting selective pressures on the enemies and compensatory responses in the plant that effectively reduce the intensities of selection for pathogenicity in the pest and against susceptibility in the plants; 4) gigantic differences in fecundity between large plants and small herbivores that compensate for real generation-time differentials; and 5) genetic mosaicism in large, arborescent plants that not only amplifies the pool of genetic variation but also multiplies the strength of natural selection by inserting another malleable layer between the gene and the individual. Each of these alternatives is discussed below.

i) The first problem concerns the inconsistent definitions of the *rate of evolution* (Maynard Smith, 1976). If it is defined as the velocity at which alleles are substituted per generation within a population, one would naturally expect the amount of evolutionary change in a long-lived species to be much less than the amount of change in a short-lived species after the same passage of time (Laird *et al.*, 1969). Historically, it has been most common to use generations as the denominator of measurements of both mutation rates (Vogel & Rathenberg, 1975) and evolutionary rates (Benveniste & Todaro, 1976). On the other hand, if evolutionary rates are defined on an absolute time scale rather than a generational time scale, then there should be no necessary relationship between the amount of evolutionary change achieved in a population and the longevity of the organisms. The controversy between the generation-time and absolute time hypotheses is far from settled, because different answers may apply to different taxa, and potentially false assumptions lurk hidden in each argument; for example: 1) constancy of the rates of cell division; 2) constancy of the strengths of selection; and 3) constancy in the amounts of available genetic variability (Antonovics, 1977).

ii) Higher chromosome number in trees (n = 11–14) compared to herbs (n = 7–9) may compensate for the difference in generation times, simply because the higher chromosome numbers should generate higher rates of recombination (Ehrendorfer, 1970). Substantial changes in chromosome number in tropical trees through dysploidy and polyploidy may also promote higher rates of mutation per locus because of the greater number of potentially mutable copies of each locus. Thus, longer-lived plants may have larger genomes, more inherent genetic variation, and larger numbers of mutable loci than shorter-lived relatives and rapidly reproducing enemies. Ehrendorfer (1970) argues that this combination effectively erases the paradoxes created by the generation effect.

iii) Effective selection for exploitative herbivory of host plants may be less strong than the selection for resistance against pests by plants. It is possible that the selective pressures on herbivores from their natural enemies are very great, and inherently antagonistic to the selective pressures generated by the plants. Hairston *et al.* (1960) and Jermy (1984) argue that the abundance of herbivores as a trophic group is controlled by predators, and that phytophagous animals are not in competition for food (but see Chapter 5). If this is so, then selection for anti-predator adaptations may take precedence over selection for responses to novel mechanisms of resistance in the plant (Gould, 1978). It could be argued, however, that enemy-regulated herbivores would be too scarce to be a potent force of selection for the origin or maintenance of resistance in the plants (Jermy, 1984; but see Crawley, 1985, and Section 10.7, above).

iv) Systematic differences may exist between the fecundities of plant and enemy that offset real disparities in the generation times of plants and their pests. In the lifetime of many forest trees, the number of seeds produced by a single individual may reach 10^9 (Kozlowski, 1971, 1973b). Assuming that plants and their enemies experience equivalent strengths of selection, and that both suffer mutation rates in the range of 10^{-5} to 10^{-6} per locus per year, then the huge fecundities in the plants convert improbable recombinants into hundreds or thousands of real mutant progeny. If effective lifetime fecundity for each individual enemy is several orders of magnitude less than that of the average plant, then the effective increase in genetic variation available to the plant by virtue of its gigantic fecundity could offset the deficit in evolutionary rate expected from a longer generation time (Williams, 1975).

Whilst the four previous explanations all assume that individual plants are the units of selection, the hypothesis of genetic mosaicism suggests that a large, deliquescent (highly branched) plant with 10^5 independently growing apical shoots is analogous to a population of 10^5 annual herbs. From a genetic point of view it should be treated as such (bearing in mind that the modules of the tree proliferate asexually, whereas the annual herbs are likely to reproduce sexually). Assuming that the rates of developmental mutation are equivalent to those in the germ line (Vogel, 1972), the amount of allelic variation available within a tree each year is expected to be the same as in the large population of herbs. However, the new evidence of somaclonal and gametoclonal variation (Section 10.5.1) suggests that the rates of recoverable mutations from somatic lines may be 2–3 orders of magnitude greater than the traditional rates of mutation of 10^{-5} to 10^{-6}. Making the additional assumptions that the spectra of mating systems and strengths of selection are similar in large woody plants and herbs, the hypothesis of genetic mosaicism predicts that the rates of evolution in herbs and trees should be commensurate, and should be scaled to absolute time not to generation time. The amplification of available genetic variation through mutation in the enlarged population of modular individuals, dovetails with the second (enhanced effective mutation) and fourth (high fecundity) hypotheses discussed above. By combining these hypotheses, one predicts that the velocity of evolution in large plants may be equal to or greater than that of their enemies, so that the burden of staying in a coevolutionary arms race may in fact lie more with the short-lived pathogens and herbivores than with the deceptively long-lived trees (as suggested by Beck & Reese, 1976).

10.8   Breeding systems in plants

The traditional concept that individual plants are genetically uniform places emphasis on cross-pollination from separate individuals (xenogamy) as the principal mechanism whereby recombinants are generated. The exquisite diversity of floral morphology, phenology and breeding systems in angiosperms frequently serves to enhance the chances of cross–pollination (Fryxell, 1957; Stebbins, 1970; see Chapter 6). Many mechanisms promote cross-pollination and discourage self-fertilization, such as asynchronous sexual expression (protandry and protogyny), temporal monoecy, dioecy, heterostyly, and self-incompatibility. The benefit traditionally attributed to cross-pollination is the production of genetically diverse progeny in the face of uncertain future environments (Stebbins, 1957; Williams, 1966).

Self-fertilizing species are usually regarded as dead-end alleys in evolutionary lineages (Stebbins, 1957; Mulcahy, 1984). Given that the genetic and developmental penalties of selfing are great (Allard, 1975: Solbrig, 1977), it is difficult to explain why selfing has arisen so often in plant evolution. It is thought that selfing has strong selective advantage only when organisms become physically isolated, as in colonizations of new habitats, or when xenogamous pollen becomes inaccesible (Antonovics, 1968; Baker, 1959; Stebbins, 1957; and see Chapter 7).

Mass-flowering in hermaphroditic, zoophilous species is a puzzle to evolutionary biologists because of the apparent inefficiency of this breeding system for out-crossing. The costs of mass-flowering seem huge: 1) as the flower visitors move about the tree, the vast majority of flowers are geitonogamously self-pollinated; 2) litres of nectar are consumed without the benefit of cross-pollination, and 3) the bulk of the pollen is either consumed or apparently squandered in geitonogamous pollinations. In self-compatible species, the massive geitonogamy is usually envisaged as resulting in excessive self-fertilization and severe inbreeding. In self-incompatible species, large quantities of geitonogamous pollen may block the deposition or tube growth of xenogamous pollen. Nevertheless, Kalin-Arroyo (1976) argued that the prevention of inbreeding selects for mechanisms of self-incompatibility over systems of self-compatibility.

Rates of outcrossing in mass-flowering trees are very low, and as few as 0.3–1.3% of flowers per day may receive xenogamous pollen (see Chapter 6). Two explanations for the adaptive significance of mass-flowering prevail: 1) the profusion of flowers acts as a large flag that attracts cross-pollinators from long distances and rewards them (Augspurger, 1980; Gentry, 1974; Stephenson, 1979; Willson & Rathke, 1974; and 2) the importance of male function (i.e. dispersing pollen) takes precedence over female function (i.e. setting seed; Willson & Burley, 1983). The large energetic costs of nectar and pollen production are allegedly worth the low xenogamous yield, because the only alternative seemed to be lower costs and *no* cross-pollination. Large surpluses of flowers relative to fruit further support these explanations (Stephenson, 1982).

In contrast, the hypothesis of genetic mosaicism predicts that mass-flowering is a breeding system of great selective value, because it provides the opportunity to recombine intra-tree genetic variation through geitonogamy. Mass-flowering therefore offers *two* avenues of recombination: 1) traditional xenogamous outcrossing; 2) geitonogamous genetic mixing. Of course, the extra benefits of mass-flowering in genetically mosaic trees can be realized only in self-compatible species. In fact, the existence of self-incompatible mass-flowering tree species could be seen as the most damaging evidence against the hypothesis of genetic mosaicism.

One expects self-incompatibility in monopodial species (palms, *Cecropia peltata*, etc.) and simple herbs (cereal grasses, mustards, etc.) because of the lack of branching and expected lack of intra-plant genetic variation. However, the frequency of self-incompatible tree species in lowland, deciduous forests of Central America is reported to be at least 54%, and the proportion of obligately xenogamous species is at least 76% (Bawa, 1974).

There is an intriguing possibility that these estimates of self-incompatibility and xenogamy are too large, because the crowns of mosaic trees have been inadequately sampled. For understandable, very practical reasons, the methodology commonly used to determine self-compatibility is hand pollination of flowers on a single, conveniently accessible branch. The selfed treatments are usually autogamous pollinations or geitonogamous pollinations between immediately adjacent flowers. By restricting the selfing treatments to flowers within the same module, the experimental protocols have maximized the chances of matching genotypically identical pollen and stigma, and thereby eliciting a judgment of self-incompatibility. If the hypothesis of genetic mosaicism is correct, there should exist within every large, deliquescent plant some modules that are compatible with other modules but incompatible with themselves. To my knowledge, this kind of genetic variability in the self-incompatibility system has never been sought systematically.

Presumably, the larger the plant, the more genetic variation there is between its reproductive modules, and the greater the amount of recombination that is accomplished by mass-flowering. Thus, it is not surprising that mass flowering is most prevalent in large canopy trees and spreading, clonal monocots, while breeding systems with sparse flowering are more common in smaller herbaceous species and understorey shrubs. I know of no previous explanation of this trend in breeding systems between herbs, shrubs and large woody species.

Genetic mosaicism has far-reaching consequences for the classification of species as self-compatible or self-incompatible. If allelic variation at compatibility loci were present between branches, and if most of the spontaneous mutations at the incompatibility loci were naught mutants rendering the self-recognition mechanisms functionless, then pollen grains from a mutant branch would be compatible with stigmata of flowers on all 'normal' branches elsewhere on the tree, and vice versa. However, pollen and stigmata on all 'normal' branches would still be highly incompatible among themselves.

There are fragmentary pieces of evidence of abnormal compatibility within otherwise self-incompatible species (East, 1927; Lester, 1971; Bawa, 1974; Stott & Campbell, 1971). Rather than dismiss these cases as irrelevant developmental abnormalities, they should be investigated as possible examples of genetic mosaicism within individual plants. The plausibility of the hypothesis of genetic mosaicism strongly recommends that the compatibility of different branches on the same tree should be tested routinely.

10.9   Conclusions

The thesis of this chapter is at an early stage of its scientific development; an idea has been suggested, both supporting and critical observations from the literature and field have been evaluated, and a hypothesis formulated. Much research lies ahead in order to test the merits of the hypothesis.

Evidence contrary to the hypothesis does exist; there are several reports in the literature where within-plant variation was sought but not discovered (e.g. Baker & Baker, 1977; Rockwood, 1973). Some intensive studies of the interaction between grazing insects and the variation in essential oil and tannin contents of *Eucalyptus* in Australia have failed to reveal any correlation of the distribution of insect damage, insect abundance, or insect growth with the concentration of these two major classes of secondary compounds (Bray, 1964; Fox & Macauley, 1977; Morrow & Fox, 1980).

First and foremost we need a thorough, quantitative assessment of the amount of genetic variation between modules of highly branched plants. The crowns of canopy trees are of particular interest, but spreading clones of meadow grasses and forbs are equally interesting subjects of investigation. Methods should include rigorous analysis of nucleotide sequences, structural proteins, and continuously-distributed characters of morphology and physiology that have a polygenic basis. Recent discoveries and advances made in tissue culture techniques (Shepard *et al.*, 1980; Evans & Sharp, 1983; Larkin *et al.*, 1984; Meins, 1983; Williams, 1984) hold great promise of uncovering the extent of spontaneous genetic variation between cells in meristems.

The genetic basis of observable intra-plant variation in phenology and morphology should be confirmed by rigorous propagation experiments. For the evolutionary implications of the theory to have any merit, it is imperative that the inheritance of intra-plant phenotypic variation through the gametophytes be demonstrated in breeding experiments. The possibility that there is allelic variation at the compatibility loci between branches within trees offers perhaps the simplest test of the importance of genetic mosaicism on the evolution of breeding systems in plants. In any event, we should recognize that ecological organisms are not necessarily the same as evolutionary individuals.

BROOM HILDA / *by Russell Myers*

Chapter 11 Photosynthesis

HAROLD A. MOONEY

11.1 Introduction

The growth of plants depends upon their capacity to incorporate atmospheric carbon into organic compounds through the use of light energy absorbed during photosynthesis. This is a two step process: 1) an initial photochemical reaction traps light energy in absorbing pigments (chlorophyll and accessory pigments), producing a reductant (NADPH) plus ATP; 2) subsequently, atmospheric CO_2 is reduced and biochemically incorporated into carbohydrate (Fig. 11.1). The CO_2 to fuel this reaction must diffuse from the atmosphere to the site of fixation within the chloroplast. Limitations to the rate of photosynthesis may occur through restrictions at each of these steps; the nature of these limitations is discussed in detail in the following sections.

11.2 Background

11.2.1 Photochemical reactions

A portion of the light energy impinging upon a leaf is absorbed by chlorophyll pigments and hence transferred to specialized reaction centres where electrons are moved against an energy gradient (Fig. 11.2). The free energy released in a series of subsequent electron transfers is utilized to phosphorylate ADP and to reduce NADP (Barber & Baker, 1985). These molecules are essential in the biochemical reduction cycle described below.

11.2.2 Biochemical reactions

In most plants, the initial reduction of CO_2 in photosynthesis depends upon the activity of a single enzyme, ribulose bisphosphate carboxylase-oxygenase. A fundamental feature of this enzyme is that both carbon dioxide and oxygen are competitive substrates for the active site. In the carboxylation reaction, atmospheric CO_2 is coupled with the 5-carbon acceptor molecule, ribulose bisphosphate (RuP_2), to form two molecules of a 3-carbon product, 3-phosphoglycerate (3-PGA).

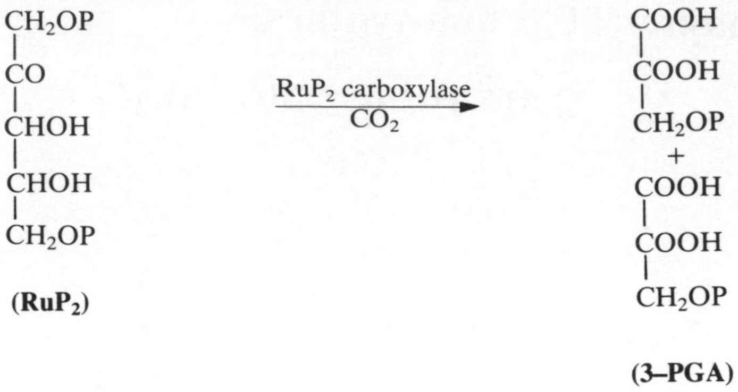
CH2OP
CO
CHOH
CHOH
CH2OP
(RuP2)
RuP2 carboxylase
CO2
COOH
COOH
CH2OP
+
COOH
COOH
CH2OP
(3–PGA)

In the oxygenation reaction, O_2 and RuP_2 combine to form a 2-carbon molecule, phosphoglycolate, and PGA.

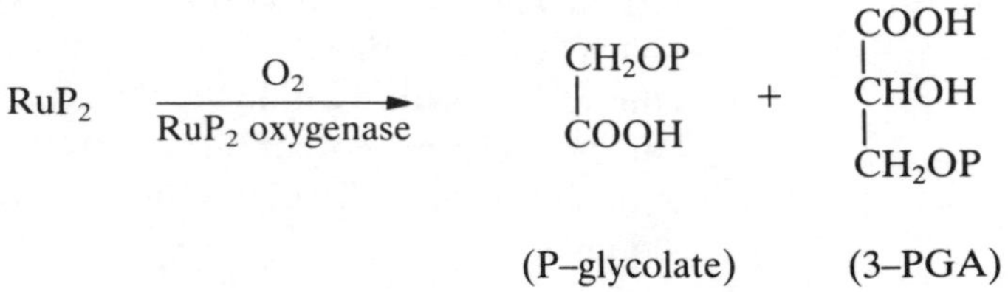
RuP2
O2
RuP2 oxygenase
CH2OP
COOH
+
COOH
CHOH
CH2OP
(P–glycolate)
(3–PGA)

RuP_2 is regenerated from glycolate in a series of reactions which involve the consumption of oxygen and the loss of carbon dioxide. This pathway is called photorespiration.

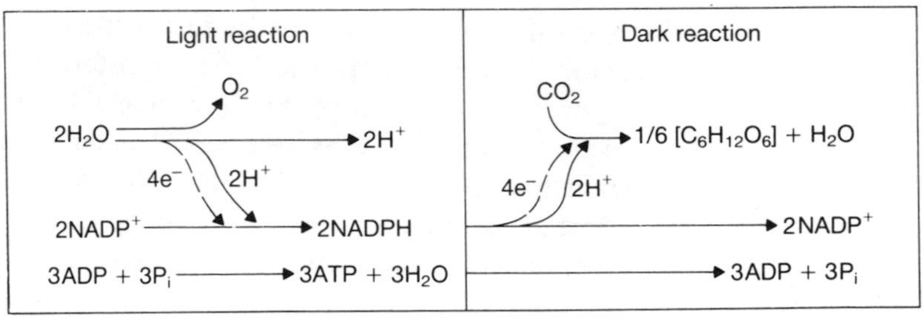
Light reaction
Dark reaction
O2
2H2O
2H+
4e−
2H+
2NADP+
2NADPH
3ADP + 3Pi
3ATP + 3H2O
CO2
1/6 [C6H12O6] + H2O
4e−
2H+
2NADP+
3ADP + 3Pi

Fig. 11.1. Coupling of photochemical and biochemical reactions in photosynthesis. Since the biochemical reactions are not dependent on light they are termed 'dark reactions'. From Miller (1979).

Photosynthesis and photorespiration are linked as shown in Fig. 11.3. This linkage acts to conserve carbon (recovering three of every four carbon atoms going to glycolate) and nitrogen (ammonia is generated during photorespiration), and to provide a photochemical sink which protects against photoinhibition under extreme environmental conditions (Osmond *et al.*, 1981). When the stomata close (e.g. under drought conditions) and photosynthesis becomes CO_2 limited, light harvesting still continues, and the resulting reductant power may become damaging to the photosynthetic machinery.

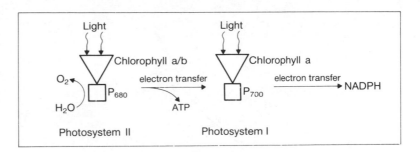

Fig. 11.2. Light reactions involved in the production of NADPH and ATP. Light is absorbed by 'antennae' chlorophyll pigments and transferred to specialized reaction centres to Photosystems I and II. These photosystems are responsive to different wavelengths with Photosystem I absorbing light of 680 nm and Photosystem II absorbing wavelengths lower than this. The electrons removed from the splitting of water are moved along this transport pathway gaining reducing potential through the light reactions and eventually producing stored energy as ATP and the strong reductant NADPH. These, in turn, are utilized to reduce CO_2 to carbohydrate during non light-requiring reactions. Plants from sun and shade habitats have leaves with different proportions of pigments, electron transfer chains and reaction centres.

The balance between the oxygenase and carboxylase reactions shifts according to atmospheric composition. At normal concentrations of atmospheric oxygen, about seven molecules of CO_2 react with RuP_2 for every two molecules of O_2. One CO_2 molecule is subsequently lost during photorespiration to give a net fixation of six molecules of CO_2. At a lowered atmospheric oxygen concentration, the CO_2 reaction predominates, resulting in: 1) an increased rate of net photosynthesis; and 2) a decrease in the compensation point (the concentration at which net CO_2 uptake balances O_2 evolution; Fig. 11.4). The compensation point is linearly related to atmospheric CO_2 concentration, as shown in Fig. 11.5.

Certain kinds of plants, the so-called C4 species, possess mechanisms for maintaining the site of carboxylation at high CO_2 concentrations, so that carbon is not lost in photorespiration (see Section 11.7).

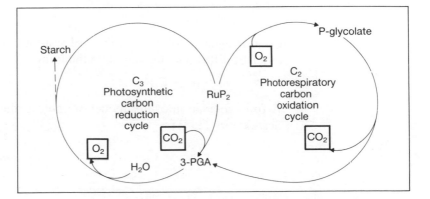

Fig. 11.3. Linkage of photosynthesis and photorespiration. Modified from Osmond *et al.* (1981).

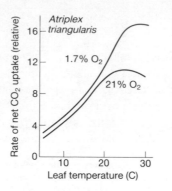

Fig. 11.4. Oxygen inhibition of net photosynthesis of *Atriplex triangularis*. From Bjorkman *et al.* (1969).

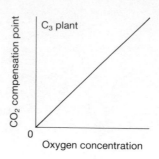

Fig. 11.5. Relationship between atmospheric oxygen concentration and photosynthetic compensation point. From Edwards & Walker (1983).

11.2.3   Diffusional limitation

The diffusion of CO_2 from the atmosphere to the site of fixation in the chloroplast is generally impeded by its passage through the stomatal pore, and by a layer of still air which covers the surface of the leaf (the boundary layer). The degree of this limitation depends upon wind speed (which determines the depth of the boundary layer), the number of stomata, and the degree to which the stomata are open, (this depends on a variety of factors including light, temperature, humidity and CO_2 levels; see Farquhar & Sharkey (1982) for a review). These relationships can be described by analogy with Ohm's Law, which states that current = voltage/resistance. In terms of photosynthesis, current (A) represents the flux of CO_2 (or the rate of photosynthesis). Voltage represents the difference in concentration in CO_2 between the bulk air (C_a) and the site of fixation (C_i). Resistance is the resistance to CO_2 diffusion in the boundary layer (R_a) and through the stomata (R_s). Thus:

$$A = \frac{C_a - C_i}{R_a + R_s}$$

Conductance, g, is the reciprocal of resistance, and can be used to describe diffusional limitations in the form

$$A = g_{CO_2} \cdot (C_a - C_i)$$

A full explanation of this model of photosynthesis can be found in Sestak *et al.* (1971).

11.3   Environmental influences on photosynthetic capacity

With this brief sketch of the process of photosynthesis as background, we are in a position to examine the influences of the physical environment on photosynthetic rates. We begin by considering the bulk of the world's plant species which utilize ribulose bisphosphate as the primary carbon acceptor in photosynthesis. These are called C3 plants because the initial product of photosynthesis is a 3-carbon compound, PGA.

11.3.1   Light

Of the total solar and terrestrial radiation impinging on a leaf, it is only the fraction lying within the band between 400 and 700 nm which is photosynthetically active (Fig. 11.6). This is referred to as photosynthetically active radiation (PAR). Photosynthetic rates of leaves increase with increasing PAR because the supply rate of reducing power increases through photochemical reactions. The rate levels off as limitations of carboxylating capacity and diffusion begin to predominate. Stomatal resistance greatly influences the maximum rates achieved at high light intensities. At very low light intensities, there is

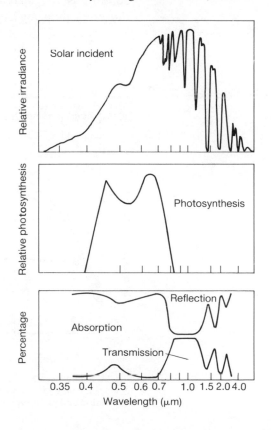

Fig. 11.6. Incoming solar radiation (top), photosynthetically-active radiation, PAR (centre), and leaf behaviour (bottom). Note the high absorptivity of PAR, and low absorptivity of the relatively high-energy, but non-usable, long wavelength radiation.

no net uptake of CO_2, since the rate of CO_2 uptake in photosynthesis is less than the rate of CO_2 evolution in respiration.

The photosynthetic light response may differ considerably between species and among leaves on the same individual plant (Fig. 11.7). The nature of these differences has been studied most thoroughly in leaves of the same plant which have been produced under either low or high light intensities ('shade leaves' or 'sun leaves'). Under these contrasting conditions, the sun leaves almost always have higher saturated photosynthetic rates and higher light compensation points (Boardman, 1977). The mechanisms underlying these differences are complex, and involve both morphological and biochemical components. Sun leaves generally have a greater density of stomata and hence a lower resistance to gas transfer. Further, they have a greater capacity for photochemical electron transport, higher respiration rates, and a higher content (and activity) of carboxylating enzymes than shade leaves.

Moving a shade leaf directly into the sun can cause damage to the photosynthetic system (photoinhibition), because the light energy trapped cannot be used fully in photosynthesis, due to the inadequate carboxylating capacity of the leaf. Over the course of several days, however, a shade leaf may be able to adjust (acclimate) to sun conditions, by increasing its photosynthetic enzyme content and thereby enhancing its capacity to utilize light energy.

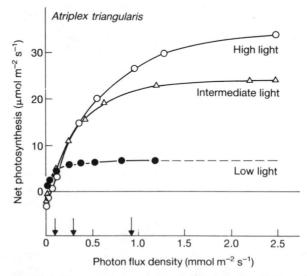

Fig. 11.7. Photosynthetic response to light intensity of leaves grown at 3 irradiance levels (indicated by arrows on the x-axis). The high-light grown plants have high respiration and photosynthetic rates. From Bjorkman *et al.* (1972).

11.3.2 Carbon dioxide

CO_2 is a primary substrate for photosynthesis. Rates of photosynthesis increase linearly with increasing CO_2 concentrations when, as at low CO_2 levels, carboxylating enzyme activity and the CO_2 receptor, RuP_2, are unlimiting (Fig. 11.8). At higher concentrations at the site of fixation, the photosynthetic rate begins to level off as the capacity of the leaf to regenerate RuP_2 fails to keep pace with the increased CO_2 supply. The regeneration of RuP_2 is dependent on photochemical activity (electron transport and photophosphorylation), which means that photosynthetic limitations at low light intensities and high CO_2 concentrations are due to similar causes.

Most leaves appear to operate at a stomatal conductance which maintains the internal leaf CO_2 concentration at the break point between CO_2 limitation and RuP_2 limitation (Farquhar & Sharkey, 1982). This means that carboxylating capacity and electron transport capacity are balanced, and results in an economical investment by the leaf in the biochemical components of these processes.

Plants which have suffered drought or been exposed to extremes of irradiation often show an alteration in the relationship between photosynthetic rate and CO_2 concentration, because of damage to one or more components of the photochemical reaction process (Powles, 1984).

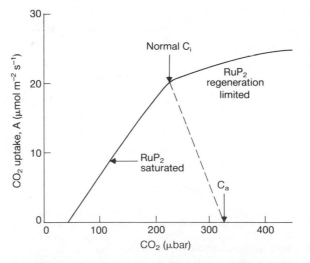

Fig. 11.8. Modelled response of chloroplasts to differing CO_2 concentrations. The arrow on the curve indicates the transition between CO_2 receptor RuP_2 being non-limiting and limiting C_i. The arrow on the x-axis indicates ambient air concentration C_a. Most plants operate at an internal CO_2 concentration at the break point, under non-limiting light and water conditions. The dashed line therefore indicates the normal CO_2 concentration gradient between air and leaf interior. From Farquhar & Sharkey (1982).

11.3.3   Temperature

Photosynthesis increases with temperature because an increase in enzymatic activity leads to an enhanced capacity to bind CO_2. At high temperatures, diffusion of CO_2 and photorespiration become limiting and the temperature response levels off. Finally, at extreme temperatures, the integrity of the photosynthetic system begins to break down and rates begin to decrease. At the highest temperatures, the decline may be irreversible. The nature of the decline depends upon the structure of the membrane lipids, and the composition of these lipids varies from species to species (Berry & Bjorkman, 1980).

Plants vary greatly in their photosynthetic response to temperature, depending upon the kind of conditions they experience in their natural environments (Fig. 11.9). For example, some desert perennials have thermal optima of more than 40°C, whereas Antarctic lichens have their optima close to freezing point. Plants which occur in the same habitat may also have different seasonal growth patterns which are associated with differences in temperature-related photosynthesis (Kemp & Williams, 1980).

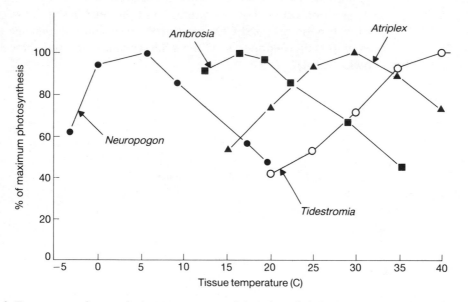

Fig. 11.9. Temperature photosynthetic response curves of plants from dissimilar habitats. Curves from left to right are for *Neuropogon acromelanus*, an Antarctic lichen (Lange & Kappen, 1972), *Ambrosia chamissonis*, a cool coastal dune plant (Mooney *et al.*, 1983), *Atriplex hymenelytra*, an evergreen desert shrub and *Tidestromia oblongifolia*, a summer active desert perennial (from Mooney *et al.*, 1976).

11.3.4   Water

The availability of water affects photosynthesis both indirectly, through effects on stomata, and directly via effects on the biochemistry of the process. Indirect effects due to stomatal closure may

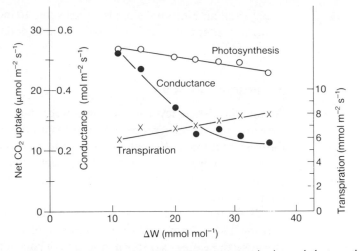

Fig. 11.10. Changes in stomatal conductance, transpiration and photosynthetic rate with changing water vapour concentration gradient for the coastal perennial. *Ambrosia chamissonis* (from Mooney *et al.*, 1983).

be induced by atmospheric drought (Fig. 11.10), by the loss of bulk leaf water (reduced leaf water potential), or possibly even by the stimulus of a hormonal signal from the root indicating soil water deficits (Davies *et al.*, 1982).

i) Water use efficiency

In order to fix carbon, plants must lose water, simply because water vapour and CO_2 travel by the same pathway. While stomata are open and CO_2 is diffusing inwards to the sites of fixation, water vapour is diffusing outwards into the atmosphere. This loss of water is directly proportional to the water vapour concentration gradient between the leaf and the atmosphere, divided by the stomatal and boundary layer resistances (just as CO_2 uptake is proportional to the CO_2 concentration gradient divided by these same resistances; see Section 11.2.3). The water vapour gradient is about 100 fold greater than the CO_2 gradient (Fig. 11.11) and usually changes markedly during the course of the day, due largely to changing temperature and the associated effects on vapour pressure. The CO_2 gradient also changes during the day as stomata open at sunrise and close at sunset. There are times of day, especially mornings, when the amount of water lost per unit of carbon fixed is rather low (i.e. water use efficiency is high). It has been proposed that plants 'manage' stomatal conductance in such a way as to optimize this relationship, and, particularly during periods of drought, plants may close their stomata during the midday period when the water loss to carbon gain would be most unfavourable (Tenhunen *et al.*, 1981). Stomatal aperture responds to a variety of external and internal signals which serve to induce changes in the osmotic content of guard cells and hence to determine the degree of opening.

ii) Integrated water use

Water use efficiency can be measured at any given moment by determining photosynthesis and transpiration simultaneously. A more important ecological measure, however, is the lifetime amount of carbon gained versus water lost. Such measures are difficult to make, but there is an indirect way of evaluating water use efficiency based on isotope analysis.

There are two stable isotopes of carbon, ^{12}C and ^{13}C. ^{13}C constitutes only about 1% of the total pool of these two isotopes. Because the isotopes differ in mass, they have different diffusivities, and the diffusivity of $^{13}CO_2$ is 4.4% less than that of $^{12}CO_2$. This results in discrimination between these isotopes both in biochemical reactions and in processes involving diffusion. The nature of this discrimination provides valuable information on plant-environment relationships (Farquhar *et al.*, 1982). For example, in mesic conditions when the stomata are fully open there will be little difference between leaf internal CO_2 concentration (C_i) and that of the air (C_a). When the stomata close during drought, the ratio of C_i/C_a will be considerably less than 1, and leaf isotope ratios will be expected to increase as in Fig. 11.12. Plant tissues which have higher isotope ratios can be assumed to have developed under conditions of water shortage.

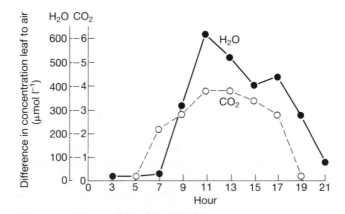

Fig. 11.11. Diurnal course of water vapour and CO_2 gradients between leaf mesophyll and the atmosphere for a maple leaf in September in Germany. Note that the differences in concentrations are a hundred-fold greater for water vapour than for CO_2 (from Küppers, 1984).

11.3.5   Nutrients

Nutrients can affect photosynthetic performance in a relatively direct fashion, since nitrogen and phosphorus are both involved in the photosynthetic reactions. Alternatively nutrients may act indirectly through their effects on the overall metabolic environment (see Chapter 12). Direct effects are most conspicuous for nitrogen. Of the nitrogen found in a leaf, a large fraction is contained in the carbon-

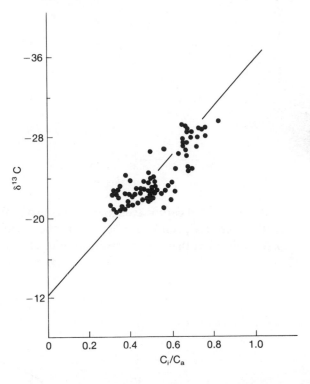

Fig. 11.12. Measured and predicted values of carbon isotope ratios for bean plants and for mangroves grown under different salinities (from Farquhar *et al.*, 1982).

fixing enzyme ribulose bisphosphate carboxylase (Fig. 11.13). It is not surprising, therefore, that there is generally a strong positive correlation between photosynthetic capacity and leaf nitrogen content (Fig. 11.14). This relationship only holds, however, if other factors such as light are not limiting. As nitrogen content is reduced,

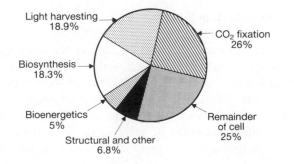

Fig. 11.13. Nitrogen apportionment in leaves. Note that about 50% of the nitrogen is devoted directly to the photo- and biochemical machinery involved in photosynthesis (from Evans, 1983).

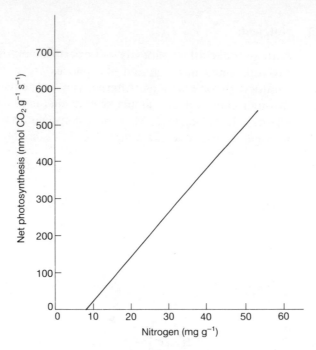

Fig. 11.14. Light-saturated photosynthetic rates as a function of leaf nitrogen content for leaves of plants from a variety of habitats. From Field & Mooney (1983).

so the amount of carbon fixed per molecule of nitrogen present in the leaf is reduced, since a fraction of the nitrogen is involved in processes other than carbon fixation (Fig. 11.15).

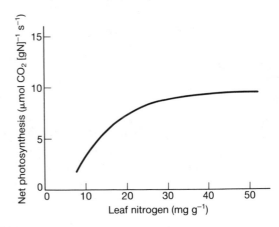

Fig. 11.15. Nitrogen use efficiency (carbon gained/unit nitrogen) versus leaf nitrogen content for a variety of plant types. At low nitrogen contents the efficiency is low (from Field & Mooney, 1986).

11.3.6   Atmospheric pollutants

Many atmospheric pollutants affect the photosynthetic capacity of leaves because they enter directly into the leaf mesophyll by the very same pathway as CO_2. The nature of their effects may be complex, and can differ from one pollutant to another. For example, SO_2 affects photosynthesis through its effect on stomatal conductance, and this effect may be positive (at least temporarily) or negative, depending upon the plant species and the concentration of the gas

(Winner, 1981; Mansfield & Freer-Smith, 1984). It may also influence carbon gain directly through effects on the photosynthetic process (Fig. 11.16). These effects are generally inhibitory at relatively low concentrations. However, under conditions of low soil sulphur availability, plant growth may be stimulated by SO_2 in the atmosphere (Noggle, 1980).

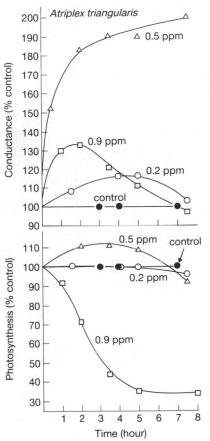

Fig. 11.16. Influence of different atmospheric concentrations of SO_2 versus exposure time on stomatal conductance (top) and photosynthetic capacity (bottom). on leaves of *Atriplex triangularis*. Note that this plant is stimulated to open stomata at these exposure levels. The high conductance at the 0.5 ppm SO_2 concentration actually enhances photosynthesis during initial exposure times. The direct depressive effects of the highest SO_2 concentration on photosynthesis results in rate reductions even though stomatal conductance is enhanced over controls. Plant species differ in the nature of their response to such pollutants (from Winner & Mooney, 1980).

11.3.7 Defoliation

Removal of some leaves on a plant, as occurs during grazing, results in a stimulation in the photosynthetic activity of the remaining leaves of comparable age (Fig. 11.17). This effect normally occurs over a period of days, and involves an increase in carboxylating efficiency, presumably due to an increased supply of nutrients or hormones to the surviving leaves. This process therefore results in a partial compensation for the lost tissue. Other aspects of plant compensation and regrowth are discussed in Chapters 8 and 12.

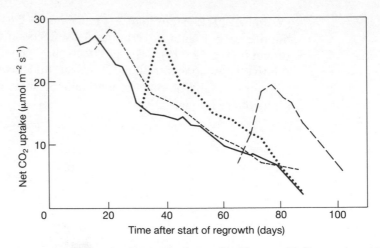

Fig. 11.17. Change in photosynthetic capacity of lucerne, *Medicago sativa*, with time, for control leaves (solid line) and for leaves remaining on plants partially defoliated at three different tissue ages (hatched lines). Leaves of the same age have higher rates of net photosynthesis on defoliated plants. From Hodgkinson (1974).

11.4   Leaf amount, type and duration

Leaf types vary greatly between plants in ways which directly influence their photosynthetic performance (see Chapter 8). Important variables include the amount of leaf tissue, the duration for which individual leaves are held on the plant, the quantity of light absorbed by these tissues, and the way in which photosynthetic capacity varies with the age of the leaf.

11.4.1   Absorptance

i) Whole leaves

As noted in Section 11.3.1, leaves generally absorb a high fraction of the usable radiation they intercept, and reflect a high fraction of the non-photosynthetic radiation of wavelengths greater than 700 nm (Fig. 11.6). In environments where radiation is abundant but water is limited, leaf reflectance may also be high in the visible wave bands (Fig. 11.18), an advantage since stomatal closure under drought conditions can lead to detrimentally high leaf temperatures (Chapter 8). In such arid environments, the loss of potential absorbed radiation is of little consequence to carbon gain, because photosynthesis is generally water rather than light-limited.

ii) Within leaves

Radiation is depleted by tissue as it passes through the leaf. Leaves are structured so that the maximum amount of radiation is absorbed by the pigment-containing chloroplasts, rather than by non-photosynthetic tissues. The chloroplast containing tissues in the upper leaf surface (palisade cells), are cylindrical in shape, with the chloroplasts pressed to the sides of the cell walls. This arrangement allows maximum transmission of light downwards, as well as the capture of scattered light.

iii) Between leaves

Just as organs within a leaf are positioned for maximum absorption of radiation, so are leaves within a plant. The general arrangement of leaves in a canopy is for those at the top to be more vertically oriented, and those at the bottom to be horizontal. With this arrangement, leaves at the top absorb enough light for high photosynthetic rates, but not so much that they are damaged or unnecessarily shade lower leaves. Lower leaves oriented horizontally maximize light interception in their light-limited positions (see Chapters 8 and 9).

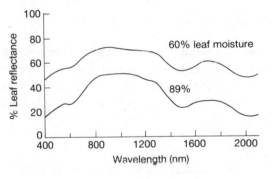

Fig. 11.18. Absorption of radiation by normally hydrated and drought stressed leaves of the desert salt bush, *Atriplex hymenelytra* (from Mooney *et al.*, 1977).

iv) Leaf movement

The amount of radiation falling on a leaf is related to leaf angle. For example, while the sun is overhead, a vertically-oriented leaf will absorb less radiation than a horizontal one. Leaves of many plant species move so that they lie perpendicular to the sun's rays throughout the day. Plants having leaves exhibiting such movement are called 'solar trackers'. This behaviour results in greater integrated daily radiation absorptance and therefore greater carbon gain for these trackers, compared with fixed-angle leaves (Fig. 11.19).

v) Leaf duration

The amount of photosynthesis which a leaf performs during its lifetime depends upon: 1) its intrinsic photosynthetic capacity; 2) the amount of limiting resources available; and 3) how long the leaf stays on the plant. The longer a leaf remains productive, the greater its

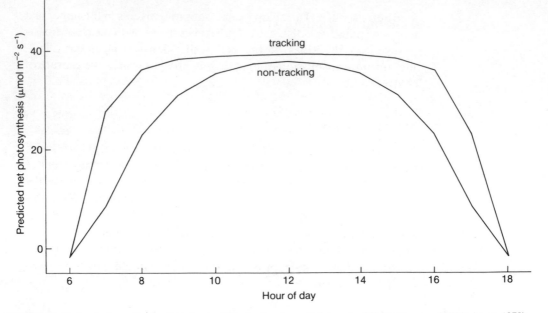

Fig. 11.19. Daily photosynthesis of a solar tracking leaf versus a stationary leaf (from Mooney & Ehleringer, 1978).

return (in terms of carbon fixed) on the resources invested to build it. Plants will generally maintain leaves as long as they are providing a positive carbon input. As a leaf is shaded by the growth of other tissues above it, respiratory losses of CO_2 will exceed photosynthetic gains, and the leaf will be abscissed. As we have seen in Chapter 8, leaves of long duration tend to be found on plants with slow growth rates.

In addition to the direct relationship between leaf duration and photosynthetic gain, there is an inverse relationship between leaf longevity and photosynthetic capacity (Fig. 11.20). Leaves of plants with fast growth rates do not live long, but they have a high photosynthetic capacity. Thus, plants with leaves of different durations (e.g. an evergreen tree and a deciduous tree) could have similar integrated carbon gain. This topic is discussed in detail in Section 12.4.2.

Fig. 11.20. Generalized relationship between leaf duration and photosynthetic capacity (from Mooney & Gulmon, 1982).

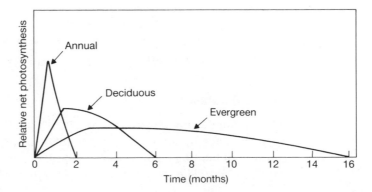

11.5   Seasonality of photosynthesis

11.5.1   Individual leaves

The photosynthetic capacity of a leaf changes with age as well as with the seasons. The leaves of most plants go through a predictable change in their photosynthetic capacity as they age (Fig. 11.21). The highest rates are attained prior to, or near the period of maximal leaf expansion, after which time fixation capacity begins to decline. It is interesting that leaves with basal growth, such as grasses, do not appear to show this decline (Sestak, 1985).

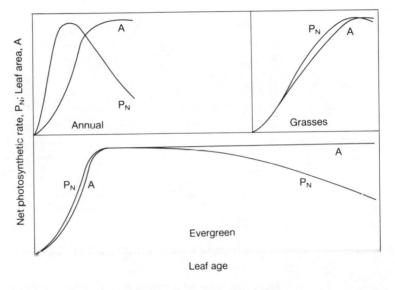

Fig. 11.21. Generalized relationships between leaf area development and photosynthetic capacity of differing plant types (from Sestak, 1985, after Mokronosov).

The rate of decline is correlated with many factors, but is most clearly related to the growth rate of the plant. Leaves at the top of a fast-growing plant rapidly overtop the older leaves. The older leaves thus become light-limited and the repayment in carbon gain per unit of nitrogen invested declines. Limiting nitrogen may then be re-allocated to the newer leaves at the apex. In slower growing plants, the life of individual leaves is prolonged, and leaves may remain active throughout different seasons or over several years. Such leaves have a changing photosynthetic capacity through time, and exhibit reduced (or no) capacity during winter or in periods of drought (Fig. 11.22). Evergreen leaves in severely cold climates, for example, may completely lose their capacity for photosynthesis in the winter, even under temporally favourable conditions (Larcher & Bauer, 1981). Evergreen leaves of plants in less severe environments, however, maintain their competence, and thus have the potential to fix carbon throughout the year.

Leaves must fix enough carbon to return not only the costs of their construction, but also the costs of their maintenance through periods which are unfavourable for photosynthesis.

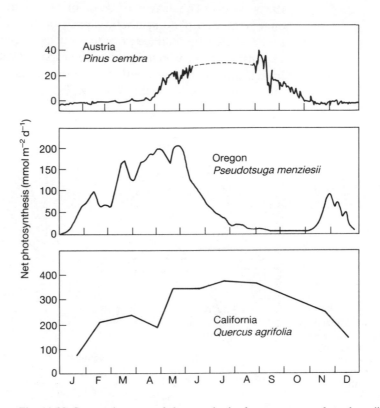

Fig. 11.22. Seasonal courses of photosynthesis of evergreen trees from three different climatic regions. Top, a timberline tree, *Pinus cembra* growing at 1980 m in the Swiss Alps. The needles have no photosynthesis during the winter months (from Hasler, unpublished). Centre, simulated net photosynthesis of *Pseudotsuga menziesii* in Oregon. Summer drought limits photosynthesis (from Emmingham, 1982). Bottom, simulated photosynthesis of the evergreen oak, *Quercus agrifolia* in central California. These plants have adequate water even during the summer drought because of a deep root system (from Hollinger, 1983). Simulations are based on site climatic features in combination with plant physiological responses to environmental parameters.

Many plants have been shown to have leaves which adjust their photosynthetic capacity with the changing seasons in an apparently adaptive manner (acclimation). The best example of this is the changing optimum temperature of photosynthesis in leaves of plants which grow in environments where temperatures fluctuate widely (Fig. 11.23). Those species which show the most complete acclimation are those which have no reduction in photosynthetic capacity through the year (although their thermal optimum may shift).

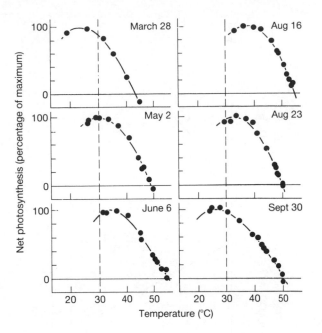

Fig. 11.23. Changes in the photosynthetic temperature optimum of irrigated apricot with seasons (Lange *et al.*, 1974).

11.5.2   Whole plants

The productive potential of a plant through the seasons depends upon the behaviour of its entire population of leaves. There is remarkably little information available on the seasonal pattern of change in leaf number on plants of different life histories. In certain perennial plants, one crop of leaves is produced per year, and all leaves appear and are lost more or less synchronously. In other species, leaves may be produced and lost throughout the growing season, with mean leaf longevity changing from one season to another. In the case of evergreen plants whose leaves last several years, leaf production may occur continuously or in simultaneous bursts. The net result of these diverse patterns in the birth and death rates of leaves, and of their changing photosynthetic capacity, is that whole plant photosynthesis shows pronounced seasonal changes (Constable & Rawson, 1980). Some of the causes and consequences of these changes in leaf abundance are described in Chapters 8 and 9.

The net photosynthesis of entire leaf canopies can be conveniently described in terms of the leaf area index (LAI); this is the area of photosynthetic surface per unit area of ground. Early in the growing season of an annual crop, for example, LAI will be much less than 1 and there is substantial bare ground. Later in the season LAI may reach values of 6 or more. The relationship between net photosynthesis and LAI is shown in Fig. 11.24. At low LAI, much light is 'wasted' by falling on bare ground. At very high LAI much photosynthate is potentially 'wasted' in the respiration of shaded leaves. At

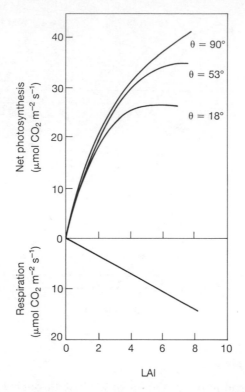

Fig. 11.24. Relationship of photosynthetic capacity of barley to leaf angle (from the ground surface) and to leaf area index (LAI). Respiration increases as the total amount of tissue increases. From Leopold & Kriedemann (1975), after Pearce *et al.*, (1967).

intermediate values of LAI, net photosynthesis reaches its maximum. This occurs at what is known as the optimal leaf area index (LAI_{opt}). Optimal leaf area occurs at higher and higher leaf areas as light intensity increases, or as leaf angles become more perpendicular to the ground surface, because more of the lower leaves remain above their compensation points.

11.6   Photosynthetic capacity and defence against herbivores

The photosynthetic capacity of a leaf and the probability of it being subject to herbivory are related through the strong positive correlations between leaf nitrogen content and photosynthetic capacity on the one hand, and leaf nitrogen content and food quality for herbivores on the other. A specific illustration of this relationship is provided by the shrub *Diplacus aurantiacus* and its principle herbivore, the butterfly *Euphydryas chalcedona*. Because the photosynthetic capacity of sun leaves is directly related to their nitrogen content (Fig. 11.14), we can employ a common axis to describe both these variables, as shown in Fig. 11.25.

On an artificial medium (and, presumably, in nature) larval growth is positively related to nitrogen content. Leaves of *Diplacus*,

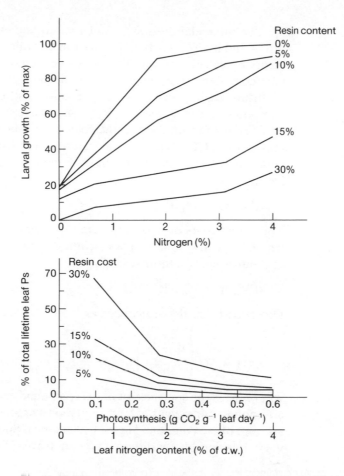

Fig. 11.25. Nitrogen-resin interaction in relation to the growth of the larvae of *Euphydryas chalcedona*, the principal herbivore of the shrub *Diplacus aurantiacus*, (from Lincoln *et al.*, 1982; top) and the fractional cost of leaf lifetime acquired carbon (from Mooney & Gulmon, 1982; bottom).

however, contain a phenolic resin called diplacol which can make up as much as 30% of leaf dry weight, and this resin effectively inhibits larval growth. Thus one can readily see that leaves could 'discourage' herbivores in two different ways; by reducing nitrogen content, or by increasing the content of resin. The economics of these two options can be assessed from a knowledge of the amount of photosynthate (or other limiting resource) needed to produce the resin. If a plant maintains a low nitrogen content it will suffer a reduced photosynthetic capacity and it will take a substantial fraction of its lifetime carbon gain in order to produce the resin in high quantities. In contrast, with high leaf nitrogen levels, the payback time is relatively short. From this kind of analysis one might conclude that the optimum leaf would have high leaf nitrogen and resin levels, since both photosynthetic capacity and herbivore protection would be greatest under these conditions.

The interrelationships between photosynthetic capacity and the direct and indirect costs of defence can be evaluated in more general terms by calculating the direct costs (in carbon units) of making a particular defensive compound, and the indirect costs in terms of loss of future carbon gain (the 'opportunity cost' of investing in that compound). The increase in dry weight of a leaf is determined by: 1) its CO_2 fixation rate (P); 2) the conversion efficiency of making dry matter from CO_2 (k); 3) the allocation of acquired assimilate to new leaf material (L); and 4) the initial amount of leaf material present (W_1):

$$dW_1/dt = W_1.P.L.k$$

The direct costs of making a defensive compound (C) can be subtracted directly from the photosynthesis term to indicate the impact of defence cost on future carbon gain:

$$dW_1/dt = W_1.(P-C).L.k$$

Direct costs are then calculated as:

$$C(t) = SC \left(\frac{d\frac{S(t)}{W(t)}}{dt} \right)$$

where SC is the specific cost of the compound expressed as weight of compound per weight of CO_2 required to produce it, and S(t) is the amount of the compound in the leaf at time t. Specific costs appear to vary considerably for different types of leaves (Table 11.1).

Table 11.1. Costs of construction of leaves and various leaf constituents presumed effective in herbivore defence (from Gulmon & Mooney, 1986).

Type	Species example	Compound	Formula	(SC) Cost $(gCO_2 g^{-1})$	(S/W)×100 Content % leaf wt.
Phenolic resin	*Diplacus aurantiacus*	diplacol	$C_{22}H_5O_7$	2.58	29
Cyanogenic glucoside	*Heteromeles arbutifolia*	prunasin	$C_{14}H_{17}NO_6$	2.79	6
Alkaloid	*Nicotiana tabacum*	nicotine	$C_{10}H_{14}N_2$	5.00	0.2–0.5
Longchain hydrocarbon	*Lycopersicum hirsutum*	2-tridecanone	$CH_3(CH_2)_{10}COCH_3$	4.78	0.9–1.7
Terpene array	*Salvia mellifera*	camphor (50%),+	$C_{10}H_{16}O+$ others	4.65	1.3
Whole leaves	various shrub species			1.93–2.69	

Growth simulations of the indirect costs of defence have been carried out using information from the herbaceous annual *Hemizonia luzulaefolia* (Fig. 11.26). This plant produces a resinous compound on cauline leaves at the beginning of reproduction, at which time the basal leaves are lost with the onset of the dry season. During the

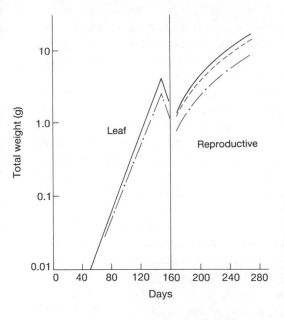

Fig. 11.26. Growth simulations for leafy tissue of the Californian annual *Hemizonia luzulaefolia*, assuming no production of resin (solid line), production throughout the life of the plant (dash/dot), or production during reproduction only (dashed). The reduction in dry weight at initial reproduction is due to basal mesophytic leaf loss associated with the annual drought (from Gulmon & Mooney, 1986).

annual drought *Hemizonia* is one of the few plants with green tissue, and it suffers enhanced herbivory in consequence. The production of resin has a different impact on biomass accumulation depending upon whether it is produced throughout the life of the plant, or only during the reproductive period (as happens in nature). Clearly, if the probability of herbivory is low during the early growth stages, the plant might increase its fitness by delaying the elaboration of the defensive resins.

11.7   Variations on the basic photosynthetic pathway

So far, we have considered the basic mode of carbon fixation via C3 photosynthesis. Several important variations on this basic metabolic theme allow improvements in photosynthetic efficiency in certain kinds of habitats. The most common variant is C4 photosynthesis, so

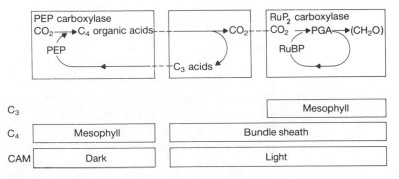

Fig. 11.27. Comparative features of C_3, C_4 and CAM photosynthesis (modified from Jones, 1983).

called because the initial products of CO_2 fixation are 4-carbon organic acids, rather than the usual 3-carbon PGA (Fig. 11.27). The carboxylating enzyme in C4 photosynthesis is phosphoenolpyruvate carboxylase (PEP carboxylase). C4 plants possess specialized cells which surround their vascular bundles, known as bundle sheaths, where chloroplasts operate in normal C3 mode (Kranz anatomy). Four-carbon products of the initial CO_2 fixation are transported from the surrounding mesophyll cells to the bundle sheath cells where the CO_2 is released and then refixed by the C3 pathway. The refixation of CO_2 in isolation in the bundle sheath acts as a CO_2 'pump', and overcomes the oxygenation reaction of RuP_2 carboxylase–oxygenase (Osmond *et al.*, 1981).

Another photosynthetic pathway, termed Crassulacean acid metabolism (CAM), occurs in certain desert plants and tropical epiphytes. It is similar to C4 photosynthesis, except that separate carboxylations take place within the same cells, displaced in time rather than in space as in C4 photosynthesis. In typical CAM plants, the stomata open at night rather than during the heat of the day. CO_2 diffusing into the leaf at night is fixed in 4-carbon organic acids through the use of stored energy (Fig. 11.27). During the day, while the stomata are closed, the stored CO_2 is refixed via the C3 pathway using light energy. An essential feature of this pathway is succulence or the possession of large vacuoles in which the organic acids can be temporarily stored. Large, columnar cacti are typical of the kind of plants which employ CAM photosynthesis.

Other CAM plants, particularly leafy succulents of the family Crassulaceae, exhibit flexibility in their mode of photosynthesis. During wet periods they fix carbon directly through the C3 mode during the day, and at night they use the CAM mode. As drought sets in, they shift entirely to the CAM mode.

11.8 Ecological consequences of different photosynthetic pathways

11.8.1 Water use efficiency

The biochemical dissimilarity of C3, C4 and CAM plants results in dissimilar physiological behaviour and this, in turn, leads to different ecological performance (Table 11.2). Because they possess an effective CO_2 pumping mechanism, C4 plants are able to saturate net photosynthesis at lower internal CO_2 concentrations than C3 plants. This means that stress-induced stomatal closure tends to have a much greater effect on photosynthesis in C3 than in C4 plants. Thus, while photosynthesis of C3 and C4 plants is differentially affected by stomatal closure, transpiration is not. This simple relationship has a profound influence on the instantaneous water use efficiency (i.e. on

Table 11.2. Comparative characteristics of the different photosynthetic pathways (modified from Jones, 1983).

	Pathway		
	C3	C4	CAM
Initial carboxylating enzyme	RuP_2	PEP	PEP, RuP_2
Tissue isotope range ($\delta\ ^{13}C$, 0/00)	-20 to -40	-10 to -20	spans C_3–C_4 range depending on fraction of daytime versus night-time fixation
Anatomy	Normal	Kranz	Succulent
Water use efficiency	Low	Medium	High
Photosynthetic capacity	Medium	High	Low
Oxygen inhibition of photosynthesis	Yes	No	Yes in day, No at night
Growth form occurrence	All	Shrubs and Herbs	Succulents
Principal geographic range	Everywhere	Open tropical areas or arid habitats	Arid regions or habitats

the amount of carbon fixed per unit water lost). At the same stomatal conductance, C3 and C4 plants will lose identical amounts of water but the C4 plant will fix more carbon and so will have a higher water use efficiency (Fig. 11.28).

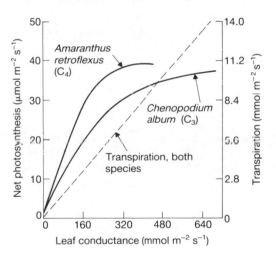

Fig. 11.28. Photosynthesis and transpiration of a C_3 and a C_4 species at different leaf conductances (from Pearcy & Ehleringer, 1984).

CAM plants have an even higher water use efficiency than C4 plants because they only open their stomata at night, when evaporative water loss is minimal (the 'inverted stomatal cycle'; see Fig. 11.29).

11.8.2   Quantum yield

Another important physiological difference between C3 and C4 plants is their differential efficiency of fixing carbon at different temperatures when light intensities are low. As we saw in Section

11.3.1, at low light intensities photosynthetic rate is directly proportional to the amount of light absorbed by the leaf (the quantum
yield). C4 plants have an intrinsically lower quantum yield because of
the extra costs associated with this pathway (they require two
additional molecules of ATP in order to regenerate PEP). This
difference is offset in favour of C4 plants at high temperatures,
however, because C4 plants lack photorespiration. This process
increases greatly with temperature, and results in a reduction of net
photosynthesis in C3 plants because previously fixed CO_2 is lost.

Although individual leaves on a plant may not be light-limited,
whole plants in natural environments generally *are* light-limited
because of carbon losses through respiration of non-photosynthetic
tissues such as stems and roots. Thus, differences in quantum yields
take on considerable significance in overall plant performance. From
the two physiological considerations just described, we would predict
that C4 plants would be most abundant in habitats where temperatures are high and where efficiency in water use is important.
Similarly, we would predict that C4 plants would not be found in
shady habitats, especially in cool climates.

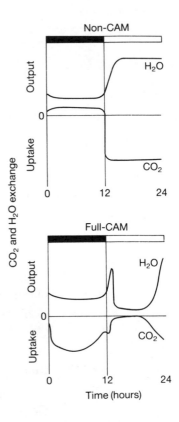

Fig. 11.29. Daily carbon fixation and water loss of a typical
CAM plant and C_3 plant (from Neales, 1979).

11.8.3   Relation to plant distribution

Because of the large difference in the physiological and ecological properties of plants with these different photosynthetic pathways, we might expect that they would have different patterns of distribution. This is true to a certain extent; for example C4 herbs predominate in tropical and subtropical grasslands which are sunny habitats and where high water use efficiency is important (Fig. 11.30). Many of the world's worst weeds such as crabgrass (*Digitaria*) and Bermuda grass (*Cynodon*) are C4 species (Holm *et al.*, 1977). C4 shrubs are common in desert and saline regions where, again high water use efficiency is paramount. It should be noted, however, that C4 metabolism is a relatively recent evolutionary innovation with multiple origins, and it is found most commonly in herbaceous plants, occasionally in woody shrubs, and almost never in trees.

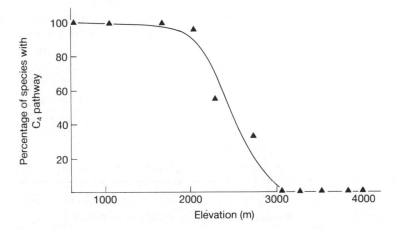

Fig. 11.30. Relative distribution of C_3 and C_4 grasses in relation to elevation on an African tropical mountain. The vegetation at the lowest elevations has all C_4 grass species and at the high elevations all C_3 (from Tieszen *et al.*, 1979). Similar distribution patterns along temperature gradients have been described in other parts of the world.

CAM plants are common only in extremely arid regions where light is not limiting to growth, and there is little above ground competition. Under these conditions plants with slow growth rates can compete successfully (see Chapter 12). CAM plants are also found as epiphytes in tree canopies in tropical regions, where there is sufficient atmospheric moisture to support metabolism even though the plants lack root systems. Submerged CAM plants have been found in lakes where intensive daytime photosynthesis by other plants depletes dissolved CO_2; in this case CAM permits CO_2 uptake at night when it is more readily available.

In addition to these broad patterns, there may be phenological growth partitioning within a plant community between species with the different photosynthetic pathways. This is seen particularly clearly in the grasslands of North America, where C3 grasses are active in the spring, while C4 grasses predominate during the hotter summer months (Boulton *et al.*, 1980).

The photosynthetic pathway can also be important in ways other than carbon balance, such as long-term stress survival. For example, during severe drought CAM plants may keep their stomata completely closed. Their thick cuticles ensure a high degree of water conservation, and light energy can be trapped during the daytime and utilized for maintenance without the incorporation of new carbon into the system (Cockburn, 1985).

11.8.4   Isotope fractionation and the photosynthetic pathway

As we saw in Section 11.3.4, mean leaf conductance can affect the fractionation of carbon isotopes in C3 plants. The leaves of plants which have experienced protracted drought will have had their stomata closed for long periods. They are therefore conductance limited and show higher (less negative) ratios than the leaves of plants which have been able to keep their stomata open. C4 plants have $^{12}C/^{13}C$ isotope ratios in the range of about -10 to -20 ‰, whereas most C3 plants have values between -20 and -40 ‰ (Fig. 11.2). Because so many CAM plants are flexible in their photosynthetic systems (e.g. employing normal C3 photosynthesis during wet periods), their values span the entire range (Table 11.2).

It is relatively straightforward to determine the isotope ratios of living and dead plant material. Isotope ratios can indicate photosynthetic pathways directly, as noted above, or can give information about the composition of past plant communities from analyses of organic matter in soil profiles. When there is a mixture of pathway types amongst the primary producers, they may even yield information about food chains, if ratios are known for the plants of a community and for the resident herbivores (Fry *et al.*, 1978).

11.9   Conclusions

Photosynthesis is a central process in the functioning of a plant. It provides the carbon skeletons and energy required to build biomass and to synthesize the wide variety of products utilized by plants in their metabolism.

The basic chemistry of photosynthesis does not vary greatly among plant species, although there are three fundamentally different biochemical pathways for the process, each with different ecological consequences. The vast majority of the world's plants operate with the C3 pathway, where CO_2 is incorporated initially into the 3-carbon product, phosphoglycerate. Plants utilizing a second pathway, Crassulacean acid metabolism (CAM), are able to fix CO_2 into organic acids during the night, refixing it during the day into carbohydrate, and utilizing light energy even though their stomata are closed. CAM results in a high ratio of carbon gained to water lost, and is found typically in desert succulent plants. A third pathway, C4 photosynthesis, also results in efficient water use. This efficiency is gained through both anatomical and biochemical features which result in the maintenance of high photosynthetic capacity even whilst the stomata are partially closed due to water stress. The C4 pathway is found commonly in tropical grasses.

Photosynthesis is very sensitive to variations in the supply rates of light and CO_2, the principal resources utilized in the process. Photosynthesis is further influenced by a wide array of environmental factors including temperature, nutrients, tissue water status, and atmospheric pollutants. These factors have influences on different time scales because of their different rates of change in natural environments. For example, leaf temperature and the quantity of radiation absorbed by a leaf change greatly during the course of a single day, whereas tissue water and nutrient status change over longer time spans.

Although plants do not differ greatly in the basic machinery utilized in photosynthesis, they do differ radically in the ways they acquire the resources needed for the process. Dissimilarities exist among plants in the amounts, display and duration of their leaves, and these affect the total amount of light intercepted and hence photosynthate accumulated. They further differ in their photosynthetic responses to various environmental factors both in the short and long term. For example, species may differ in the amount of photosynthesis they perform at a given light level, as well as in the way they respond (acclimate) to a long-term change in the light environment. These differences are often the result (and may be the cause in some cases) of dissimilar patterns of resource acquisition by plants. Such differences between species presumably play a role in permitting their coexistence.

The photosynthetic capacity of plants is directly linked to their ability to acquire water, light and nutrients, and the process itself serves as an integrator of 'success' in a given habitat. Photosynthetic capacity, however, may also be related directly to potential rates of herbivory, since leaves which have high photosynthetic rates generally have high leaf protein contents, and this makes them additionally attractive to herbivores.

Chapter 12 Acquisition and Utilization of Resources

ALASTAIR H. FITTER

12.1 Plant resources

Plant resources are chemically simpler than those required by animals. Plants use solar energy to manufacture their organic requirements from CO_2, H_2O and a dozen or so ions, but this metabolic simplicity leads to ecological complexity. Animals consume food items which have been pre-packaged by earlier metabolism, so that the ecological factors governing their abundance, distribution and accessibility in space and time can be clearly defined. In contrast, the availability to plants of CO_2, solar radiation, water and dissolved soil ions, varies according to controls which are unique to each resource. Thus, one of the major problems facing a plant is the acquisition of these resources, and their utilization in such a way as to maximize fitness. This chapter examines the problems associated with resource acquisition and utilization in relation to the evolution of plant life histories.

12.1.1 Response curves and limiting resources

The effect of increasing the supply of any resource on plant growth is asymptotic, often sigmoid (Fig. 12.1). Above some minimal level where growth is negligible, there is a continuously declining response, up to a point beyond which additional supplies lead to no further increases in growth, and may even cause decreases due to toxicity. The sigmoid nature of some response curves is the result of control processes at very low supply rates: for example, at low radiant energy fluxes, sigmoid curves are due to a reduction in respiration rates such that photosynthetic gains approximately balance respiratory losses (Mahmoud & Grime, 1974), and at low potassium supply they are the result of the existence of a threshold level of potassium in root cells, which must be exceeded if potassium is to be transported to the shoots (Glass, 1978).

The asymptote reached on a response curve, however, is neither a species nor a process characteristic; the asymptote depends upon the levels of other factors (Fig. 12.2), and therefore upon which factor is limiting growth. The concept of a limiting resource has a long history, but was first quantified by Blackman (1905) and by Mitscherlich

(1913) who proposed the equation:

$$Y = A(1-e^{-cX}) \tag{12.1}$$

where Y is the yield of a crop, X is the level of supply of a resource, A is the maximum yield obtainable, and c is a coefficient which describes the steepness of the ascending part of the curve. In this formulation A is defined by the prevailing levels of other resources. When two resources are in short supply, there may still be a response to the augmentation of one, but the value of A in equation 12.1 will be lower.

The identification of limiting resources is crucial to an understanding of resource limitation and acquisition, because only those utilization patterns which lead to improvements in the acquisition of

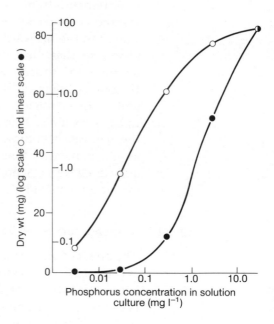

Fig. 12.1. Yield of the grass *Cenchrus ciliaris* in solution culture as a function of phosphorus concentration. The form of the curve is sigmoid, whether or not the yield is expressed as a logarithm. Data from Christie & Moorby (1975).

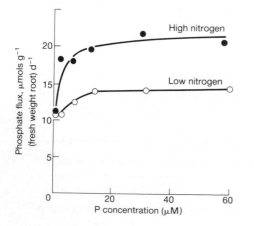

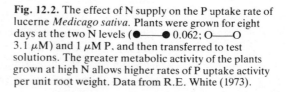

Fig. 12.2. The effect of N supply on the P uptake rate of lucerne *Medicago sativa*. Plants were grown for eight days at the two N levels (●———● 0.062; O———O 3.1 μM) and 1 μM P. and then transferred to test solutions. The greater metabolic activity of the plants grown at high N allows higher rates of P uptake activity per unit root weight. Data from R.E. White (1973).

limiting resources will increase fitness. The recognition that several resources may simultaneously limit growth does not negate this point; it merely emphasizes the complexity of the control problem which the plant must solve. Not only may different resources simultaneously affect one process, but these resources may require quite different activities for their acquisition; (e.g. radiation is intercepted by leaves, and nitrogen is taken up by roots). Further problems arise if the resources demand *conflicting* activities, for example, when conservation of water requires closing the stomata, through which the CO_2 required for photosynthesis diffuses to the chloroplasts.

These examples stress the role of various resources on a single process. It is entirely possible for the different processes in various parts of the plant to be limited by different resources (e.g. leaf growth by nitrogen and root growth by carbon fixation). Balancing these conflicting processes is central to the control of plant growth.

Finally, it is normal for plant processes to suffer limitations from environmental factors other than resources. Temperature is the most important of such limiting factors, along with various toxins (both soil- and airborne), and the activities of herbivores, mycorrhizas and other organisms. Water is a marginal case, since the vast bulk of that taken up is not used strictly as a resource but acts to transport materials to the shoots and to cool the leaves.

12.1.2 Which resources limit growth?

It has been traditional to regard plant growth as energy-limited (Caldwell, 1979), and Mooney & Gulman (1979) have argued that carbon is a unique primary resource for plants. In this view all other resources are of lower order, and control the fundamental activity,

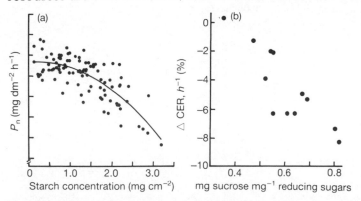

Fig. 12.3. Effect of leaf sugar concentration on photosynthetic rate, indicating sink limitation. (a) Net photosynthetic rate, P_n, of the lowermost trifoliate leaves of 21-day old soybean *Glycine max* after 13 h in light, as a function of leaf starch concentration. After Nafziger & Koller (1976). (b) Reduction in carbon exchange rate, as a percentage of the initial value (ΔCER %), and as a function of the ratio of sucrose to other reducing sugars in soybean leaves. Petioles were chilled to inhibit sucrose export. From Hanson & Yeh (1979).

which is carbon gain. In contrast, Harper (1977) suggests that plants are generally carbon-sufficient, and that other resources are generally limiting (Harper & Ogden, 1970). This view is supported by the work of Lambers (1982) who has demonstrated the occurrence of a 'luxury' pathway of respiration in roots, which suggests that root activity is unlikely to be limited by carbon supply. At a physiological level this dichotomy depends upon whether photosynthesis is seen as source- or sink-limited (Chapter 8). Various experimental techniques have been used to simulate changes in sink strength, most of which have involved altering levels of carbohydrate in photosynthesizing leaves, for example by the use of excised leaves (Azcon-Bieto, 1983), by chilling petioles to inhibit translocation (Hanson & Yeh, 1979) or by pre-treating leaves with various CO_2 levels (Nafziger & Koller, 1976). Herbivore feeding can lead to increased photosynthetic rate by creating a sink for carbohydrate (Crawley, 1983), and many plant parasites act in the same way. In most cases there is a close negative correlation between leaf sugar content and carbon exchange rate (Fig. 12.3). The relationship may take several days to develop, implying a hormonal or similar mechanism, rather than direct feedback (Geiger, 1976).

It seems, therefore, that photosynthetic rates are normally determined by sink activity (Gifford & Evans, 1981), except where levels of photosynthetically active radiation (PAR) are sub-optimal. This being so, patterns of allocation of carbon may give limited insight into the importance of different organs and functions in the control of fitness. When the allocation of carbon and various mineral elements is compared, there are often very large differences (Abrahamson & Caswell, 1981; Lovett Doust, 1980). Differences in

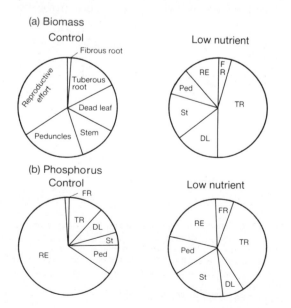

Fig. 12.4. Patterns of (a) biomass and (b) phosphorus allocation by the umbellifer *Smyrnium olusatrum* under high and low nutrient supply rates. The reproductive allocation differences are particularly marked for phosphorus. After Lovett Doust (1980).

nitrogen and phosphate allocation are often more pronounced than those of carbon, and tend to be more closely related to known constraints (Fig. 12.4).

12.2 Acquisition of resources

All the resources used by plants reach the plant by physical transport processes: radiation in the case of PAR; diffusion in the case of CO_2 and many ions in the soil (notably $H_2PO_4^-$); and convection in the case of water and other ions in the soil, particularly those present at high concentrations. With the exception of PAR which is simply intercepted by plant tissues, the supply of all other resources is created by absorption. It is the removal of water from soil, a process ultimately driven by evaporation from the leaves, that brings about convective transport of water and its dissolved ions through the soil. Where this convective supply of ions is less than the rate of absorption at the root surface, depletion there will create concentration gradients in the soil and initiate diffusion down those gradients (Fitter & Hay, 1981). For all resources other than PAR, it is *use* of a resource that creates supply.

12.2.1 Mechanisms of resource capture

Since mechanisms of interception of PAR and absorption of CO_2 have been dealt with in Chapter 11, only absorption of water and minerals are discussed here. Uptake of water is entirely passive and depends upon the existence of a water potential gradient between root cells and the soil, generated by evaporation from leaves. Since the root cell walls are freely permeable to water as far as the suberized endodermis (and as far as the xylem in very young, unsuberized roots), the surface area for uptake is equivalent to that of all the root cortical cells. Actual transport into the symplasm may occur anywhere in the cortex, although hydraulic considerations make transport through the cell walls (apoplastic transport) the more likely (Newman, 1974). This water movement produces convective transport of dissolved ions to the root surface, where the cations may either be absorbed by negatively charged cell wall materials or move passively into the symplasm down the electrochemical gradient (root cortical cells have a standing potential of -60 to -150 mV); anions are actively absorbed against that gradient.

Since soil water permeates the whole cortex and can therefore be taken up into the symplasm by any of the cortical cells, whereas many ions tend to move into the symplasm only at, or near the epidermis, the effective surface area for absorption of ions is much smaller than that for water. The convective ion flux F is given simply by the water flux at the root surface, V, and the concentration of ions in the soil solution C_1:

$$F = V \cdot C_1 \tag{12.2}$$

When, for a particular ion, F is less that the rate of absorption by root cells, depletion at the root surface will lead to diffusion of the ion from the bulk soil. The diffusion coefficient for this depends upon the moisture content of the soil (θ), the tortuosity of the diffusion pathway (f) and the reactivity of the ion with the soil (soil buffer power, dC/dC_1):

$$D^* = D \cdot \theta \cdot f \cdot \frac{dC_1}{dC} \tag{12.3}$$

where D^* and D are the diffusion coefficient for the ion in soil solution and free solution respectively (Nye & Tinker, 1977). Note that buffer power enters the equation as its reciprocal. The effect of these relationships is that the availability of a particular ion in soil is determined by the following factors:

i) **Soil buffer power (dC/dC_1):** this controls the soil solution concentration (C_1) in relation to total soil concentration (C) and so the level of convective supply. In consequence ions which are strongly buffered in any soil will tend to have low values of C_1 (unless the total concentration, C, is very high) and hence low rates of both convective and diffusive supply. The extreme example is phosphate, where uptake is nearly always diffusion-limited. If buffer power is very weak, as is the case for nitrate (where $C = C_1$), convection may be inadequate if C_1 is low. In this event, rapid diffusion may lead to exhaustion of soil nitrate stocks.

ii) **Soil water content:** the demand for water by leaves, largely to maintain energy balances, interacts with soil water content to control the rate of water uptake and hence convective supply of ions (Eqn. 12.2). Where soil water content is sufficiently low to limit convective supply, rates of diffusion are also depressed, because of the effect on diffusion coefficients (Eqn. 12.3).

iii) **Soil structure:** the compaction of the soil and the nature of the soil-aggregates determine both the size and distribution of soil pores, affecting both the hydraulic conductivity of the soil and the length of the diffusion pathway (f in Eqn. 12.3).

These soil factors control water and ion availability per unit root area or volume. Thus the actual levels of resource capture are strongly influenced by the total amount of root and its three-dimensional distribution within the soil.

12.2.2   Limiting resources

Although the factors outlined in the previous section control soil resource availability, there are differences between individual resources in the relative importance of these factors. Nitrogen and phosphorus frequently limit plant growth under natural conditions, but their availability and the demand for them by plants respond to distinct constraints.

Nitrogen

There is no large mineral reserve of nitrogen in soil, and it is present mainly as organic compounds in various stages of mineralization. Once nitrogen is released from soil organic matter or leaf litter as NH_4^+, it may be: 1) adsorbed by clay minerals and other soil components; 2) leached out of the root zone; 3) converted to NO_3^- by nitrifying bacteria; or 4) taken up by roots or soil microbes.

Very little nitrogen remains in solution. If it is converted to NO_3^-, leaching becomes much more likely, because NO_3^- ions are not adsorbed by soil and more remain in solution. Under anaerobic conditions, NO_3^- may be converted to N_2 gas (denitrification) by bacteria such as *Pseudomonas* spp., which can use NO_3^- as an alternative electron acceptor.

Soils may therefore contain both NO_3^- and NH_4^+ ions, but since nitrification is inhibited at low pH (pH < 5), nitrate ions tend to be less common as pH falls, and are apparently almost absent from acid peat. Plants can in general use both NH_4^+ and NO_3^- ions as nitrogen sources, although NO_3^- must be reduced to NH_4^+ before incorporation into organic compounds, with consequent energy cost.

If nitrogen is taken up predominantly as NH_4^+, then this normally involves the concomitant extrusion of H^+ ions to achieve electrochemical balance, resulting in acidification of the soil around the roots. Equally, nitrification of NH_4^+ in (or added to) soil involves H^+ production. Consequently, many plants adapted to calcareous soils do poorly if offered a low pH environment with nitrogen mainly as NH_4^+. The effects of NO_3^- and NH_4^+ uptake and their interaction with other factors are extremely complex (Rorison *et al.*, 1983).

If nitrate is taken up, it must be reduced by the plant to NH_4^+ prior to use. This reduction is carried out by an enzyme, *nitrate reductase*, which is inducible in most plants (but not in heathers and a few other species characteristic of acid peats; Dirr *et al.*, 1973). Where soil nitrate levels are high, nitrate reductase levels are usually correspondingly high (Havill *et al.*, 1974). Nitrate reduction takes place in the leaves or stems of some plants, but additionally or alternatively in the roots of others. Since herbivores cannot use inorganic nitrogen, it is possible that nitrate reduction in leaves, by ensuring that there is no organic nitrogen in the xylem, may act as a defence mechanism which reduces the food value of the xylem fluid.

Nitrogen metabolism is certainly central to the chemical defence systems that plants possess against insect herbivores. Bernays (1983) lists cyanogenesis, alkaloids, non-protein amino acids, glucosinolates and proteins (including proteinase inhibitors) as major classes of chemical defence compounds which have large nitrogen contents. The resource costs of these systems are reviewed in Section 12.3.1.

As a result of experimental applications of NO_3^- fertilizers to natural communities 1) the individual plants may take up nitrogen but not increase their growth dramatically, so that tissue nitrogen concentrations increase (e.g. *Cytisus scoparius*); 2) the plants may respond by growing so rapidly that their tissue nitrogen concentrations do not rise (e.g. *Ulex europeus*); or 3) the individuals may not take up the fertilizer at all. These different responses have important consequences for the insect herbivores feeding on these plants, especially when their reproduction is limited by the quantity or quality of available amino acids in leaf cells or phloem sap (Prestidge & McNeill, 1983). Insect outbreaks in forests may be triggered by increases in plant amino acid levels which occur following drought stress (T.C.R. White, 1973) or after air pollution with oxides of nitrogen (Port & Thompson, 1980).

Phosphorus

In contrast to nitrogen, the phosphorus dissolved in soil water is strongly buffered against depletion by a large reservoir of inorganic, adsorbed phosphorus. In some soils, organic phosphorus is also present in large quantities, but it is often resistant to mineralization, and cycles slowly. The affinity of soil components (particularly hydrated aluminium oxides) for phosphorus is so great that concentrations in solution are very low, and leaching is unimportant. Variation between soils in phosphorus availability largely reflects the buffer power of the soil (Section 12.2.1), and the conventional wisdom that pH is the main factor causing variation in phosphorus availability can be misleading.

Soil pH does, however, affect the form in which orthophosphate ions appear in soil solution. The principle ion species is $H_2PO_4^-$, but in alkaline soils the HPO_4^{2-} ion occurs as well; as far as is known this does not affect plant uptake. Once in the plant, phosphate ions are rapidly esterified, but the roots in particular maintain a pool of inorganic phosphorus (P_i), with which organic phosphorus forms are in equilibrium. The pool of P_i in the roots is important in controlling the translocation of phosphorus to the shoots, which appears to occur mostly as organic forms in the phloem, rather than as P_i in the xylem.

The availability of phosphorus in soils varies seasonally,

especially where organic phosphorus occurs in large quantities (Gupta & Rorison, 1975; Veresoglou & Fitter, 1984). Some plants may be able to take advantage of this by 'luxury' uptake in excess of immediate needs at times of relatively high availability. Such excess phosphorus is stored as P_i in the roots (Nassery, 1971); it has been suggested (Jeffrey, 1964) that plants in extremely phosphorus deficient soils in Australia may store phosphorus as polyphosphates (condensed chains of orthophosphate ions), but the advantages of this are unclear, and the general existence of polyphosphates in plant tissues has not been established.

A more general response to phosphorus deficiency appears to depend upon 1) the adoption of mycorrhizal symbioses (Section 12.3.1); 2) low growth rates (Section 12.4.1); and 3) increased leaf longevity and the evergreen habit (Section 12.4.2).

12.2.3   Capture of limiting resources

Plant growth is extremely plastic, and the relative numbers and sizes of different modules vary greatly (Chapter 9). It is generally accepted that these variations are related to the acquisition of limiting resources, at least in part. For example, plants grown at low rates of supply of any mineral ion tend to have high root weight ratios (RWR; root weight divided by total plant weight), whilst those experiencing low levels of PAR have low RWRs, having allocated more to leaf growth (Fig. 12.5a). Intuitively these responses should lead to a balanced growth path, since allocation of materials to the organ concerned with acquisition of the limiting resource should alleviate

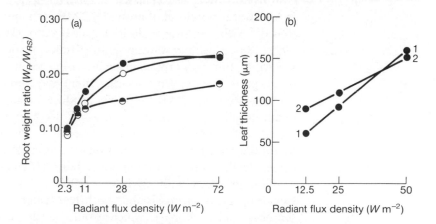

Fig. 12.5. (a) Effect of radiant flux density on root weight ratio of three species varying in shade resistance, *Galinsoga parviflora* (O) being the least, *Urtica dioica* (◑) next, and *Scrophularia nodosa* (●) the most resistant. Plants are grown in growth chambers under 16 h days at 20/15°C and 65% relative humidity (from Corre, 1983). (b) Thickness of the second leaf pair of (1) a sun plant, *Galinsoga parviflora*; and (2) a shade plant, *Stachys sylvatica* at three radiant flux densities. Note the thinner leaves of both species at low flux densities, and the lack of development of those of *Galinsoga* at the lowest level.

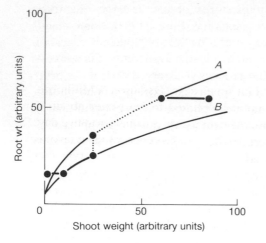

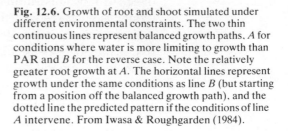

Fig. 12.6. Growth of root and shoot simulated under different environmental constraints. The two thin continuous lines represent balanced growth paths, *A* for conditions where water is more limiting to growth than PAR and *B* for the reverse case. Note the relatively greater root growth at *A*. The horizontal lines represent growth under the same conditions as line *B* (but starting from a position off the balanced growth path), and the dotted line the predicted pattern if the conditions of line *A* intervene. From Iwasa & Roughgarden (1984).

that limitation. In theory, the optimal growth pattern can be shown to be achieved by simultaneous growth of root and shoot along a balanced growth path determined by environmental conditions.

Changes in environmental conditions are met by allocation of growth resources to the organ that is capable of alleviating the limitation imposed by the new conditions (Iwasa & Roughgarden, 1984; Fig. 12.6). In other words, for a given stage in plant development, the environment defines an optimal RWR, and when the environment changes, the plant channels more (or less) of its resources to roots in order to achieve the new optimum RWR.

The physiological evidence for such precise switching is indirect. Cell expansion is very sensitive to changes in turgor, and reductions in leaf water potential of around 0.2 MPa can inhibit expansion and 0.3–0.4 MPa can stop it altogether (Acevedo, Hsiao & Henderson, 1971). Since low leaf water potentials reflect inadequate supply from the roots, this may represent such a switching mechanism (although in adapted species, osmotic adjustment may permit maintenance of turgor). Reduced PAR levels can produce a comparable effect. For example, nitrate uptake by ryegrass showed a periodicity related to radiation flux and CO_2 uptake, with a small reduction (around 10%) in uptake occurring immediately on reduction of radiation flux, and a larger decline starting after about 12 hours (Clement *et al.*, 1978).

The problem with envisaging either of these possibilities as switching mechanisms is that both leaf water potential and PAR undergo daily changes which would mask genuine changes in resource availability. It is more likely that levels of both PAR and water produce graded responses of root and shoot growth. Certainly for minerals such as phosphorus and potassium, where control appears to rely on the maintenance of threshold concentrations in root cells (Glass, 1978; Loneragan & Asher, 1967; Clarkson,

Sanderson & Scattergood, 1978), rates of shoot growth relative to the roots are quantitatively dependent upon the excess nutrient transported to shoots. More severe changes, such as defoliation, can bring about immediate cessation of root growth. Following defoliation in *Dactylis glomerata*, root respiration was reduced by over half and root extension almost ceased for seven days (Davidson & Milthorpe, 1966). In ryegrass nitrate uptake declined to zero over about 12 hours, and then recovered slowly over a period from two to five days (Clement *et al.*, 1978). Partial defoliation, however, whilst it favours shoot growth over root growth, does not stop the latter altogether (Brouwer, 1963).

Plant age

In annual plants, and in the non-storage components of perennials, there is a downward trend in root weight ratio with time. In annuals this is the result of high allocation to root tissue early in life, and in perennials it results from peak root growth occurring before peak shoot growth (at least in most seasonal environments). In perennials this early peak of root growth (e.g. Garwood, 1967; Veresoglou & Fitter, 1984) involves the expenditure of stored resources, whereas in annuals it is due either to the diversion of seed-borne resources or to a switch in allocation of new photosynthate. This pattern of development probably reflects the need to establish a reliable water supply, since alone amongst the resources acquired by plants, water flux must be maintained whenever demand exists. Also, minerals obtained by the root system in early growth can be redistributed later, to a much greater extent than is possible for structural carbon fixed by leaves (Williams, 1948). Barley, for example, can take up all the phosphorus it requires until maturity in the first six weeks of growth (Brenchley, 1929), wheat in five (Boatwright & Viets, 1966), and tomatoes in the period from 15–25 days after germination (Awerbuch, 1975).

The redistribution of minerals may be the more important of these two causes. Taking typical figures (Briggs, 1967) for water flux into plant roots (0.6×10^{-5} cm^3 cm^{-2} s^{-1}), root radius (0.03 cm) and transpiration rate (20 mg cm^{-2} h^{-1} ~ 5.6×10^{-6} cm^3 cm^{-2} s^{-1}), one can show that approximately 5 cm of root are required to supply 1 cm^2 of transpiring leaf surface. In practice this value is easily exceeded, and since the ratio of root length to leaf area varies with species, plant age and nutrient supply, it appears unlikely that transpiration is limited by the absorbing surface during early development. The growth pattern of allocating resources first to roots and then to shoots, therefore most probably reflects the ability of plants to store and retranslocate nutrients, and the fact that PAR is not so amenable to this.

Imbalance of supply

Under favourable conditions, the balanced growth path of Fig. 12.6 reflects early allocation of resources to root growth, followed by an allometric relationship in which the ratio of shoot to root relative growth rates is constant but greater than 1. Where particular resources limit growth, the balanced path may be shifted in favour of roots or shoots, but the overall pattern remains the same; this result can be shown experimentally (Atkinson, 1973) and can be derived from theory (Mooney & Gulman, 1979; Iwasa & Roughgarden, 1984).

Other morphological responses may also occur. It is common to find that leaf thickness (either measured directly (Corré, 1983; Fig. 12.5b) or indirectly by the specific leaf area, A_L/W_L (the area of leaf divided by leaf dry weight); Newton, 1963; Evans & Hughes, 1961; Fitter & Ashmore, 1974; McLaren & Smith, 1978; Jurik *et al.*, 1979), is lower in shaded leaves. It also changes to a greater extent than other allocation indexes like leaf weight ratio (W_L/W_{RS}; the fraction of total plant weight in leaves); this may be because smaller cells (Wild, 1979), or more usually fewer layers of cells are produced, so that the mesophyll surface area (A_{mes}) is reduced relative to leaf area. This ratio, A_{mes}/A_L, correlates well with maximum photosynthetic rate in a range of species typical of well illuminated environments (Nobel, 1977), but not in more shade resistant species (Chabot & Chabot, 1977), which suggests that although high A_{mes}/A_L values tend to be associated with the high light-saturated photosynthetic rates characteristic of sun plants, the relationship is not direct. High A_{mes}/A_L values merely provide the potential for further sites for CO_2 fixation; in other words there must also be greater investment in the biochemical machinery, which in allocation terms is largely the enzyme RuBP carboxylase (Björkman, 1981). Shade-resistant plants do not normally make such an investment.

Where soil-based resources are limiting, the analogous change to that in specific leaf area is in the specific root length, L_R/W_R, which is related to root diameter in the same way that A_L/W_L is related to leaf thickness. Again this seems more sensitive to changes in water and mineral nutrient availability than is root weight ratio, W_R/W_{RS}. It is also true that plants which are capable of large growth responses to increases in nutrient supply tend to have more plastic root systems in terms of root diameter (Christie & Moorby, 1975; Robinson & Rorison, 1983; Fitter, 1985), individual roots becoming finer as nutrient supply declines.

The analogy with responses to PAR is precise; sun plants which are responsive to high PAR levels have more plastic leaf morphology, and again, low resource levels lead to lower allocation of biomass per unit absorbing surface or length.

Enforced changes

Morphological changes in response to altered resource supply rates allow the resumption of a balanced growth path. Plants may also suffer enforced deviations from this path following defoliation by grazing or mowing (and, more rarely, root pruning). Removal of leaves has two main consequences: 1) the remaining leaves show large increases in photosynthetic rate (Maggs, 1965; Hodgkinson, 1974), as a result of the increased sink strength or developing leaves; 2) sink strength of roots is reduced relative to these new leaves and roots may actually become a carbon source (Bokhari, 1977).

After defoliation, both *Trifolium repens* (Ennik, 1966; Evans, 1971) and *Lolium perenne* (Ennik, 1966) undergo compensating shoot growth. Evans (1971) found that it took from 11 to 17 days for rates of clover root elongation to return to control levels after a single defoliation at 7.5, 5.0 or 2.5 cm above ground level. After the recovery of root growth was complete the shoot:root ratio was the same in all treatments (2.43–2.56), despite very large differences between cutting treatments in root growth rate. After the most

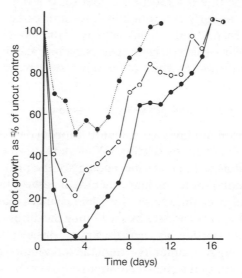

Fig. 12.7. The effect of a single defoliation to three different heights (●···● 7.5 cm, o—o 5 cm and ●—● 2.5 cm) on root growth of ryegrass *Lolium perenne*. Defoliation occurred at day 0. From Evans (1971).

severe treatment root growth was brought to an almost complete standstill for several days (Fig. 12.7). In other words, changes in root and shoot growth rates were complementary.

As with all root-shoot interactions, there are powerful feedback effects in any such treatment. Defoliation causes transport of minerals (particularly phosphate) and carbohydrate reserves from root to shoot, and this may stimulate absorption rates in the roots, which are normally controlled by internal nutrient levels (inorganic phosphorus in the case of phosphate; Chapin, Follett & O'Connor,

1982). Defoliation of *Eriophorum vaginatum* growing under nutrient-limited conditions reduced root phosphorus and carbohydrate levels, but not root nitrogen levels, and consequently stimulated phosphorus absorption activity despite reducing root growth rate (Cnapin & Slack, 1979). In the same study, *Carex aquatilis*, which has a large, perennial root system in contrast to the smaller, almost annual system of *E. vaginatum*, was less responsive to defoliation, and was able to finance leaf regrowth from stored reserves over several cycles of defoliation.

Severe defoliation may, however, have the opposite effect: cutting *Dactylis glomerata* to 2.5 cm above ground caused transport of more carbon to regenerating leaves than was available as soluble carbohydrate. This implies a breakdown of other substrates, probably protein, for respiratory and synthetic use, which was associated with a large reduction in phosphorus uptake rates, such that regrowth became phosphorus-limited after about six days (Davidson & Milthorpe, 1966).

Both root and shoot activity are therefore strongly sink-limited when resource supply rates are adequate. As we have seen, photosynthetic rate is a function of leaf sugar levels, and nutrient uptake rate is similarly controlled by both root carbohydrate and nutrient levels. Increase of the sink activity of the shoot (for nutrients) or the roots (for carbon) triggers compensating activity which permits the growth responses which ultimately restore balanced growth.

Morphological and physiological responses

Responses to limiting resource supplies, whether shoot or root based, may be largely morphological; that is they may depend upon alterations to the gross morphology of the plant (allocation to roots and shoots), or to the fine structure of parts of the plants (changes in leaf and root thickness). Such fine changes may affect the rates of physiological processes by shortening the diffusion pathways for CO_2 from the outside air (leaf thickness: distance from atmosphere to CO_2-fixing site) or for ions from the soil solution (root thickness: root density per unit weight). Alternatively there may be modifications to the physiology and biochemistry of the processes themselves.

Levels of RuBP carboxylase in leaves are closely related to both the PAR level experienced during leaf development and to the light-saturated photosynthetic rate (Björkman, 1968; Singh, Ogren & Widholm, 1974). Equally, levels of mineral ion supply profoundly influence the characteristics of the ion uptake system for nitrate (Smith, 1973; Doddema, Telkamp & Otten, 1979), phosphate (Cartwright, 1972) and potassium (Glass, 1978). Nevertheless, where the relative importance of morphological and physiological adjustments has been compared, it is generally the morphological responses that prove to be dominant (Chapin, 1980). In crop plants, selection for greater response to high levels of resources has been taking place, first unconsciously, then consciously, over thousands of years. Strikingly, the relative growth rates and carbon dioxide exchange rates of crop ancestors are not systematically different from modern cultivars (Evans & Dunstone, 1970; Duncan & Hesketh, 1968), and most differences in final yield relate to the *proportion* of photosynthate allocated to the harvested tissues (e.g. grain weight as a fraction of total shoot weight in cereals; Snaydon, 1984). Overall, development of leaf area is a more significant determinant of crop growth rate than is the rate of photosynthesis per unit leaf area (Gifford & Jenkins, 1981).

12.3   Allocation of resources

So far, allocation has been discussed in terms of the root-shoot balance of carbon, largely because this is where most studies have concentrated (despite the caveats of various workers, e.g. Harper & Ogden, 1970; Caswell & Abrahamson, 1981). During vegetative growth it is the root-shoot term which dominates the allocation process, and determines the ability of the plant to garner further resources. There are, however, many other processes which make demands on plant resources.

12.3.1   Resource utilization

Capital and maintenance costs

Calculations of biomass distribution indicate the relative value of different organs to the plant on a capital investment basis only. The true cost of any organ must also take into account the continuing cost of maintaining it. Such respiratory costs include resynthesis of labile cell components, transport activities, repair processes, defence, and all the normal metabolic functions of the organ. In *Trifolium repens*, McCree (1970) found that whole plant respiration was a dual function of photosynthetic rate and plant weight—about 25% of carbon fixed was respired in addition to about 1.5% of the total dry weight per day (Fig. 12.8). These figures compare favourably with the estimates of around 28% of carbon fixed and 30 mg sugar/g dry weight/day given as an overall average by Penning de Vries (1975).

Respiratory rates are closely linked to metabolic and growth activity, as shown by the large increases caused by uncoupling agents such as dinitrophenol, by the high values found in actively dividing meristematic tissue, and by the concomitant changes of respiratory rate and growth rate in an organ (Beevers, 1970). In organs whose resource gaining activity is lowered by resource limitation or inhibit-

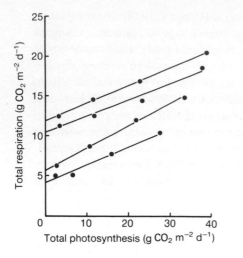

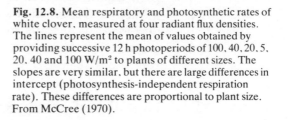

Fig. 12.8. Mean respiratory and photosynthetic rates of white clover, measured at four radiant flux densities. The lines represent the mean of values obtained by providing successive 12 h photoperiods of 100, 40, 20, 5, 20, 40 and 100 W/m² to plants of different sizes. The slopes are very similar, but there are large differences in intercept (photosynthesis-independent respiration rate). These differences are proportional to plant size. From McCree (1970).

ing factors (like temperature or toxins), respiratory rates are commonly reduced by a factor of 10 (7.0–0.7 mg CO_2/dm²/h in shaded white clover; McCree & Troughton, 1966). In extreme cases, some plants are able to lower respiration rates almost to zero, enabling them to survive long periods in total darkness; for example *Deschampsia flexuosa* survived this treatment for 227 days (Hutchinson, 1967).

Respiration rates in non-photosynthetic organs (except meristems) are generally much lower than in leaves because of the absence of the component associated with carbon fixation, but show similar correlations with metabolic activity. Although maintenance respiration is therefore an important fate for photosynthate, it is under close metabolic control and is linked to resource capture rates.

Supporting structures

Although numerous plant organs act as supporting structures (for example stems, petioles, peduncles, tendrils), they are normally photosynthetic and, at least in terms of carbon economy, function as leaves. In resource allocation studies they are sometimes lumped together with leaves in a photosynthetic component (although peduncles are often added to reproductive effort). Root systems also serve a vital anchoring role, which is confounded with the roles of resource capture and transport. The supporting function of these various organs is described in Chapter 9.

Reproductive structures

Reproductive effort (RE) is sometimes defined to include only sexual reproductive effort (SRE), and sometimes to include vegetative reproductive effort (VRE) as well. SRE variously covers just seeds, whole fruits, fruits and floral structures, or fruits, flowers, peduncles and stems together. Stewart & Thompson (1981) included the whole stem in rosette species (*Carex flacca, Leontodon hispidus*) or the stem above the topmost leaf in *Centaurea nigra* and *Poterium sanguisorba*. Since stems generally have a high biomass, these variations in definition make comparisons difficult.

There is no clear distinction between vegetative growth and vegetative reproduction (or clonal growth, as it is sometimes known). Grasses, for example, grow by continuously forming new shoots or tillers, which become self-sufficient in all resources by developing leaves and roots after an initial period of dependence on older tillers. In rhizomatous and stoloniferous species all gradations occur from wholly dependent tillers acting as sinks to wholly independent tillers acting as sources, or even physically separated from other tillers.

Reproduction may achieve a number of results: 1) production of new genotypes (genets) by sexual reproduction; 2) increase in biomass of the parent individual; 3) dispersal of new individuals to new sites, beyond the range of interference with the parent; 4) formation of resistant phases in the life cycle either sexually (seeds) or asexually (corms, bulbs, etc., and seeds in apomictic species). For vegetative reproduction, therefore, the main allocation functions are dispersal and resistance. Dispersal is normally the result of stolon or rhizome growth (although occasionally achieved by bulbils and other discrete organs, as in *Polygonum viviparum*). Stolons are short-lived (usually surviving less than one growing season) and typically above-ground; they provide a channel for transport of resources from parent to offspring. Rhizomes are perennial, subterranean and additionally perform a storage function; they are frequently the predominant organ during the unfavourable season. The resource costs of these structures therefore differ. Stolons, being short-lived, have low capital and maintenance costs, but may still occupy up to 20% of summer biomass in species like *Viola pallens* (Newell, Solbrig & Kincaid, 1981). In a study of several *Viola* species N. Setters (pers. comm.) found peak values of around 40% of summer biomass in rhizomes in *V. lutea, V. palustris* and *V. hirta*. In *V. odorata* the maximum was around 20%, but that species had stolons as well which accounted for a further 10–15%.

The distribution of radioactively labelled ^{14}C can be used to estimate short-term transfers of carbohydrates. In *Dupontia fisheri* up to 15% of ^{14}C assimilated was exported from vegetative tillers, but only 5% from flowering tillers (Alessio & Tieszen, 1978). In *Elymus repens*, another rhizomatous plant, 33–35% was exported from mature tillers, but only 6–8% from the youngest tiller though this could be increased to 40% by defoliating other tillers, and to 48% by shading them (Rogan & Smith, 1974).

Rates of carbon export from leaves seem relatively constant, irrespective of the reproductive method employed by the species (Hofstra & Nelson, 1969). In mature *Ranunculus repens,* stolons may account for 50% total biomass, whereas *R. bulbosus,* which has no stolons, allocates around 25% biomass to the corm and rather more than this to seed production under most conditions (Ginzo & Lovell, 1974). Export of ^{14}C from rosette leaves of each species is similar, at around 35–40%, except in young plants. Leaves on the ramets of the stolons of *R. repens* may export more, 45–60%, which travels to the stolon, the new ramets and the stolon apex. Such export values are typical of many mature leaves; the particular significance of rhizomes and stolons is that they permit exploration and exploitation of relatively remote patches of environment by long-distance transport of resources from the parent.

Sexual reproduction involves very different costs from those of stolon and rhizome growth. Whereas production of new vegetative ramets is physiologically indistinguishable from growth of an 'individual', sexual reproduction requires diversion of a meristem away from growth functions. Where all meristems typically become active, this 'competition for meristems' may be as significant as that for resources (Watson, 1984).

Estimates of resource allocation to sexual reproduction (SRE) vary, but values of around 20–30% of net annual assimilation for annuals are generally accepted (Harper, 1977). Schulze (1983) quotes harvest indices (seed yield as a percentage of above-ground biomass) ranging from 14–40% for *Helianthus annuus* to 39–56% for *Oryza sativa.* Allowing for root biomass, these figures are comparable. It is generally accepted that herbaceous perennials have lower SRE (e.g. Warner & Platt, 1976; Bostock & Benton, 1979; Primack *et al.,* 1981), but calculations for SRE for *Solidago pauciflosculosa* on a lifetime basis, allowing for carry over of vegetative components from year to year, give a figure of 30%, very similar to that for annuals (Pritts & Hancock, 1983). A figure of around 30% for overall reproductive effort seems quite general, and where SRE and VRE have been recorded together they often total approximately this figure; for instance, in *Tussilago farfara* SRE was 3–8% and VRE 4–23% (Ogden, 1974). The actual resource costs of sexual reproduction include the seeds and associated dispersal or protective structures as well as the pollination system, including pollen and (for entomophilous species) petals and nectar (Heinrich, 1981). Insect pollination thus has two costs; the direct cost of attractants and rewards, and also the loss of photosynthetic activity resulting from investment in non-green structures. Flowers, however, are not wholly dependent on the rest of the plant for carbon, since peduncles, sepals, bracts and other structures may photosynthesize actively: ear photosynthesis may contribute one-third of the carbon required for grain-filling in cereals (Biscoe *et al.*, 1975), and floral photosynthesis reaches 65% of carbon SRE in *Acer platanoides*, though only as little as 2% in *Quercus macrocarpa* (Bazzaz *et al.*, 1979).

Symbiosis

The two most important and widespread plant symbioses are root nodules and mycorrhizas. Each operates on the basis of transfer of resources; carbon from the plant in return for nitrogen in the case of nodules (Sprent, 1979), and phosphate (and possibly other ions and water) in the case of mycorrhizas (Harley & Smith, 1983). The almost universal occurrence of mycorrhizas confirms suspicions that carbon is less limiting than phosphate in many situations, certainly to uninfected plants.

   These carbon costs, particularly of nitrogen fixation, may be high. Minchin & Pate (1974) found that as much as 57% of the carbon translocated to actively fixing pea roots in a 24 hour period was used as skeletons for nitrogen fixation, and another 39% was respired, at least some of which must have been associated with fixation. Mycorrhizal costs are generally lower; recent studies estimate that 8% (Snellgrove *et al.*, 1982) and 6–10% (Koch & Johnson, 1984) of the plant's total photosynthate is diverted to mycorrhizal fungi. These figures should be compared to the losses of carbon experienced normally by plants growing in soil, in a form generally known as 'exudate'. A number of figures have been published (Barber & Martin, 1976; Martin, 1977; Martin & Kemp, 1980; Sauerbeck & Johnen, 1977), but recent studies with wheat and barley suggest that figures as high as 40% of total net fixed carbon may appear in soil in insoluble form (Whipps, 1984); less is lost at lower temperatures, because less carbon is translocated to roots. Typically, however, around two-thirds of ^{14}C translocated to roots appeared in soil, either as organic compounds or CO_2, but the bulk of this is probably from the death of roots, and only 3–15% is likely to be exudation (Newman, 1985). It is not clear whether root nodule and mycorrhizal utilization of carbon is additional to, or alternative to these losses, but it is interesting that current figures for mycorrhizal costs (6–8 or 6–10%) are comparable to estimates for exudation when root death is excluded (3–15%).

Defence

Plant defence against herbivores is of five main kinds: i) '*non-apparency*', in the sense of Feeny (1976); i.e. rarity in space or brevity of site occupation in time. This mechanism has no obvious resource cost, but imposes constraints on life history ii) *nutritional inadequacy* (see Southwood, 1973): plants with particularly low nitrogen contents may be avoided by grazers; this imposes low rates of resource acquisition by reducing physiological activity and growth rates. This is not a defence likely to work in the long-term against specialist herbivores, as these would simply eat more of the plant to compensate for its low quality (Moran & Hamilton, 1980), though it will reduce their growth rates and so may affect their population density. iii) *physical defences*, such as thick cuticles, spines and thorns, which have an easily calculated carbon cost (though few seem to be published), and a demonstrable benefit (e.g. Gilbert, 1971; Singh *et al.*, 1971). The cost in terms of other resources is usually low, although there may be some loss of photosynthetic activity. iv) *symbioses*, as in the classic case of the ant-acacias of Central America (Janzen, 1979). The cost here is again easily measurable: it is the resource content of the extra-floral nectaries and protein-bodies that the plant supplies. v) *chemical defences*, these provide the most taxing problem in determining costs (Bernays, 1983, and Coley *et al.*, 1985, provide good reviews).

Most secondary compounds are present in extremely small amounts although some, such as tannins, may only be effective at concentrations as high as around 2% plant weight (Feeny, 1975). Even where compounds are present at low concentrations, a high rate of turnover could mean that they were expensive in energy terms. Equally, a particular secondary compound may be relatively rich in a scarce mineral resource, making it disproportionately expensive. Several related defence compounds have been shown to be negatively correlated in concentration, suggesting that allocation of resources to one precludes allocation to the other (Romeo & Bell, 1974; Russell & Berryman, 1976).

Chew & Rodman (1979) have attempted to calculate minimum biosynthetic costs of a cyanogenic glycoside, a glucosinolate and a non-protein amino acid, each of which is synthesized from isoleucine. The costs, per molecule, are given in Table 12.1. They point out that the substrate cost of the isoleucine precursor is comparable to that of the derived compounds and it seems likely that, unless these compounds turn over rapidly, they represent a relatively small cost. This may well account for their ubiquity. On the other hand, tannins are very widely distributed and have a biosynthetic cost as much as 120 times greater than that for the production of an equivalent deterrent effect (Swain, 1979). The high cost of tannins may also suggest that turnover *is* an important factor in the metabolic costing of other toxins (but, since tannins represent a different *kind* of defence, and are effective against different sorts of herbivores, comparative costing is fraught with difficulties; see Crawley, 1983).

That chemical defences do have a cost is admirably illustrated by the fact that ant-acacias lack the cyanogenic glucosides found in many non-ant-acacias (Rehr *et al.*, 1973). Presumably there has been selection against such 'useless' defences in those species which are effectively protected by ants. Equally, the dramatic defoliation of such plants when their ant defences are removed points to the benefits of these apparently low-cost defence systems.

Table 12.1. Biosynthetic cost of manufacturing one molecule of each of three secondary compounds. From Chew & Rodman, 1979.

Secondary compound	Molecules of			
	isoleucine	ATP	$NADH_2$	glucose
lotaustralin	1	1	1	1
2-butylglucosinolate	1	12	5	1
2-amino-4-methylhex-4-enoic acid	1		−1 or 2	½

12.3.2   Allocation and environment

Allocation and acquisition patterns are important components of the response of plants to environmental conditions. Much of what is normally referred to as phenotypic plasticity (Bradshaw, 1965) rests on differences in these patterns. The importance of resource supply levels has already been pointed out (Section 12.2.2) in relation to spatial differences in environment, which may select either for plasticity or for genotypic differences in allocation and acquisition patterns. Clearly an individual must be able to withstand all the environmental conditions it is likely to encounter, so that temporal instability may impose plastic responses on plants, if the time scale is similar to that of the plant's life cycle.

Hickman (1975) studied the annual endemic herb *Polygonum cascadense*, growing on open mountain slopes, in habitats varying in the predictability of soil water availability. In the least predictable environment, in which plant growth was most reduced, SRE approached 60% compared to under 40% at the other extreme (Fig. 12.9). There were no differences in allocation between two populations grown under glasshouse conditions, so that all the differences were plastic in origin. This contrasts with the rather more constant patterns found in *Taraxacum* sect. *Vulgaria* (Gadgil & Solbrig, 1972) and various *Solidago* spp. (Abrahamson & Gadgil, 1973), and suggests that plasticity of allocation may be selected for in relation to the unpredictability of the environment.

This plasticity may also depend upon the severity of the extremes experienced; in Harper and Ogden's work on *Senecio vulgaris*, allocation was unaltered in the 'medium stress' treatment, where biomass was 15% of that in the 'low stress' treatment, but radically different under 'high stress', where biomass was reduced below 2%. In Hickman's work the biomass of the plants in the four most severe habitats (as a percentage of those in the most favourable) were 6, 7, 17 and 17%; in the poorest conditions (the first three), SRE was significantly higher than in the mesic habitat. Allocation patterns must not therefore be viewed as species characteristics, but as both adaptive and adaptable response patterns.

Plasticity of allocation means that field studies based on observations in a single year must be treated with caution. Soulé & Werner (1981) found large differences in SRE between populations of *Potentilla recta* in one year, which were negatively correlated with successional development of the habitat; in the second year of the study, no such correlation existed and there were no differences between populations!

12.4   Efficiency of resource use

An important aspect of resource ecology concerns the relationship between the costs of an organ and its output, as measured in terms of plant fitness. Analyses of cost and benefits are normally conducted in terms of carbon, since it is the most easily measured compound, and represents the basic constructional resource. Others (particularly nitrogen and phosphorus) may be more important in functional terms. The balance between cost and benefit can be affected in three ways: 1) by changes in costs—the amount of resources required to construct and run the organ; 2) by changes in the activity of the organ; and 3) by changes in the longevity of the organ.

12.4.1   Resource concentrations and activity

The carbon costs of organ construction vary in well known and predictable ways, as already discussed for specific leaf area and specific root length (Section 12.2.2). Costs of functioning may be expressed in other ways, in terms of nutrient concentrations for example, to reflect efficiency of metabolic activity, rates of turnover of enzymes, etc. The nutrient concentrations of leaves vary widely, but they are highest in young leaves and at high nutrient supply rates. Expressed on a dry weight basis, leaf nitrogen concentration is typically 20–40 mg/g in young leaves, falling to around 5–10 mg/g in leaves immediately prior to abscission. Leaf phosphorus varies from 2–5 mg/g in young leaves to less than 1, and leaf potassium from 5–10 to around 2–5. This decline is partly due to export during senescence, which is why leaf calcium concentration normally increases (calcium being immobile in phloen), but in the early part of the season, these differences reflect the proportion of structural tissue of low nutrient concentration, rather than differences in the availability of nutrients to metabolic centres. Since the primary enzyme of CO_2 fixation,

RuBP carboxylase, is usually the major component of leaf protein, nitrogen level may have a potentially profound influence on photosynthetic rate, as may inorganic phosphate levels (Herrold, 1980). Phosphorus deficiency is often associated with increased mesophyll resistance, a sign of reduced carboxylation activity (Longstreth & Nobel, 1980; Terry & Ulrich, 1973).

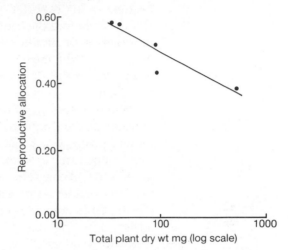

Fig. 12.9. The decline of sexual reproductive effort (total seed output/total weight) with increasing plant weight in *Polygonum cascadense*. Plant weight can be taken here as an index of environmental favourability. From Hickman (1975).

There is a general correlation between leaf nitrogen and photosynthetic rate (Chapter 12), although it is often the case that, for a given leaf nitrogen content, shorter-lived leaves tend to have greater photosynthetic rates. This may be because of the diversion of nitrogen to other activities (defence, resistance to environmental extremes) in long lived leaves (Grubb, 1984; Coley *et al.*, 1985).

12.4.2 Longevity of organs

Other things being equal, the longer a plant part persists, the greater the returns it can make to the plant. In practice, reduced operational effectiveness due to local resource exhaustion and predator or pathogen damage, may mean that the cost of maintaining the part, and the loss of use of the resources that could be re-allocated from it, outweigh the benefits from its continued function. Most of the plant-controlled abscission processes appear to obey this simple economic model.

The life spans of plant parts vary from ephemeral (root hairs and anthers) to the complete life of the plant (tree trunks and main roots); others lie in between. Both the shoot and root systems of perennial plants may contain permanent members, concerned with the placement of absorbing organs and transport, and short-lived ones, the leaves and fine roots. The control of the life span of the ephemeral tissues is internal, but responds to environmental stimuli. In the case of roots the stimuli are predominantly soil temperature and moisture (Head, 1973), while for leaves they are photoperiod and water availability (Addicott & Lyon, 1973).

If carbon can be regarded as the primary resource, then cost-benefit calculations for leaves are relatively straightforward. The costs comprise the amount of carbon imported into the leaf before it becomes a net exporter, and the benefit is the total net export. If a leaf is lost before the benefit exceeds the cost, it will have been a net drain on the plant; after that point it is in profit. The critical variables here are: 1) the initial construction cost; 2) the net photosynthetic rate; and 3) the duration of the favourable conditions after leaf initiation.

Where rates of carbon gain are limited by resources or temperature, increasing potential photosynthetic rates will have little impact on leaf profit, which can only be improved by increasing life span. This is illustrated in a simulation by Chabot & Hicks (1982), based on measured leaf characteristics of *Fragaria virginiana* (Fig. 12.10). For the open-habitat leaf, which has a much greater maximum photosynthetic rate and a lower area, the greatest improvement would be achieved by doubling the rate of carbon gain; for the larger shaded leaf, with an already low photosynthetic rate, increased leaf life span is the most productive option.

Leaf characteristics		Open	Forest
Maximum net photosynthesis ($mgCO_2 \cdot dm^{-2} \cdot hr^{-1}$)		10.3	2.2
Specific leaf weight ($mg \cdot cm^{-2}$)		6.5	2.6
Leaf area (cm^2)		7.5	22.0
Life span (days)		77	85
Computed leaf profit (mg CO_2)			
Based on measured values —		290	220
Double maximum Net photosynthesis – – – –		441	294
Increase duration of maximum net photosynthesis by 1/3 of life span ·······		392	231
Increase life span by 1/3		394	325

Fig. 12.10. Estimates of carbon gain over the whole life span of individual leaves of *Fragaria virginiana*. Figures are based on actual data and extrapolated to various changed conditions. From Chabot & Hicks (1980).

It follows from this that we should expect to find a trend towards increased leaf longevity as resources or other limiting factors progressively reduce achievable photosynthetic rates. In contrast, since maximum photosynthetic rate appears invariably to decline with leaf age, whenever high rates *are* sustainable, greatest carbon gain will be achieved by maintaining a population of active, young leaves.

In some environments, the duration of one favourable season for photosynthesis may be inadequate for a leaf to achieve profitability, necessitating retention over more than one growing season (i.e. evergreenness). It is notable that evergreenness is associated not just with high latitude and altitude, but also with low soil nutrient status, as in mires (Reader, 1978) and tundra (Shaver & Chapin, 1980), with low soil moisture availability (Gray, 1983), or with combinations of these (Goldberg, 1982). All these conditions increase the time required before profitability is achieved.

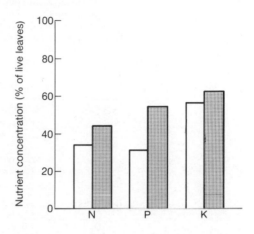

Fig. 12.11. Concentration of N, P and K in abscissed leaves of evergreen *Ceanothus megacarpus* (open columns) and the drought-deciduous shrub *Salvia leucophylla* (hatched), expressed as a percentage of that in live leaves. Note the higher retention of nutrients, particularly P, by the evergreen. From Gray (1983).

Increase of leaf life span, however, increases the probability of encountering severe climatic conditions, or attack by herbivores and pathogens. Countering these involves different leaf structures, with greater investment in structural tissue, reduced palatability, and increased chemical defence. These in turn may increase the initial structural costs, and reduce photosynthetic activity, by increasing stomatal and mesophyll resistances (Mooney & Dunn, 1970; Coley *et al.*, 1985).

Although this analysis can explain the widespread relationship between low soil nutrient levels and the evergreen habit, it is also revealing to examine the problem from the standpoint of limiting mineral resources. Leaf-fall involves a loss of minerals from the plant, even though some are retranslocated prior to abscission. The rate of leaf-fall must therefore be balanced against the potential nutrient capture rate. Evergreens almost invariably have lower nutrient levels in both live and dead leaves than herbaceous species, and are more efficient at withdrawing nitrogen, phosphorus and potassium before abscission (Fig. 12.11). This internal cycling permits greater efficiency of carbon fixation per unit mineral resource (Small, 1972; Table 12.2).

A number of studies of leaf longevity have highlighted the central role of phosphorus deficiency, which has a strong association with xeromorphy and evergreenness (Beadle, 1966; Goldberg, 1982; Gray, 1983). Even in mesic species, phosphorus deficiency has a pronounced effect on transpiration rate by increasing stomatal resistance (Nelson & Safir, 1982; Allen *et al.*, 1981), and it may be that on phosphorus deficient soils, reduced water efficiency results in the need for longer-lived leaves, and hence evergreenness.

Table 12.2. Estimated potential yield of photosynthate (g) per g N in a range of evergreen and deciduous species from mires and non-wetland habitats. Estimates were based on the leaf maximum photosynthetic rate per unit dry weight, leaf longevity, retention of N by the plant before leaf fall, and N concentration of live leaves. From Small, 1972.

	Mire		Non-mire	
	Species	g/g	Species	g/g
Evergreens	*Ledum groenlandicum*	82	*Picea glauca*	60
	Chamaedaphne calyculata	66	*Pinus strobus*	23
	Andromeda glaucophylla	52	*Thuja occidentalis*	17
	Kalmia angustifolia	45		
	K. polifolia	74		
	Picea mariana	23		
	Mean	57	Mean	33
Deciduous	*Aronia melanocarpa*	25	*Spiraea alba*	23
	Gaylussacia baccata	37	*S. tomentosa*	22
	Vaccinium myrtilloides	23	*Acer rubrum*	13
	Larix laricina	22	*Alnus rugosa*	11
	Betula pumila	9	*Rubus hispidus*	19
	B. populifolia	23	*Salix fragilis*	11
	Nemopanthus mucronata	30	*Ulmus americana*	12
	Viburnum cassinoides	23	*Comptonia peregrina*	14
			Populus tremuloides	10
			Ilex verticillata	9
	Mean	23	Mean	14

12.5 Life histories

The longevities of plant parts and of entire plants are linked: conditions which favour evergreenness are unsuitable for short life cycles, because the same economic model applies. A genotype will only persist if it can successfully transmit offspring to the next generation, and it must therefore either produce those offspring within a single favourable growth period and then die, or else be able to survive the unfavourable period. In this sense, and because they make demands upon limiting resources, reproduction and growth are alternatives (Harper, 1977).

12.5.1 Reproduction and growth

No reproductive activity is wholly self-sustaining, although green parts of many flowers do contribute 10–30% of reproductive effort measured in terms of carbon (Bazzaz *et al.*, 1979; see Section 12.3.1).

Reproduction is therefore bound to reduce growth rate by the use of resources, some of which are likely to be growth-limiting, and it may limit the amount of resources that can be stored for use in a subsequent growth period. Further, reproduction diverts meristems away from the development of productive tissue (Watson, 1984). There is often a compensatory reduction in seed production where new ramets are produced vegetatively: for example, *Elymus repens* produced 30 seeds and 215 rhizome buds in its first year, whereas *E. caninus* produced 258 seeds (Tripathi & Harper, 1973). Similarly *Viola soraria* growing in New England produced 0.78 fruits per plant and no stolons, while *V. blanda* averaged 0.10 fruits and 0.35 stolons per plant (Newell *et al.*, 1981).

These rough equivalences in numbers may not reflect equal demands for resources, but it is clear that diversion of resources to flowering reduces their availability for vegetative activity, and the negative correlation between annual tree growth and fruit production is well known to foresters. Law (1979) clearly showed that size in the second season of growth of the annual or short-lived perennial grass *Poa annua* was inversely related to the reproductive effort in the first growing season (Fig. 12.12), and there was some evidence for a similar effect of reproduction on mortality.

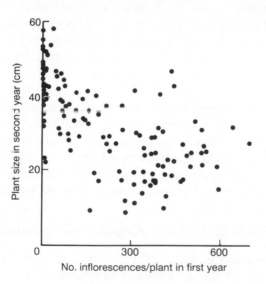

Fig. 12.12. The relationship between the number of inflorescences per plant in the first year of growth and plant size in the second year in *Poa annua*. Each point represents the mean of several progeny of a single mother. From Law (1979).

Monocarpy and reproductive thresholds

Since reproduction is antagonistic to growth and survival of the parent, it is not surprising to find that many plants are monocarpic: they undergo genetically programmed senescence and death after flowering. This permits diversion of all available resources into seed production. In monocarpic plants, flowering may occur at any time from a few weeks after germination (as in desert ephemerals; Went, 1955), to several decades later, as in *Yucca* and *Agave* species (Schaeffer & Schaeffer, 1979) and in *Frasera speciosa* (Inouye & Taylor, 1980). This apparently wasteful behaviour, which involves the sacrifice of up to sixty years accumulated tissues, has been interpreted in terms of competition for pollinators (Heinrich, 1981; see Chapter 6) and as a means of escape from seedling predators (see Chapter 5).

Failure to commit resources to reproduction in the early stages of the life cycle is commonly observed in forest trees, where diversion of resources into height growth can be seen as adaptive. Equally, plants suffering from resource deprivation may fail to flower; for example, *Aster acuminatus* grows in patches in forest clearings, and flowers only when light intensity and plant size exceed critical thresholds (Pitelka *et al.*, 1980). VRE of both flowering and non-flowering plants of *A. acuminatus* was more or less constant (at 30–50%), but SRE ranged from 0–13%. The high ratio of VRE to SRE in this species indicates the importance of persistence by rhizome storage as its basic pattern of survival.

In other herbaceous perennials, it is often SRE that is constant and VRE that responds to changes in plant size, as in strawberry *Fragaria virginiana* (Holler & Abrahamson, 1977) and *Tussilago farfara* (Ogden, 1974). Some species do not possess the facility for both sexual and vegetative reproduction, but in those that do it seems to be generally true that one method is fixed and the other plastic in terms of resource allocation, and that the latter acts as a luxury method of reproduction when resources are plentiful. This contrasts with the widely held view that it should be sexual reproduction that is favoured when environmental conditions are difficult. In fact, the conditions favouring sexual reproduction can only be interpreted in relation to life history; where habitats are sufficiently stable to make prolonged occupancy of a site feasible (as in *Aster acuminatus*), vegetative reproduction is likely to be the fixed method.

12.5.2   Life histories, patterns of allocation and vegetation

The great majority of the earth's land surface is dominated by perennial plants. Annuals and ephemerals are characteristic of sites where dispersal is of major significance, either in space, where the ground is disturbed, or in time, where conditions become temporarily unfavourable for survival, as in deserts. The establishment of annuals in closed communities is extremely rare, since their allocation patterns makes it difficult for them to compete successfully for resources with established perennials.

It is instructive to examine a notable exception to this rule, the hemi-parasites in the Scrophulariaceae (*Euphrasia*, *Rhinanthus*,

Pedicularis, *Melampyrum*). Unlike the chlorophyll-free total parasites such as *Orobanche* and *Lathraea*, which are mostly perennials, these hemi-parasites have normal photosynthetic shoots and are mainly annuals, yet they grow in closed grasslands. Their seeds are relatively large and after germination they grow by using seed reserves until the root chances upon that of another plant. The hemi-parasite root then invades the host root, and makes a xylem-to-xylem connection across which water and ions can move. In this way the hemi-parasite does away with the need to allocate resources to roots, retaining them in the shoots, and thereby becoming more competitive with neighbouring plants, some of which may be its hosts.

Since these annual hemi-parasites have no rootstock, they have no means of overwintering other than as seeds, and so can devote a very high proportion of their resources to seed production. It may also be that, since the host roots tend to die off in the winter to be replaced by new growth in the spring, they are responding to a seasonal resource (roots) in much the same way that a desert ephemeral responds to water.

Other annuals which do well in essentially perennial communities (e.g. hedgerow plants like *Galium aparine* and *Vicia sepium*) grow tall by scrambling up their longer-lived neighbours; by diverting few resources to structural materials in self-supporting stems they can compete with the perennials.

It is an essential characteristic of plants with short life cycles that they concentrate the allocation of resources into reproduction. This occurs at the expense of root growth (annuals typically have low root weight ratios), of the development of symbioses (mycorrhizal associations are generally poorly developed for example), and of defence compounds. Such a developmental pattern permits the rapid production of seeds from the time of germination and hence offers the potential for rapid population growth—a high potential rate of population increase (r in the classic logistic equation describing population growth). Conversely the diversion of resources to root growth, storage organs, symbioses and defence compounds has the opposite effect, but increases the ability of the individual to survive and to monopolize space. This is a developmental pattern that will be favoured where environments are relatively stable or predictable. Hence the well-known dichotomy between r-selection and K-selection can be seen as acting primarily on the allocation of resources.

The idea of r and K-selection has penetrated deeply into biological language because of its apparently universal applicability. As a simple linear axis of differentiation it has provided considerable insight into the selection pressures acting on organisms. As an analytical tool, however, its utility is limited; all too often it is used almost as a synonym for more mundane terms such as large and small, long-lived and short-lived. The frequent clashes in texts on succession between statements like 'pioneer species are r-selected' and 'early stages of primary succession are dominated by lichens', which are clearly incompatible, demonstrate the confusion to which the concept can lead.

A more generally applicable model is proposed by Grime (1979); this has three poles, namely competition, stress and frequency of disturbance. In essence it is a two-axis ordination of selection pressures, as opposed to the single-axis model of r and K-selection. The first axis describes environmental favourability, the second disturbance (Fig. 12.13). The square is reduced to a triangle by virtue of the fact that the corner characterized by extremely unfavourable environments (stress) and high disturbance is uninhabitable—mobile sand and the surfaces of busy roads might be examples. Grime's model allows resolution of the succession conundrum, since the early stages of secondary succession are characterized by high disturbance, (and hence are colonized by Grime's 'ruderals', which are in effect r-selected species); those of primary successions are characterized by environmental unfavourability ('stressed' in Grime's terminology), and correspond perhaps to a particular class of K-selected species. Old communities are generally dominated by competitive pressure as soils mature and biomass increases, allowing establishment of K-selected or 'competitive' species (see Chapters 1 and 2).

The addition of the second axis in Grime's model makes it more generally applicable than than of r- and K-selection, and an analogous concept—that of 'adversity (A) selection'—has recently been suggested in studies of soil animals (Greenslade, 1983; Usher, 1985). The triangular model also lends itself readily to an interpretation in terms of resource allocation. The disturbance axis selects for species of short life cycles in the most unstable habitats, and hence for high reproductive effort. In this respect it scarcely differs from r-selection. It is the other axis, environmental favourability, which provides the greater contrast. In the most favourable environments (defined in terms of physical and chemical environmental factors to avoid circularity: Terborgh, 1973; Fitter & Hay, 1981), productivity is high, resource depletion is greatest and competitive interactions are most fully developed. Where disturbance is low, opportunities for colonization are few, and site occupancy is favoured. Plants of these environments therefore tend to be large (which is particularly beneficial in competition for PAR), long-lived, and to have few, large seeds. Plants of unfavourable environments, in contrast, tend to be much smaller (since resource competition is no longer so important) and to have particularly high root weight ratios—in tundra as much as 90% of biomass can be below ground—with a predominant allocation of resources to storage functions.

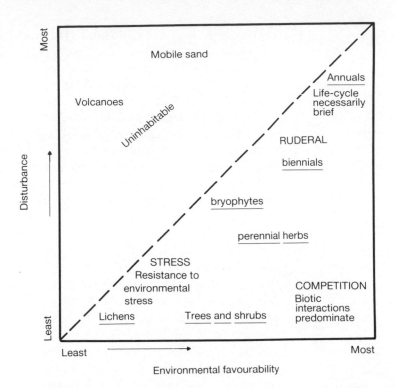

Fig. 12.13. Diagram illustrating the derivation of Grime's (1979) triangular scheme as a reduction of a two-axis ordination of disturbance and environmental favourability. half of the space being uninhabitable. Capital letters represent Grime's three 'primary strategies' (ruderals, competitors and stress-tolerators), underlined are the suggested general position of various life forms (after Grime), and roman type shows the dominant selective forces.

Models such as these are not intended as classifications which allow all plant species to be neatly pigeon-holed, and they contrast with the traditional schemes of Raunkiaer (1907), Warming (1909) and Walter (1973), based upon perennation and flowering frequency (see Chapter 8). Nevertheless, they point to the existence of general patterns of behaviour, often termed strategies (Harper, 1977; Grime, 1979), where similar selection pressures acting upon plants growing in similar conditions have resulted in the evolution of similar patterns of resource acquisition and utilization. If we are to understand the mechanisms underlying the structure and functioning of plant communities, and to make sense of both the repetitiveness of vegetation patterns which the synoptic ecologist sees as he surveys communities in many parts of the world, and the endless differences in detail which the analytical ecologist perceives at his scale of enquiry, an understanding of the control of resource acquisition and utilization will be essential.

References

Abbott D.L. (1959) The effects of seed removal on the growth of apple fruitlets. *Report of the Long Ashton Research Station* 1958, 52–56.

Abbott D.L. (1960) The bourse shoot as a factor in the growth of apple fruits. *Annals of Applied Biology* **48**, 434–438.

Abbott D.L. (1965) The effect of severe pruning on the fruiting of young apple trees. *Report of the Long Ashton Research Station* 1964, 106–112.

Abbott H.G. & Quink T.F. (1970) Ecology of eastern white pine seed caches made by small forest mammals. *Ecology* **51**, 271–278.

Aberdeen J.E.C. (1958) The effect of quadrat size, plant size and plant distribution on frequency estimates in plant ecology. *Australian Journal of Botany* **6**, 47–59.

Abrahamson W.G. (1980) Demography and vegetative reproduction. In *Demography and Evolution in Plant Populations* (Ed. O.T. Solbrig), pp. 89–106. Blackwell Scientific Publications, Oxford.

Abrahamson W.G. & Caswell H. (1982) On the comparative allocation of biomass, energy and nutrients in plants. *Ecology* **63**, 982–991.

Abrams P. (1976) Environmental variability and niche overlap. *Mathematical Bioscience* **28**, 357–372.

Abul-Faith A.H. & Bazzaz F.A. (1979) The biology of *Ambrosia trifida* L. I. Influence of species removal on the organization of the plant community. *New Phytologist* **83**, 813–816.

Acevedo E., Hsiao T.C. & Henderson D.W. (1971) Immediate and subsequent growth responses of maize leaves to changes in water status. *Plant Physiology* **49**, 631–636.

Ackerman J.D. & Mesler M.R. (1979) Pollination biology of *Listera cordata* (Orchidaceae). *American Journal of Botany* **66**, 820–824.

Adams M.W., Ellingboe A.H. & Rossman E.C. (1971) Biological uniformity and disease epidemics. *Bioscience* **21**, 1067–1070.

Addicott F.T. & Lyon J.L. (1973) Physiological ecology of abscission. In *Shedding of Plant Parts* (Ed. T.T. Kozlowski), pp. 85–124. Academic Press, London.

Addicott J.F. (1979) On the population biology of aphids. *American Naturalist* **114**, 762–763.

Aikman D.P. & Watkinson A.R. (1980) A model for growth and self-thinning in even-aged monocultures of plants. *Annals of Botany* **45**, 419–427.

Alessio M.L. & Tieszen L.L. (1978) Translocation and allocation of ^{14}C-photosynthate by *Dupontia fisheri*. In *Vegetation and Production Ecology of an Alaskan Arctic Tundra* (Ed. L. Tieszen). Ecological Studies, **29**, 393–413. Springer, Berlin.

Alexander H.M. & Burdon J.J. (1984) The effect of disease induced by *Albugo candida* (white rust) and *Peronospera parasitica* (downy mildew) on the survival and reproduction of *Capsella bursa-pastoris* (Shepherd's purse). *Oecologia* **64**, 314–318.

Alexander I.J. (1983) The significance of ectomycorrhizas in the nitrogen cycle. In *Nitrogen as an Ecological Factor* (Eds J.A. Lee, S. McNeill & I.H. Rorison), pp. 69–93. Blackwell Scientific Publications, Oxford.

Allard R.W. (1975) The mating system and microevolution. *Genetics* **79** (Suppl.), 115–126.

Allen M.F., Smith W.K., Moore T.S. & Christensen M. (1981) Comparative water relations and photosynthesis of mycorrhizal and non-mycorrhizal *Bouteloua gracilis* H.B.K. Lag ex Stend. *New Phytologist* **88**, 684–693.

Al-Mufti M.M., Sydes C.L., Furness S.B., Grime J.P. & Band S.R. (1977) A quantitative analysis of shoot phenology and dominance in herbaceous vegetation. *Journal of Ecology* **65**, 759–791.

Alstad D.N. & Edmunds G.F. (1983) Selection, outbreeding depression, and the sex ratio of scale insects. *Science* **220**, 93–95.

Anagnostakis S.L. (1982) Biological control of chestnut blight. *Science* **215**, 466–471.

Anderson D.J. (1971) Pattern in desert perennials. *Journal of Ecology* **59**, 555–560.

Anderson R.M. & May R.M. (1981) The population dynamics of microparasites and their invertebrate hosts. *Philosophical Transactions of the Royal Society, London. Series B* **291**, 451–524.

Andrew C.S. (1978) Legumes and acid soil. In *Limitations and Potentials for Biological Nitrogen Fixation in the Tropics*. (Eds J. Dubereiner, R. Burris & A. Hollaender), pp. 135–160. Plenum, New York.

Anikster Y. & Wahl I. (1980) Coevolution of the rust fungi on Graminae and Liliaceae and their hosts. *Annual Review Phytopathology* **17**, 367–403.

Antolin M.F. & Strobeck C. (1985) The population genetics of somatic mutation in plants. *American Naturalist* **126**, 52–62.

Antonovics J. (1968) Evolution of closely adjacent plant populations V. Evolution of self-fertility. *Heredity* **23**, 219–238.

Antonovics J. (1977) The input from population genetics, "The New Ecological Genetics." *Systematic Botany* **1**, 233–245.

Antonovics J. (1978) The population genetics of mixtures. In *Plant Relations in Pastures*. (Ed. J.R. Wilson), pp. 233–253. CSIRO, Melbourne.

Antonovics J., Bradshaw A.D. & Turner R.G. (1971) Heavy metal tolerance in plants. *Advances in Ecological Research* **71**, 1–85.

Antonovics J. & Fowler N.L. (1985) Analysis of frequency and density effects on growth in mixtures of *Salvia splendens* and *Linum grandiflorus* using hexagonal fan designs. *Journal of Ecology* **73**, 219–234.

Antonovics J. & Levin D.A. (1980) The ecological and genetical consequences of density dependent regulation in plants. *Annual Review of Ecology and Systematics* **11**, 411–452.

Armesto J.J. & Pickett S.T.A. (1985) Experiments on disturbance in old-field plant communities: Impact on species richness and abundance. *Ecology* **66**, 230–240.

Armstrong W. (1982) Waterlogged soils. In *Environment and Plant Ecology*, pp. 290–330. John Wiley, Chichester.

Arrhenius O. (1921) Species and area. *Journal of Ecology* **9**, 95–99.

Ashenden T.W. & Mansfield T.A. (1978) Extreme pollution sensitivity of grasses when SO_2 and NO_2 are present in the atmosphere together. *Nature* **273**, 142–143.

Ashton P.S. (1977) A contribution of rainforest research to evolutionary theory. *Annals of the Missouri Botanical Garden* **64**, 694–705.

Asker S. (1980) Gametophytic apomixis: elements and genetic regulation. *Hereditas* **93**, 277–293.

Atanasoff D. (1964) Universality of interferon formation. *Phytopathology Zeitschrift* **50**, 336–358.

Atkinson D. (1973) Some general effects of phosphorus deficiency on growth and development. *New Phytologist* **72**, 101–111.

Atkinson W.D. & Shorrocks B. (1981) Competition on a divided and ephemeral resource: a simulation model. *Journal of Animal Ecology* **50**, 461–471.

Aubreville A. (1971) Regenerative patterns in the closed forest of the Ivory Coast. In *World Vegetation Types* (Ed. S.R. Eyre), pp. 41–55. Columbia University Press, New York.

Augspurger C.K. (1980) Mass-flowering of a tropical shrub (*Hybanthus prunifolius*): influence on pollinator attraction and movement. *Evolution* **34**, 475–488.

Augspurger C.K. (1981) Reproductive synchrony of a tropical shrub: experimental studies on effects of pollinators and seed predators on *Hybanthus prunifolius* (Violaceae). *Ecology* **62**, 775–788.

Augspurger C.K. (1983a) Offspring recruitment around tropical trees: changes in cohort distance with time. *Oikos* **40**, 189–196.

Augspurger C.K. (1983b) Seed dispersal of the tropical tree, *Platypodium elegans*, and the escape of its seedlings from fungal pathogens. *Journal of Ecology* **71**, 759–772.

Augspurger C.K. & Kelly C.K. (1984) Pathogen mortality of tropical tree seedlings: experimental studies of the effects of dispersal distance, seedling density, and light conditions. *Oecologia* **61**, 211–217.

Awerbuch T.E. (1975) The utilisation of phosphate in tomato plants at different growth stages. *Plant and Soil* **43**, 443–450.

Azion-Bieto J. (1983) Inhibition of photosynthesis by carbohydrates in wheat (*Triticum aestivum* cultiver Gabo) leaves. *Plant Physiology* **73**, 681–686.

Bach C.E. (1980) Effects of plant density and diversity on the population dynamics of a specialist herbivore, the striped cucumber beetle, *Acalymma vittata* (Fab.) *Ecology* **61**, 1515–1530.

Bakelaar R. & Odum E. (1978) Community and population level responses to fertilization in an old-field ecosystem. *Ecology* **59**, 660–671.

Baker G.A. & O'Dowd D.J. (1982) Effects of parent plant density on the production of achene type in the annual *Hypochoeris glabra*. *Journal of Ecology* **70**, 201–215.

Baker H.G. (1959) Reproductive methods as factors in speciation in flowering plants. *Cold Spring Harbor Symposium on Quantitative Biology* **24**, 177–199.

Baker H.G. & Baker I. (1977) Intraspecific constancy of floral nectar amino acid complements. *Botanical Gazette* **138**, 183–191.

Baker H.G. & Baker I. (1983a) Floral nectar sugar constituents in relation to pollinator type. In *Handbook of Experimental Pollination Biology* (Ed C.E. Jones & R.J. Little), pp. 117–141. Van Nostrand Reinhold, New York.

Baker H.G. and Baker I. (1983b) A brief historical review of the chemistry of floral nectar. In *The Biology of Nectaries* (Eds B. Bentley & T. Elias), pp. 126–152. Columbia, New York.

Baker H.G. & Hurd P. (1968) Intrafloral ecology. *Annual Review of Entomology* **13**, 385–414.

Baldwin I.T. & Schulz J.C. (1983) Rapid changes in tree leaf chemistry induced by damage: evidence for communication between plants. *Science* **221**, 277–279.

Barber D.A. & Martin J.K. (1976) The release of organic substances by cereal roots into soil. *New Phytologist* **76**, 69–80.

Barber J. & Baker N.R. (Eds) (1985) *Photosynthesis and the Environment. Topics in Photosynthesis*. Volume 6. Elsevier, Amsterdam.

Barbour M.G. (1973) Desert dogma re-examined: root/shoot production and plant spacing. *American Midland Naturalist* **89**, 41–57.

Barkham J.P. (1980) Population dynamics of the wild daffodil (*Narcissus pseudo-narcissus*) I. Clonal growth, seed reproduction, mortality and the effects of density. *Journal of Ecology* **68**, 607–633.

Barlow H.W.B. (1964) An interim report on a long-term experiment to assess the effect of cropping on apple tree growth. *Report of the East Malling Research Station* 1963, 84–93.

Barlow H.W.B. (1966) The effect of cropping on the number and kinds of shoots on four apple varieties. *Report of the East Malling Research Station* 1963, 84–93.

Barrett J. (1985) The gene-for-gene hypothesis: parable or paradigm? In *Ecology and Genetics of Host-Parasite Interactions* (Ed D. Rollinson & R.M. Anderson), pp. 215–225. Academic Press, London

Bartholomew B. (1970) Bare zones between California shrub and grassland communities: the role of animals. *Science* **170**, 1210–1212.

Bartlett M.S. (1975) *The Statistical Analysis of Spatial Pattern*. Chapman and Hall, London.

Bartram W. (1955) *Travels of William Bartram* (Ed. M. Van Doren). Dover Publications, New York.

Bateman A.J. (1947) Contamination of seed crops. III. Relation with isolation distance. *Heredity* **1**, 303–336.

Bates G.H. (1935) The vegetation of footpaths, sidewalks, cart tracks and gateways. *Journal of Ecology* **23**, 470–487.

Bawa K.S. (1974) Breeding systems of tree species of a lowland tropical community. *Evolution* **28**, 85–92.

Bayer R.J. & Stebbins G.L. (1983) Distribution of sexual and apomictic populations of *Antennaria parlinii*. *Evolution* **37**, 555–561.

Bazzaz F.A., Carlson R.W. & Harper J.L. (1979) Contribution to reproductive effort by photosynthesis of flowers and fruits. *Nature* **279**, 554–555.

Bazzaz F.A. & Harper J.L. (1976) Relationship between plant weight and numbers in mixed populations of *Sinapis arvensis* (L.) Rabenh. & *Lepidium sativum* L. *Journal of Applied Ecology* **13**, 211–216.

Bazzaz F.A. & Harper J.L. (1977) Demographic analysis of the growth of *Linum usitatissimum*. *New Phytologist* **78**, 193–208.

Bazzaz F.A., Levin D.A. & Schmierbach M.R. (1982) Differential survival of genetic variants in crowded populations of *Phlox*. *Journal of Applied Ecology* **19**, 891–900.

Beadle N.C.W. (1966) Soil phosphate and its role in molding segments of the Australian flora and vegetation, with special reference to xeromorphy and sclerophylly. *Ecology* **47**, 992–1007.

Beals E.W. & Cope J.B. (1964) Vegetation and soils in an eastern Indiana woods. *Ecology* **45**, 777–792.

Beard J.S. (1944) Climax vegetation in tropical America. *Ecology* **25**, 127–158.

Beard J.S. (1955) The classification of tropical American vegetation-types. *Ecology* **36**, 89–100.

Beard J.S. (1983) Ecological control of the vegetation of South-western Australia: moisture versus nutrients. In *Mediterranean-Type Ecosystems* (Eds F.J. Kruger, D.T. Mitchell & J.U.M. Jarvis). Springer–Verlag, Berlin.

Beck S.D. & Reese J.C. (1976) Insect–plant interactions: nutrition and metabolism. *Recent Advances in Phytochemistry* **10**, 41–92.

Becking R.W. (1957) The Zürich–Montpellier School of Phytosociology. *Botanical Review* **23**, 411–488.

Beddington J.R., Free C.H. & Lawton J.H. (1978) Characteristics of successful natural enemies in models of biological control of insect pests. *Nature* **273**, 513–519.

Beever H. (1970) Respiration in plants and its regulation. In *Prediction and Measurement of Photosynthetic Activity* (Ed. I. Setlik), pp. 209–214. Pudoc, Wageningen.

Begon M. (1984) Density and individual fitness: asymmetric competition. In *Evolutionary Ecology* (Ed. B. Shorrocks), pp. 179–194. Blackwell, Oxford.

Begon M.E., Harper J.L. & Townsend C.R. (1986) *Ecology: Individuals, Populations and Communities*. Blackwell Scientific Publications, Oxford.

Begon M. & Mortimer M. (1981) *Population ecology: a unified study of animals and plants*. Blackwell Scientific Publications, Oxford.

Bell A.D., Roberts D. & Smith A. (1979) Branching patterns: the simulation of plant architecture. *Journal of Theoretical Biology* **81**, 351–375.

Bell A.D. & Tomlinson P.B. (1980) Adaptive architecture in rhizomatous plants. *Botanical Journal of the Linnean Society* **80**, 125–160.

Bender E.A., Case T.J. & Gilpin M.E. (1984) Peturbation experiments in community ecology: theory and practice. *Ecology* **65**, 1–13.

Benedict J.B. (1984) Rates of tree-island migration, Colorado Rocky Mountains, USA. *Ecology* **65**, 820–823.

Bentley S. & Whittaker J.B. (1979) Effects of grazing by a chrysomelid beetle *Gastrophysa viridula* on competition between *Rumex obtusifolius* and *R. crispus*. *Journal of Ecology* **68**, 671–674.

Benveniste R.E. & Todaro G.J. (1976) Evolution of Type C viral genes: evidence for an Asian origin of man. *Nature* **261**, 101–108.

Benzing D.H. (1983) Vascular epiphytes: a survey with special reference to their interactions with other organisms. In *Tropical Rain Forest: Ecology and Management* (Eds S.L. Sutton, T.C. Whitmore & A.C. Chadwick), pp. 11–24. Blackwell Scientific Publications, Oxford.

Bernays, E.A. (1983) Nitrogen in defence against insects. In *Nitrogen as an Ecological Factor*, pp. 321–344. British Ecological Society Symposium 22. Blackwell Scientific Publications, Oxford.

Bernays E.A. & Chapman R.F. (1978) Plant chemistry and acridoid feeding behaviour. In *Biochemical Aspects of Plant and Animal Coevolution*, pp. 99–141. Academic Press, London.

Berry J. & Bjorkman O. (1980) Photosynthetic response and adaptation to temperature in higher plants. *Annual Review of Plant Physiology* **31**, 491–543.

Bertin R.I. (1982a) Floral biology, hummingbird pollination and fruit production of trumpet creeper (*Campsis radicans*, Bignonicaceae). *American Journal of Botany* **69**, 122–134.

Bertin R.I. (1982b) Paternity and fruit production in trumpet creeper (*Campsis radicans*). *American Naturalist* **119**, 694–709.

Bierhorst D.W. (1977) On the stem apex, leaf initiation and early ontogeny in Filicalean ferns. *American Journal Botany* **64**, 125–152.

Bierzychudek P. (1981) Pollinator limitation of plant reproductive effort. *American Naturalist* **117**, 838–840.

Bierzychudek P. (1982) The demography of Jack-in-the-pulpit, a forest perennial that changes sex. *Ecological Monographs* **52**, 335–351.

Biscoe P.V., Gallagher J.N., Littleton E.J., Monteith J.L. & Scott K.K. (1975) Barley and its environment. IV: Sources of assimilate for the grain. *Journal of Applied Ecology* **12**, 295–318.

Bishop G.F. & Davy A.J. (1984) Significance of rabbits for the population regulation of *Hieracium pilosella* in Breckland. *Journal of Ecology* **72**, 273–284.

Bishop G.F., Davy A.J. & Jefferies R.L. (1978) Demography of *Hieracium pilosella* in a Breck grassland. *Journal of Ecology* **66**, 615–629.

Björkman O. (1968) Further studies on differentiation of photosynthetic properties in sun and shade ecotypes of *Solidago virgaurea*. *Physiologia Plantarum* **21**, 84–99.

Björkman O. (1981) Responses to different quantum flux densities. In *Encyclopaedia of Plant Physiology, New Series, Volume 12A* (Eds O.L. Lange, P.S. Nobel, C.B. Osmond & H. Ziegler), Chapter 3, pp. 57–107. Springer, Berlin.

Björkman O., Boardman N.K., Anderson J.M., Thorne S.W., Goodchild D.J. & Pyliotis N.A. (1972) Effect of light intensity during growth of *Atriplex patula* on the capacity of photosynthetic reactions, chloroplast components and structure. *Carnegie Institution Year Book* **71**, 115–135.

Björkman O., Gauhl E. & Nobs M.A. (1969) Comparative studies of *Atriplex* species with and without B–carboxylation and their first generation hybrid. *Carnegie Institution Year Book* **68**, 620–633.

Black J.N. (1957) The early vegetative growth of three strains of subterranean clover (*Trifolium subterraneum* L.) in relation to size of seed. *Australian Journal of Agricultural Research* **8**, 1–14.

Black J.N. (1959) Seed size in herbage legumes. *Herbage Abstracts* **29**, 235–341.

Blackman F.F. (1905) Optima and limiting factors. *Annals of Botany* **19**, 281–295.

Blasco D. (1971) Composion y distribucion del fitoplancton en la region del afloramiento de las costas peruanas. *Investigationes Pesques* **35**, 61–112.

Bliss L.C. (1956) A comparison of plant development in microenvironments of arctic and alpine tundras. *Ecological Monographs* **26**, 303–307.

Bliss L.C. (1985) Alpine. In *Physiological Ecology of North American Plant Communities* (Eds B.F. Chabot & H.A. Mooney), pp. 41–65. Chapman & Hall, New York.

Boardman N.K. (1977) Comparative photosynthesis of sun and shade plants. *Annual Review of Plant Physiology* **28**, 355–377.

Boatwright G.O. & Viets F.G. (1966) Phosphorus absorption during various growth stages of spring wheat and intermediate wheatgrass. *Agronomy Journal* **58**, 185–188.

Bock W.J., Balda R.P. & Vander Wall S.B. (1973) Morphology of the sublingual pouch and tongue musculature in Clark's Nutcracker. *Auk* **90**, 491–519.

Boerner R.E.J. (1982) Fire and nutrient cycling in temperate ecosystems. *Bio Science* **32**, 187–192.

Boerner R.E.J. (1984) Foliar nutrient dynamics and nutrient use efficiency of four deciduous tree species in relation to site fertility. *Journal of Applied Ecology* **21**, 1029–1040.

Bokhari V.G. (1977) Regrowth of western wheatgrass utilising ^{14}C–labelled assimilates stored in below-ground parts. *Plant and Soil* **48**, 115–127.

Bond G. (1976) The results of the IBP survey of root nodule formation in non-leguminous angiosperms. In *Symbiotic Nitrogen Fixation in Plants* (Ed. P.S. Nutman), pp. 443–474. Cambridge University Press, Cambridge.

Bond W. (1983) Alpha diversity of southern cape fynbos. In *Mediterranean-Type Ecosystems* (Eds F.J. Kruger, D.T. Mitchell & J.U.M. Jarvis), pp. 337–356. Springer-Verlag, Berlin.

Bossema I. (1968) Recovery of acorns in the European jay (*Garrulus g. glandarius* L.) *Verhandelingen van der Koninklijke Vlaamse Academie voor Wetenschappen*, **71**, 1–5.

Bostock S.J. & Benton R.A. (1979) The reproductive strategies of five perennial Compositae. *Journal of Ecology* **67**, 91–107.

Boucher D.H. (Ed.) (1985) *The Biology of Mutualism: Ecology and Evolution.* Croom Helm, London.

Boulton T.W., Harrison A.T. & Smith B.N. (1980) Distribution of biomass of species differing in photosynthetic pathway along an altitudinal transect in southeastern Wyoming grassland. *Oecologia* **45**, 287–298.

Bowen G.D. (1980) Mycorrhizal roles in tropical plants and ecosystems. In *Tropical Mycorrhiza Research* (Ed. P. Mikola), pp. 165–190. Clarendon Press, Oxford.

Box T.W. (1961) Relationships between plants and soils of four range plant communities in south Texas. *Ecology* **42**, 794–810.

Braakhekke W.G. (1980) On coexistence: a causal approach to diversity and stability of grassland vegetation. *Verslagen van Lanabouwkundige Onder zoekingen* **902**, 1–164.

Bradford D.F. & Smith C.C. (1977) Seed predation and seed number in *Schellea* palm fruits. *Ecology* **58**, 667–673.

Bradshaw A.D. (1965) Evolutionary significance of phenotype plasticity in plants. *Advances in Genetics* **13**, 115–155.

Bradshaw A.D. (1973) Environment and phenotypic plasticity. *Brookhaven Symposium in Biology* **25**, 75–94.

Bradshaw A.D. (1976) Pollution and Evolution. In *Effects of Air Pollutants on Plants* (Ed. T.A. Mansfield), pp. 135–159, Cambridge University Press, Cambridge.

Bradshaw A.D. (1984a) The importance of evolutionary ideas in ecology and vice versa. In *Evolutionary Ecology* (Ed. B. Shorrocks), pp. 1–25. Blackwell Scientific Publications, Oxford.

Bradshaw A.D. (1984b) Ecological significance of genetic variation between populations. In *Perspectives on Plant Population Ecology* (Eds R. Dirzo & J. Sarukhán), pp. 213–228. Sinauer Associates, Sunderland, Massachusetts.

Bradshaw A.D. & Chadwick M.J. (1980) *The Restoration of Land.* Blackwell Scientific Publications, Oxford.

Bradshaw A.D. & McNeilly T. *Evolution and Pollution.* Arnold, London.

Bradshaw M.E. (1981) Monitoring grassland plants in Upper Teesdale, England. In *The Biological Aspects of Rare Plant Conservation* (Ed. H. Synge), pp. 241–251. John Wiley & Sons, Chichester.

Braun-Blanquet J. (1982) *Plant Sociology: the Study of Plant Communities.* McGraw Hill, New York.

Bray J.R. (1964) Primary consumption of three forest canopies. *Ecology* **45**, 165–167.

Breemen N. van (1985) Acidification and decline of Central European forests. *Nature* **315**, 16.

Breese E.L. (1959) Selection for differing degrees of out-breeding in *Nicotiana rustica*. *Annals of Botany* **23**, 331–344.

Breese E.L., Hayward M.D. & Thomas A.C. (1965) Somatic selection in perennial ryegrass, *Heredity* **20**, 367–379.

Brenchley W.E. (1929) Phosphate requirement of barley at different periods of growth. *Annals of Botany* **43**, 89–110.

Briggs G.E. (1967) *Movement of Water in Plants*. Blackwell Scientific Publications, Oxford.

Brokaw N.V.L. (1982a) Treefalls: frequency, timing, and consequences. In *The Ecology of a Tropical Forest: Seasonal Rhythms and Long-term Changes* (Eds E.G. Leigh, Jnr., A.S. Rand & D.M. Windsor), pp. 101–108. Smithsonian Institution Press, Washington D.C.

Brokaw N.V.L. (1982b) The definition of treefall gaps and its effect on measures of forest dynamics. *Biotropica* **14**, 158–160.

Brokaw N.V.L. (1985a) Treefalls, regrowth, and community structure in tropical forests. In *Natural Disturbance: The Patch Dynamics Perspective* (Eds S.T. Pickett & P.S. White), Chapter 4. Academic Press, New York.

Brokaw N.V.L. (1985b) Gap-phase regeneration in a tropical forest. *Ecology* **66**, 682–687.

Brouwer R. (1963) Some aspects of the equilibrium between overground and underground plant parts. *Jaarb. Inst. Biol. Scheikd Onderzoek (IBS)*, pp. 31–39.

Brown A.H.D. (1979) Enzyme polymorphism in plant populations. *Theoretical Population Biology* **15**, 1–42.

Brown B.A. & Clegg M.T. (1984) Influence of flower color polymorphism on genetic transmission in a natural population of the common morning glory, *Ipomoea purpurea*. *Evolution* **38**, 796–803.

Brown J.H. & Kodric-Brown A. (1979) Convergence, competition, and mimicry in a temperate community of hummingbird-pollinated flowers. *Ecology* **60**, 1022–1035.

Brues C.T. (1946) *Insect Dietary*. Harvard University Press, Cambridge, Mass.

Buchman S.L. (1983) Buzz pollination in angiosperms. In *Handbook of Experimental Pollination Biology* (Eds C.E. Jones & R.J. Little), pp. 73–113. Van Nostrand Reinhold, New York.

Bülow-Olsen A., Sackville-Hamilton N.R. & Hutchings M.J. (1984) A study of growth form in genets of *Trifolium repens* L. as affected by intra- and interplant contacts. *Oecologia* **61**, 383–387.

Bulmer M.G. (1985) Selection for iteroparity in a variable environment. *American Naturalist* **126**, 63–71.

Burbidge A.H. & Whelan R.J. (1982) Seed dispersal in a cycad, *Macrozamia riedlei*. *Australian Journal of Ecology* **7**, 63–67.

Burdon J.J. (1980) Intra-specific diversity in a natural population of *Trifolium repens*. *Journal of Ecology* **68**, 717–735.

Burdon J.J. & Chilvers G.A. (1975) Epidemiology of damping-off disease (*Phythium irregulare*) in relation to density of *Lepidium sativum* seedlings. *Annals of Applied Biology* **81**, 135–143.

Burdon J.J. & Chilvers G.A. (1977) Preliminary studies on a native Australian eucalypt forest invaded by exotic pines. *Oecologia* **31**, 1–12.

Burdon J.J., Groves R.H. & Cullen J.M. (1981) The impact of biological control on the distribution and abundance of *Chondrilla juncea* in south-eastern Australia. *Journal of Applied Ecology* **18**, 957–966.

Burdon J.J., Groves R.H., Kaye P.E. & Speer S.S. (1984) Competition in mixtures of susceptible and resistant genotypes of *Chondrilla juncea* differentially infected with rust. *Oecologia* **64**, 199–203.

Burdon J.J., Marshall D.R. & Brown A.H.D. (1983) Demographic and genetic changes in populations of *Echium plantagineum*. *Journal of Ecology* **71**, 667–679.

Burdon J.J. & Shattock R.C. (1980) Disease in plant communities. *Applied Biology* **5**, 145–219.

Burgess T.M. & Webster R. (1980) Optimal interpolation and isarithmic mapping of soil properties. II. Block Kriging. *Journal of Soil Science* **31**, 333–341.

Burggraaf-van Nierop Y.D. & Meijden E. van der (1984) The influence of rabbit scrapes on dune vegetation. *Biological Conservation* **30**, 133–146.

Buss L.W. (1983) Evolution, development, and the units of selection. *Proceedings of the National Academy Science USA* **80**, 1387–1391.

Butcher R.E. (1983) *Studies on interference between weeds and peas*. Ph.D. Thesis, University of East Anglia.

Byth K. & Ripley B.D. (1980) On sampling spatial patterns by distance methods. *Biometrics* **36**, 279–284.

Cahn M.A. & Harper J.L. (1976a) The biology of the leaf mark polymorphism in *Trifolium repens* L. I. Distribution of phenotypes at a local scale. *Heredity* **37**, 309–325.

Cahn M.A. & Harper J.L. (1976b) The biology of the leaf mark polymorphism in *Trifolium repens* L. II. Evidence for the selection of leaf marks by rumen fistulated sheep. *Heredity* **37**, 327–333.

Cain S.A. (1947) Characteristics of natural areas and factors in their development. *Ecological Monographs* **17**, 185–200.

Calder W.A. (1984) *Size, Function and Life History*. Harvard University Press, Cambridge, Massachusetts.

Caldwell M.M. (1979) Root structure: the considerable cost of belowground function. In *Topics in Plant Population Biology* (Eds O.T. Solbrig, S. Jain, G.N. Johnson & P.H. Raven), pp. 408–427. Macmillan, London.

Caldwell M.M., Eissenstat D.M., Richards J.H. & Allen M.F. Competition for phosphorus: differential uptake from dual isotype-labelled soil interspaces between shrub and grass. *Science* **229**, 384–386.

Callaghan T.V. & Collins N.J. (1981) Life cycles, population dynamics and the growth of tundra plants. In *Tundra Ecosystems: A Comparative Analysis* (Eds L.C. Bliss, O.W. Heal & J.J. Moore), pp. 257–284. Cambridge University Press, Cambridge.

Cammell M.E. & Way M.J. (1983) Aphid Pests. In *The Faba Bean (Vicia faba L.). A Basis for Improvement* (Ed. P.D. Hebblethwaite), pp. 315–346. Butterworths, London.

Campbell A.I. & Jetzer P. (1972) Clonal variations in apple. *Report of the Long Ashton Research Station 1971*, 37–38.

Campbell E. (1927) Wild legumes and soil fertility. *Ecology* **8**, 480–483.

Cannell M.G.R., Rothery P. & Ford E.D. (1984) Competition within stands of *Picea sitchensis* and *Pinus contorta*. *Annals of Botany* **53**, 349–362.

Carpenter F.L. (1983) Pollination energetics in avian communities: simple concepts and complex realities. In *Handbook of Experimental Pollination Biology* (Eds C.E. Jones & R.J. Little), pp. 215–234. Van Nostrand Reinhold, New York.

Cartwright B. (1972) The effect of phosphate deficiency on the kinetics of phosphate absorption by sterile excised barley roots, and some factors affecting the ion uptake efficiency of roots. *Communications in Soil Science and Plant Analysis* **3**, 313–322.

Chabot B.F. & Chabot J.F. (1977) Effects of light and temperature on leaf anatomy and photosynthesis in *Fragaria vesca*. *Oecologia* **26**, 363–377.

Chabot B.F. & Hicks D.J. (1982) The ecology of leaf life spans. *Annual Review of Ecology and Systematics* **13**, 229–259.

Chandler G.E. & Anderson J.W. (1976) Studies of the nutrition and growth of *Drosera* spp. with reference to the carnivorous habit. *New Phytologist* **76**, 129–141.

Chapin F.S. (1980) The mineral nutrition of wild plants. *Annual Review of Ecology and Systematics* **11**, 233–260.

Chapin F.S., Follett J.M. & O'Connor K.F. (1982) Growth, phosphate absorption and phosphorus chemical fractions in two *Chionochloa* species. *Journal of Ecology* **10**, 305–32.

Chapin F.S. & Shaver G.R. (1985) Individualistic growth response of tundra plant species to environmental manipulations in the field. *Ecology* **66**, 564–576.

Chapin F.S. & Slack M. (1979) Effect of defoliation upon root growth, phosphate absorption and inspiration in nutrient-limited tundra graminoids. *Oecologia* **42**, 67–80.

Chapman V.J. (1977) *Wet Coastal Ecosystems. Ecosystems of the World*, Volume I. Elsevier, Amsterdam.

Charles-Edwards P.A. (1984) On the ordered development of plants 2. Self-thinning in plant communities. *Annals of Botany* **53**, 709–714.

Charlesworth B. (1980) *Evolution in Age-Structured Populations.* Cambridge University Press, Cambridge.

Charnov E.L. (1982) *The Theory of Sex Allocation.* Princeton University Press.

Cheliak W.M. & Dancik B.P. (1982) Genic diversity of natural populations of a clone-forming tree *Populus tremuloides. Canadian Journal of Genetics and Cytology* **24**, 611–616.

Chesson P.L. & Warner K.R. (1981) Environmental variability promotes coexistence in lottery competitive systems. *American Naturalist* **117**, 923–943.

Chester K.S. (1933) The problem of acquired physiological immunity in plants. *Quarterly Review of Biology* **8**, 129–154, 275–324.

Chew F.S. & Rodman J.E. (1979) Plant resources for chemical defence. In *Herbivores, their Interaction with Secondary Plant Metabolites* (Eds G.A. Rosenthal & D.H. Jansen), pp. 271–307. Academic Press, London.

Chiariello N., Hickman J.C. & Mooney H.A. (1982) Endomycorrhizal role for interspecific transfer of phosphorus in a community of annual plants. *Science*, **217**, 941–943.

Christie E.K. & Moorby J. (1975) Physiological responses of arid grasses. I. The influence of phosphorus supply on growth and phosphorus absorption. *Australian Journal of Agricultural Research* **26**, 423–436.

Cittadino E. (1980) Ecology and the professionalization of Botany in America, 1890–1905. *Studies in the History of Biology* **4**, 171–198.

Clark D.A. & Clark D.B. (1984) Spacing dynamics of a tropical rain forest tree: evaluation of the Janzen-Connell model. *American Naturalist* **124**, 769–788.

Clark J.R. & Benforado J. (Eds) (1981) *Wetlands of Bottomland Hardwood Forests.* Elsevier, New York.

Clarke G.A.E. & Knox R.B. (1979) Plants and immunity. *Developmental Immunology* **3**, 571–589.

Clarkson D.T. & Hanson J.B. (1980) The mineral nutrition of higher plants. *Annual Review of Plant Physiology* **31**, 239–298.

Clarkson D.T. Sanderson J. & Scattergood C.B. (1978) Influence of phosphate stress on phosphate absorption and translocation by various parts of the root system of *Hordeum vulgare. Planta* **139**, 47–53.

Clausen J. & Hiesey W.M. (1958) Experimental studies on the nature of species. IV. Genetic structure of ecological races. *Carnegie Institution Washington Publications* **615**.

Clay K. (1982) Environmental and genetic determinants of cleistogamy in a natural population of the grass *Danthonia spicata. Evolution* **36**, 734–741.

Clegg M.T. (1980) Measuring plant mating systems. *Bioscience* **30**, 814–818.

Clegg M.T. & Brown A.D.H. (1983) The founding of plant populations. In *Genetics and Conservation* (Eds C.M. Schonewald-Cox, S.M. Chambers, B. MacBryde & L. Thomas), pp. 216–228. Benjamin/Cummings Publishing Company, London.

Clement C.R., Hopper M.J. & Jones L.H.P. (1978) The uptake of nitrate by *Lolium perenne* from flowing nutrient solution. I. Effect of NO_3^- concentration. *Journal of Experimental Botany* **29**, 453–464.

Clements F.E. (1916) Plant Succession. An analysis of the development of vegetation. *Carnegie Institute, Washington, Publication* **242**, Washington D.C.

Clements F.E. (1928) *Plant Succession and Indicators.* H.W. Wilson, New York.

Clowes F.A.L. (1961) *Apical Meristems.* p. 249. Blackwell Scientific Publications, Oxford.

Clymo R.S. (1962) An experimental approach to part of the calcicole problem. *Journal of Ecology* **50**, 701–731.

Clymo R.S. (1973) The growth of *Sphagnum:* some effects of environment. *Journal of Ecology* **61**, 849–869.

Cockburn W. (1985) Variation in photosynthetic acid metabolism in vascular plants: CAM and related phenomena. *New Phytologist* **101**, 3–24.

Cody M.L. (1966) A general theory of clutch size. *Evolution* **20**, 174–184.

Cody M.L. (1984) Branching patterns in columnar cacti. In *Being Alive on Land* (Eds N.S. Margaris, M. Arianoustou-Faraggitaki & W.C. Oechel), pp. 201–233. Junk, The Hague.

Cody M.L. & Mooney H.A. (1978) Convergence versus non-convergence in Mediterranean-climate ecosystems. *Annual Review of Ecology and Systematics* **9**, 265–321.

Cohen D. (1966) Optimizing reproduction in a randomly varying environment. *Journal of Theoretical Biology* **12**, 119–129.

Cohen D. (1971) Maximizing final yield when growth is limited by time or by limiting resources. *Journal of Theoretical Biology* **33**, 299–307.

Cohen J.E. (1978) *Food Webs and Niche Space.* Princeton University Press, New Jersey.

Coleman D.C., Coe C.V. & Elliot E.T. (1984) Decomposition, organic matter turn-over and nutrient dynamics in agroecosystems. In *Agricultural Ecosystems* (Eds R. Lowrance, B.R. Stinner & G.T. House), pp. 83–104. John Wiley, New York.

Coles J.F. & Fowler D.P. (1976) Inbreeding in neighbouring trees in two white spruce populations. *Silvae Genetica* **25**, 29–34.

Coles J.S. & Campbell A.I. (1965) Compatibility of "Bristol Cross" pear with some quince rootstocks. *Report of the Long Ashton Station 1964*, 113–116.

Coley P.D. (1983) Herbivory and defensive characteristics of tree species in a lowland tropical forest. *Ecological Monographs* **53**, 209–233.

Coley P.D. Bryant J.P. & Chapin F.S. (1985) Resource availability and plant antiherbivore defence. *Science* **230**, 895–899.

Comstock R.E., Kelleher T. & Morrow E.G. (1958) Genetic variation in an asexual species of garden strawberry. *Genetics* **43**, 634–646.

Connell J.H. (1971) On the role of natural enemies in preventing competitive exclusion in some marine animals and in rain forest trees. *Proceedings of the Advanced Study Institute on Dynamics of Numbers in Populations (Oosterbeek, 1970)*, pp. 298–312. Pudoc, Wageningen.

Connell J.H. (1979) Tropical rain forests and coral reefs as open non-equilibrium systems. In *Population Dynamics: 20th Symposium of the British Ecological Society* (Eds R.M. Anderson, B.D. Turner & L.R. Taylor), pp. 141–163. Blackwell Scientific Publications, Oxford.

Connell J.H. & Slatyer R.O. (1977) Mechanisms of succession in natural communities and their role in community stability and organization. *American Naturalist* **111**, 1119–1144.

Connell J.H., Tracey J.G. & Webb L.J. (1984) Compensatory recruitment, growth, and mortality as factors maintaining rain forest tree diversity. *Ecological Monographs* **54**, 141–164.

Connor E.F. & McCoy E.D. (1979) The statistics and biology of the species-area relationship. *American Naturalist* **113**, 791–833.

Constable G.A. & Rawson H.M. (1980) Carbon production and utilization in cotton: inferences from a carbon budget. *Australian Journal of Plant Physiology* **7**, 539–553.

Cook R.E. (1979) Patterns of juvenile mortality and recruitment in plants. In *Topics in Plant Population Biology* (Eds O.T. Solbrig, S. Jain, G.B. Johnson & P.H. Raven), pp. 207–231. Macmillan, London.

Cook R. (1980) The Biology of Seeds in the Soil. In *Demography and Evolution in Plant Populations* (Ed. O. Solbrig), pp. 107–130. University of California, Berkeley.

Cook R.E. (1983) Clonal plant populations. *American Scientist* **71**, 244–253.

Cook S.A. & Johnson M.P. (1968) Adaptation to heterogenous environments. I. Variation in heterophylly in *Ranunculus flammula* L. *Evolution* **22**, 496–516.

Cooper W.S. (1923) The recent ecological history of Glacier Bay, Alaska. *Ecology* **4**, 93–128; 223–246; 355–365.

Cooper W.S. (1926) The fundamentals of vegetation change. *Ecology* **7**, 391–413.

Cooper W.S. (1939) A fourth expedition to Glacier Bay, Alaska. *Ecology* **20**, 130–155.

Cornell H.V. (1983) Why and how gall wasps form galls: Cynipids as genetic engineers? *Antenna* **7**, 53–58.

Corré W.J. (1983) Growth and morphogenesis of sun and shade plants. I. The influence of light intensity. *Acta botanica Neerlandica* **32**, 49–62.

Cottam G. & Curtis J.T. (1949) A method for making rapid surveys of woodlands, by means of pairs of randomly selected trees. *Ecology* **30**, 101–104.

Cowles H.C. (1899) The ecological relations of the vegetation on the sand dunes of Lake Michigan. *Botanical Gazette* **27**, 95–117; 167–202; 281–308; 361–391.

Crawford T.J. (1984a) The estimation of neighborhood parameters for plant populations. *Heredity* **52**, 273–283.

Crawford T.J. (1984b) What is a population? In *Evolutionary Ecology* (Ed. B. Shorrocks), pp. 135–173. Blackwell Scientific Publications, Oxford.

Crawley M.J. (1983) *Herbivory: The Dynamics of Animal-Plant Interactions*. Blackwell Scientific Publications, Oxford.

Crawley M.J. (1985) Reduction of oak fecundity by low density herbivore populations. *Nature* **314**, 163–164.

Crawley M.J. (1986) What makes a community invasible? In *Colonization, Succession and Stability* (Eds A.J. Gray, M.J. Crawley, & P.J. Edwards). Blackwell Scientific Publications, Oxford.

Crawley M.J. & Nachapong M. (1985) The establishment of seedlings from primary and regrowth seeds of ragwort (*Senecio jacobaea*). *Journal of Ecology* **73**, 255–261.

Crisp M.D. & Lange R.T. (1976) Age structure, distribution and survival under grazing of the arid-zone shrub *Acacia burkittii*. *Oikos* **27**, 86–92.

Crocker R.L. & Major J. (1955) Soil development in relation to vegetation and surface age at Glacier Bay, Alaska. *Journal of Ecology* **43**, 427–448.

Cruden R.W. (1976) Intraspecific variation in pollen-ovule ratios and nectar secretion—preliminary evidence of ecotypic adaptation. *Annals of the Missouri Botanical Garden* **63**, 277–289.

Cruden R.W. (1977) Pollen-ovule ratios: a conservative indicator of the breeding systems in the flowering plants. *Evolution* **31**, 32–46.

Cruden R.W. & Hermann-Parker S.M. (1979) Butterfly pollination of *Caesalpinia pulcherrima*, with observations on a psychophilous syndrome. *Journal of Ecology* **67**, 155–168.

Cullen J.M. & Groves R.H. (1977) The population biology of *Chondrilla juncea* L. in Australia. *Proceedings of the Ecological Society of Australia* **10**, 121–134.

Curtis J.T. (1959) *The vegetation of Wisconsin. An Ordination of Plant Communities*. University of Wisconsin Press, Madison.

Curtis J.T. & McIntosh R.P. (1951) An upland forest continuum in the prairie–forest border region of Wisconsin. *Ecology* **32**, 476–496.

Darwin C. (1859) *On the Origin of Species by Means of Natural Selection*. John Murray, London.

Datta A.N., Maiti S.N. & Basak S.L. (1982) Outcrossing and isolation requirement in jute (*Corchorus olitorius* L.) *Euphytica* **31**, 97–101.

Davidson D.W. (1985) An experimental study of diffuse competition in harvester ants. *American Naturalist* **125**, 500–506.

Davidson D.W. & Morton S.R. (1981a) Competition for dispersal in ant-dispersed plants. *Science* **213**, 1259–1261.

Davidson D.W. & Morton S.R. (1981b) Myrmecochory in some plants (F. *Chenopodiaceae*) of the Australian Arid Zone. *Oecologia* **50**, 357–366.

Davidson J.L. & Milthorpe F.L. (1966) The effect of defoliation on the carbon balance in *Dactylis glomerata. Annals of Botany* **30**, 186–198.

Davidson D.W., Samson D.A. & Inouye R.S. (1985) Granivory in the Chihuahuan desert: interactions within and between trophic levels. *Ecology* **66**, 486–502.

Davies W.J., Rodriguez J.L. & Fiscus E.L. (1982) Stomatal behavior and water movement through roots of wheat plants treated with abscissic acid. *Plant, Cell and Environment* **5**, 485–493.

Davis L.D. (1957) Flowering and alternate bearing. *Proceedings of the American Society of Horticultural Science* **70**, 545–556.

Davis M.B. (1981) Quaternary history and the stability of plant communities. In *Forest Succession: Concepts and Application* (Eds D.C. West, H.H. Shugart & D.B. Botkin), pp. 132–153. Springer-Verlag, Berlin.

Davy A.J. & Jefferies R.L. (1981) Approaches to the monitoring of rare plant populations. In *The Biological Aspects of Rare Plant Conservation* (Ed. H. Synge), pp. 219–232. John Wiley & Sons, Chichester.

Dawkins R. (1976) *The Selfish Gene*. Oxford University Press, Oxford.

Dawkins R. & Krebs J.R. (1979) Arms races within and between species. *Proceedings of the Royal Society of London Series B* **205**, 489–511.

Day P.R. (1973) Genetic variability of crops. *Annual Review Phytopathology* **11**, 293–312.

De Angelis D.L., Travis C.C. & Post W.M. (1979) Persistence and stability of seed-dispersed species in a patchy environment. *Theoretical Population Biology* **16**, 107–125.

Del Moral R. & Muller C.H. (1969) Fog drip: a mechanism of toxin transport from *Eucalyptus globosus. Bulletin of the Torrey Botanical Club* **96**, 467–475.

Dement W.A. & Mooney H.A. (1974) Seasonal variation in the production of tannins and cyanogenic glycosides in the chaparral shrub, *Heteromeles abutifolia. Oecologia* **15**, 65–76.

Dempster J.P. (1971) The population ecology of the cinnabar moth, *Tyria jacobaeae* L. (Lepidoptera, Arctiidae). *Oecologia* **7**, 26–67.

Dempster J.P. (1975) *Animal Population Ecology*. Academic Press, London.

Dempster J.P. & Lakhani K.H. (1979) A population model for cinnabar moth and its food plant, ragwort. *Journal of Animal Ecology* **48**, 143–164.

Denslow J.S. (1980a) Gap partitioning among tropical rain-forest trees. *Biotropica* **12** (Suppl.), 47–55.

Denslow J.S. (1980b) Patterns of plant species diversity during succession under different disturbance regimes. *Oecologia* **46**, 18–21.

Derman H. (1960) Nature of plant sports. *American Horticultural Magazine* **39**, 123–173.

De Steven D. (1981) Abundance and survival of a seed-infesting weevil, *Pseudanthonomus hamamelidis* (Coleoptera: Curculionidae), on its variable-fruiting host plant, witch-hazel (*Hamamelis virginiana*). *Ecological Entomology* **6**, 387–396.

De Steven D. (1982) Seed production and seed mortality in a temperate witch-hazel forest shrub (*Hamamelis virginiana*). *Journal of Ecology* **70**, 437–443.

Diggle P.J. (1983) *Statistical Analysis of Spatial Point Patterns*. Academic Press, London.

Dingle H. & Hegmann J.P. (Eds) (1982) *Evolution and Genetics of Life Histories*. Springer-Verlag, Berlin.

Dinoor A. & Eshed N. (1984) The role and importance of pathogens in natural plant communities. *Annual Review of Phytopathology* **22**, 443–466.

Dirr M.A., Barker A.N. & Maynard D.M. (1973) Extraction of nitrate reductase from leaves of *Ericaceae. Phytochemistry* **12**, 1261–1264.

Dirzo R. (1984) Herbivory: a phytocentric overview. In *Perspectives on Plant Popu-*

lation Ecology (Eds R. Dirzo & J. Sarukhán), pp. 141–165. Sinauer Associates Inc., Sunderland, Massachusetts.

Dirzo R. & Harper J.L. (1982a) Experimental studies on slug-plant interactions. III. Differences in the acceptability of individual plants of *Trifolium repens* to slugs and snails. *Journal of Ecology* **70**, 101–117.

Dirzo R. & Harper J.L. (1982b) Experimental studies on slug-plant interactions. IV. The performance of cyanogenic and acyanogenic morphs of *Trifolium repens* in the field. *Journal of Ecology* **70**, 119–138.

Dix R. & Smeins F. (1967) The prairie, meadow and marsh vegetation of Nelson County, North Dakota. *Canadian Journal of Botany* **45**, 21–58.

Dobkin D.S. (1984) Flowering patterns of long-lived *Heliconia* inflorescences: implications for visiting and resident nectarivores. *Oecologia* **64**, 245–254.

Docters van Leeuwen W.M. (1929) Krakatau's new flora. *Proceedings of the Fourth Pacific Science Congress (Batavia) Part 2*, pp. 56–71g.

Dodd A.P. (1940) *The Biological Campaign against Prickly Pear*. Commonwealth Prickly Pear Board, pp. 1–177. Government Printer, Brisbane.

Doddema H., Telkamp G.P. & Otten H. (1979) Uptake of nitrate by mutants of *Arabidopsis thaliana*, disturbed in uptake or reduction of nitrate. *Plant Physiology* **45**, 297–346.

Dohmen G.P., McNeill S. & Bell J.N.B. (1984) Air pollution increases *Aphis fabae* pest potential. *Nature* **307**, 52–53.

Domée B. (1981) Roles du milieu et du genotype dans le regime de la reproducion de *Thymus vulgaris*. *Acta Oecologia* **2**, 137–147.

Downs A.A. & McQuilkin W.E. (1944) Seed production of southern Appalachian oaks. *Journal of Forestry* **42**, 913–920.

Dressler R.L. (1981) *The Orchids*. Harvard University Press, Cambridge, Massachusetts.

Drury W.H. & Nisbet I.C. (1973) Succession. *Journal of the Arnold Aboretum* **54**, 331–368.

Dugdale R. (1972) Chemical oceanography and primary productivity in upwelling regions. *Geoforum* **11**, 47–61.

Duggan A.E. (1985) Pre-dispersal seed predation by *Anthocaris cardamines* (Pieridae) in the population dynamics of the perennial *Cardamine pratensis* (Brassicaceae). *Oikos* **44**, 99–106.

Duncan W.G. & Hesketh J.D. (1968) Net photosynthetic rates, relative growth rates and leaf numbers of 22 races of maize grown at eight temperatures. *Crop Science* **8**, 670–674.

Dyson W.G. & Herbin G.A. (1968) Studies on plant cuticular waxes. IV. Leaf alkanes as a taxonomic discriminant for cypresses grown in Kenya. *Phytochemistry* **9**, 585–589.

East E.M. (1908) Suggestions concerning certain bud variation. *The Plant World* **11**, 77–83.

East E.M. (1927) Peculiar genetic results due to genetic gametophytic factors. *Hereditas* **9**, 49–58.

Edmunds G.F. & Alstad D.N. (1978) Coevolution in insect herbivores and conifers. *Science* **199**, 941–945.

Edmunds G.F. & Alstad D.N. (1982a) Effects of air pollutants on insect populations. *Annual Review of Entomology* **27**, 369–384.

Edmunds G.F. & Alstad D.N. (1982b) Responses of black pineleaf scales to host plant variability. In *Insect Life History Patterns and Habitat and Geographic Variation* (Eds R. Denno & H. Dingle), pp. 29–38. Springer-Verlag, New York.

Edwards G. & Walker D. (1983) *C3, C4: Mechanisms, and Cellular and Environmental Regulation, of Photosynthesis*. Blackwell Scientific Publications, Oxford.

Egbert T.A. & McMaster G.S. (1981) Regular pattern of desert shrubs: a sampling artefact? *Journal of Ecology* **69**, 559–564.

Egler F.E. (1947) Arid southeast Oahu vegetation, Hawaii. *Ecological Monographs* **17**, 383–435.

Egler F.E. (1954) Vegetation science concepts: I. Initial floristic composition—a factor in old-field vegetation development. *Vegetatio* **4**, 412–417.

Egler F.E. (1976) *Nature of vegetation. Its management and mismanagement.* Connecticut Conservation Association, Bridgewater, Connecticut.

Ehleringer J.R. & Mooney H.A. (1978) Leaf hairs: effects on physiological activity and adaptive value to a desert shrub. *Oecologia* **37**, 183–200.

Ehrendorfer F. (1970) Evolutionary patterns and strategies in seed plants. *Taxon* **19**, 185–195.

Ehrlich P.R. & Raven P.H. (1964) Butterflies and plants: a study in coevolution. *Evolution* **18**, 586–608.

Eldridge K.G. & Griffin A.R. (1983) Selfing effects in *Eucalyptus regnans*. *Silvae Genetica* **32**, 216–221.

Ellenberg H. (1953) Physiologisches und ökologisches Verhalten derselben Pflanzenarten. *Bericht der Deutschen Botanischen Gesellschaft* **65**, 351–361.

Ellingboe A.H. (1961) Somatic recombination in *Puccinia graminis* var *tritici*. *Phytopathology* **51**, 13–15.

Ellingboe A.H. (1968) Inoculum production and infection by foliage pathogens. *Annual Review Phytopathology* **6**, 317–330.

Ellingboe A.H. (1972) Genetics and physiology of primary infection by *Erysiphe graminis*. *Phytopathology* **62**, 401–406.

Ellingboe A.H. (1978) A genetic analysis of host-parasite interactions. In *The Powdery Mildews* (Ed. D.M. Spencer), pp. 159–181. Academic Press, London.

Ellingboe A.H. & Raper J.R. (1962) Somatic mutation in *Schizophyllum commune*. *Genetics* **17**, 85–98.

Elliot J.M. (1977) *Some methods for the statistical analysis of samples of benthic invertebrates*, 2nd edn. Freshwater Biological Association Scientific Publication Number 25.

Ellis W.M., Keymer R.J. & Jones D.A. (1977) On the polymorphism of cyanogenesis in *Lotus corniculatus* L. VIII Ecological Studies in Anglesey. *Heredity* **39**, 45–65.

Ellner S. & Shmida A. (1981) Why are adaptations for long-range seed dispersal rare in desert plants? *Oecologia* **51**, 133–144.

Emmingham W.H. (1982) Ecological indexes as a means of evaluating climate, species distribution and primary production. In *Analysis of Coniferous Forest Ecosystems in the Western United States* (Ed. R.L. Edmonds). Hutchinson, Ross Publishing Company, Stroudsburg, Pennsylvania.

Ennik G.C. (1966) Influence of clipping and soil fumigation on shoot and root production of perennial ryegrass and white clover. *Jaarboek der Institut Biol. schiek. Onderz. Landb Gewass 1966*, 11–18.

Epling C. & Dobzhansky Th. (1942) Microgeographic races in *Linanthus parryae*. *Genetics* **27**, 317–332.

Epling C., Lewis H. & Bell F.M. (1960) The breeding group and seed storage: a study in population dynamics. *Evolution* **14**, 238–255.

Esau K. (1977) *Anatomy of Seed Plants*, 2nd edn. John Wiley & Sons, New York.

Etherington J.R. (1982) *Environment and Plant Ecology*, 2nd edn. John Wiley, Chichester.

Evans D.A. & Sharp W.R. (1983) Single gene mutations in tomato plants regenerated from issue culture. *Science* **221**, 949–951.

Evans D.A., Sharp W.R. & Medina-Filho H.P. (1984) Somaclonal and gametoclonal variation. *American Journal of Botany* **71**, 759–774.

Evans G.C. (1972) *The Quantitative Analysis of Plant Growth*. Blackwell Scientific Publications, Oxford.

Evans G.C. & Hughes A.P. (1961) Plant growth and the aerial environment. I. Effect of artificial shading on *Impatiens parviflora*. *New Phytologist* **60**, 150–180.

Evans J.R. (1983) Photosynthesis and nitrogen partitioning in leaves of *T. aestivium* and related species. PhD thesis, Australian National University, Canberra.

Evans L.T. & Dunstone R.L. (1970) Some physiological aspects of evolution in wheat. *Australian Journal of Biological Science* **23**, 725–741.

Evans P.S. (1971) Root growth of *Lolium perenne* L. II. Effects of defoliation and shading. *New Zealand Journal of Agricultural Research* **14**, 552–562.

Faegri K. & Pijl L. van der (1979) *The Principles of Pollination Ecology*, 3rd edn. Pergamon Press, New York.

Farquhar G.D., Ball M.C., Caemmerer S. von & Roksandic Z. (1982) Effect of salinity and humidity on $\sigma^{13}C$ value of halophytes—evidence for diffusional isotope fractionation determined by the ratio of intercellular/atmospheric partial pressure of CO_2 under different environmental conditions. *Oecologia* **52**, 121–124.

Farquhar G.D. & Sharkey T.D. (1982) Stomatal conductance and photosynthesis. *Annual Review of Plant Physiology* **33**, 317–345.

Feeny P. (1975) Biochemical coevolution between plants and their insect herbivores. In *Coevolution of Plants and Animals* (Eds L.E. Gilbert & P.H. Raven), pp. 3–19. University of Texas Press, Austin.

Feeny P. (1976) Plant apparency and chemical defence. In *Recent Advances in Phytochemistry* **10**, *Biochemical Interactions between Plants and Insects* (Eds J.W. Wallace & R.L. Mansell), pp. 1–40. Plenum Press, New York.

Feinsinger P. (1983) Coevolution and pollination. In *Coevolution* (Eds D.J. Futuyma & M. Slatkin), pp. 282–310. Sinauer, Sunderland, Massachusetts.

Fenner M. (1978) Susceptibility to shade in seedlings of colonising and closed turf species. *New Phytologist* **81**, 739–744.

Field C. & Mooney H.A. (1986) The photosynthesis-nitrogen relationship in wild plants. In *On the Economy of Plant Form and Function* (Ed. T.J. Givnish). Cambridge University Press, Cambridge.

Finegan B. (1984) Forest Succession. *Nature* **312**, 109–114.

Firbank L.G. & Watkinson A.R. (1985) On the analysis of competition within two-species mixtures of plants. *Journal of Applied Ecology* **22**, 503–517.

Firbank L.G., Manlove R.J., Mortimer A.M. & Putwain P.D. (1984) The management of grass weeds in cereal crops, a population biology approach. *Proceedings of the 7th International Symposium on Weed Biology, Ecology and Systematics*, 375–384.

Fischer A. (1960) Latitudinal variations in organic diversity. *Evolution* **14**, 64–81.

Fisher J.B. & Hibbs D.E. (1982) Plasticity of tree architecture: specific and ecological variations found in Aubréville model. *American Journal of Botany* **69**, 690–702.

Fitter A.H. (1985) The functional significance of root morphology and root system architecture. In *Ecological Interactions in Soil* (Eds A.H. Fitter, D. Atkinson, D.J. Read & M.B. Usher) British Ecological Society Special Publication **4**, pp. 87–106. Blackwell Scientific Publications, Oxford.

Fitter A.H. & Ashmore C.J. (1974) Response of two *Veronica* species to a simulated woodland light climate. *New Phytologist* **73**, 997–1001.

Fitter A.H. & Hay R.K.M. (1981) *Environmental Physiology of Plants*. Academic Press, London.

Flor H.H. (1971) Current status of the gene-for-gene concept. *Annual Review Phytopathology* **9**, 275–296.

Flower-Ellis J.G.K. & Persson H. (1980) Investigation of structural properties and dynamics of Scots pine stands. *Ecological Bulletin (Stockholm)* **32**, 125–138.

Fogel R. (1980) Mycorrhizae and nutrient cycling in natural forest ecosystems. *New Phytologist* **86**, 199–212.

Fogle H.W. & Faust M. (1975) Ultrastructure of nectarine fruit surfaces. *Journal of American Society of Horticultural Science* **100**, 74–77.

Fonteyn P.J. & Mahall B.E. (1978) Competition among desert perennials. *Nature* **275**, 544–545.

Fonteyn P.J. & Mahall B.E. (1981) An experimental analysis of structure in a desert plant community. *Journal of Ecology* **69**, 883–896.

Foote L.E. & Jackobs J.A. (1966) Soil factors and the occurrence of partridge pea (*Cassia fasciculata* Michx.) in Illinois. *Ecology* **47**, 968–974.

Ford E.D. (1975) Competition and stand structure in some even-aged plant mono-cultures. *Journal of Ecology* **63**, 311–333.

Ford E.D. & Newbould P.J. (1970) Stand structure and dry weight production through the sweet chestnut (*Castanea sativa* Mill.) coppice cycle. *Journal of Ecology* **58**, 275–296.

Foster R.B. (1982) The seasonal rhythm of fruit-fall on Barro Colorado Island. In: *The Ecology of a Tropical Forest* (Eds E.G. Leigh, A.S. Rand & D.M. Windsor), pp. 151–172. Smithsonian Press, Washington.

Foster R.B. & Brokaw N.V.L. (1982) Structure and history of the vegetation of Barro Colorado Island. In *The Ecology of a Tropical Forest: Seasonal Rhythms and Long-term Changes* (Eds E.G. Leigh Jr., A.S. Rand & D.M. Windsor), pp. 67–81. Smithsonian Institution Press, Washington D.C.

Fowler N. (1981) Competition and coexistence in a North Carolina grassland. II. The effects of the experimental removal of species. *Journal of Ecology* **69**, 843–854.

Fowler N.L. (1984) The role of germination date, spatial arrangement and neighbourhood effects in competitive interactions in *Linum*. *Journal of Ecology* **72**, 307–318.

Fowler S.V. & Lawton J.H. (1985) Rapidly induced defences and talking trees: the devils advocate position. *American Naturalist* **126**, 181–195.

Fox L.R. & Macauley B.J. (1977) Insect grazing on *Eucalyptus* in response to variation in leaf tanning and nitrogen. *Oecologia* **29**, 145–162.

Foy C.L., Hurtt W. & Hale M.G. (1971) Root exudation and plant growth regulators. In *Biochemical Interactions Among Plants* (Eds U.S. National Committee for IBP), pp. 75–85. National Academy of Sciences, Washington D.C.

Fraenkel G.S. (1959) The *raison d'être* of secondary plant substances. *Science* **129**, 1466–1470.

Frankie G.W. (1976) Pollination of widely dispersed trees by animals in Central America, with an emphasis on bee pollination systems. In *Tropical Trees: Variation, Breeding and Conservation* (Eds J. Burley & B.T. Styles), pp. 151–159. Academic Press, New York.

Frankie G.W., Haber W.A., Opler P.A. & Bawa K.S. (1983) Characteristics and organization of the large bee pollination system in the Costa Rican dry forest. In *Handbook of Experimental Pollination Biology* (Eds C.E. Jones & R.J. Little), pp. 411–447. Van Nostrand Reinhold, New York.

Free J.B. (1970) *Insect Pollination of Crops*. Academic Press, New York.

Freeman D.C., Klikoff L.G. & Harper K.T. (1976) Differential resource utilisation by the sexes of dioecious plants. *Science* **193**, 597–599.

Frey K.J. & Browning J.A. (1971) Breeding crop plants for disease resistance. In *Mutation Breeding for Disease Resistance*, pp. 45–54. International Atomic Energy Agency, Vienna.

Frisch K. von (1967) *The Dance Language and Orientation of Bees*. Belknap Press of Harvard University Press, Cambridge, Massachusetts.

Fry B., Joern A. & Parker P.L. (1978) Grasshopper food web analysis: use of carbon isotope ratios to examine feeding relationships among terrestrial herbivores. *Ecology* **59**, 498–506.

Fryxell P.A. (1957) Modes of reproduction in higher plants. *Botanical Review* **23**, 135–233.

Fuller W., Hance C.E. & Hutchings M.J. (1983) Within-season fluctuations in mean fruit weight in *Leondonton hispidus* L. *Annals of Botany* **51**, 545–549.

Gadgil M.D. & Solbrig O.T. (1972) The concept of *r*- and *K*-selection: evidence from wild flowers and some theoretical considerations. *American Naturalist* **106**, 14–31.

Galston A.W. & Davies P.J. (1970) Control mechanisms in plant development. Prentice Hall, Englewood Cliffs, New Jersey.

Ganders F.R. (1979) The biology of heterostyly. *New Zealand Journal of Botany* **17**, 607–636.

Garwood E.A. (1967) Seasonal variation in appearance and growth of grass roots. *Journal of the British Grassland Society* **22**, 121–130.

Garwood N.C. (1982) Seasonal rhythm of seed germination in a semideciduous tropical forest. In *The Ecology of a Tropical Forest: Seasonal Rhythms and*

Long-term Changes (Eds E.G. Leigh, A.S. Rand & D.M. Windsor), pp. 173–188. Smithsonian Press, Washington.

Garwood N.C. (1983) Seed germination in a seasonal tropical forest in Panama: a community study. *Ecological Monographs* **53**, 159–181.

Garwood N.C., Janos D.P. & Brokaw N.V.L. (1979) Earthquake-caused landslides: a major disturbance to tropical forests. *Science* **205**, 997–999.

Gatsuk E., Smirnova O.V., Vorontzova L.I., Zaugolnova L.B. & Zhukova L.A. (1980) Age-states of plants of various growth forms: a review. *Journal of Ecology* **68**, 675–696.

Gauch H.G. (1982) *Multivariate Analysis in Community Ecology.* Cambridge University Press, Cambridge.

Gautier-Hion A., Duplantier J.-M., Quris R., Feer F., Sourd C., Decoux J.-P., Dubost G., Emmons L., Erard C., Hecketsweiler P., Moungazi A., Roussilhon C. & Thiollay J.-M. (1985) Fruit characters as a basis of fruit choice and seed dispersal in a tropical forest vertebrate community. *Oecologia* **65**, 324–337.

Geiger D.R. (1976) Effects of translocation and assimilate demand on photosynthesis. *Canadian Journal of Botany* **54**, 2337–2345.

Geiger R. (1950) *The Climate Near the Ground.* Harvard University Press, Cambridge, Massachusetts.

Gentry A.H. (1974) Flowering phenology and diversity in tropical Bignoniaceae. *Biotropica* **6**, 64–68.

Gerhold H.D. (1966) In quest of insect-resistant forest trees. In *Breeding Pest-Resistant Trees* (Eds H.D. Gerhold, E.J. Schreiner, R.E. McDermott & J.A. Winieski), pp. 305–318. Pergamon Press, Oxford.

Geritz S.A.H., de Jong T.J. & Klinkhamer P.G.L. (1984) The efficacy of dispersal in relation to safe-site area and seed production. *Oecologia* **62**, 219–221.

Gifford R.M. & Evans L.T. (1981) Photosynthesis, carbon partitioning and yield. *Annual Review of Plant Physiology* **32**, 485–509.

Gifford R.M. & Jenkins C.L. (1981) Prospects of applying knowledge of photosynthesis toward improving crop production. In *Photosynthesis: CO₂ Assimilation and Plant Productivity* (Ed. Gorindjee), Vol. 2. Academic Press, New York.

Gilbert L.E. (1971) Butterfly: plant coevolution: has *Passiflora adenopoda* won the selectional race with Heliconid butterflies? *Science* **172**, 585–586.

Gill D.E. (1983) Within tree variation in fruit production and seed set in *Capparis odoratissima* Jacq. in Costa Rica. *Brenesia* **21**, 33–40.

Gill D.E. & Halverson T.G. (1984) Fitness variation among branches within trees. In *Evolutionary Ecology* (Ed. B. Shorrocks), pp. 105–116. Blackwell Scientific Publications, Oxford.

Ginzo H.D., Collantes M. & Caso O.H. (1982) Fertilization of a native grassland in the 'Depresion del Rio Salado', Province of Buenos Aires: herbage dry matter accumulation and botanical composition. *Journal of Range Management* **35**, 35–42.

Ginzo H.D. & Lovell P.H. (1974) Aspects of the comparative physiology of *Ranunculus bulbosus* L. and *Ranunculus repens* L. II. Carbon dioxide assimilation and distribution of photosynthates. *Annals of Botany* **37**, 765–776.

Givnish T.J. (1978) On the adaptive significance of compound leaves, with particular reference to tropical trees. In *Tropical Trees as Living Systems* (Eds P.B. Tomlinson & H. Zimmermann). Cambridge University Press, Cambridge.

Givnish T.J. (1979) On the adaptive significance of leaf form. In *Topics in Plant Population Biology* (Eds O.T. Solbrig, S. Jain, G.B. Johnson & P.H. Raven), pp. 375–407. Macmillan, London.

Givnish T.J. (1982) On the adaptive significance of leaf height in forest herbs. *American Naturalist* **120**, 353–381.

Givnish T.J. (1985) Mechanical constraints on growth form in forest herbs. In *On the Economy of Plant Form and Function* (Ed. T.J. Givnish). Cambridge University Press, Cambridge.

Givnish T.J. (1986) Biochemical constraints on self-thinning in plant populations. *Journal of Theoretical Biology* **119**, 139–146.

Givnish T.J., Terborgh J.W. & Waller D.M. (1986) Plant form, temporal community structure, and species richness in forest herbs of the Virginia Piedmont. In review, *Ecological Monographs*.

Glass A.D.M. (1978) Regulation of potassium influx into intact roots of barley by internal potassium levels. *Canadian Journal of Botany* **56**, 1759–1764.

Gleason H.A. (1917) The structure and development of the plant association. *Bulletin of the Torrey Botanical Club* **44**, 463–481.

Gleason H.A. (1926) The individualistic concept of plant association. *Bulletin of the Torrey Botanical Club* **53**, 7–26.

Gleason H.A. (1927) Further views on the succession concept. *Ecology* **8**, 299–326.

Godwin H. (1975) *The History of the British Flora. A Factual for Phytogeography*, 2nd edn. Cambridge University Press, Cambridge.

Goldberg D.E. (1982) The distribution of evergreen and deciduous trees, relative to soil type: an example from the Sierra Madre, Mexico, and a general model. *Ecology* **63**, 942–951.

Goldberg D.E. & Werner P.A. (1983) Equivalence of competitors in plant communities: a null hypothesis and a field experimental approach. *American Journal of Botany* **70**, 1098–1104.

Goldblatt P. (1978) An analysis of the flora of Southern Africa: its characteristics, relationships and origins. *Annals of the Missouri Botanical Garden* **65**, 369–436.

Goodall D.W. (1952) Quantitative aspects of plant distribution. *Biological Reviews* **27**, 194–245.

Goodall D.W. (1965) Plotess tests of interspecific association. *Journal of Ecology* **53**, 197–210.

Goodrum P.D., Reid V.H. & Boyd C.E. (1971) Acorn yields, characteristics, and management criteria of oaks for wildlife. *Journal of Wildlife Management* **35**, 520–532.

Gordan A.G. & Gorham E. (1963) Ecological aspects of air pollution from an iron sintering plant at Wawa, Ontario. *Canadian Journal of Botany* **41**, 1063–1078.

Gottlieb L.D. (1977) Genotypic similarity of large and small individuals in a natural population of the annual plant *Stephanomeria exigua* ssp. *coronaria* (Compositae). *Journal of Ecology* **65**, 127–134.

Gottlieb L.D. (1981) Electrophoretic evidence and plant populations. *Progress in Phytochemistry* **7**, 1–46.

Gottlieb L.D. (1984a) Genetics and morphological evolution in plants. *American Naturalist* **123**, 681–709.

Gottlieb L.D. (1984b) Electrophoretic analysis of the phylogeny of the self-pollinating populations of *Clarkia xanthiana*. *Plant Systematics & Evolution* **147**, 91–102.

Gould S.J. (1977) *Ontogeny and Phylogeny*. Belknap Press of Harvard University, Cambridge, Massachusetts.

Gould S.J. & Lewontin R. (1979) The spandrels of San Marco and the Panglossian paradigm: a critique of the adaptionist programme. In *The Evolution of Adaptation by Natural Selection* (Eds J. Maynard Smith & R. Holliday). The Royal Society, London.

Gourly J.H. & Howlett F.S. (1941) *Modern Fruit Production*. Macmillan Company, New York.

Grace J. (1977) *Plant Response to Wind*. Academic Press, London.

Grace J., Ford E.D. & Gravis P.G. (Eds) (1981) *Plants and their Atmospheric Environment*. Blackwell Scientific Publications, Oxford.

Grace J.B. & Wetzel R.G. (1981) Habitat partitioning and competitive displacement in cattails (*Typha*): experimental field studies. *American Naturalist* **118**, 463–474.

Graham B.F. & Bormann F.H. (1966) Natural root grafts. *Botanical Review* **32**, 255–292.

Graham C.F. & Wareing P.F. (1984) *Developmental Control in Animals and Plants*, 2nd edn. Blackwell Scientific Publications, Oxford.

Grant M.C. & Mitton J.B. (1979) Elevational gradients in adult sex ratios and sexual differentiation in vegetative growth rates of *Populus tremuloides* Michx. *Evolution* **33**, 914–918.

Grant V. (1963) *The Origin of Adaptations.* Columbia University Press, New York.

Grant V. (1971) *Plant Speciation.* Columbia University Press, New York.

Grant V. & Grant K.A. (1965) *Flower Pollination in the Phlox Family.* Columbia University Press, New York.

Gray A.I., Crawley M.J. & Edwards P.J. (Eds) (1986) *Colonization, Succession and Stability.* Blackwell Scientific Publications, Oxford.

Gray J.T. (1983) Nutrient use by evergreen and deciduous shrubs in southern California. I. Community nutrient cycling and nutrient use efficiency. *Journal of Ecology* **71**, 21–41.

Green K.A., Zasada J.E. & Van Cleve K. (1971) An albino aspen sucker. *Forest Science* **17**, 272.

Green T.R. & Ryan C.A. (1972) Wound-induced proteinase inhibition in plant leaves: a possible defense mechanism against insects. *Science* **175**, 776–777.

Greenhill G. (1881) Determination of the greatest height consistent with stability that a vertical pole or mast can be made, and of the greatest height to which a tree of given proportions can grow. *Proceedings of the Cambridge Philosophical Society* **4**, 65–73.

Greenslade P.J.M. (1983) Adversity selection and the habitat templet. *American Naturalist* **122**, 325–365.

Greenway H. & Munns R. (1980) Mechanisms of salt tolerance in non-halophytes. *Annual Review of Plant Physiology* **31**, 149–190.

Greig-Smith P. (1983) *Quantitative Plant Ecology*, 3rd edn. Blackwell Scientific Publications, Oxford.

Griffen A.R. (1980) Floral phenology of a stand of mountain ash (*Eucalyptus regnans* F. Muell.) in Gippsland, Victoria. *Australian Journal of Botany* **28**, 393–404.

Grime J.P. (1973) Competitive exclusion in herbaceous vegetation. *Nature* **242**, 344–347.

Grime J.P. (1979) *Plant Strategies and Vegetation Processes.* John Wiley, Chichester.

Grime J.P. & Curtis A.V. (1976) The interaction of drought and mineral nutrient stress in calcareous grassland. *Journal of Ecology* **64**, 975–988.

Grime J.P. & Hodgson J.G. (1969) An investigation of the ecological significance of lime-chlorosis by means of large-scale comparative experiments. In *Ecological Aspects of the Mineral Nutrition of Plants* (Ed. I.H. Rorison), pp. 67–100. Blackwell Scientific Publications, Oxford.

Grime J.P. & Hunt R. (1975) Relative growth rate: its range and adaptive significance in a local flora. *Journal of Ecology* **63**, 393–422.

Grime J.P., Mason G., Curtis A.V., Rodman J., Band S.R., Mowforth M.A.G., Neal A.M. & Shaw S. (1981) A comparative study of germination characteristics in a local flora. *Journal of Ecology* **69**, 1017–1059.

Gross K.L. (1980) Colonization of *Verbascum thapsus* (Mullein) in an old field in Michigan: the effects of vegetation. *Journal of Ecology* **68**, 919–928.

Gross K.L. (1981) Predictions of fate from rosette size in four 'biennial' plant species: *Verbascum thapsus, Oenothera biennis, Daucus carota* and *Tragopogon dubius. Oecologia* **48**, 209–213.

Gross K.L. & Werner P.A. (1982) Colonising abilities of "biennial" plant species in relation to ground cover: implications for their distributions in a successional sere. *Ecology* **63**, 921–931.

Gross R.S. & Werner P.A. (1983) Relationships among flowering phenology, insect visitors, and seed-set of individuals: experimental studies on four co-occurring species of goldenrod (*Solidago*: Compositae). *Ecological Monographs* **53**, 95–117.

Grubb P.J. (1977) The maintenance of species-richness in plant communities: the importance of the regeneration niche. *Biological Reviews* **52**, 107–145.

Grubb P.J. (1984) Some growth points in investigative plant ecology. In *Trends in Ecological Research for the 1980s* (Eds J.H. Cooley & F.B. Golley). Plenum, New York.

Grubb P.J. (1986) The ecology of establishment. In *Ecology and Landscape Design* (Eds A.D. Bradshaw, D.A. Goode & E. Thorpe). Blackwell Scientific Publications, Oxford (in press).

Grubb P.J., Green H.E. & Merrifield R.C.J. (1969) The ecology of chalk heath: its relevance to the clacicole—calcifuge and soil acidification problems. *Journal of Ecology* **57**, 175–211.

Grubb P.J., Kelly D. & Mitchley J. (1982) The control of relative abundance in communities of herbaceous plants. In *The Plant Community as a Working Mechanism* (Ed. E.I. Newman), pp. 79–97. Blackwell Scientific Publications, Oxford.

Gulmon S.L. & Mooney H.A. (1986) Costs of defense on plant productivity. In *On the Economy of Plant Form and Function* (Ed. T.J. Givnish), Cambridge University Press, Cambridge.

Gulmon S.L., Rundel P.W., Ehleringer J.R. & Mooney H.A. (1979) Spatial relationships and competition in a Chilean desert cactus. *Oecologia* **44**, 40–43.

Gupta P.L. & Rorison I.H. (1975) Seasonal differences in the availability of nutrients down a podzolic profile. *Journal of Ecology* **63**, 521–534.

Gutierrez M.G. & Sprague G.F. (1959) Randomness of mating in isolated polycross plantings of maize. *Genetics* **44**, 1075–1082.

Hackett C. (1965) Ecological aspects of the nutrition of *Deschampsia flexuosa* (L.) Trin. II. The effects of Al, Ca, Fe, K, Mn, P and pH on the growth of seedlings and established plants. *Journal of Ecology* **53**, 315–333.

Hagerup O. (1951) Pollination in the Faroes—in spite of rain and poverty of insects. *Biologiske Meddelelser* **18**, 1–48.

Hair J.B. (1956) Subsexual reproduction in *Agropyron*. *Heredity* **10**, 129–160.

Hairston N.G., Smith F.E. & Slobodkin L.B. (1960) Community structure, population control, and competition. *American Naturalist* **94**, 421–425.

Hallé F. & Oldemann R.A.A. (1970) *Essai sur l'architecture et dynamique de croissance des arbres tropicaux*. Masson, Paris.

Hallé F., Oldemann R.A.A. & Tomlinson P.B. (1978) *Tropical Trees and Forests: an Architectural Analysis*. Springer-Verlag, Berlin.

Haller M.H. (1933) The relation of leaf area and position to quantity of fruit and bud differentiation in apples. *U.S. Department Agricultural Technology Bulletin* **338**.

Haller M.H. & Magness J.R. (1925) The relation of leaf area to the growth and composition of apples. *Proceedings American Society of Horticultural Science* **22**, 189–196.

Hamilton W.D. (1964) The genetical evolution of social behaviour. I & II. *Journal of Theoretical Biology* **7**, 1–52.

Hamilton W.D. (1966) The moulding of senescence. *Journal of Theoretical Biology* **12**, 12–45.

Hamilton W.D. & May R.M. (1977) Dispersal in stable habitats. *Nature* **269**, 578–581.

Hamrick J.L. (1983) The distribution of genetic variation within and among natural plant populations. In *Genetics and Conservation* (Eds C.M. Schonewald-Cox, S.M. Chambers, B. MacBryde & L. Thomas), pp. 335–348. Benjamin/Cummings Publishing Company, London.

Hamrick J.L., Linhart Y.B. & Mitton J.B. (1979) Relationships between life history characteristics and electrophoretically-detected genetic variation in plants. *Annual Review of Ecology & Systematics* **10**, 173–200.

Handel S.N. (1985) The intrusion of clonal growth patterns on plant breeding systems. *American Naturalist* **125**, 367–384.

Handel S.N., Fisch S.B. & Schatz G.E. (1981) Ants disperse a majority of herbs in a mesic forest community in New York State. *Bulletin of the Torrey Botanical Club* **108**, 430–437.

Hanover J.W. (1966) Environmental variation in the monoterpenes of *Pinus monticola* Dougl. *Phytochemistry* **5**, 713–717.

Hanover J.W. (1975) Physiology of tree resistance to insects. *Annual Review Entomology* **20**, 75–95.

Hansen S.R. & Hubbell S.P. (1980) Single-nutrient microbial competition: qualitative agreement between experimental and theoretically forecast outcomes. *Science* **207**, 1491–1493.

Hanski I. (1983) Coexistence of competitors in patchy environments. *Ecology* **64**, 493–500.

Hanski I. (1986) Colonization of ephemeral habitats. In *Colonization, Succession and Stability* (Eds A.J. Gray, M.J. Crawley, & P.J. Edwards). Blackwell Scientific Publications, Oxford (in press).

Hanson W.D. & Yeh R.Y. (1979) Genotypic differences for reduction in carbon exchange rates as associated with assimilate accumulation in soybean leaves. *Crop Science* **19**, 54–58.

Hara T. (1984) A stochastic model and the moment dynamics of the growth and size distribution in plant populations. *Journal of Theoretical Biology* **109**, 173–190.

Harberd D.J. (1961) Observations on population structure and longevity of *Festuca rubra* L. *New Phytologist* **60**, 184–206.

Harberd D.J. (1963) Observations on natural clones of *Trifolium repens* L. *New Phytologist* **62**, 198–204.

Harberd D.J. (1967) Observations on natural clones of *Holcus mollis*. *New Phytologist* **66**, 401–408.

Harborne J.B. (1982) *Introduction to Ecological Biochemistry*. Academic Press, London.

Hardin G. (1968) *Exploring New Ethics for Survival. The Voyage of the Spaceship Beagle*. Viking, London.

Hare R.C. (1966) Physiology of resistance to fungal diseases in plants. *Botanical Review* **32**, 95–137.

Harlan H.V. & Martini M.L. (1938) The effect of natural selection on a mixture of barley varieties. *Journal of Agricultural Research* **57**, 189–199.

Harley J.L. & Smith S.E. (1983) *Mycorrhizal Symbioses*. Academic Press, London.

Harper J.L. (1957) The ecological significance of dormancy and its importance in weed control. *Proceedings of the 4th International Congress on Crop Production*. Hamburg, pp. 415–420.

Harper J.L. (1961) Approaches to the study of competition. In *Mechanisms in Biological Competition* (Ed. F.L. Milthorpe), pp. 1–39. *Symposium of the Society for Experimental Biology* 15.

Harper J.L. (1967) A Darwinian approach to plant ecology. *Journal of Ecology* **55**, 247–270.

Harper J.L. (1975) Review of 'Allelopathy' by E.L. Rice. *Quarterly Review of Biology* **50**, 493–495.

Harper J.L. (1977) *Population Biology of Plants*. Academic Press, London.

Harper J.L. (1980) Plant demography and ecological theory. *Oikos* **35**, 244–253.

Harper J.L. (1981) The concept of population in modular organisms. In *Theoretical Ecology: Principles and Applications*, 2nd edn. (Ed. R.M. May), pp. 53–77. Blackwell Scientific Publications, Oxford.

Harper J.L. (1982) After description. In *The Plant Community as a Working Mechanism* (Ed. E.I. Newman), pp. 11–25. Blackwell Scientific Publications, Oxford.

Harper J.L. (1983) A Darwinian plant ecology. In *Evolution from Molecules to Man* (Ed. D.S. Bendall), pp. 323–345. Cambridge University Press, Cambridge.

Harper J.L. & Bell A.D. (1979) The population dynamics of growth form in organisms with modular construction. In *Population Dynamics* (Eds R.M. Anderson *et al.*), pp. 29–52. Blackwell Scientific Publications, Oxford.

Harper J.L., Lovell P.H. & Moore K.G. (1970) The shapes and sizes of seeds. *Annual Review of Ecology and Systematics*, **1**, 327–356.

Harper J.L. & Ogden J. (1970) The reproductive strategy of higher plants. I. The concept of strategy with special reference to *Senecio vulgaris* L. *Journal of Ecology* **58**, 681–698.

Harper J.L. & White J. (1971) The dynamics of plant populations. In *Dynamics of Populations* (Eds P.J. den Boer & G.R. Gradwell), pp. 44–63. Centre for Agricultural Publishing and Documentation, Wageningen, The Netherlands.

Harper J.L., Williams J.T. & Sagar G.R. (1965) The behaviour of seeds in soil. Part I.

The heterogeneity of soil surfaces and its role in determining the establishment of plants from seed. *Journal of Ecology* **53**, 273–286.

Harris P. (1974) The impact of the cinnabar moth on ragwort in eastern and western Canada and its implication for biological control strategy. *Commonwealth Institute Biological Control, Trinidad. Miscellaneous Publications* **105**, 119–123.

Harris P. (1984) *Carduus nutans* L., nodding thistle and *C. acanthoides* L., plumeless thistle (Compositae). In *Biological Control Programmes against Insects and Weeds in Canada 1969–1980* (Eds J.S. Kellener & M.A. Hulme), pp. 115–126. Commonwealth Agricultural Bureaux, London.

Hart R. (1977) Why are biennials so few? *American Naturalist* **111**, 792–799.

Hartgerink A.P. & Bazzaz F.A. (1984) Seedling-scale environmental heterogeneity influences individual fitness and population structure. *Ecology* **65**, 198–206.

Hartley C. (1939) The clonal variety of tree planting: asset or liability? *Phytopathology* **29**, 9.

Hartmann H.T. & Kester D.E. (1975) *Plant Propagation*. Prentice-Hall, Englewood Cliffs, New Jersey.

Hartshorn G.S. (1975) A matrix model of tree population dynamics. In *Tropical Ecological Systems*: trends in terrestrial and aquatic research (Eds F.B. Golley & E. Medina), pp. 41–51. Springer-Verlag, Berlin.

Hartshorn G.S. (1978) Treefalls and tropical forest dynamics. In *Tropical Trees as Living Systems* (Eds P.B. Tomlinson & M.H. Zimmerman), Chapter 26. Cambridge University Press, Cambridge.

Hartshorn G.S. (1980) Neotropical forest dynamics. *Biotropica* **12**, (Suppl.) pp. 23–30.

Harvey P.H., Colwell R.K., Silvertown J.W. & May R.M. (1983) Null models in ecology. *Annual Review of Ecology and Systematics* **14**, 189–211.

Hassell M.P. (1975) Density-dependence in single species populations. *Journal of Animal Ecology* **44**, 283–295.

Hassell M.P. (1978) *The Dynamics of Arthropod Predator–Prey Systems*. Princeton University Press, Princeton.

Hastings A. (1980) Disturbance, coexistence, history and competition for space. *Theoretical Population Biology* **18**, 363–373.

Haukioja E. (1980) On the role of plant defences in the fluctuation of herbivore populations. *Oikos* **35**, 202–213.

Haukioja E. & Niemelä P. (1976) Does birch defend itself actively against herbivores? *Reports of the Kevo Subarctic Research Station* **13**, 44–47.

Haukioja E. & Niemelä P. (1977) Retarded growth of a geometrid larva after mechanical damage to leaves of its host tree. *Annals Zoologia Fenn Lauija* **14**, 48–52.

Haukioja E. & Niemelä P. (1979) Birch leaves as a resource for herbivores: seasonal occurrence of increased resistance in foliage after mechanical damage of adjacent leaves. *Oecologia* **39**, 151–159.

Havill D.C., Lee J.A. & Stewart G.R. (1974) Nitrate utilisation by species from acidic and calcareous soils. *New Phytologist* **73**, 1221–1232.

Hawksworth D.L. (1973) Mapping Studies. In *Air Pollution and Lichens* (Eds B.W. Ferry, M.S. Baddeley & D.L. Hawksworth), pp. 38–76. Athlone Press, London.

Head G.C. (1973) Shedding of roots. In *Shedding of Plant Parts* (Ed. T.T. Kozlowski), pp. 237–294. Academic Press, New York.

Heichel G.H. & Turner N.C. (1976) Phenology and leaf growth of defoliated hardwood trees. In *Perspectives in Forest Entomology* (Eds J.F. Anderson & H.K. Kaya), pp. 31–40. Academic Press, New York.

Heichel G.H. & Turner N.C. (1984) Branch growth and leaf numbers of red maple (*Acer rubrum* L.) and red oak (*Quercus rubra* L.): response to defoliation. *Oecologia* **62**, 1–6.

Heinrich B. (1979) *Bumblebee Economics*, pp. 153–155. Harvard University Press, Cambridge, Massachusetts.

Heinrich B. (1981) The energetics of pollination. *Annals of the Missouri Botanic Garden* **60**, 370–378.

Heithaus E.R. (1982) Coevolution between bats and plants. In *Ecology of Bats* (Ed. T.H. Kunz), pp. 327–367. Plenum Press, New York.

Herrera C.M. (1984a) Determinants of plant-animal coevolution: the case of mutualistic vertebrate seed dispersal systems. *Oikos* (in press).

Herrera C.M. (1984b) A study of avian frugivores, bird-dispersed plants, and their interaction in Mediterranean scrublands. *Ecological Monographs* **54**, 1–23.

Herrold A. (1980) Regulation of photosynthesis by sink activity—the missing link. *New Phytologist* **86**, 131–144.

Heslop-Harrison J. (1982) The reproductive versatility of flowering plants: an overview. In *Strategies of Plant Reproduction* (Ed. W.J. Meudt), pp. 3–18. Allenheld, Osmun Publishing, London.

Hett J.M. (1971) A dynamic analysis of age in sugar maple seedlings. *Ecology* **52**, 1071–1074.

Hickman J. (1975) Environmental unpredictability and plastic energy allocation strategies in the annual *Polygonum cascadense* (Polygonaceae). *Journal of Ecology* **63**, 689–701.

Hils M.H. & Vankat J.L. (1982) Species removals from a first-year old field plant community. *Ecology* **63**, 705–711.

Hladik C.M. (1967) Surface relative du tractus digestif de quelques primates. Morphologie des villosites intestinales et correlations avec la regime alimentaire. *Mammalia* **31**, 120–147.

Hodgkinson K.C. (1974) Influence of partial defoliation on photosynthesis, photorespiration and transpiration by lucerne leaves of different ages. *Australian Journal of Plant Physiology* **1**, 561–578.

Hofstra G. & Nelson C.D. (1969) A comparative study of translocation of assimilated ^{14}C from leaves of different species. *Planta* **88**, 103–112.

Holdridge L., Grenke W., Hatheway W., Liang T. & Tosi Jr. J. (1971) *Forest Environments in Tropical Life Zones: A Pilot Study*. Pergamon Press, Oxford.

Hole F.D. (1976) *Soils of Wisconsin*. University of Wisconsin Press, Madison.

Hollinger D.Y. (1983) *Photosynthesis, water relations and herbivory in co-occurring deciduous and evergreen oaks*. PhD Thesis, Stanford University.

Holm L.G., Plucknett D.L., Pancho J.V. & Herberger J.P. (1977) *The World's Worst Weeds: Distribution and Biology*. The University Press of Hawaii, Honolulu.

Holm N.P. & Armstrong D. (1981) Role of nutrient limitation and competition in controlling the populations of *Asterionella formosa* and *Microcystis aeruginosa* in semi-continuous culture. *Limnology and Oceanography* **26**, 622–635.

Holmes R.T., Schul J.C. & Notunagle P. (1979) Bird predation on forest insects: an exclosure experiment. *Science* **206**, 462–463.

Holmes W. (Ed.) (1980) *Grass: its Production and Utilization*. Blackwell Scientific Publications, Oxford.

Holt R.D. (1977) Predation, apparent competition and the structure of prey communities. *Theoretical Population Biology* **11**, 197–229.

Holt R.D. & Pickering J. (1985) Infectious diseases and species coexistence: a model of Lotka-Volterra form. *American Naturalist* **126**, 196–211.

Holter L.C. & Abrahamson W.G. (1977). Seed and vegetative reproduction in relation to density in *Fragaria virginiana* (Rosaceae). *American Journal of Botany* **64**, 1003–7.

Honda H. (1971) Description of the form of trees by the parameters of the tree-like body: effects of the branching angle and branch length in the shape of the tree-like body. *Journal of Theoretical Biology* **31**, 331–338.

Honda H. & Fisher J.B. (1978) Tree branch angle maximizing effective leaf area. *Science* **199**, 888–890.

Honda H. & Fisher J.B. (1979) Ratio of tree branch lengths: the equitable distribution of leaf clusters on branches. *Proceedings of the National Academy of Sciences* **76**, 3875–3879.

Honda H., Tomlinson P.B., Fisher J.B. (1981) Computer simulation of branch inter-action and regulation by unequal flow rates in botanical trees. *American Journal of Botany* **68**, 569–585.

Honda H., Tomlinson P.B. & Fisher J.B. (1982) Two geometrical models of branching of botanical trees. *Annals of Botany* **49**, 1–11.

Hooker A.L. & Saxena K.M.S. (1971) Genetics of disease resistance in plants. *Annual Review of Genetics* **5**, 407–424.

Horn C.L. (1940) Existence of only one variety of cultivated mangosteen explained by asexually formed "seed". *Science* **92**, 237.

Horn H.S. (1971) *The Adaptive Geometry of Trees*. Princeton University Press, New Jersey.

Horn H.S. (1981) Some causes of variety in patterns of secondary succession. In *Forest Succession, Concepts and Application* (Eds D.C. West, H.H. Stugart & D.R. Botkin). Springer-Verlag, Berlin.

Horn H.S. (1976) Successions In *Theoretical Ecology*, 1st edn. (Ed. R.M. May), pp. 187–204. Blackwell Scientific Publications, Oxford.

Horn H.S. (1979) Adaptation from the perspective of optimality. In *Topics in Plant Population Biology* (Eds O.T. Solbrig, S. Jain, G.B. Johnson & P.H. Raven). Columbia University Press.

Horn H.S. & MacArthur R.H. (1972) Competition among fugitive species in a harlequin environment. *Ecology* **53**, 749–752.

Howard-Williams C. (1970) The ecology of *Becium homblei* in central Africa with special reference to metalliferous soils. *Journal of Ecology* **58**, 745–763.

Howe H.F. (1980) Monkey dispersal and waste of a neotropical tree. *Ecology* **61**, 944–959.

Howe H.F. (1983) Annual variation in a neotropical seed-dispersal system. In *Tropical Rain Forest: Ecology and Management* (Eds S.L. Sutton, T.C. Whitmore & A.C. Chadwick), pp. 211–227. Blackwell Scientific Publications, Oxford.

Howe H.F. (1984) Constraints on the evolution of mutualisms. *American Naturalist* **123**, 764–777.

Howe H.F. (1985) Gomphothere fruits: a critique. *American Naturalist* **125**, 853–865.

Howe H.F. (1986) Seed dispersal by fruit-eating birds and mammals In *Seed Dispersal* (Ed. D. Murray). Academic Press, Sydney.

Howe H.F. & Estabrook G.F. (1977) On intraspecific competition for avian dispersers in tropical trees. *American Naturalist* **111**, 817–832.

Howe H.F. & Richter W. (1982) Effects of seed size on seedling size in *Virola surinamensis*; a within and between tree analysis. *Oecologia* **53**, 347–351.

Howe H.F., Schupp E.W. & Westley L.C. (1985) Early consequences of seed dispersal for a neotropical tree (*Virola surinamensis*). *Ecology* (in press).

Howe H.F. & Smallwood J. (1982) Ecology of seed dispersal. *Annual Review of Ecology and Systematics* **13**, 201–228.

Howe H.F. & Vande Kerckhove G.A. (1979) Fecundity and seed dispersal of a tropical tree. *Ecology* **60**, 180–189.

Howe H.F. & Vande Kerckhove G.A. (1980) Nutmeg dispersal by tropical birds. *Science* **210**, 925–927.

Howe H.F. & Vande Kerckhove G.A. (1981) Removal of wild nutmeg (*Virola surinamensis*) crops by birds. *Ecology* **62**, 1093–1106.

Hsu S.B., Hubbell S.P. & Waltman P. (1977) A mathematical theory for single-nutrient competition in continuous cultures of microorganisms. *SIAM Journal of Applied Mathematics* **32**, 366–383.

Hubbell S.P. (1979) Tree dispersion, abundance and diversity in a tropical dry forest. *Science* **203**, 1299–1309.

Hubbell S.P. (1980) Seed predation and the coexistence of tree species in tropical forests. *Oikos* **35**, 214–229.

Hubbell S.P. (1984) Seed predation and tropical tree species diversity: *redressee*. *Oikos* (in press).

Hubbell S.P. (1984) Methodologies for the study of the origin and maintenance of tree diversity in tropical rain forest. In *The Significance of Species Diversity in Tropical*

Forest Ecosystems (Eds G. Maury-Lechon, G.M. Hadley & T. Younes). Biology International (IUBS) **6**, 8–13.

Hubbell S.P. & Foster R.B. (1983) Diversity of canopy trees in a neotropical forest and implications for conservation. In *Tropical Rain Forest Ecology and Management* (Eds S.L. Sutton, T.C. Whitmore & A.C. Chadwick), pp. 25–41. Blackwell Scientific Publications, Oxford.

Hubbell S.P. & Foster R.B. (1986a) Biology, chance and history, and the structure of tropical tree communities. In *Community Ecology* (Eds J.M. Diamond & T.J. Case), Chapter 19, pp. 314–324. Harper & Row, New York.

Hubbell S.P. & Foster R.B. (1986b) The spatial context of regeneration in a neotropical forest. In *Colonization, Succession and Stability* (Eds A.J. Gray, M.J. Crawley, & P.J. Edwards). Blackwell Scientific Publications, Oxford (in press).

Hughes T.P. (1984) Population dynamics based on individual size rather than age: a general model with a reef coral example. *American Naturalist* **123**, 778–795.

Hughes M.K., Lepp N.W. & Phipps D.A. (1980) Aerial heavy metal pollution and terrestrial ecosystems. *Advances in Ecological Research* **11**, 217–327.

Hunt R. (1978a) *Plant Growth Analysis*. Edward Arnold, London.

Hunt R. (1978b) Demography versus plant growth analysis. *New Phytologist* **80**, 269–272.

Hurlbert S.H. (1984) Pseudoreplication and the design of ecological field experiments. *Ecological Monographs* **54**, 187–211.

Huston M. (1979) A general hypothesis of species diversity. *American Naturalist* **113**, 81–101.

Huston M. (1980) Soil nutrients and tree species richness in Costa Rican forests. *Journal of Biogeography* **7**, 147–157.

Hutchings M.J. (1978) Standing crop and pattern in pure stands of *Mercurialis perennis* and *Rubus fruticosus* in a mixed deciduous woodland. *Oikos* **31**, 351–357.

Hutchings M.J. (1979) Weight density relationships in ramet populations of clonal perennial herbs, with special reference to the −3/2 power law. *Journal of Ecology* **67**, 21–33.

Hutchings M.J. (1985) Plant population biology. In *Methods in Plant Ecology*, 2nd edn. (Eds P.D. Moore & S.B. Chapman). Blackwell Scientific Publications, Oxford (in Press).

Hutchings M.J. & Barkham J.P. (1976) An investigation of shoot interactions in *Mercurialis perennis* L., a rhizomatous perennial herb. *Journal of Ecology* **64**, 723–743.

Hutchings M.J. & Bradbury I.K. (1986) Ecological perspectives on clonal perennial herbs. *BioScience* **36**, 178–182.

Hutchinson G.E. (1951) Copepodology for the ornithologist. *Ecology* **32**, 571–577.

Hutchinson G.E. (1957) Concluding remarks. In *Population Studies: Animal Ecology and Demography. Cold Spring Harbor Symposium on Quantitative Biology* **22**, 415–427. Long Island Biological Association, New York.

Hutchinson T.C. (1967) Comparative studies of the ability of species to withstand prolonged periods of darkness. *Journal of Ecology* **55**, 291–299.

Imai Y. (1935) The mechanism of bud variation. *American Naturalist* **69**, 587–595.

Inouye D.W. & Taylor J.R. (1980) Variation in generation time in *Frasera speciosa* (Gentianaceae), a long-lived perennial monocarp. *Oecologia* **47**, 171–174.

Inouye R.S. (1980) Density-dependent germination response by seeds of desert annuals. *Oecologia* **46**, 235–238.

Islam Z. & Crawley M.J. (1983) Compensation and regrowth in ragwort (*Senecio jacobaea*) attacked by cinnabar moth (*Tyria jacobaeae*). *Journal of Ecology* **71**, 829–843.

Iwasa Y. & Roughgarden J. (1984) Shoot/root balance of plants: optimal growth of a system with many vegetative organs. *Theoretical Population Biology* **25**, 78–105.

Jackson M.B. & Drew M.C. (1984) Effects of flooding on growth and metabolism of herbaceous plants. In *Flooding and Plant Growth* (Ed. T.T. Kozlowski), pp. 47–128. Academic Press, London.

Jager H.J. & Grill D. (1975) Einfluss von SO_2 und HF auf freie Aminosauren der Fichte. (*Picea abies* (L.) Karsten). *European Journal of Forest Pathology* 5, 279–286.

Jain S.K. (1979) Adaptive strategies: polymorphism, plasticity and homeostasis. In *Topics in Plant Population Biology* (Eds O.T. Solbrig, S. Jain, G.B. Johnson & P.H. Raven), pp. 160–187. Columbia University Press, New York.

Janson C. (1983) Adaptation of fruit morphology to dispersal agents in a neotropical forest. *Science* 219, 187–189.

Janzen D.H. (1969) Seed-eaters versus seed size, number, toxicity and dispersal. *Evolution* 23, 1–27.

Janzen D.H. (1970) Herbivores and the number of tree species in tropical forests. *American Naturalist* 104, 501–528.

Janzen D.H. (1976a) Why bamboo wait so long to flower. *Annual Review of Ecology and Systematics* 7, 347–391.

Janzen D.H. (1976b) Effect of defoliation on fruit bearing branches of the Kentucky coffee tree, *Gymnocladus dioicus* (Leguminosae). *American Naturalist* 95, 474–478.

Janzen D.H. (1977) What are dandelions and aphids? *American Naturalist* 111, 586–589.

Janzen D.H. (1979) Reply to Addicott. *American Naturalist* 114, 763–764.

Janzen D.H. (1980) When is it coevolution? *Evolution* 34, 611–612.

Janzen D.H. (1982) Differential seed survival and passage rates in cows and horses. *Oikos* 38, 150–156.

Janzen D.H. (1985) The natural history of mutualisms. In *The Biology of Mutualism* (Ed. D.H. Boucher), pp. 40–99. Croom Helm, London.

Jefferies R.L., Davy A.J. & Rudinik T. (1981) Population biology of the salt marsh annual *Salicornia europaea* agg. *Journal of Ecology* 69, 17–32.

Jefferies R.L. & Gottlieb L.D. (1983) Genetic variation within and between populations of the asexual plant *Puccinellia* (hybrid) *phryganodes* (Trin.) Scribner and Merr. *Canadian Journal of Botany* 61, 774–779.

Jeffrey D.W. (1964) The formation of polyphosphate in *Banksia ornata*, an Australian heath plant. *Australian Journal of Biological Sciences* 17, 845–854.

Jenny H. (1980) *Soil genesis with ecological perspectives*. Springer-Verlag, New York.

Jermy T. (1984) Evolution of insect/host plant relationships. *American Naturalist* 124, 609–630.

Jewell F.F. (1966) Inheritance of rust resistance in southern pines. In *Breeding Pest-Resistant Trees* (Eds H.D. Gerhold, E.J. Schreiner, R.E. McDermott & J.A. Winieski), pp. 107–109. Pergamon Press, Oxford.

Joenje W. (1978) *Plant colonization and succession on embanked sandflats: a case study in the Lauwerszeepolder*. Ph.D. Thesis, University of Groningen.

Johnson I.R. & Parsons A.J. (1985) A theoretical analysis of grass growth under grazing. *Journal of Theoretical Biology* 112, 345–367.

Johnson L.K. & Hubbell S.P. (1974) Aggression and competition among stingless bees: field studies. *Ecology* 55, 120–127.

Johnson T. (1961) Man-guided evolution in plant rusts. *Science* 133, 357–362.

Jones H.G. (1983) *Plants and microclimate*. Cambridge University Press, Cambridge.

Jones L.H. (1961) Aluminium uptake and toxicity in plants. *Plant and Soil* 13, 297–310.

Jones R.N. (1975) B-chromosome systems in flowering plants and animal species. *International Review of Cytology* 40, 1–100.

Juloka-Sulonen E.L. (1983) Vegetation succession of abandoned hay fields in central Finland: a quantitative approach. *Communicationes Instituti Forestalis Fenmae* 112.

Jurik T.W., Chabot J.F. & Chabot B.F. (1979) Ontogeny of photosynthetic performance in *Fragaria virginiana* under changing light regimes. *Plant Physiology* 63, 542–547.

Kalin Arroyo M.T. (1976) Geitonogamy in animal pollinated tropical angiosperms. A stimulus for the evolution of self-incompatibility. *Taxon* 25, 543–548.

Kay Q.O.N. (1982) Intraspecific discrimination by pollinators and its role in evolution. In *Pollination and Evolution* (Eds J.A. Armstrong, J.M. Powell & A.J. Richards), pp. 9–28. Royal Botanic Gardens, Sydney.

Kays S. & Harper J.L. (1974) The regulation of plant and tiller density in a grass sward. *Journal of Ecology* **62**, 97–105.

Keddy P.A. (1981) Experimental demography of the sand-dune annual *Cakile edentula*, growing along an environmental gradient in Nova Scotia. *Journal of Ecology* **69**, 615–630.

Keddy P.A. (1982) Population ecology on an environmental gradient: *Cakile edentula* on a sand dune. *Oecologia* **52**, 348–355.

Keeley J.E. (1979) Population differentiation along a flood frequency gradient: physiological adaptation to flooding in *Nyssa sylvatica*. *Ecological Monographs* **49**, 98–108.

Keever C. (1950) Causes of succession on old fields of the Piedmont, North Carolina. *Ecological Monographs* **20**, 229–250.

Kelly D. (1985) Why are biennials so maligned? *American Naturalist* **125**, 473–479.

Kemp P.R. & Williams G.J. III (1980) A physiological basis for niche separation between *Agropyron smithii* (C_3) and *Bouteloua gracilis* (C_4). *Ecology* **61**, 846–858.

Kerner A. (1894) *The Natural History of Plants. Their Forms, Growth, Reproduction and Distribution*. Blackie and Son, London.

Kernick M.D. (1961) Seed production of specific crops. In *Agricultural and Horticultural Seeds*, pp. 181–547. FAO Agricultural Studies No. 55.

Kershaw K.A. (1963) Pattern in vegetation and its causality. *Ecology* **44**, 377–388.

Kershaw K.A. (1973) *Quantitative and Dynamic Plant Ecology*, 2nd edn. Arnold, London.

Kevan P.G. (1983) Floral colors through the insect eye: What they are and what they mean In *Handbook of Experimental Pollination Biology* (Eds C.E. Jones & R.J. Little), pp. 3–30. Van Nostrand Reinhold, New York.

Kevan P.G. & H.G. Baker (1983) Insects as flower visitors and pollinators. *Annual Review of Entomology* **28**, 407–453.

Kilham S.S. & Kilham P. (1982) The importance of resource supply rates in determining phytoplankton community. In *Trophic Dynamics of Aquatic Ecosystems* (Eds D.C. Meyers & J.R. Strickler). AAAS Symposium.

King R.W., Wardlaw I.F. & Evans L.T. (1967) Effects of assimilate utilization and photosynthetic rate in wheat. *Planta* **77**, 261–276.

King T.J. (1975) Inhibition of seed germination under leaf canopies in *Arenaria serpyllifolia*, *Veronica arvensis* and *Cerastium holosteoides*. *New Phytologist* **75**, 87–90.

King T.J. & Woodell S.R.J. (1984) Are regular patterns in desert shrubs artefacts of sampling? *Journal of Ecology* **72**, 295–298.

Kirchner T. (1977) The effects of resource enrichment on the diversity of plants and arthropods in a shortgrass prairie. *Ecology* **58**, 1334–1344.

Kirkpatrick M. (1984) Demographic models based on size, not age, for organisms with indeterminate growth. *Ecology* **65**, 1874–1884.

Klekowski E.J. Jr. (1984) Mutational load in clonal plants: a study of two fern species. *Evolution* **38**, 417–426.

Klekowski E.J. Jr., & Karinova-Fukshansky N.K. (1984) Shoot apical meristems and mutation: fixation of selectively neutral cell genotypes. *American Journal of Botany* **71**, 22–27.

Klinkowski, M. (1970) Catastrophic plant diseases. *Annual Reviews of Phytopathology* **8**, 37–60.

Knapp R. (Ed.) (1974) *Vegetation Dynamics*. Handbook of Vegetation Science, Volume 8. Junk, The Hague.

Knowles P. & Grant M.C. (1983) Age and size structure analyses of Engelmann spruce, Ponderosa pine, Lodgepole pine and Limber pine in Colorado. *Ecology* **64**, 1–9.

Knox R.B. (1984) Pollen pistil interactions. *Encyclopaedia of Plant Physiology* **17**, 508–608.

Koch K.E. & Johnson C.R. (1984) Photosynthate partitioning in split-root seedlings with mycorrhizal and non-mycorrhizal root systems. *Plant Physiology* **75**, 26–30.

Koyama H. & Kira T. (1956) Intraspecific competition among higher plants. VIII. Frequency distribution of individual plant weight as affected by the interaction between plants. *Journal of the Institute of Polytechnics, Osaka City University* **7**, 73–94.

Koziol M.J. & Whatley F.R. (Eds) (1984) *Gaseous Air Pollutants and Plant Metabolism*. Butterworths, London.

Kozlowski T.T. (1971) *Growth and Development of Trees*, Volume II. Academic Press, New York.

Kozlowski T.T. (Ed.) (1972) *Water Deficits and Plant Growth*. Academic Press, London.

Kozlowski T.T. (1973) Extent and significance of shedding of plant parts. In *Shedding of Plant Parts* (Ed. T.T. Kozlowski), pp. 1–44. Academic Press, New York.

Kozlowski T.T. (Ed.) (1984) *Flooding and Plant Growth*. Academic Press, London.

Kozlowski T.T. & Ahlgren C.E. (Eds) (1974) *Fire and Ecosystems*. Academic Press, New York.

Kozlowski T.T. & Keller T. (1966) Food relations of woody plants. *Botanical Review* **32**, 293–382.

Kramer P.J. & Kozlowski T.T. (1960) *Physiology of Trees*. McGraw-Hill, New York.

Krefting L.W. & Roe E.I. (1949) The role of some birds and mammals in seed germination. *Ecological Monographs* **19**, 269–286.

Kruger F.J. (1982) Prescribing fire frequencies in Cape fynbos in relation to plant demography. In *Dynamics and Management of Mediterranean-Type Ecosystems* (Eds C.E. Conrad & W.C. Oechel). *USDA Forest Service General Technical Report PSW* **58**, 483–489.

Kruger F.J. & Taylor H.C. (1979) Plant species diversity in Cape Fynbos: gamma and delta diversity. *Vegetation* **41**, 85–93.

Kulman H.M. (1971) Effects of insect defoliation on growth and mortality of trees. *Annual Review Entomology* **16**, 289–324.

Küppers M. (1984) Carbon relations and competition between woody species in a Central European hedgerow. I. Photosynthetic characteristics. *Oecologia* **64**, 332–343.

Kuroiwa S. (1960) Ecological and physiological studies on the vegetation of Mt. Shimagare. V. Intraspecific competition and productivity difference among tree classes in the *Abies* stand. *Botanical Magazine, Tokyo* **73**, 165–174.

Laessle A.M. (1965) Spacing and competition in natural stands of sand pine. *Ecology* **46**, 65–72.

Laird C.D., McConaught B.L. & McCarthy B.J. (1969) Rate of fixation of nucleotide substitutions in evolution. *Nature* **224**, 149–154.

Lambers H. (1982) Cyanide resistant respiration: a non-phosphorylating electron transport pathway acting as an energy overflow. *Physiologia Plantarum* **55**, 478–485.

Lande R. & Schemske D.W. (1985) The evolution of self-fertilization and inbreeding depression in plants. I. Genetic models. *Evolution* **39**, 24–40.

Lange O. & Kappen L. (1972) Photosynthesis of lichens from Antarctica. *Antarctic Research Series* **20**, 83–95. American Geophysical Union.

Lange O., Schulze E.D., Evanari M., Kappen L. & Buschbom U. (1974) The temperature-related photosynthetic capacity of plants under desert conditions. I. Seasonal changes of the photosynthetic response to temperature. *Oecologia* **17**, 97–110.

Larcher W. & Bauer H. (1981) Ecological significance of resistance to low temperature. In *Encyclopedia of Plant Physiology. New Series, Volume 12A Physiological Plant Ecology I* (Eds O.L. Lange, P.S. Nobel, C.B. Osmond & H. Ziegler). Springer-Verlag, Berlin, Heidelberg.

Larkin P.J., Ryan S.A., Brettell R.I.S. & Snowcroft W.R. (1984) Heritable soma-clonal variation in wheat. *Theoretical and Applied Genetics* **67**, 443–455.

Law R. (1975) *Colonization and the evolution of life histories in Poa annua.* Ph.D. Thesis, University of Liverpool.

Law R. (1979) The cost of reproduction in annual meadow grass. *American Naturalist* **113**, 3–16.

Law R. (1981) The dynamics of a colonizing population of *Poa annua. Ecology* **62**, 1267–1277.

Law R. (1983) A model for the dynamics of a plant population containing individuals classified by age and size. *Ecology* **64**, 224–230.

Lawes J. & Gilbert J. (1880) Agricultural, botanical and chemical results of experiments on the mixed herbage of permanent grassland, conducted for many years in succession on the same land. I. *Philosophical Transactions of the Royal Society* **171**, 189–514.

Lawrence D.B. (1958) Glaciers and vegetation in Southeastern Alaska. *American Scientist* **46**, 89–122.

Lawrence D.B. (1979) Primary versus secondary succession at Glacier Bay National Monument, southeastern Alaska. In *Proceedings of the First Conference on Scientific Research in the National Parks* (Ed. Robert M. Linn), pp. 213–224.

Lawrence D.B., Schoenike R.E., Quispel A. & Bond G. (1967) The role of *Dryas drummondi* in vegetation development following ice recession at Glacier Bay, Alaska, with special reference to its nitrogen fixation by root nodules. *Journal of Ecology* **55**, 793–813.

Lawrence M.J. (1984) The genetical analysis of ecological traits. In *Evolutionary Ecology* (Ed. B. Shorrocks), pp. 27–63. Blackwell Scientific Publications, Oxford.

Lawrence W.H. & Rediske J.H. (1962) Fate of sown Douglas-fir seed. *Forestry Science* **8**, 210–218.

Laws R.M., Parker I.S.C. & Johnstone R.C.B. (1975) *Elephants and their Habitat.* Clarendon Press, Oxford.

Lawton J.H. (1983) Plant architecture and the diversity of phytophagous insects. *Annual Review of Entomology* **28**, 23–39.

Lawton J.H. & Hassel M.P. (1984) Interspecific competition in insects. In *Ecological Entomology* (Eds C.B. Huffaker & R.L. Rabb), pp. 451–495. John Wiley, New York.

Lechowicz M.J. (1984) Why do temperate deciduous trees leaf out at different times? Adaptation and ecology of forest communities. *American Naturalist* **124**, 821–842.

Lee R. (1978) *Forest Microclimatology.* Columbia University Press, New York.

Lefkovitch L.P. (1965) The study of population growth in organisms grouped by stages. *Biometrics* **21**, 1–18.

Lefroy R.B.D. (1982) *The supply and utilisation of phosphorus in the control of plant growth.* D.Phil. Thesis, University of York.

Lehman J.T. (1982) Grazing, nutrient release, and their impacts on the structure of phytoplankton communities. In *Trophic Dynamics of Aquatic Ecosystems* (Eds D.C. Meyers & J.R. Strickler), pp. 49–74. American Association for the Advancement of Science Symposium.

Leigh E.G. (1975) Structure and climate in tropical rain forest. *Annual Review of Ecology and Systematics* **6**, 67–86.

Leigh E.G., Rand A.S. & Windsor D.M. (Eds) (1982) *The Ecology of a Tropical Forest: Seasonal Rhythms and Long-Term Changes.* Smithsonian Press, Washington.

Lenz F. (1967) Relationships between the vegetative and reproductive growth of Washington navel orange cuttings (*Citrus sinensis* (L.) Osbeck). *Journal of Horticultural Science* **42**, 31–39.

Leon J.A. (1985) Germination strategies. In *Evolution: Essays in Honour of John Maynard Smith* (Eds P.J. Greenwood, P.H. Harvey & M. Slatkin), pp. 129–142. Cambridge University Press, Cambridge.

Leopold A.C. & Kriedemann P.E. (1975) *Plant Growth and Development*. 2nd edn. McGraw-Hill, New York.

Lertzman K.P. & Gass C.L. (1983) Alternative methods of pollen transfer. In *Handbook of Experimental Pollination Biology* (Eds C.E. Jones & R.J. Little), pp. 474–489. Van Nostrand-Reinhold, New York.

Leslie P.H. (1945) On the use of matrices in certain population mathematics. *Biometrika* **33**, 183–212.

Leslie P.H. (1948) Some further notes on the use of matrices in population mathematics. *Biometrika* **35**, 213–245.

Lester D.T. (1971) Self-compatibility and inbreeding depression in elm. *Forest Science* **17**, 321–322.

Leverich W.J. & Levin D.A. (1979) Age-specific survivorship and reproduction in *Phlox drummondii*. *American Naturalist* **113**, 881–903.

Levey D.J., Moermond T.C. & Denslow J.S. (1984) Fruit choice in neotropical birds: the effect of distance between fruits on preference patterns. *Ecology* **65**, 844–850.

Levin D.A. (1972) Plant density, cleistogamy, and self-fertilization in natural populations of *Lithospermum caroliniense*. *American Journal of Botany* **59**, 71–77.

Levin D.A. (1975) Gametophytic selection in *Phlox*. In *Gamete Competition in Plants and Animals* (Ed. D.L. Mulcahy), pp. 207–217. North-Holland Publishing Company, Oxford.

Levin D.A. (1978a) The origin of isolating mechanisms in flowering plants. *Evolutionary Biology* **11**, 185–317.

Levin D.A. (1978b) Genetic variation in annual *Phlox:* Self-compatible versus self-incompatible species. *Evolution* **32**, 245–263.

Levin D.A. (1981) Dispersal versus gene flow in plants. *Annals of the Missouri Botanical Gardens* **68**, 232–253.

Levin D.A. (1983) An immigration-hybridization episode in *Phlox*. *Evolution* **37**, 575–582.

Levin D.A. (1984) Inbreeding depression and proximity-dependent crossing success in *Phlox drummondii*. *Evolution* **38**, 116–127.

Levin D.A. (1984b) Genetic variation and divergence in a disjunct *Phlox*. *Evolution* **38**, 223–225.

Levin D.A. & Kerster H.W. (1969) Density-dependent gene dispersal in *Liatris*. *American Naturalist* **103**, 61–74.

Levin D.A. & Kerster H.W. (1974) Gene flow in seed plants. *Evolutionary Biology* **7**, 139–220.

Levin D.A. & Watkins L. (1985) Assortative mating in *Phlox*. *Heredity* (in press).

Levin D.A. & Wilson J.B. (1978) The genetic implications of ecological adaptations in plants. In *Structure and Functioning of Plant Populations* (Eds A.H.J. Freysen & J.W. Woldendorp), pp. 75–100. North-Holland Publishing Company, Amsterdam.

Levin S.A. (1974) Dispersion and population interactions. *American Naturalist* **108**, 207–228.

Levins R. (1969) Dormancy as an adaptive strategy. *Symposium of the Society for Experimental Biology* **23**, 1–10.

Levins R. & Culver D. (1971) Regional coexistence of species and competition between rare species. *Proceedings of the National Academy of Sciences of the USA* **68**, 1246–1248.

Levins R. & Lewontin R. (1980) Dialectics and reductionism in ecology. In *Conceptual Issues in Ecology* (Ed. E. Saarinen), pp. 107–138. D. Reidel, London.

Levitt J. (1972) *Responses of Plants to Environmental Stresses*. Academic Press, London.

Levitt J. (1978) An overview of freezing injury and survival, and its interrelationships with other stresses. In *Plant Cold Hardiness and Freezing Stress* (Eds P.H. Li & A. Sakai). Academic Press, London.

Lewis A.C. (1979) Feeding preference for diseased and wilted sunflower in the grasshopper *Malanopus differentialis*. *Entomologia Experimentalis et Applicata* **26**, 202–207.

Lewis W.H. (1970) Chromosomal drift, a new phenomenon in plants. *Science* **168**, 1115–1116.

Lewis W.H. (1976) Temporal adaptation correlated with ploidy in *Claytonia virginiana*. *Systematic Botany* **1**, 340–347.

Lewis W.H., Oliver R.L. & Luikart T.K. (1971) Multiple genotypes in individuals of *Claytonia virginica*. *Science* **172**, 564–565.

Lewontin R.C. (1970) The units of selection. *Annual Review Ecology and Systematics* **1**, 1–11.

Lewontin R.C. (1974) The analysis of variance and the analysis of causes. *American Journal of Human Genetics* **26**, 400–411.

Libby W.J. (1974) A summary statement on the 1973 vegetative propagation meeting in Rotorua, New Zealand. *New Zealand Journal of Forest Science* **4**, 454–458.

Libby W.L., Stettler R.F. & Seitz F.W. (1969) Forest genetics and forest-tree breeding. *Annual Review Genetics* **3**, 469–494.

Libby W.J., McCutchan B.G. & Miller C.I. (1981) Inbreeding depression in selfs of redwood. *Silvae Genetica* **30**, 15–25.

Liddle M.J., Budd C.S.J. & Hutchings M.J. (1982) Population dynamics and neighbourhood effects in establishing swards of *Festuca rubra*. *Oikos* **38**, 52–59.

Lincoln D.E., Newton T.S., Ehrlich P.R. & Williams K.S. (1982) Coevolution of the checkerspot butterfly *Euphydras chalcedona* and its larval food plant *Diplacus aurantiacus*: larvae response to protein and leaf resin. *Oecologia* **52**, 216–223.

Lindsey A.A. (1961) Vegetation of the drainage-aeration classes of northern Indiana soils in 1830. *Ecology* **42**, 432–436.

Linhart Y.B. (1976) Density-dependent seed germination strategies in colonising versus non-colonising species. *Journal of Ecology* **64**, 375–380.

Linhart Y.B. & Feinsinger P. (1980) Plant-hummingbird interactions: effects of island size and degree of specialization on pollination. *Journal of Ecology* **68**, 745–760.

Linsley E.G., MacSwain J.W., Raven P.H. & Thorp R.W. (1973) Comparative Behavior of Bees and Onagraceae. V. *University of California Publications in Entomology* **71**, 1–68.

Lloyd M. (1967) Mean crowding. *Journal of Animal Ecology* **36**, 1–30.

Loneragan J.F. & Asher C.J. (1967) Response of plants to phosphorus concentrations in solution culture. II. Rate of phosphorus absorption and its relation to growth. *Soil Science* **103**, 311–318.

Longstreth D.J. & Nobel P.S. (1980) Nutrient influences on leaf photosynthesis. Effects of N, P and K for *Gossypium hirsutum*. *Plant Physiology* **65**, 541–543.

Lonsdale W.M. & Watkinson A.R. (1982) Light and self-thinning. *New Phytologist* **90**, 431–445.

Lonsdale W.M. & Watkinson A.R. (1983) Plant geometry and self-thinning. *Journal of Ecology* **71**, 285–297.

Lord E.M. (1981) Cleistogamy: a tool for the study of floral morphogenesis, function and evolution. *Botanical Review* **47**, 421–449.

Loveless M.D. & Hamrick J.L. (1984) Ecological determinants of genetic structure in plant populations. *Annual Review of Ecology and Systematics* **15**, 65–95.

Lovett Doust J. (1980) Experimental manipulation of patterns of resource allocation in the growth cycle and reproduction of *Smyrnium olusatrum*. *Biological Journal of the Linnean Society* **13**, 155–166.

Lovett Doust L. (1981a) Intraclonal variation and competition in *Ranunculus repens*. *New Phytologist* **89**, 495–502.

Lovett Doust L. (1981b) Population dynamics and local specialization in a clonal perennial (*Ranunculus repens*) I. The dynamics of ramets in contrasting habitats. *Journal of Ecology* **69**, 743–755.

Lovric L.M. & Lovric A.Z. (1984) Morpho-anatomical syndromes in phytoindicators of extreme stormy habitats in the northeastern Adriatic. In *Being Alive on Land* (N.S. Margaris, M. Arianoustou-Faraggitaki & W.C. Oechel), pp. 41–49. Junk, The Hague.

Lyman J.C. & Ellstrand N.C. (1984) Clonal diversity in *Taraxacum officinale* (Compositae), an apomict. *Heredity* **53**, 1–10.

Maarel E. van der & Leertouwer J. (1967) Variation in vegetation and species diversity along a local environmental gradient. *Acta Botanica Neerlandica* **16**, 211–221.

MacArthur R.H. (1972) *Geographical Ecology. Patterns in the Distribution of Species.* Harper & Ross, New York.

MacArthur R.H. & MacArthur J. (1961) On bird species diversity. *Ecology* **42**, 594–598.

MacArthur R.H. & Wilson E.O. (1967) *The Theory of Island Biogeography.* Princeton University Press, New Jersey.

MacDonald N. & Watkinson A.R. (1981) Models of an annual plant population with a seed bank. *Journal of Theoretical Biology* **93**, 643–653.

Macevicz S. & Oster G. (1976) Modelling social insect populations. II. Optimal reproductive strategies in annual eusocial insect colonies. *Behavioral Ecology and Sociobiology* **1**, 265–282.

Mack R.N. (1981) Invasion of *Bromus tectorum* L. into western North America an ecological chronicle. *Agro-Ecosystems* **7**, 145–165.

Mack R.N. & Harper J.L. (1977) Interference in dune annuals: spatial pattern and neighbourhood effects. *Journal of Ecology* **65**, 345–363.

Mack R.N. & Pyke D.A. (1983) The demography of *Bromus tectorum:* variation in time and space. *Journal of Ecology* **71**, 69–93.

Mack R.N. & Pyke D.A. (1984) The demography of *Bromus tectorum*: the role of microclimate, grazing and disease. *Journal of Ecology* **72**, 731–748.

Mack R.N. & Thompson J.N. (1982) Evolution in steppe with few large, hooved mammals. *American Naturalist* **119**, 757–773.

Maggs D.H. (1963) The reduction in growth in apple trees brought about by fruiting. *Journal of Horticultural Science* **38**, 119–129.

Maggs D.H. (1965) Growth rates in relation to assimilate supply and demand. II. The effect of particular leaves and growing regions in determining dry matter distribution in young apple trees. *Journal of Experimental Botany* **16**, 387–404.

Magness J.R. (1928) Relation of leaf area to size and quality in apples. *Proceedings American Society of Horticultural Science* **25**, 285–288.

Mahmoud A. & Grime J.P. (1974) A comparison of negative relative growth rates in shaded seedlings. *New Phytologist* **73**, 1215–1220.

Maiti S.N., Datta A.N. & Basak S.L. (1981) Outcrossing and isolation requirement in kenaf (*Hibiscus cannabinus* L.), *Euphytica* **30**, 739–745.

Mallik A.U., Hobbs R.J. & Legg C.J. (1984) Seed dynamics in *Calluna-Arctostaphylos* heath in northeastern Scotland. *Journal of Ecology* **72**, 855–871.

Malloch D.W., Pirozynski K.A. & Raven P.H. (1980) Ecological and evolutionary significance of mycorrhizal symbioses in vascular plants (a review). *Proceedings of the National Academy of Sciences of the USA* **77**, 2113–2118.

Malmberg C. & Smith H. (1982) Relationship between plant weight and density in mixed populations of *Medicago sativa* and *Trifolium pratense. Oikos* **38**, 365–368.

Manasse R.S. & Howe H.F. (1983) Competition among tropical trees for dispersal agents: influences of neighbors. *Oecologia* **59**, 185–190.

Mansfield T.A. & Freer-Smith P.H. (1984) The role of stomata in resistance mechanisms. In *Gaseous Air Pollutants and Plant Metabolism* (Eds M.J. Koziol & F.R. Whatley), pp. 131–146. Butterworths, London.

Marks P.L. (1974) The role of pin cherry (*Prunus pensylvanica* L.) in the maintenance of stability in northern hardwood ecosystems. *Ecological Monographs* **44**, 73–88.

Marks P.L. (1983) On the origin of the field plants of the northeastern United States. *American Naturalist* **122**, 210–228.

Marrs R.H., Roberts R.D., Skeffington R.A. & Bradshaw A.D. (1983) Nitrogen and the development of ecosystems. In *Nitrogen as an Ecological Factor* (Eds J.A. Lee, S. McNeill & I.H. Rorison), pp. 113–136.

Martin J.K. (1977) Factors influencing the loss of organic carbon from wheat roots. *Soil Biology and Biochemistry* **9**, 1–7.

Martin J.K. & Kemp J.R. (1980) Carbon loss from roots of wheat cultivars. *Soil Biology and Biochemistry* **12**, 551–4.

Mason H.L. (1947) Evolution in certain floristic associations in western North America. *Ecological Monographs* **17**, 201–210.

Maxwell F.G. & Jennings P.R. (Eds) (1980) *Breeding Plants Resistant to Insects*. John Wiley, New York.

May P. & Antcliffe A.J. (1963) The effect of shading on fruitfulness and yielding in the sultana. *Journal of Horticultural Science* **38**, 85–94.

May R.M. (1975) Patterns of species abundance and diversity. In *Ecology and Evolution of Communities* (Eds M.L. Cody & J.M. Diamond), pp. 81–120. Harvard University Press, Cambridge, Massachusetts.

May R.M. (Ed.) (1981) *Theoretical Ecology. Principles and Applications*, 2nd edn. Blackwell Scientific Publications, Oxford.

May R.M. (1981) Models for two interacting populations. In *Theoretical Ecology* (Ed. R.M. May), pp. 78–104. Blackwell Scientific Publications, Oxford.

May R.M. (1985) Evolutionary ecology and John Maynard Smith. In *Evolution: Essays in Honour of John Maynard Smith* (Eds P.J. Greenwood, P.H. Harvey & M. Slatkin), pp. 107–116. Cambridge University Press, Cambridge.

May R.M. & MacArthur R.H. (1972) Niche overlap as a function of environmental variability. *Proceedings of the National Academy of Sciences of the USA* **69**, 1109–1113.

Maynard Smith J. (1976) What determines the rate of evolution? *American Naturalist* **110**, 331–338.

Maynard Smith J. (1982) *Evolution and the Theory of Games*. Cambridge University Press, Cambridge.

Maynard Smith J. & Price G.R. (1973) The logic of animal conflict. *Nature* **246**, 15–18.

McArthur E.D. (1977) Environmentally induced changes in sex expression in *Atriplex canescens*. *Heredity* **38**, 97–103.

McCree K.J. (1970) An equation for the rate of respiration of white clover plants grown under controlled conditions. In *Prediction and Measurement of Photosynthetic Activity* (Ed. I. Setlik), pp. 221–230. Pudoc, Wageningen.

McCree K.J. & Troughton J.H. (1966) Prediction of growth rate at different light levels from measured photosynthesis and respiration rates. *Plant Physiology* **41**, 559–566.

McGraw J.B. & Antonovics J. (1983) Experimental ecology of *Dryas octopetala* ecotypes. II. A demographic model of growth, branching and fecundity. *Journal of Ecology* **71**, 899–912.

McGuinness K.A. (1984) Equations and explanations in the study of species area curves. *Biological Reviews* **59**, 423–440.

McKendrick J.D., Ott V.J. & Mitchell G.A. (1979) Effects of nitrogen and phosphorus fertilisation on carbohydrate and nutrient levels in *Dupontia fisheri* and *Arctagrostis latifolia*. In *Vegetation and Production Ecology of an Alaskan Arctic Tundra* (Ed. L. Tieszea), pp. 509–537, *Ecological Studies* **29**. Springer-Verlag, New York.

McKey D. (1975) The ecology of coevolved seed dispersal systems. In *Coevolution in Animals and Plants* (Eds L.E. Gilbert & P.H. Raven), pp. 159–191. University of Texas Press, Austin.

McLaren J.S. & Smith H. (1978) Phytochrome control of the growth and development of *Rumex obtusifolius* under simulated canopy light environments. *Plant Cell and Environment* **1**, 61–68.

McMahon T. (1975) The mechanical design of trees. *Scientific American* **233**, 92–102.

McMahon T.A. & Kronauer R.E. (1976) Tree structures: deducing the principle of mechanical design. *Journal of Theoretical Biology* **59**, 443–466.

McNaughton S.J. (1968) Autotoxic feedback in relation to germination and seedling growth in *Typha latifolia*. *Ecology* **49**, 367–369.

McNaughton S.J. (1979) Grassland-herbivore dynamics. In *Serengeti: Dynamics of an Ecosystem* (Eds A.R.E. Sinclair & M. Norton-Griffiths), pp. 46–81. University of Chicago Press, Chicago.

McNeilly T. (1981) Ecotypic differentiation in *Poa annua*: interpopulation differences in response to competition and cutting. *New Phytologist* **88**, 539–547.

McPherson J.K. & Muller C.H. (1969) Allelopathic effects of *Adenostoma fasciculatum* "chamise", in the Californian chaparral. *Ecological Monographs* **39**, 177–198.

Meagher T.R. (1980) Population biology of *Chamaelirium luteum*, a dioecious lily. I. Spatial distributions of males and females. *Evolution* **34**, 1127–1137.

Meijden E. van der (1979) Herbivore exploitation of a fugitive plant species: local survival and extinction of the cinnabar moth and ragwort in a heterogeneous environment. *Oecologia* **42**, 307–323.

Meins F. Jr. (1983) Heritable variation in plant cell culture. *Annual Review of Plant Physiology* **34**, 327–346.

Mellinger M. & McNaughton S. (1975) Structure and function of successional vascular plant communities in central New York. *Ecological Monographs* **34**, 161–182.

Menges E.S. (1986) Adaptive allocation and geometry of a perennial forest herb. *American Naturalist* (in press).

Metcalf C.L. & Flint W.P. (1951) *Destructive and Useful Insects. Their Habitat and Control.* McGraw-Hill, New York.

Milewski A.F. & Bond W.J. (1982) Convergence of myrmecochoryin Mediterranean Australia and Southern Africa. In *Ant-Plant Interactions in Australia* (Ed. R.C. Buckley), pp. 89–98. Junk, The Hague.

Milewski A.V. (1983) A comparison of ecosystems in Mediterranean Australia and Southern Africa: Nutrient-poor sites at the Barrens and the Caledon Coast. *Annual Review of Ecology and Systematics* **14**, 57–76.

Millington W.F. & Chaney W.R. (1973) Shedding of shoots and branches. In *Shedding of Plant Parts* (Ed. T.T. Kozlowski), pp. 149–204. Academic Press, New York.

Miller K.R. (1979) The photosynthetic membrane. *Scientific American* **241**, 102–113.

Milton W. (1934) The effect of controlled grazing and manuring on natural hill pastures. *Welsh Journal of Agriculture* **10**, 192–211.

Milton W. (1940) The effect of manuring, grazing and cutting on the yield, botanical and chemical composition of natural hill pastures. *Journal of Ecology* **28**, 326–356.

Milton W. (1947) The yield, botanical and chemical composition of natural hill herbage under manuring, controlled grazing and hay conditions. I. Yield and botanical. *Journal of Ecology* **35**, 65–89.

Minchin F.R. & Pate J.S. (1974) Diurnal functioning of the legume root nodule. *Journal of Experimental Botany* **14**, 483–495.

Mitscherlisch E.A. (1913) *Bodenkunde für Landund Forstuirte.* Berlin.

Moermond T.C. & Denslow J.S. (1985) Neotropical Avian Frugivores: Patterns of Behavior Morphology, and Nutrition With Consequences for Fruit Selection. In *Neotropical Ornithology* (Eds P.A. Buckley, M.S. Foster, E.S. Morton, R.S. Ridgely & N.G. Smith), A.O.U. Monographs.

Mohler C.L., Marks P.L. & Sprugel D.G. (1978) Stand structure and allometry of trees during self-thinning of pure stands. *Journal of Ecology* **66**, 599–614.

Molisch H. (1938) *The Longevity of Plants.* E.H. Fulling, New York.

Monro J. (1967) The exploitation and conservation of resources by populations of insects. *Journal of Animal Ecology* **36**, 531–547.

Mooney H.A. (Ed.) (1977) Convergent Evolution in Chile and California: Mediterranean Climate Ecosystems. *US/IBP Synthesis Series* No. **5**. Dowden, Hutchinson & Ross, Stroudsburg, Pennsylvania.

Mooney H.A., Bjorkman O., Ehleringer J. & Berry J. (1976) Photosynthetic capacity of *in situ* Death Valley plants. *Carnegie Institution Year Book* **75**, 410–413.

Mooney H.A., Bonnicksen T.M., Christensen N.L., Lothan J.E. & Reiners W.A. (Eds) (1981) Fire Regimes and Ecotype Properties. *USDA Forest Service General Technical Reports (Washington) GTR-WO-26.*

Mooney H.A. & Conrad C.E. (Eds) (1977) Environmental Consequences of Fire and Fuel Management in Mediterranean Ecosystems. *General Technical Reports of the US Forest Service (Washington) GTR-WO-3.*

Mooney H.A. & Dunn E.L. (1970) Convergent evolution of Mediterranean-climate evergreen sclerophyllous shrubs. *Evolution* **24**, 292–303.

Mooney H.A. & Ehleringer J. (1978) The carbon gain benefits of solar tracking in a desert annual. *Plant, Cell and Environment* **1**, 307–311.

Mooney H.A., Ehleringer J. & Bjorkman O. (1977) The energy balance of leaves of the evergreen shrub, *Atriplex hymenelytra. Oecologia* **29**, 301–310.

Mooney H.A., Field C., Williams W.E., Berry J.A. & Bjorkman O. (1983) Photosynthetic characteristics of plants of a Californian cool coastal environment. *Oecologia* **57**, 38–42.

Mooney H.A. & Godron M. (Eds) (1983) *Disturbance and Ecosystems.* Springer-Verlag, Berlin.

Mooney H.A. & Gulmon S.L. (1979) Environmental and evolutionary constraints on the photosynthetic characteristics of higher plants. In *Topics in Plant Population Biology* (Eds O.T. Solbrig, S. Jain, G.B. Johnson & P.H. Raven), pp. 316–337. Macmillan, London.

Mooney H.A. & Gulmon S.L. (1982) Constraints on leaf structure and function in reference to herbivory. *BioScience* **32**, 198–206.

Moran N. & Hamilton W.D. (1980) Low nutritive quality as defence against herbivores. *Journal of Theoretical Biology* **86**, 247–254.

Moran P.A.P. (1962) *The Statistical Process of Evolutionary Theory.* Oxford University Press, London.

Moravec. J. (1973) The determination of the minimal area of phytocenoses. *Folia Geobotanica et Phytotaxonomica* **8**, 23–47.

Morris M.G. (1981) Responses of grassland invertebrates to management by cutting. III. Adverse effects on Auchenorhyncha. *Journal of Applied Ecology* **18**, 107–123.

Morrow P.A. & Fox L.R. (1980) Effects of variation in *Eucalyptus* essential oil yield on insect growth and grazing damage. *Oecologia* **45**, 209–219.

Morrow P.A. & La Marche V.C. (1978) Tree ring evidence for chronic insect suppression of productivity in subalpine *Eucalyptus. Science* **201**, 1224–1226.

Mulcahy D.L. (1984) Self-incompatibility, self-compatibility, and Stebbins Rule. In *Pollination '84* (Eds E.G. Williams & R.B. Knox), pp. 48–52. The School of Botany, University of Melbourne.

Mulcahy D.L., Curtis P.S. & Snow A.A. (1983) Pollen competition in a natural population. In *Handbook of Experimental Pollination Biology* (Eds C.E. Jones & R.J. Little), pp. 330–337. Von Nostrand Reinhold, New York.

Muller C.H. & Del Moral R. (1966) Soil toxicity induced by terpenes from *Salvia leucophylla. Bulletin of the Torrey Botanical Club* **93**, 130–137.

Murton R.K., Isaacson A.J. & Westwood N.J. (1966) The relationship between wood pigeons and their clover food supply and the mechanisms of population control. *Journal of Applied Ecology* **3**, 55–93.

Nafziger E.D. & Koller H.R. (1976) Influence of leaf starch concentration on CO_2 assimilation in soybean. *Plant Physiology* **57**, 560–563.

Nassery H. (1971) Phosphate absorption by plants from habitats of different phosphate status. III. Phosphate fractions in the roots of intact plants. *New Phytologist* **70**, 949–951.

Naveh Z. (1975) The evolutionary significance of fire in the Mediterranean region. *Vegetatio* **29**, 199–208.

Naveh Z. & Whittaker R.H. (1980) Structural and floristic diversity of shrublands and woodlands in Northern Israel and other Mediterranean areas. *Vegetatio* **41**, 171–190.

Neales T.F. (1975) The gas exchange of patterns of CAM plants. In *Environmental and Biological Control of Photosynthesis* (Ed. R. Marcelle), pp. 299–310. Junk, The Hague.

Nelsen C.E. & Safir G.R. (1982) Increased drought tolerance of mycorrhizal onion plants caused by improved phosphorus nutrition. *Planta* **154**, 407–413.

Nelson D. & Goering J. (1978) Assimilation of silicic acid by phytoplankton in the Baja California and northwest Africa upwelling systems. *Limnology and Oceanography* **23**, 508–517.

Nettancourt D. de (1977) *Incompatibility in Angiosperms.* Springer-Verlag, New York.

Newbould P.J. (1967) *Methods for Estimating the Primary Production of Forests.* IBP Handbook No. 2. Blackwell Scientific Publications, Oxford.

Newell S.J., Solbrig O.T. & Kincaid D.T. (1981) Studies on the population biology of the genus *Viola.* III. The demography of *Viola blanda* and *Viola pallens. Journal of Ecology* **69,** 997–1016.

Newman E.I. (1974) Root and soil water relations. In *The Plant Root and its Environment* (Ed. E.W. Carson), pp. 363–440. University Press of Virginia, Charlottesville.

Newman E.I. (1985) The rhizosphere: carbon sources and microbial populations. In *Ecological Interactions in Soil. Plants, Microbes and Animals* (Eds A.H. Fitter, D. Atkinson, D.J. Read & M.B. Usher), pp. 107–121. Blackwell Scientific Publications, Oxford.

Newman E.I. & Rovira A.D. (1975) Allelopathy among some British grassland species. *Journal of Ecology* **63,** 727–737.

Newton P. (1963) Studies on the expansion of the leaf surface. II. The influence of light intensity and photoperiod. *Journal of Experimental Botany* **14,** 458–482.

Ng F.S.P. (1978) Strategies of Establishment in Malaysian Forest Trees. In *Tropical Trees as Living Systems* (Eds P.B. Tomlinson & M.H. Zimmermann), pp. 129–162. Cambridge University Press, Cambridge.

Nieuwhof M. (1963) Pollination and contamination of *Brassica oleracea* L. *Euphytica* **12,** 17–26.

Niklas K.J. & Kerchner V. (1984) Mechanical and photosynthetic constraints on the evolution of plant shape. *Paleobiology* (in press).

Nip-van der Voort J., Hengeveld R. & Haeck J. (1979) Immigration rates of plant species in three Dutch polders. *Journal of Biogeography* **6,** 301–308.

Nobel P.S. (1977) Internal leaf area and cellular CO_2 resistance: photosynthetic implications of variations with growth conditions and plant species. *Physiologia Plantarum* **40,** 137–144.

Noble J.C., Bell A.D. & Harper J.L. (1979) The population biology of plants with clonal growth. I. The morphology and structural demography of *Carex arenaria. Journal of Ecology* **67,** 983–1008.

Noble I.R. & Slatyer R.O. (1980) The use of vital attributes to predict successional changes in plant communities subject to recurrent disturbances. *Vegetatio* **43,** 5–21.

Noggle J.C. (1980) Sulfur accumulation by plants: the role of gaseous sulfur in crop nutrition. In *Atmospheric Sulfur Deposition: Environmental Impact and Health Effects* (Eds D.S. Shriner, C.R. Richmond & S.E. Lindberg), pp. 289–297. Ann Arbor Science Publications, Ann Arbor, Michigan.

Nyahoza F., Marshall C. & Sagar G.R. (1973) The interrelationships between tillers and rhizomes of *Poa pratensis* L.—an autoradiographic study. *Weed Research* **13,** 304–309.

Nye P.H. & Tinker P.B. (1977) *Solute Movement in the Soil-Root System.* Blackwell Scientific Publications, Oxford.

Obeid M., Machin D. & Harper J.L. (1967) Influence of density on plant to plant variations in fiber flax, *Linum usitatissimum. Crop Science* **7,** 471–473.

O'Dowd D.J. & Hay M.E. (1980) Mutualism between harvester ants and a desert ephemeral: seed escape from rodents. *Ecology* **61,** 531–540.

Ogden J. (1970) Plant population structure and productivity. *Proceedings of the New Zealand Ecological Society* **17,** 1–9.

Ogden J. (1974) The reproductive strategy of higher plants. II. The reproductive strategy of *Tussilago farfara. Journal of Ecology* **62,** 291–324.

Ogilvie R.T. & von Rudloff E. (1968) Chemosystematic studies in the genus *Picea* (Pinaceae). IV. The introgression of white and Engelmann spruce as found along the Bow River. *Canadian Journal of Botany* **46,** 901–908.

Oinonen E. (1967) Sporal regeneration of bracken in Finland in the light of dimensions and age of its clones. *Acta Forestalia Fennica* **83** 3–96.

Oldeman R.A.A. (1978) Architecture and energy exchange of dicotyledenous trees in the forest. In *Tropical Trees as Living Systems* (Eds P.B. Tomlinson & M.H. Zimmerman), Chapter 23. Cambridge University Press, Cambridge.

Olsen C. (1923) Studies on the hydrogen ion concentration of the soil and its significance to the vegetation, especially to the natural distribution of plants. *Compte rendu des travaux du Laboratoire de Carlsberg* **15**, 1–166.

Olsen C. (1953) The significance of concentration for the rate of ion absorption by higher plants in water culture. IV. The influence of hydrogen ion concentration. *Physiologia Plantarum* **6**, 848.

Olson J.S. (1958) Rates of succession and soil changes on southern Lake Michigan sand dunes. *Botanical Gazette* **119**, 125–169.

Opler P.A., Frankie G.W. & Baker H.G. (1980) Comparative phenological studies of treelet and shrub species in tropical wet and dry forests in the lowlands of Costa Rica. *Journal of Ecology* **68**, 167–188.

Orians G.H. (1982) The influence of tree-falls in tropical forests in tree species richness. *Tropical Ecology* **23**, 255–279.

Orians G.H. & Solbrig O.T. (1977) A cost-income model of leaves and roots with special reference to arid and semi-arid areas. *American Naturalist* **111**, 677–690.

Osmond C.B., Winter K. & Zeigler H. (1981) Functional significance of different pathways of CO_2 fixation in photosynthesis. In *Encyclopedia of Plant Physiology. New Series*, Vol. 12B (Eds A. Pirson & M.H. Zimmermann), pp. 480–547. Springer Verlag, Berlin.

Pacala S.W. & Roughgarden J. (1982) Spatial heterogeneity and interspecific competition. *Theoretical Population Biology* **21**, 92–113.

Pacala S.W. & Silander J.A. (1985) Neighborhood models of plant population dynamics. I. Single species models of annuals. *American Naturalist* **125**, 385–411.

Paces T. (1985) Sources of acidification in Central Europe estimated from elemental budgets in small basins. *Nature* **315**, 31–36.

Paige K.N. & Whitham T.G. (1985) Individual and population shifts in flower color by scarlet gilia: a mechanism for pollinator tracking. *Science* **227**, 315–317.

Paine R.T. (1966) Food web complexity and species diversity. *American Naturalist* **100**, 65–75.

Painter R.H. (1958) Resistance of plants to insects. *Annual Review Entomology* **3**, 267–290.

Painter R.H. (1966) Lessons to be learned from past experience in breeding plants for insect resistance. In *Breeding Pest Resistant Trees* (Eds H.D. Gerhold, E.J. Schreiner, R.E. McDermott & J.A. Winieski), pp. 349–355. Pergamon Press, Oxford.

Pamblad I.G. (1968) Competition in experimental studies on population of weeds with emphasis on the regulation of population size. *Ecology* **49**, 26–34.

Paltridge F.W. & Denholm J.V. (1974) Plant yield and the switch from vegetative to reproductive growth. *Journal of Theoretical Biology* **44**, 23–34.

Parker A.J. & Peet R.K. (1984) Size and age structure of coniferous forests. *Ecology* **65**, 1685–1689.

Parker J. (1977) Phenolics in black oak bark and leaves. *Journal of Chemical Ecology* **3**, 489–496.

Parker M.A. (1985) Size-dependent herbivore attack and the demography of an arid grassland shrub. *Ecology* **66**, 850–860.

Parker M.A. & Root R.B. (1981) Insect herbivores limit habitat distribution of a native composite, *Machaeranthera canescens*. *Ecology* **62**, 1390–1392.

Parkhurst D.F. & Loucks O.L. (1972) Optimal leaf size in relation to environment. *Journal of Ecology* **60**, 505–537.

Paterniani E. & Short A.C. (1974) Effective maize pollen dispersal in the field. *Euphytica* **23**, 129–134.

Patrick R. (1963) The structure of diatom communities under varying ecological conditions. *Annals of the New York Academy of Science* **108**, 359–365.

Patrick R. (1967) The effect of varying amounts and ratios of nitrogen and phosphate on algae blooms. *Proceedings of the 21st Annual Industrial Waste Conference*, pp. 41–51. Purdue University, Indiana.

Patten B. (1962) Species diversity in net phytoplankton of Raritan Bay. *Journal of Marine Research* **20**, 57–75.

Pearce R.B., Brown R.H. & Blaser R.E. (1967) Photosynthesis in plant communities as influenced by leaf angle. *Crop Science* **7**, 321–324.

Pearcy R.W. & Ehleringer J. (1984) Comparative ecophysiology of C_3 and C_4 plants. *Plant, Cell and Environment* **7**, 1–13.

Peart D.R. (1982) *Experimental analysis of succession in a grassland at Sea Ranch, California.* Ph.D. Thesis, University of California, Davis.

Penning de Vries F.W.T. (1975) The cost of maintenance processes in plant cells. *Annals of Botany* **39**, 77–92.

Pennington W. (1969) *The History of British Vegetation.* English Universities Press, London.

Pergande T. (1901) The life history of two species of plant-lice. *Technical Series* No. 9, pp. 7–44. Department of Agriculture, Government Printing Office, Washington D.C.

Person C. (1966) Genetic polymorphism in parasitic systems. *Nature* **212**, 266–267.

Person C. (1967) Genetic aspects of parasitism. *Canadian Journal of Botany* **45**, 1193–1204.

Person C., Samborski D.J. & Rohringer R. (1962) The gene-for-gene concept. *Nature* **194**, 561–562.

Person C. & Sidhu G. (1971) Genetics of host-parasite interrelationships. In *Mutation Breeding for Disease Resistance*, pp. 31–37. International Energy Agency, Vienna.

Phillips D.L. & MacMahon J.A. (1981) Competition and spacing patterns in desert shrubs. *Journal of Ecology* **69**, 97–115.

Pianka E.R. (1970) On r- and K-selection. *American Naturalist* **104**, 592–597.

Pianka E.R. (1981) Competition and niche theory. In *Theoretical Ecology* (Ed. R.M. May), pp. 167–176. Blackwell Scientific Publications, Oxford.

Pickard W.F. (1983) Three interpretations of the self-thinning rule. *Annals of Botany* **51**, 749–757.

Pickett S.T.A. (1982) Population patterns through twenty years of oldfield succession. *Vegetatio* **49**, 15–59.

Pickett S.T.A. & White P.S. (Eds) (1985) *The Ecology of Natural Disturbance as Patch Dynamics.* Academic Press, New York.

Pielou E.C. (1961) Segregation and symmetry in two-species populations as studied by nearest-neighbour relationships. *Journal of Ecology* **49**, 255–269.

Pielou E.C. (1977) *Mathematical Ecology.* John Wiley, New York.

Pigott C.D. (1983) Regeneration of oak-birch woodland following exclusion of sheep. *Journal of Ecology* **71**, 629–646.

Pigott C.D. & Taylor K. (1964) The distribution of some woodland herbs in relation to the supply of nitrogen and phosphorus in the soil. *Journal of Ecology* **52**(Suppl.), 175–185.

Pijl L. van der (1972) *Principles of Dispersal In Higher Plants*, 2nd edn. Springer-Verlag, Berlin.

Pimm S.L. (1982) *Food Webs.* Chapman & Hall, London.

Pinero D., Martinez-Ramos M. & Sarukhán J. (1984) A population model of *Astrocaryum mexicanum* and a sensitivity analysis of its finite rate of increase. *Journal of Ecology* **72**, 977–991.

Pitelka L.F. (1984) Application of the $-3/2$ power law to clonal herbs. *American Naturalist* **123**, 442–449.

Pitelka L.F. & Ashmun J.W. (1985) The physiology and ecology of connections between ramets in clonal plants. In *The Population Biology and Evolution of Clonal Organisms* (Eds J. Jackson, L. Buss & R. Cook). Yale University Press.

Pitelka L.F., Stanton D.S. & Peckenham M.O. (1980) Effects of light and density on resource allocation in a forest herb, *Aster acuminatus* (Compositae). *American Journal of Botany* **67**, 942–948.

Pitelka L.F., Thayer M.E. & Hansen S.B. (1983) Variation in achene weight in *Aster acuminatus*. *Canadian Journal of Botany* **61**, 1415–1420.

Platt W.J. & Weiss I.M. (1977) Resource partitioning and competition within guild of fugitive prairie plants. *American Naturalist* **111**, 479–513.

Poole R.W. & Rathcke B.J. (1979) Regularity, randomness, and aggregation in flowering phenologies. *Science* **203**, 470–471.

Poore M.E.D. (1955) The use of phytosociological methods in ecological investigations. III. Practical application. *Journal of Ecology* **43**, 606–651.

Port G.R. & Thompson J.R. (1980) Outbreaks of insect herbivores on plants along motorways in the United Kingdom. *Journal of Applied Ecology* **17**, 649–656.

Posnette A.F. & Cropley R. (1962) Further studies on a selection of "Williams Bon Chretin" pear compatible with quince A rootstocks. *Journal of Horticultural Science* **37**, 291–294.

Post W.M., Pastor J., Zinke P.J. & Stangenberger A.G. (1985) Global patterns of soil nitrogen storage. *Nature* **317**, 613–616.

Powles S.B. (1984) Photoinhibition of photosynthesis induced by visible light. *Annual Review Plant Physiology* **35**, 15–44.

Prestidge R.A. & McNeill S. (1983) The role of nitrogen in the ecology of grassland *Auchenorrhyncha*. In *Nitrogen as an Ecological Factor* (Eds J.A. Lee, S. McNeill & I.H. Rorison), pp. 257–281. Blackwell Scientific Publications, Oxford.

Preston F.W. (1962) The canonical distribution of commoness and rarity. Part I. *Ecology* **43**, 185–215.

Price W.C. (1940) Acquired immunity from plant virus diseases. *Quarterly Review of Biology* **15**, 338–361.

Primack R.B., Rittenhouse A.R. & August P.V. (1981) Components of reproductive effort and yield in goldenrods. *American Journal of Botany* **68**, 855–858.

Pritts M.P. & Hancock J.F. (1983) Seasonal and lifetime allocation patterns of the woody goldenrod, *Solidago pauciflosculosa* Michaux (Compositae). *American Journal of Botany* **70**, 216–221.

Procter J. & Woodell S.R.J. (1975) The ecology of serpentine soils. *Advances in Ecological Research* **9**, 256–366.

Proctor M. & Yeo P. (1972) *The Pollination of Flowers*. Taplinger, New York.

Putwain P.D. & Harper J.L. (1970) Studies in the dynamics of plant populations. III. The influence of associated species on populations of *Rumex acetosella* L. and *Rumex acetosa* L. in grassland. *Journal of Ecology* **58**, 251–264.

Putz F.E. (1984) The natural history of lianes on Barro Colorado Island, Panama. *Ecology* **65**, 1713–1724.

Putz F.E., Coley P.D., Lu K., Montalvo A. & Aiella A. (1983) Uprooting and snapping of trees, structural causes and ecological consequences. *Canadian Journal of Forestry Research* **13**, 1011–1020.

Quinlan J.D. & Preston A.P. (1968) Effects of thinning blossom and fruitlets on growth and cropping of Sunset apple. *Journal of Horticultural Science* **43**, 373–381.

Quinlan G.J.D. & Preston A.P. (1971) The influence of shoot competition on fruit retention and cropping of apple trees. *Journal of Horticultural Science* **46**, 525–534.

Rabinovitch-Vin A. (1983) Influence of nutrients on the composition and distribution of plant communities in Mediterranean-Type ecosystems in Israel. In *Mediterranean-Type Ecosystems* (Eds F.J. Kruger, D.T. Mitchell & J.U.M. Jarvis), pp. 74–85. Springer-Verlag.

Rabinowitz D. (1979) Bimodal distributions of seedling weight in relation to density of *Festuca paradoxa* Desv. *Nature* **277**, 297–298.

Rabinowitz D. (1981) Seven forms of rarity. In *The Biology of Rare Plant Conservation* (Ed. H. Synge), pp. 205–217. John Wiley, Chichester.

Rabinowitz D. & Rapp J.K. (1980) Seed rain in a North American tall grass prairie. *Journal of Applied Ecology* **17**, 793–802.

Rabinowitz D. & Rapp J.K. (1981) Dispersal abilities of seven sparse and common grasses from a Missouri prairie. *American Journal of Botany* **68**, 616–624.

Rabinowitz D., Rapp J.K., Sork V.L., Rathcke B.J., Reese G.A. & Weaver J.C. (1981) Phenological properties of wind- and insect-pollinated prairie plants. *Ecology* **62**, 49–56.

Rabotnov T.A. (1978) On coenopopulations of plants reproducing by seeds. In *Structure and Functioning of Plant Populations* (Eds A.H.J. Freysen & J.W. Woldendorp), pp. 1–26. North-Holland Publishing Company, Amsterdam.

Ranwell D.S. (1959) Newborough Warren, Anglesey. I. The dune system and dune slack habitat. *Journal of Ecology* **47**, 571–601.

Ranwell D.S. (1972) *Ecology of Salt Marshes and Sand Dunes.* Chapman & Hall, London.

Rapport D.J. (1971) An optimization model of food selection. *American Naturalist* **105**, 575–578.

Rathcke B. (1983) Competition and facilitation among plants for pollination. In *Pollination Biology* (Ed. L. Real), pp. 305–329. Academic Press, New York.

Raunkiaer C. (1934) *The Life Forms of Plants and Statistical Plant Geography.* Clarendon Press, Oxford.

Raven P.H., Evert R.F. & Curtis H. (1981) *Biology of Plants*, 3rd edn. Worth Publishers, Inc., New York.

Read D.J., Francis R. & Finlay R.D. (1985) Mycorrhizal mycelia and nutrient cycling in plant communities. In *Ecological Interactions in Soil. Plants, Microbes and Animals* (Eds A.H. Fitter, D. Atkinson, D.J. Read & M.B. Usher), pp. 193–217. Blackwell Scientific Publications, Oxford.

Reader R.J. (1978) Contribution of overwintering leaves to the growth of three broad-leaved evergreen shrubs belonging to the Ericaceae family. *Canadian Journal of Botany* **56**, 1247–1261.

Real L. (1983) *Pollination Biology.* Academic Press, London.

Regal P. (1982) Pollination by wind and animals: Ecology of geographic patterns. *Annual Review of Ecology and Systematics* **13**, 497–524.

Rehr S.S., Feeny P.P. & Janzen D.H. (1973) Chemical defense in Central American non-ant acacias. *Journal of Animal Ecology* **42**, 405–416.

Reiners W.A., Worley I.A. & Lawrence D.B. (1971) Plant diversity in a chronosequence at Glacier Bay, Alaska. *Ecology* **52**, 55–69.

Rhoades D.F. (1982) Responses in alder and willow to attack by tent caterpillars and webworms: evidence for phenomenal sensitivity of willow. In *Plant Resistance to Insects* (Ed. P. Hedin), pp. 55–68. American Chemical Society, Washington D.C.

Rhoades D.F. (1985) Offensive-defensive interactions between herbivores and plants: their relevance in herbivore population dynamics and ecological theory. *American Naturalist* **125**, 205–238.

Rhoades D.F. & Cates R.G. (1976) Toward a general theory of plant antiherbivore chemistry. In *Biochemical Interaction Between Plants and Insects* (Eds J.W. Wallace & R.L. Mansell), pp. 168–213. [*Recent Advances in Phytochemistry* **10**, 168–213.]

Rice B. & Westoby M. (1983) Plant species richness at the 0.1 hectare scale in Australian vegetation compared to other continents. *Vegetatio* **52**, 129–140.

Rice E.L. (1984) *Allelopathy*, 2nd edn. Academic Press, London.

Rice W.R. (1983) Parent-offspring pathogen transmission: a selective agent promoting sexual reproduction. *American Naturalist* **121**, 187–203.

Richards P.W. (1983) The three dimensional structure of tropical rain forest. In *Tropical Rain Forest: Ecology and Management* (Eds S.L. Sutton, T.C. Whitmore & A.C. Chadwick), pp. 3–10. Blackwell Scientific Publications, Oxford.

Ricklefs R.E. (1977) Environmental heterogeneity and plant species diversity: a hypothesis. *American Naturalist* **111**, 376–381.

Risch S. (1980) The population dynamics of several herbivorous beetles in a tropical agroecosystem: the effect of intercropping corn, beans and squash in Costa Rica. *Journal of Applied Ecology* **17**, 593–612.

Risch S. & Boucher D. (1976) What ecologists look for. *Bulletin of the Ecological Society of America* **57**, 8–9.

Ritland K. & Jain S. (1984) The comparative life histories of two annual *Limnanthes* species in a temporally variable environment. *American Naturalist* **124**, 656–679.

Ritz K. & Newman E.I. (1984) Movement of ^{32}P between intact grassland plants of the same age. *Oikos* **43**, 138–142.

Roberts D.R. (1970) Within tree variation of monoterpene hydrocarbon composition of slash pine oleoresin. *Phytochemistry* **9**, 809–815.

Roberts E.H. (1972) Dormancy: a factor affecting seed survival in soil. In *Viability of Seeds* (Ed. E.H. Roberts), pp. 321–359. Syracuse University Press, Syracuse.

Roberts H.A. (1981) Seed banks in soils. *Advances in Applied Biology* **6**, 1–55.

Roberts H.A. & Feast P.M. (1973) Changes in the numbers of viable weed seeds in soil under different regimes. *Weed Research* **13**, 298–303.

Robertson G.P. & Vitousek P.M. (1981) Nitrification potentials in primary and secondary succession. *Ecology* **62**, 376–386.

Robinson D. & Rorison I.H. (1983) A Comparison of the responses of *Lolium perenne* L., *Holcus lanatus* L. and *Deschampsia flexuosa* (L.) Trin. to a localised supply of nitrogen. *New Phytologist* **94**, 263–273.

Rockwood L.L. (1973) The effect of defoliation on seed production of six Costa Rican tree species. *Ecology* **54**, 1363–1369.

Rogan P.G. & Smith D.L. (1974) Patterns of translocation of ^{14}C-labelled assimilates during vegetative growth of *Agropyron repens* (L.) Beans. *Zeitschrift fur Pflanzenphysiologie* **73**, 405–414.

Romeo J.T. & Bell E.A. (1974) Distribution of amino-acids and certain alkaloids in *Erythrina* spp. *Lloydia* **37**, 543–568.

Root R.B. (1973) Organization of a plant–arthropod association in simple and diverse habitats: the fauna of collards (*Brassica oleracea*). *Ecological Monographs* **43**, 95–124.

Rorison I.H., Peterkin J.H. & Clarkson D.T. (1983) Nitrogen source, temperature and the growth of herbaceous plants. In *Nitrogen as an Ecological Factor* (Eds J.A. Lee, S. McNeill & I.H. Rorison), pp. 189–209. British Ecological Society Symposium 22. Blackwell Scientific Publications, Oxford.

Rosenthal G.A. & Janzen D.H. (Eds) (1979) *Herbivores. Their Interaction with Secondary Plant Metabolites*. Academic Press, New York.

Ross M.A. & Harper J.L. (1972) Occupation of biological space during seedling establishment. *Journal of Ecology* **60**, 77–78.

Roth I. (1984) *Stratification of Tropical Forests as seen in Leaf Structure*. Junk, The Hague.

Roughgarden J. (1976) Resource partitioning among competing species: a co-evolutionary approach. *Theoretical Population Biology* **9**, 388–424.

Rudloff E. von (1962b) Gas-liquid chromatography of terpenes. V. The volatile oils of the leaves of black, white, and Colorado spruce. *Journal of Technical Association Pulp Paper Industry* **45**, 181–184.

Rudloff E. von (1968) Gas-liquid chromatography of terpenes. XVI. The volatile oil of the leaves of *Juniperus ashei* Buchholz. *Canadian Journal of Chemistry* **46**, 679–683.

Rudloff E. von (1969) Scope and limitations of gas chromatography of terpenes in chemosystematic studies. *Recent Advances in Phytochemistry* **2**, 127–159.

Rudloff E. von (1972b) Seasonal variation in the composition of the volatile oils of the leaves, buds, and twigs of white spruce (*Picea glauca*). *Canadian Journal of Botany* **50**, 1595–1603.

Russell C.E. & Berryman A.A. (1976) Host resistance to the fir engraver beetle. Monoterpene composition of *Abies grandis* pitch blisters and fungus-infected woods. *Canadian Journal of Botany* **54**, 14–18.

Sagar G.R. (1974) On the ecology of weed control. In *Biology in Pest and Disease Control* (Eds D. Price-Jones & M.E. Solomon), pp. 42–56. *BES Symposium* **13**. Blackwell Scientific Publications, Oxford.

Sagar G.R. & Mortimer A.M. (1976) An approach to the study of the population dynamics of plants with special reference to weeds. *Applied Biology* **1**, 1–47.

Sakai A.K. & Oden N.L. (1983) Spatial pattern of sex expression in silver maple (*Acer saccharinum* L.): Morisita's index and spatial autocorrelation. *American Naturalist* **122**, 489–508.

Sale P.F. (1977) Maintenance of high diversity in coral reef fish communities. *American Naturalist* **111**, 337–359.

Salisbury E.J. (1921) The significance of the calcicolous habit. *Journal of Ecology* **8**, 202–215.

Salisbury E.J. (1942) *The Reproductive Capacity of Plants: Studies in Quantitative Biology.* G. Bell & Sons, London.

Salisbury F.B. & Ross C. (1985) *Plant Physiology*, 3rd edn. Wadsworth Publishing Co., California.

Salzman A.G. (1985) Habitat selection in a clonal plant. *Science* **228**, 603–604.

Sarukhán J. (1974) Studies on plant demography: *Ranunculus repens* L., *R. bulbosus* L., and *R. acris* L. II. Reproductive strategies and seed population dynamics. *Journal of Ecology* **62**, 151–177.

Sarukhán J. (1980) Demographic problems in tropical systems. In *Demography and Evolution in Plant Populations* (Ed. O.T. Solbrig), pp. 161–188. Blackwell Scientific Publications, Oxford.

Sarukhán J. & Gadgil M. (1974) Studies on plant demography: *Ranunculus repens* L., *R. bulbosus* L. and *R. acris* L. III. A mathematical model incorporating multiple models of reproduction. *Journal of Ecology* **62**, 921–936.

Sarukhán J. & Harper J.L. (1973) Studies on plant demography; *Ranunculus repens* L., *R. bulbosus* L. and *R. acris* L. I. Population flux and survivorship. *Journal of Ecology* **61**, 675–716.

Sarukhán J., Martinez-Ramos M. & Piñero D. (1984) The analysis of demographic variability at the individual level and its population consequences. In *Perspectives on Plant Population Ecology* (Eds R. Dirzo & J. Sarukhán), pp. 83–106. Sinauer, Sunderland, Massachusetts.

Sauerbeck D. & Johnen B.G. (1977) Root formation and decomposition during plant growth. In *Proceedings of the International Symposium on Soil Organic Matter Studies*, Vol. 1, pp. 141–148. IAEA, Vienna.

Schaal B.A. (1980a) Measurement of gene flow in *Lupinus texensis*. *Nature* **284**, 450–451.

Schaal B.A. (1980b) Reproductive capacity and seed size in *Lupinus texensis*. *American Journal of Botany* **67**, 703–709.

Schaal B.A. & Leverich W.J. (1981) The demographic consequences of two-stage life cycles: survivorship and the time of reproduction. *American Naturalist* **118**, 135–138.

Schaeffer W.M. & Gadgil M.D. (1975) Selection for optimal life histories in plants. In *The Ecology and Evolution of Communities* (Eds. M. Cody & J. Diamond), pp. 142–157. Harvard University Press, Cambridge, Massachusetts.

Schaeffer W.M. & Schaeffer M.V. (1979) The adaptive significance of variations in reproductive habit in Agaraceae. II. Pollinator foraging behaviour and selection for increased reproductive expenditure. *Ecology* **60**, 1051–1069.

Schelske C.L. & Stoermer E. (1971) Eutrophication, silica depletion, and predicted changes in algal quality in Lake Michigan. *Science* **173**, 423–442.

Schemske D.W. (1977) Flowering phenology and seed set in *Claytonia virginica* (Portulaceae). *Bulletin of the Torrey Botanical Club* **104**, 254–263.

Schemske D.W. (1978) Evolution of reproductive characteristics in *Impatiens* (Balsaminaceae): the significance of cleistogamy and chasmogamy. *Ecology* **59**, 596–613.

Schemske D.W. (1981) Floral convergence and pollinator sharing in two bee-pollinated tropical herbs. *Ecology* **62**, 946–954.

Schemske D.W. (1983a) Breeding system and habitat effects on fitness components in three neotropical *Costus* (Zingiberaceae) *Evolution* **37**, 523–539.

Schemske D.W. (1983b) Limits to Specialization and Coevolution in Plant-animal Mutualisms. In *Coevolution* (Ed. M. Nitecki), pp. 67–110. University of Chicago, Chicago.

Schemske D.W. & Horvitz C.C. (1984) Variation among floral visitors in pollination ability: a precondition for mutualism specialization. *Science* **225**, 519–521.

Schemske D.W. & Lande R. (1985) The evolution of self-fertilization and inbreeding depression in plants. II. Empirical observations. *Evolution* **39**, 41–52.

Schimper A.F.W. (1903) *Plant Geography upon a Physiological Basis.* Translated by W.R. Fisher, Clarendon Press, Oxford.

Schindler D. (1977) Evolution of phosphorus limitation in lakes. *Science* **195**, 260–262.

Schlesinger W.H. & Gill D.S. (1978) Demographic studies of the chaparral shrub, *Ceanothus megacarpus*, in the Sant Ynez mountains, California. *Ecology* **59**, 1256–1263.

Schmitt J. (1980) Pollinator foraging behavior and gene dispersal in *Senecio* (Compositae). *Evolution* **34**, 934–943.

Schoen D.J. (1982) Genetic variation and the breeding system of *Gilia achilleifolia*. *Evolution* **36**, 361–370.

Schoen D.J. (1983) Relative fitness of selfed and outcrossed progeny in *Gilia achilleifolia* (Polemoniaceae). *Evolution* **37**, 292–301.

Schulze E.D. (1983) Plant life forms and their carbon, water and nutrient relations. In *Encyclopaedia of Plant Physiology*, Volume 12B, pp. 615–676. Springer, Berlin.

Seaward M.R.D. (Ed.) *Lichen Ecology.* Academic Press, London.

Service P.M. (1984) Genotypic interactions in an aphid-host plant relationship: *Uroleucon rudbeckiae* and *Rudbeckia lacinata*. *Oecologia* **61**, 271–276.

Service P.M. & Lenski R.E. (1982) Aphid genotypes, plant phenotypes, and genetic diversity: a demographic analysis of experimental data. *Evolution* **36**, 1276–1282.

Sestak Z. (Ed.) (1985) *Photosynthesis During Leaf Development.* T:VS 11. Junk, The Hague.

Sestak Z., Catsky J. & Jarvis P.G. (1971) Plant photosynthetic production. *Manual of Methods.* Junk, The Hague.

Sharitz R.R. & McCormick J.F. (1973) Population dynamics of two competing annual species. *Ecology* **54**, 723–740.

Shaver G.R., Chapin F.S. (1980) Response to fertilization by various plant growth forms in an Alaskan (USA) tundra: nutrient accumulation and growth. *Ecology* **61**, 662–675.

Shaver G.R., Chapin F.S. & Billings W.D. (1979) Ecotypic differentiation in *Carex aquatilis* on ice wedge polygons in the Alaskan coastal tundra. *Journal of Ecology* **67**, 1025–1045.

Shepard J.F., Bidney D. & Shahin E. (1980) Potato protoplasts in crop improvement. *Science* **208**, 17–24.

Shinozaki K. & Kira T. (1956) Intraspecific competition among higher plants. VII. Logistic theory of the C-D effect. *Journal of the Institute of Polytechnics, Osaka City University, Series D* **7**, 35–72.

Shinozaki K., Yoda K., Hozumi K. & Kira T. (1964a) A quantitative analysis of plant form—the pipe model theory. I. Basic analyses. *Japanese Journal of Ecology* **14**, 97–105.

Shinozaki K., Yoda K., Hozumi K. & Kira T. (1964b) A quantitative analysis of plant form—the pipe model theory. II. Further evidence of the theory and its application in forest ecology. *Japanese Journal of Ecology* **14**, 133–139.

Shmida A., Evanari M. & Noy-Meir I. (1984) Hot desert ecosystems; an integrated view. In *Hot Desert Ecosystems* (Eds M. Evanari, I. Noy-Meir & D.W. Goodall). Elsevier Publications, Holland.

Shmida A. & Whittaker R.H. (1984) Convergence and non-convergence of Mediterranean type communities in the Old and the New World. In *Being Alive on Land* (Eds N.S. Margaris, M. Arianoustou-Faraggitati & W.C. Oechel), pp. 5–11. Junk, The Hague.

Sibly R. & Calow P. (1982) Asexual reproduction in protozoa and invertebrates. *Journal of Theoretical Biology* **96**, 401–424.

Sibly R. & Calow P. (1983) An integrated approach to life-cycle evolution using selective landscapes. *Journal of Theoretical Biology* **102**, 527–547.

Silander J.A. & Antonovics J. (1982) Analysis of interspecific interactions in coastal plant community—a peturbation approach. *Nature* **298**, 557–560.

Silvertown J.W. (1980) The dynamics of a grassland ecosystem: Botanical equilibrium in the park grass experiment. *Journal of Applied Ecology* **17**, 491–504.

Silvertown J.W. (1982) *Introduction to Plant Population Ecology*. Longman, London.

Silvertown J.W. (1983) Why are biennials sometimes not so few? *American Naturalist* **121**, 448–453.

Silvertown J.W. (1985) When plants play the field. In *Evolution: Essays in Honour of John Maynard Smith* (Eds P.J. Greenwood, P.H. Harvey & M. Slatkin), pp. 143–153. Cambridge University Press, Cambridge.

Simkin T. & Fiske R.S. (1983) *Krakatau 1883. The Volcanic Eruption and its Effects*. Smithsonian Institute Press, Washington D.C.

Simpson B.B. & Neff J.L. (1983) Evolution and diversity of floral rewards. In *Handbook of Experimental Pollination Biology* (Eds C.E. Jones & R.J. Little), pp. 142–159. Von Nostrand Reinhold, New York.

Singh B.B., Hadley H.H. & Bernard R.L. (1971) Morphology of pubescence in soybeans and its relation to plant vigour. *Crop Science* **11**, 13–16.

Singh M., Ogren W.L. & Widholm J.M. *et al.* (1974) Photosynthetic characteristics of several C_δ and C_ϕ plant species grown under different light intensities. *Crop Science* **14**, 563–566.

Slatkin M. (1974) Competition and regional coexistence. *Ecology* **55**, 128–134.

Slatyer R.O. (1967) *Plant-Water Relationships*. Academic Press, London.

Small E. (1972) Photosynthetic rates in relation to nitrogen recycling as an adaptation to nutrient deficiency in peat bog plants. *Canadian Journal of Botany* **50**, 2227–2233.

Smayda T. (1975) Net phytoplankton and the greater than 20 micron phytoplankton size fraction in upwelling waters off Baja California. *Fisheries Bulletin* **73**, 38–50.

Smith B.H. (1983) Demography of *Floerkia proserpinacoides,* a forest floor annual. I. Density dependent growth and mortality. *Journal of Ecology* **71**, 391–404.

Smith B.H. (1984) The optimal design of a herbaceous body. *American Naturalist* **123**, 197–211.

Smith C.C. (1975) The coevolution of plants and seed predators. In *Coevolution of Animals and Plants* (Eds L.E. Gilbert & P.H. Raven), pp. 53–77. University of Texas Press, Austin.

Smith C.C. & Follmer D. (1972) Food preferences of squirrels. *Ecology* **53**, 82–91.

Smith F.A. (1973) The internal control of nitrate uptake into excised barley roots with differing salt contents. *New Phytologist* **72**, 769–782.

Smith F.G. & Thorneberry G.O. (1951) The tetrazolium test and seed viability. *Proceedings of the Association of Official Seed Analysts of North America* **41**, 105–108.

Smith J.N. (1962) Detoxification mechanisms. *Annual Review of Entomology* **7**, 465–480.

Smith R.I.L. (1984) Terrestrial plant biology of the sub-Antarctic and Antarctic. In *Antarctic Biology,* Vol. 1 (Ed. R.M. Laws), pp. 61–162. Academic Press, London.

Smith V.H. (1983) Low nitrogen to phosphorus ratio favour dominance by blue-green algae in lake phytoplankton. *Science* **221**, 669–671.

Smith-Gill S.J. (1983) Developmental plasticity: developmental conversion versus phenotypic modulation. *American Zoologist* **23**, 47–55.

Snaydon R.W. (1962) Micro-distribution of *Trifolium repens* L. and its relation soils factors. *Journal of Ecology* **50**, 133–143.

Snaydon R.W. (1984) Plant demography in an agricultural context. In *Perspectives on Plant Population Ecology* (Eds R. Dirzo & J. Sarukhán), pp. 389–407. Sinauer, Sunderland, Massachusetts.

Snellgrove R.C., Splitstoesser W.E., Stibley D.P. & Tinker P.B. (1982) The distribution of carbon and the demand of the fungal symbiont in leek plants with vesicular-arbuscular mycorrhizas. *New Phytologist* **92**, 75–87.

Snow B.K. & Snow D.W. (1984) Long-term defence of fruit by mistle thrushes *Turdus viscivorus*. *Ibis* **126**, 39–49.

Soane I.D. & Watkinson A.R. (1979) Clonal variation in populations of *Ranunculus repens*. *New Phytologist* **82**, 537–573.

Sokal R.R. & Rohlf J.E. (1981) *Biometry*. (2nd Edn) Freeman, San Francisco.

Solbrig O.T. (1977) Plant population biology: an overview. *Systematic Botany* **1**, 202–208.

Solbrig O.T. (1981) Studies on the population biology of the genus *Viola*. II. The effect of plant size on fitness in *Viola sororia*. *Evolution* **35**, 1080–1093.

Solbrig O.T. & Simpson B.B. (1974) Components of regulation of a population of dandelions in Michigan. *Journal of Ecology* **62**, 473–486.

Soó R. (1964–1973) *A maggur flóra és vegetáció rendszertani-növényföldrajzi kézikonyue*. Vols. 1–5. Akadémian Kiadó, Budapest.

Sork V.L. (1985) Germination response in a large-seeded neotropical tree species, *Gustavia superba* (Lecythidaceae). *Biotropica* (in press).

Soule J.D. & Werner P.A. (1981) Patterns of resource allocation in plants, with special reference to *Potentilla recta*. *Bulletin of the Torrey Botanical Club* **108**, 311–319.

Southwood T.R.E. (1973) The insect-plant relationship—an evolutionary perspective. In *Insect-Plant Relationships* (Ed. H.F. van Emden). *Royal Entomological Society Symposium* **6**, 3–30.

Specht R.L. (1963) Dark Island Heath (Ninety Mile Plain, South Australia). VII. The effect of fertilizers on composition and growth, 1950–60. *Australian Journal of Botany* **11**, 62–66.

Specht R.L. & Rayson P. (1957) Dark Island Heath (Ninety Mile Plain, South Australia) I. Definition of the ecosystem. *Australian Journal of Botany* **5**, 52–85.

Spitters C.J.T. (1983) An alternative approach to the analysis of mixed cropping experiments. I. Estimation of competition effects. *Netherlands Journal of Agricultural Science* **31**, 1–11.

Sporne K.R. (1965) *The Morphology of Gymnosperms*. Hutchinson, London.

Sprent J.I. (1979) *The Biology of Nitrogen-Fixing Organisms*. McGraw Hill, New York.

Sprent J.I. (1983) Adaptive variation in legume nodule physiology resulting from host-rhizobial interactions. In *Nitrogen as an Ecological Factor*. (Eds J.A. Lee, S. McNeill & I.H. Rorison), pp. 29–42. Blackwell Scientific Publications, Oxford.

Sprugel D.G. (1976) Dynamic structure of wave-generated *Abies balsamea* forests in the north-eastern United States. *Journal of Ecology* **64**, 889–911.

Stalfelt M.G. (1972) *Plant Ecology: Plants, the Soil and Man*. Translated by M.S. Jarvis & P.G. Jarvis. Longman, London.

Stamp N.E. & Lucas J.R. (1983) Ecological correlates of explosive seed dispersal. *Oecologia* **59**, 272–278.

Stapanian M.A. & Smith C.C. (1978) A model for scatterhoarding: coevolution of fox squirrels and black walnuts. *Ecology* **59**, 884–896.

Stearns S.C. (1976) Life-history tactics: a review of ideas. *Quarterly Review of Biology* **51**, 3–47.

Stearns S.C. (1977) The evolution of life history traits: A critique of the theory and a review of the data. *Annual Review of Ecology and Systematics* **8**, 145–171.

Stearns S.C. & Crandall R.E. (1981) Bet-hedging and persistence as adaptations of colonizers. In Evolution Today. *Proceedings of 2nd International Congress on Systematics and Evolution 1980*, 371–384. Vancouver B.C.

Stebbins G.L. (1957) Self-fertilization and population variability in higher plants. *American Naturalist* **91**, 337–354.

Stebbins G.L. (1970) Adaptive radiation of reproductive characteristics in angiosperms. I. Pollination mechanisms. *Annual Review Ecology and Systematics* **1**, 307–326.

Stebbins G.L. (1974) *Flowering Plants: Evolution Above The Species Level*. Harvard University Press, Cambridge, Massachusetts.

Steeman-Neilsen E. (1954) On organic production in the oceans. *Journal de Conseil (Conseil Permanent International pour l'exploration de la Mer)* Vol. 19, No. 1.

Steingraeber D.A. (1982a) Phenotypic plasticity of branching pattern in sugar maple (*Acer saccharum*). *American Journal of Botany* **69**, 638–640.

Steingraeber D.A. (1982b) Heterophylly and neoformation of leaves in sugar maple (*Acer saccharum*). *American Journal of Botany* **69**, 1277–1282.

Steingraeber D.A., Kascht L.J. & Franck D.H. (1979) Variation of shoot morphology and bifurcation ratio in sugar maple (*Acer saccharum*) saplings. *American Journal of Botany* **66**, 441–445.

Stephenson A.G. (1979) Evolutionary examination of the floral display of *Catalpa speciosa* (Bignoniaceae). *Evolution* **33**, 1200–1209.

Stephenson A.G. (1980) Fruit set, herbivory, fruit abortion, and the fruiting strategy of *Catalpa speciosa* (Bignoniaceae). *Ecology* **6**, 57–64.

Stephenson A.G. (1981) Flower and fruit abortion: proximate causes and ultimate function. *Annual Review Ecology and Systematics* **12**, 253–281.

Stephenson A.G. (1982) When does outcrossing occur in a mass-flowering plant? *Evolution* **36**, 762–767.

Stern A.C., Boubel R.W., Turner D.B. & Fox D.L. (1984) *Fundamentals of Air Pollution*, 2nd Edn. Academic Press, New York.

Stern K. (1974) *The Genetics of Forest Ecosystems*. Springer-Verlag, New York.

Stevens P.S. (1974) *Patterns in Nature*. Atlantic Monthly Press.

Stewart A.J.A. & Thompson K. (1981) Reproductive strategies of six herbaceous species in relation to a successional sequence. *Oecologia* **52**, 269–272.

Stewart R.N. (1978) Ontogeny of the primary body in chimeral forms of higher plants. In *The Clonal Basis of Development* (Eds S. Subtelny & I.M. Sussex). pp. 131–160. Prentice-Hall, New Jersey.

Stocker G.C. & Irvine A.K. (1983) Seed dispersal by cassowaries (*Casuarius casuarius*) in North Queensland's rainforests. *Biotropica* **15**, 1970–1976.

Stott K.G. & Campbell A.I. (1971) Variation in self-compatibility in Cox clones. *Report of Long Ashton Research Station 1970*, **23.**

Streeter D.T. (1974) Ecological aspects of oak woodland conservation. In *The British Oak: Its History and Natural History* (Eds M.G. Morris & F.H. Perring). pp. 341–354. Classey, Berkshire.

Strong D.R., Lawton J.H. & Southwood T.R.E. (1984) *Insects on Plants. Community Patterns and Mechanisms*. Blackwell Scientific Publications, Oxford.

Stumpf P.K. & Conn E.E. (Eds) (1982) *Biochemistry of Plants: A Comprehensive Treatise. Vol. 7 Secondary Plant Products*. Academic Press, London.

Suighara G. (1981) S = CA^z z = 1/4: A reply to Connor and McCoy. *American Naturalist* **117**, 790–793.

Suighara G. (1984) Graph Theory, homology and food webs. *Proceedings of Symposia in Applied Mathematics* **30**, 83–101.

Summerhayes V.S. (1951) *Wild orchids of Britain*. Collins New Naturalist, London.

Sutton S.L., Whitmore T.C. & Chadwick A.C. (Eds) (1983) *Tropical Rain Forest: Ecology and Management*. Blackwell Scientific Publications, Oxford.

Sutton O.G. (1947) The theoretical distribution of airborne pollution from factory chimneys. *Quarterly Journal of the Royal Meteorological Society* **73**, 426–436.

Swain A. (1979) Tannin and lignins. In *Herbivores, their Interaction with Secondary Plant Metabolites* (Eds G.A. Rosenthal & D.H. Janzen), pp. 657–682. Academic Press, London.

Sweet G.B. (1973) Shedding of reproductive structures in forest trees. In *Shedding of Plant Parts* (Ed. T.T. Kozlowski), pp. 341–382. Academic Press, New York.

Sydes C.L. (1984) A comparative study of leaf demography in limestone grassland. *Journal of Ecology* **72**, 331–345.

Sydes C.L. & Grime J.P. (1984) A comparative study of root development using a simulated rock crevice. *Journal of Ecology* **72**, 937–946.

Sykes J.T. (1975) Tree crops. In *Crop Genetic Resources for Today and Tomorrow*, Vol. 2 (Eds O.H. Franakel & J.G. Hawkes), pp. 123–137. Cambridge University Press, Cambridge.

Symonides E. (1983) Population size regulation as a result of intra-population inter-actions. I. Effect of density on the survival and development of individuals of *Erophila verna* (L.) *Ekologia Polska* **31,** 839–881.

Tansley A.G. (1935) The use and abuse of vegetational concepts and terms. *Ecology* **16,** 284–307.

Tansley A.G. (1939) *The British Islands and their Vegetation.* Cambridge University Press, Cambridge.

Templeton A.R. & Levin D.A. (1979) Evolutionary consequences of seed pools. *American Naturalist* **114,** 232–249.

Tenhunen J.D., Lange O.L., Pereira J.S., Losch R. & Catarino F. (1981) Midday stomatal closure in *Arbutus unedo* leaves. In *Components of Productivity of Mediterranean-climate Regions—Basic and Applied Aspects* (Eds N.S. Margaris & H.A. Mooney) pp. 61–69. Junk, The Hague.

Terborgh J. (1973) On the notion of favourableness in plant ecology. *American Naturalist* **107,** 481–501.

Terborgh J. (1983) *Five New World Primates: A Study in Comparative Ecology.* Princeton University Press, Princeton.

Terry N. & Ulrich A. (1973) Effects of phosphorus deficiency on the photosynthesis and respiration of leaves of sugar beet. *Plant Physiology* **51,** 43–47.

Thompson F. (1918) *The Mistress of Vision.* Douglas Pepler, Ditchling.

Thompson J.N. (1984) Variation among individual seed masses in *Lomatium grayi* (Umbelliferae) under controlled conditions: magnitude and partitioning of the variance. *Ecology* **65,** 626–631.

Thompson J.N. (1985) Post-dispersal seed predation in *Lomatium* spp. (Umbel-liferae): variation among individuals and species. *Ecology* **66,** 1608–1616.

Thompson K. & Grime J.P. (1979) Seasonal variation in the seed banks of herbaceous species in ten contrasting habitats. *Journal of Ecology* **67,** 893–921.

Thresh J.M. (Ed.) (1981) *Pests, Pathogens and Vegetation.* Pitman, Boston.

Thurston J. (1969) The effect of liming and fertilizers on the botanical composition of permanent grassland, and on the yield of hay. In *Ecological Aspects of the Mineral Nutrition of Plants* (Ed. I. Rorison), pp. 3–10. Blackwell Scientific Publications, Oxford.

Tieszen L.L., Senyimba M.M., Imbamba S.K. & Troughton J.H. (1979) The dis-tribution of C_3 and C_4 grasses and carbon isotope discrimination along an altitudinal and moisture gradient in Kenya. *Oecologia* **37,** 337–350.

Tilman D. (1977) Resource competition between planktonic algae: an experimental and theoretical approach. *Ecology* **58,** 338–348.

Tilman D. (1980) Resources: a graphical-mechanistic approach to competition and predation. *American Naturalist* **116,** 362–393.

Tilman D. (1981) Experimental tests of resource competition theory using four species of Lake Michigan algae. *Ecology* **62,** 802–815.

Tilman D. (1982) *Resource Competition and Community Structure.* Princeton Uni-versity Press, Princeton.

Tilman D. (1984) Plant dominance along an experimental nutrient gradient. *Ecology* **65,** 1445–1453.

Tilman D. (1985) The resource ratio hypothesis of succession. *American Naturalist* **125,** 827–852.

Tilman D. (1986) Evolution and differentiation in terrestrial plant communities: the importance of the soil resource: light gradient. In *Community Ecology.* (Eds J. Diamond & T. Case), pp. 359–380. Harper and Row, New York.

Tilman D. & Kiesling R. (1983) Freshwater algal ecology: Taxonomic tradeoffs in the temperature dependence of nutrient competitive abilities. In *Current Perspectives in Microbial Ecology* (Eds M.J. Klug & C.A. Reddy). American Society of Microbiology, Washington D.C.

Tilman D., Kilham S.S. & Kilham P. (1982) Phytoplankton community ecology: the role of limiting nutrients. *Annual Review of Ecology and Systematics* **13,** 349–372.

Titman D. (1976) Ecological competition between algae: experimental confirmation of resource-based competition theory. *Science* **192,** 463–465.

Torssell B.W.R., Rose C.W. & Cunningham B.R. (1975) Population dynamics of an annual pasture in a dry monsoonal climate. *Proceedings of the Ecological Society of Australia* **9**, 157–162.

Treshow M. (Ed.) (1984) *Air Pollution and Plant Life*. John Wiley, Chichester.

Tripathi R.S. & Harper J.L. (1973) The comparative biology of *Agropyron repens* L. (Beaur.) and *A. caninum* L. (Beaur.) I. The growth of mixed populations established from tillers and from seeds. *Journal of Ecology* **61**, 353–368.

Turkington R. (1983) Leaf and flower demography of *Trifolium repens* L. I. Growth in mixture with grasses. *New Phytologist* **93**, 599–616.

Turkington R. & Aarssen L.W. (1984) Local-scale differentiation as a result of competitive interactions. In *Perspectives on Plant Population Ecology* (Eds R. Dirzo & J. Sarukhán), pp. 107–127. Sinauer, Sunderland, Massachusetts.

Turkington R. & Harper J.L. (1979) The growth, distribution and neighbour relationships of *Trifolium repens* in a permanent pasture. IV. Fine-scale biotic differentiation. *Journal of Ecology* **67**, 245–254.

Turner M.D. & Rabinowitz D. (1983) Factors affecting frequency distributions of plant mass: the absence of dominance and suppression in competing monocultures of *Festuca paradoxa*. *Ecology* **64**, 469–475.

Turrill W.B. (1929) *The Plant Life of the Balkan Peninsula*. Clarendon, Oxford.

Underwood A.J. & Densley E.J. (1984) Paradigms, explanations and generalizations in models for the structure of intertidal communities on rocky shores. In *Ecological Communities: Conceptual Issues and Evidence* (Eds D.R. Strong, D. Simberloff, L.G. Abele & A.B. Thistle), pp. 151–180. Princeton University Press, Princeton New Jersey.

Upadhya M.D. & Yee J. (1968) Isozyme polymorphism in flowering plants. VII. Isozyme variations in tissues of barley seedling. *Phytochemistry* **7**, 937–943.

Uphof J.C.T. (1938) Cleistogamic flowers. *Botanical Review* **4**, 21–49.

Usberti J.A. & Jain S.K. (1978) Variation in *Panicum maximum*: a comparison of sexual and asexual populations. *Botanical Gazette* **139**, 112–116.

Usher M.B. (1966) A matrix approach to the management of renewable resources, with special reference to selection forests. *Journal of Applied Ecology* **3**, 355–367.

Usher M.B. (1972) Developments in the Leslie matrix model. In *Mathematical Models in Ecology* (Ed. J.N.R. Jeffers), pp. 29–60. Blackwell Scientific Publications, Oxford.

Usher M.B. (1973) *Biological Management and Conservation*. Chapman & Hall, London.

Usher M.B. (1985) Population and community dynamics in the soil ecosystem. In *Ecological Interactions in Soil* (Eds A.H. Fitter, D. Atkinson, D.J. Read & M.B. Usher), pp. 243–266. *British Ecological Society Special Publication* **4**. Blackwell Scientific Publications, Oxford.

Usher M.B. (1986) Modelling successional processes in ecosystems. In *Colonization, Succession and Stability* (Eds A.J. Gray, M.J. Crawley & P.J. Edwards), Blackwell Scientific Publications, Oxford (in press).

Valdeyron G., Dommee B. & Vernet Ph. (1977) Self-fertilization in male-fertile plants of a gynodioecious species: *Thymus vulgaris* L. *Heredity* **39**, 243–249.

Valk A.G. van der (1981) Succession in wetlands: a Gleasonian approach. *Ecology* **62**, 688–696.

Vandermeer J.H. (1975) On the construction of the population projection matrix for a population grouped in unequal stages. *Biometrics* **31**, 239–242.

Vandermeer J.H. (1984) The evolution of mutualism. In *Evolutionary Ecology* (Ed. B. Shorrocks), pp. 221–232. Blackwell Scientific Publications, Oxford.

Vasquez-Yanes C. & Smith H. (1982) Phytochrome control of seed germination in the tropical rain forest pioneer trees *Cecropia obtusifolia* and *Piper auritum* and its ecological significance. *New Phytologist* **92**, 177–485.

Veleminsky J. & Gichner T. (1978) DNA repair in mutagen-injured higher plants. *Mutation Research* **55**, 71–84.

Venable D.L. (1984) Using intraspecific variation to study the ecological significance and evolution of plant life histories. In *Perspectives on Plant Population*

Ecology (Eds R. Dirzo & J. Sarukhán), pp. 166–178. Sinauer, Sunderland, Massachusetts.

Venable D.L. & Lawlor L. (1980) Delayed germination and dispersal in desert annuals: escape in space and time. *Oecologia* **46**, 272–282.

Veresoglou D.S. & Fitter A.H. (1984) Spatial and temporal patterns of growth and nutrient uptake of five co-existing grasses. *Journal of Ecology* **72**, 259–272.

Vitousek P.M. (1982) Nutrient cycling and nutrient use efficiency. *American Naturalist* **119**, 553–572.

Vitousek P.M. & Matson P.A. (1984) Mechanisms of nitrogen retention in forest ecosystems: A field experiment. *Science,* **255**, 51–52.

Vogel F. (1972) Spontaneous mutation. International Titisee Workshop. *Humangenetik* **16**, 1–180.

Vogel F. & Rathenberg R. (1975) Spontaneous mutation in man. *Advances in Human Genetics* **5**, 223–318.

Wainwright S.J. (1984) Adaptations of plants to flooding with salt water. In *Flooding and Plant Growth* (Ed. T.T. Kozlowski), pp. 295–343. Academic Press, London.

Waite S. (1980) *Autecology and population biology of* Plantago coronopus L. *at coastal sites in Sussex.* D.Phil. Thesis, University of Sussex.

Waite S. & Hutchings M.J. (1978) The effects of sowing density, salinity and substrate upon the germination of seeds of *Plantago coronopus* L. *New Phytologist* **81**, 341–348.

Waite S. & Hutchings M.J. (1979) A comparative study of establishment of *Plantago coronopus* L. from seeds sown randomly and in clumps. *New Phytologist* **82**, 575–583.

Waller D.M. (1980) Environmental determinants of outcrossing in *Impatiens capensis* (Balsaminaceae). *Evolution* **34**, 747–761.

Waller D.M. & Steingraeber D. (1986) Branching and modular growth: Theoretical models and empirical patterns. In *The Population Biology and Evolution of Clonal Organisms* (Eds J.B.C. Jackson, L.W. Buss & R.E. Cook).

Waloff B. & Richards O.W. (1977) The effect of insect fauna on growth, mortality and natality of broom, *Sarothamnus scoparius. Journal of Applied Ecology* **14**, 787–798.

Wapshere A.J., Hasan S., Wabba W.K. & Caresche L. (1974) The ecology of *Chondrilla juncea* in the western Mediterranean. *Journal of Applied Ecology* **11**, 783–799.

Warming E. (1909) *Oecology of Plants. An Introduction to the Study of Plant-Communities.* Clarendon Press, Oxford.

Waser N.M. (1978) Competition for hummingbird pollination and sequential flowering in two Colorado wildflowers. *Ecology* **59**, 934–944.

Waser N.M. (1983) Competition for pollination and floral character differences among sympatric plant species: a review of the evidence. In *Handbook of Experimental Pollination Biology* (Eds C.E. Jones & R.J. Little), pp. 277–293. Van Nostrand Reinhold, New York.

Waser N.M. & Price M.V. (1983) Optimal and actual outcrossing in plants, and the nature of plant pollinator interaction. In *Handbook of Experimental Pollination Biology* (Eds C.E. Jones & R.J. Little), pp. 342–359. Van Nostrand Reinhold, New York.

Waser N.M., Vickery R.K. & Price M.V. (1982) Patterns of seed dispersal and population differentiation. *Mimulus guttatus. Evolution* **36**, 753–761.

Watkinson A.R. (1978a) The demography of a sand dune annual: *Vulpia fasciculata.* II. The dynamics of seed populations. *Journal of Ecology* **66**, 35–44.

Watkinson A.R. (1978b) The demography of a sand dune annual, *Vulpia fasciculata.* III. The dispersal of seeds. *Journal of Ecology* **66**, 483–498.

Watkinson A.R. (1980) Density-dependence in single species populations of plants. *Journal of Theoretical Biology* **83**, 345–357.

Watkinson A.R. (1981a) The population ecology of winter annuals. In *The Biological Aspects of Rare Plant Conservation* (Ed. H. Synge), pp. 253–264. John Wiley, Chichester.

Watkinson A.R. (1981b) Interference in pure and mixed populations of *Agrostemma githago* L. *Journal of Applied Ecology* **18**, 967–976.

Watkinson A.R. (1984) Yield-density relationships: the influence of resource availability on growth and self-thinning in populations of *Vulpia fasciculata*. *Annals of Botany* **53**, 469–482.

Watkinson A.R. (1985a) Plant responses to crowding. In *Studies on Plant Demography: A Festschrift for John L. Harper* (Ed. J. White), pp. 275–289. Academic Press, London.

Watkinson A.R. (1985b) On the abundance of plants along an environmental gradient. *Journal of Ecology* **73**, 569–578.

Watkinson A.R. & Davy A.J. (1985) Population biology of salt marsh and sand dune annuals. *Vegetatio* **62**, 487–497.

Watkinson A.R. & Harper J.L. (1978) The demography of a sand dune annual: *Vulpia fasciculata*. I. The natural regeneration of populations. *Journal of Ecology* **66**, 15–33.

Watson M.A. (1984) Developmental constraints: effect on population growth and patterns of resource allocation in a clonal plant. *American Naturalist* **123**, 411–426.

Watson M.A. & Casper B.B. (1984) Morphogenetic constraints on patterns of carbon distribution in plants. *Annual Review of Ecology and Systematics* **15**, 233–258.

Watson P.J. (1969) Evolution in closely adjacent plant populations. VI. An entomophilous species, *Potentilla erecta*, in two contrasting habitats. *Heredity* **24**, 407–422.

Watt A.S. (1947) Pattern and process in the plant community. *Journal of Ecology* **35**, 1–22.

Watt A.S. (1955) Bracken versus heather, a study in plant sociology. *Journal of Ecology* **43**, 490–506.

Watt A.S. (1971) Factors controlling the floristic composition of some plant communities in Breckland. In *The Scientific Management of Animal and Plant Communities for Conservation* (Eds E. Duffey & A.S. Watt), pp. 137–152. Blackwell Scientific Publications, Oxford.

Watt A.S. (1981) A comparison of grazed and ungrazed grassland in East Anglian Breckland. *Journal of Ecology* **69**, 499–508; 509–536.

Weinberger J.H. (1931) The relation of leaf area to size and quality of peaches. *Proceedings American Society of Horticultural Science* **28**, 18–22.

Weinberger J.H. & Cullinan F.P. (1932) Further studies on the relation between leaf area and size of fruit, chemical composition and fruit bud formation in Elberta peaches. *Proceedings American Society of Horticultural Science* **29**, 23–27.

Weiner J. (1985) Size hierarchies in experimental populations of annual plants. *Ecology* **6**, 743–752.

Weiner J. & Solbrig O.T. (1984) The meaning and measurement of size hierarchies in plant populations. *Oecologia* **61**, 334–336.

Weins D. (1984) Ovule survivorship, brood size, life history, breeding systems, and reproductive success in plants. *Oecologia* **64**, 47–53.

Wellington W.G. & Trimble R.M. (1984) Weather. In *Ecological Entomology* (Eds C.B. Huffaker & R.L. Rabb), pp. 399–425. John Wiley, New York.

Wells T.C.E. (1981) Population ecology of terrestrial orchids. In *The Biological Aspects of Rare Plant Conservation* (Ed. H. Synge), pp. 281–295. John Wiley, Chichester.

Went F.W. (1953) The effect of temperature on plant growth. *Annual Review of Plant Physiology* **4**, 347–362.

Went F.W. (1955) The ecology of desert plants. *Scientific American* **192**, 68–75.

Went F.W., Juhren G. & Juhren M.C. (1952) Fire and biotic factors affecting germination. *Ecology* **33**, 351–364.

Werner P.A. (1975) Predictions of fate from rosette size in teasel (*Dipsacus fullonum* L.). *Oecologia* **20**, 197–201.

Werner P.A. (1977) Colonization success of a biennial plant species: experimental field studies of species cohabitation and replacement. *Ecology* **58**, 840–849.

Werner P.A. & Caswell H. (1977) Population growth rates and age versus stage distribution models for teasel (*Dipsacus sylvestris* Huds.) *Ecology* **58**, 1103–1111.

Werner P.A. & Platt W.J. (1976). Ecological relationships of co-occurring goldenrods (*Solidago*; Compositae). *American Naturalist* **110**, 959–971.

Westoby M. (1982) Frequency distributions of plant size during competitive growth of stands: the operation of distribution-modifying functions. *Annals of Botany* **50**, 733–735.

Westoby M. (1984) The self-thinning rule. *Advances in Ecological Research* **14**, 167–225.

Westoby M. & Howell J. (1981) Self-thinning: the effect of shading on glasshouse populations of silver beet (*Beta vulgaris*). *Journal of Ecology* **69**, 359–365.

Westoby T. & Rice B. (1981) A note on combining two methods of dispersal for distance. *Australian Journal of Ecology* **6**, 189–192.

Wheeler W.M. (1911) The ant-colony as an organism. *Journal of Morphology* **22**, 307–325.

Wheelwright N.T., Haber W.A., Murray K.G. & Guindon C. (1984) Tropical fruit-eating birds and their food plants: a survey of a Costa Rican lower montane forest. *Biotropica* **16**, 173–191.

Wheelwright N.T. & Orians G. (1982) Seed dispersal by animals: contrasts with pollen dispersal, problems of terminology, and constraints on coevolution. *American Naturalist*, **119**, 402–413.

Whipps J.M. (1984) Environmental factors affecting the loss of carbon from the roots of wheat and barley seedlings. *Journal of Experimental Botany* **35**, 767–773.

White J. (1980) Demographic factors in populations of plants. In *Demography and Evolution in Plant Populations* (Ed. O.T. Solbrig), pp. 21–48. Blackwell Scientific Publications, Oxford.

White J. (1981) The allometric interpretation of the self-thinning rule. *Journal of Theoretical Biology* **89**, 475–500.

White J. (1984) Plant metamerism. In *Perspectives on Plant Population Ecology* (Eds R. Dirzo and J. Sarukhán). Sinauer, Sunderland, Massachusetts.

White J. & Harper J.L. (1970) Correlated changes in plant size and number in plant populations. *Journal of Ecology* **58**, 467–485.

White P.S. (1984) The architecture of Devil's Walking Stick, *Aralia spinosa* L. (Araliaceae). *Journal of the Arnold Arboretum* **65**, 403–418.

White R.E. (1973) Studies on mineral ion absorption by plants. II. The interaction between metabolic activity and the rate of phosphorus uptake. *Plant and Soil* **38**, 509–523.

White T.C.R. (1974) A hypothesis to explain outbreaks of looper caterpillars, with special reference to populations of *Selido suavis* in a plantation of *Pinus radiata* in New Zealand. *Oecologia* **16**, 279–301.

Whitham T.G. (1981) Individual trees as heterogeneous environments: adaptation to herbivory or epigenetic noise? In *Insect Life History Patterns: Habitat and Geographic Variation.* (Eds R.F. Denno & H. Dingle), pp. 9–27. Springer-Verlag, New York.

Whitham T.G. (1983) Host manipulation of parasites: within-plant variation as a defense against rapidly evolving pests. In *Variable Plants and Herbivores in Natural and Managed Systems* (Eds R.F. Denno & S. McClure), pp. 15–41. Academic Press, New York.

Whitham T.G. & Slobodchikoff C.N. (1981) Evolution by individuals, plant-herbivore interactions, and mosaics of genetic variability: the adaptive significance of somatic mutations in plants. *Oecologia* **49**, 287–292.

Whitham T.G., Williams A.G. & Robinson A.M. (1984) The variation principle: individual plants as temporal and spatial mosaics of resistance to rapidly evolving pests. In *Novel Approaches to Interactive Systems* (Eds P.W. Price, C.N. Slobodchikoff & W.S. Gaud), pp. 15–52. John Wiley, New York.

Whitmore T.C. (1975) *Tropical Rain Forests of the Far East*. Clarendon Press, Oxford.

Whitmore T.C. (1978) Gaps in the forest canopy. In *Tropical Trees as Living Systems* (Eds P.B. Tomlinson & M.H. Zimmerman), Chapter 27. Cambridge University Press, Cambridge.

Whitmore T.C. (1985) Forest Succession. *Nature* 315, 692.

Whittaker J.B. (1982) The effect of grazing by a chrysomelid beetle, *Gastrophysa viridula* on growth and survival of *Rumex crispus* on a shingle bank. *Journal of Ecology* 70, 291–296.

Whittaker R.H. (1956) Vegetation of the Great Smoky Mountains. *Ecological Monographs* 26, 1–80.

Whittaker R.H. (1967) Gradient analysis of vegetation. *Biological Reviews* 42, 207–264.

Whittaker R.H. (1975) *Communities and Ecosystems*, 2nd edn. Collier MacMillan, London.

Whittaker R.H. & Feeney P. (1971) Allelochemics: chemical interactions between species. *Science* (1971), 757–770.

Whittaker R.H., Likens G.E., Bormann F.H., Eaton J.S. & Siccama T.G. (1979) The Hubbard Brock Ecosystem Study: forest nutrient cycling and element behaviour. *Ecology* 60, 203–220.

Whittaker R.H. & Niering W.A. (1975) Vegetation of the Santa Catalina mountains, Arizona. V. Biomass, production, and diversity along the elevation gradient. *Ecology* 56, 771–790.

Wiebes J.T. (1979) Co-evolution of figs and their insect pollinators. *Annual Review of Ecology and Systematics* 10, 1–12.

Wild A. (1979) Physiologie der Photosynthese höherer Pflanzen. Die Ampassung an Lichtbedingungen. *Berichte der Deutsche Botanische Gesellschaft* 92, 341–364.

Williams E.G. (1984) Tissue culture and reproductive biology. In *Pollination '84* (Eds E.G. Williams & R.B. Knox), pp. 30–42. University of Melbourne, Melbourne.

Williams G.C. (1957) Pleiotropy, natural selection, and the evolution of senescence. *Evolution* 11, 398–411.

Williams G.C. (1966) *Adaptation and Natural Selection*. Princeton University Press, Princeton, New Jersey.

Williams G.C. (1975) *Sex and Evolution*. Princeton University Press, Princeton, New Jersey.

Williams L. (1964) Possible relationships between plankton-diatom species numbers and water-quality estimates. *Ecology* 45, 809–823.

Williams O.B. (1981) Monitoring changes in populations of desert plants. In *Biological Aspects of Rare Plant Conservation* (Ed. H. Synge), pp. 223–240. John Wiley, Chichester.

Williams R.D. & Evans G. (1935) The efficiency of spatial isolation in maintaining the purity of red clover. *Welsh Journal of Agriculture* 11, 164–171.

Williams R.F. (1948) The effects of phosphorus supply on the rates of intake of phosphorus and nitrogen and upon certain aspects of phosphorus metabolism in gramineous plants. *Australian Journal of Scientific Research B* 1, 333–361.

Williamson M.H. (1972) *The Analysis of Biological Populations*. Arnold, London.

Willis A. (1963) Braunton Burrows: the effects on the vegetation of the addition of mineral nutrients to the dune soils. *Journal of Ecology* 51, 353–374.

Willis A. & Yemm E. (1961) Braunton Burrows: mineral nutrient status of the dune soils. *Journal of Ecology* 49, 377–390.

Willson M.F. (1983) *Plant Reproductive Ecology*. John Wiley, New York.

Willson M.F. & Burley N. (1983) *Mate Choice in Plants: Tactics, Mechanisms and Consequences*. Princeton University Press, New Jersey.

Willson M.F. & Rathke B.J. (1974) Adative design of the floral display in *Asclepias syriaca* L. *American Midland Naturalist* 92, 47–57.

Wilson B.F. (1984) *The Growing Tree* (revised edn.). University of Massachusetts Press, Amherst.

Wilson D.E. & Janzen D.H. (1972) Predation on *Scheelea* palm seeds by bruchid beetles: seed density and distance from the parent palm. *Ecology* 53, 954–959.

Wilson E.O. (1971) *The Insect Societies*. The Belknap Press of Harvard University Press, Cambridge, Massachusetts.

Winkler A.J. (1930) The relation of number of leaves to size and quality of table grapes. *Proceedings American Society of Horticultural Science* 27, 158–160.

Winner W.E. (1981) The effect of SO_2 on photosynthesis and stomatal behaviour of Mediterranean-climate shrubs and trees. In *Components of Productivity of Mediterranean-climate Regions—Basic and Applied Aspects* (Eds N.S. Margaris & H.A. Mooney). Junk, The Hague.

Winner, W.E. and H.A. Mooney (1980) Ecology of SO_2 resistance: III. Metabolic changes of C_3 and C_4 Atriplex species due to SO_2 fumigation. *Oecologia* 46, 196–200.

Wit C.T. de (1960) On competition. *Verslagen van landbouwkundige onderzoekingen* 66, 1–82.

Woodell S.R.J., Mooney H.A. & Hill A.J. (1969) The behaviour of *Larrea divaricata* (creosote bush) in response to rainfall in California. *Journal of Ecology* 57, 37–44.

Worley I.A. (1973) The "Black Crust" phenomenon in upper Glacier Bay, Alaska. *Northwest Science* 47, 20–29.

Wright A.J. (1981) The analysis for yield-density relationships in binary mixtures using inverse polynomials. *Journal of Agricultural Science, Cambridge* 96, 561–567.

Wright S. (1969) Evolution and genetics of populations. Vol. 2. *The Theory of Gene Frequencies*. University of Chicago Press, Chicago.

Wright S.T.C. (1956) Studies of fruit development in relation to plant hormones. III. Auxins in relation to fruit morphogenesis and fruit drop in the black currant *Ribes nigrum* L. *Journal Horticultural Science* 31, 196–211.

Yarwood C.E. (1967) Response to parasites. *Annual Review Plant Physiology* 18, 419–438.

Yeaton R.I. & Cody M.L. (1976) Competition and spacing in plant communities: the northern Mohave desert. *Journal of Ecology* 64, 689–696.

Yeaton R.I., Travis J. & Gilinsky E. (1977) Competition and spacing in plant communities: the Arizona upland association. *Journal of Ecology* 65, 587–595.

Yoda K., Kira T., Ogawa H. & Hozumi K. (1963) Self-thinning in over-crowded pure stands under cultivated and natural conditions. *Journal of Biology, Osaka City University* 14, 107–129.

Yost R.S., Behara G. & Fox R.L. (1982) Geostatistical analysis of soil chemical properties of large land areas. I. Semi-variograms. *Soil Science Society of America Journal* 46, 1028–1032.

Young V. (1934) Plant distribution as influenced by soil heterogeneity in Cranberry Lake region of the Adirondack Mountains. *Ecology* 15, 154–196.

Zavarin E., Cobb F.W., Bergot J. & Barber H.W. (1971) Variation of the *Pinus ponderosa* needle oil with season and needle age. *Phytochemistry* 10, 3107–3114.

Zedler J. & Zedler P. (1969) Association of species and their relationship to micro-topography within old fields. *Ecology* 50, 432–442.

Zimmermann M.H. (1978) Hydraulic architecture of some diffuse-porous trees. *Canadian Journal of Botany* 56, 2286–2295.

Zimmermann M.H. & Brown C.L. (1971) *Trees: Structure and Function*. Springer-Verlag, New York.

Zohary M. (1950) Evolutionary trends in the fruiting head of compositae. *Evolution* 4, 103–109.

Zucker W.V. (1982) How aphids choose leaves: the roles of phenolics in host selection by a galling aphid. *Ecology* 63, 972–981.

Author Index

This index lists all individual contributors to the papers and articles cited in the text of this book. The names are arranged alphabetically and can be traced to the relevant part of the book using the section references alongside each one.

Subject Index

Page numbers in italics denote major text entries.